CROSS-BRIDGE MECHANISM IN MUSCLE CONTRACTION

Explanation of Plates

Plate 1
(First row)   S. Okinaga/ H.E. Huxley, H. Sugi/ G.H. Pollack
(Second row) S. Winegrad, G.H. Pollack, G.J. Steiger/ S. Ebashi, R. Natori, M. Endo
(Third row)   S. Winegrad/ D.C.S. White/ J. Dawson, E.M. Bartels, K. Yamada, Y. Umazume/ D.L. Brutsaert
(Fourth row) T. Iwazumi, K.A.P. Edman/ E.M. Bartels, A. Drake, H.E.D.J. ter Kers, H.E. Huxley, M.I.M. Noble/ A. Oplatka

Plate 2
(First row)   T. Blangé/ J. Dawson, D.R. Wilkie, H. Shimizu, H. Mashima, R.T. Tregear, T. Sakai/ M.J. Kushmerick
(Second row) M.M. Dewey/ D.R. Wilkie/ H. Shimizu/ R.C. Woledge
(Third row)   T.L. Hill/ During the session/ J.C. Rüegg
(Fourth row) H. Sugi/ Dinner party at Happo-en/ M.I.M. Noble

Plate 1

Plate 2

CROSS-BRIDGE MECHANISM IN MUSCLE CONTRACTION

Proceedings of the International Symposium
on the Current Problems of Sliding Filament Model
and Muscle Mechanics
Tokyo, Japan, 13th-15th September, 1978

Edited by
Haruo Sugi and Gerald H. Pollack

UNIVERSITY PARK PRESS
Baltimore

Sponsored by Teikyo University

UNIVERSITY PARK PRESS
Baltimore

Library of Congress Cataloging in Publication Data

International Symposium on the Current Problems
  of Sliding Filament Model and Muscle Mechanics,
  Tokyo, 1978.
  Cross-bridge mechanism in muscle contraction.

  Includes index.
  1. Muscle contraction—Congresses. I. Sugi,
Haruo, 1933– II. Pollack, Gerald H.
III. Teikyō Daigaku. IV. Title.
QP321.I59 1978   591.1′852   79-19121
ISBN 0-8391-1481-8

© UNIVERSITY OF TOKYO PRESS, 1979
UTP 3047-68925-5149

Printed in Japan

All rights reserved. No part of this publication may be reproduced
or transmitted in any form or by any means, electronic or mechanical,
including photocopy, recording, or any information storage and
retrieval system, without permission in writing from the publisher.

Originally published by
UNIVERSITY OF TOKYO PRESS

Preface

Although more than twenty years have passed since the monumental discovery of a relative sliding motion between two sets of interdigitating filaments during length changes in striated muscle by H. E. Huxley, A. F. Huxley, and their co-workers, a question which still remains to be answered is, what makes the filaments slide? Among a number of possible causes for the relative sliding between the thick and thin filaments, the attention of most investigators has been focused on the cyclic interaction of the projections on the thick filaments—*i.e.* the cross-bridges—with the sites on the thin filaments to produce force and motion. In spite of considerable progress in the studies of muscle contraction, including mechanics, energetics, ultrastructural changes and enzyme kinetics of actomyosin ATPase, the cross-bridge mechanism is still a matter for speculation and controversy.

The purpose of this symposium is to thoroughly discuss various problems related to the current sliding filament/cross-bridge hypotheses. Since the validity of any contraction model should be tested on intact muscles or on other preparations in which three-dimensional filament-lattice structure is preserved, most papers presented are related to experiments on intact, skinned or glycerinated striated muscles. The results are discussed mainly in connection with the contraction models put forward by A. F. Huxley and A. F. Huxley and Simmons, which are central in the field of muscle physiology. In this connection, it is a great pity that we could not have Professor A. F. Huxley at this meeting, but it is our great pleasure to have Dr. H. E. Huxley, one of the founders of the current sliding filament hypotheses, with us.

This symposium consists of eight sessions. The following is an outline of the organization of the sessions.

Session I. Sarcomere dynamics. Studies on striated muscle at the level of sarcomeres are very important, since a sarcomere constitutes the functional unit in muscle contraction. Contrary to the general view that the length of the filaments does not change during contraction, Dewey *et al.* presents evidence that the thick filaments in *Limulus* striated muscle can shorten during contraction, suggesting a mechanism to produce force and motion other than the cross-bridges. By means of laser light diffraction technique, Pollack *et al.* demonstrate the stepwise shortening of sarcomeres in vertebrate cardiac and skeletal muscles; this suggests a synchronization of cross-bridge activity against the general concept of cross-bridges as independent force-generators.

Natori *et al.* show that overstretched skinned frog muscle fibers provide suitable material for studying the properties of the thick and thin filaments separately. Using the same preparations, Fujime *et al.* measure the flexibility of the thin filaments by analyzing optical changes when an electric field is applied.

Session II. Instantaneous elasticity and stiffness. The instantaneous elasticity or the stiffness of active muscle, as examined by tension changes in response to quick length changes (or *vice versa*), has generally been taken as a measure of cross-bridge elasticity. Pollack *et al.* describe a method of measuring the instantaneous elasticity in complex cardiac muscle tissues by use of laser light diffraction. Sugi demonstrates a uniform distribution of a highly compliant series elasticity along the length of single crayfish muscle fibers, indicating that it may originate mainly from structures other than the cross-bridges. He also shows a localized buckling at the released end of a frog muscle fiber during a quick release, suggesting that the tension changes in response to excessively quick length changes may not give correct information about the cross-bridges. Gott discusses the problems of measuring the instantaneous elasticity with quick length changes applied at one end, based on his multiple segment analysis of muscle responses. Rüegg *et al.* examine the stiffness of skinned or glycerinated cardiac, skeletal and insect fibrillar muscles in a variety of experimental conditions, including the replacement of ATP by AMP-PNP, to obtain information about cross-bridge properties. Another way of approaching cross-bridge properties is the so-called "sinusoidal analysis" technique, in which the tension changes in response to small sinusoidal length oscillations of varying frequency are studied. Kawai and Sagawa *et al.* apply this method to skinned skeletal muscle fibers and to intact cardiac muscles, respectively, and the visco-elastic elements inferred are discussed in connection with the cross-bridge properties.

Session III. Isotonic and isometric transients. The transient length changes after a quick change in load (isotonic transients) or the transient tension changes after a quick change in length (isometric transients) have been taken to reflect the kinetic properties of cross-bridge turnover. Though the isometric transients in insect fibrillar muscle is known to include a marked delayed rise in tension (stretch activation), its character is obscured by the presence of resting tension. In this connection, White *et al.* show that the mechanical properties of relaxed fibrillar muscle act functionally in parallel with the cross-bridges. Blangé and Stienen study the effect of metabolic inhibitors on the isometric transients in vertebrate skeletal muscle, showing that the quick tension recovery disappears after metabolic inhibition. Tsuchiya and Sugi's experiments are concerned with the isotonic transients in frog muscle fibers with special reference to marked oscillatory length changes following quick increases in load. Brutsaert *et al.* also study the isotonic transients in cardiac muscle in detail, including the time course of isotonic lengthening. Steiger's paper deals with the determination of rate constants on the time course of isometric transients and on the results of sinusoidal analysis in cardiac muscle.

Session IV. Length-tension relation. The well-known linear descending limb of the length-tension diagram in striated muscles has been taken to support the current sliding filament models in which the cross-bridges act as independent force-generators. ter Keurs *et al.* show, however, that the descending limb of the length-tension curve in single frog fibers is not linear at all, and that their results are not due to the dispersion in sarcomere length, thus challenging the concept of independent force-generators. Edman *et al.* demonstrate the enhancement of mechanical performance by stretch in single frog fibers; since the number of cross-bridges decreases with decreasing overlap between the filaments, this phenomenon is not readily explained by the current sliding filament models. Using rapid cooling contractures, Sakai also shows a non-linear descending limb of the length-tension relation in frog toe muscle. Maruyama reports the presence of a network of connectin, an elastic protein descovered by him, around the myofibrils in muscle, serving as a parallel elastic component. Winegrad and McClellan's paper deals with the control of contractile force by a cyclic AMP dependent mechanism in cardiac muscle.

Session V. Force-velocity relation. The independence of the shortening velocity at zero load (V_{max}) on the amount of overlap between the filaments has also been taken to support the concept of independent force-generators, though the measurement of V_{max} is technically difficult. Edman describes a new method of measuring the shortening velocity at zero load in single frog muscle fibers, and finds that it is length-independent over a wide range of sarcomere lengths but increases steeply at sarcomere lengths above 2.7 μm where resting tension introduces a compressive force. Endo studies the effect of viscosity of the surrounding medium on the mechanical performance of skinned frog muscle fibers in order to give information about the effect of viscosity on cross-bridge kinetics. Mashima studies the force-velocity relation in tetanized cardiac muscle under a variety of conditions, and determines the dynamic constants according to the Hill equation.

Session VI. X-ray diffraction studies. X-ray diffraction studies on resting, active and rigor muscles are extremely useful in obtaining information about the ultrastructural changes during contraction. The time resolution of X-ray diffraction studies has now been markedly improved by use of very intense X-ray sources such as synchrotron radiation. H. E. Huxley describes attempts to obtain direct evidence about the actual behavior of the cross-bridges by means of time-resolved X-ray diffraction of contracting frog skeletal muscle, including the recent finding that the 143 Å periodicity of the thick filaments increases during contraction. Another approach to the cross-bridge mechanism being made by Tregear *et al.* is to study the X-ray diffraction patterns from insect fibrillar muscle under various chemical conditions, with special reference to the effect of unhydrolyzable ATP analogs. Amemiya *et al.* show, with the time-resolved X-ray diffraction technique, that the intensity ratio of two equatorial reflections ($I_{1,0}/I_{1,1}$), which is generally taken as a measure of attached cross-bridges, does not change by stretching active frog muscle, and that the intensity ratio is smaller during an isotonic twitch under a small load than during an isometric twitch. The papers of Namba *et al.* and Maéda are related to the detailed structural analysis of the thin filaments in crab muscles in rigor to give information about the mode of attachment of the cross-bridges to the thin filaments. Their results indicate that crab muscle provides suitable material for studying the ultrastructure of the thin filaments.

Session VII. Energetics. Since the classic work of A.V. Hill, muscle energetics has constituted an important field of muscle physiology, providing information about how ATP splitting is coupled with cross-bridge performance. Woledge and Curtin review the problems on muscle energetics, and report their recent work on the effect of stretch on the rate of energy output in active frog skeletal muscle. Kodama and Yamada construct a model to explain the rate of energy output and the rate of ATP splitting in shortening muscle, based on enzyme kinetics of actomyosin ATPase. Hoh demonstrates the presence of myosin isoenzymes in cardiac and skeletal muscles, and shows that their profile in fast-twitch and slow-twitch muscles can be reversed by nerve cross-union. Kushmerick studies the relation between the high-energy phosphate utilization during contraction and the subsequent metabolic recovery reactions. Wilkie *et al.* describe their pioneering work in simultaneously recording the ^{31}P nuclear magnetic resonance spectra and the mechanical activity in intact skeletal muscles, and measure the time course of change in metabolite levels during contraction to give information about cross-bridge cycling.

Session VIII. Theories of contraction. For the future development of studies of cross-bridge mechanisms, it is very useful to construct new theories of muscle contraction which, since they can be tested experimentally, stimulate the interest of investigators. Hill describes a simplified theory which accounts for the results of Huxley and Simmons in terms of transitions between attached states of the cross-bridges. Shimizu illustrates his three-state contraction model, emphasizing the dynamic cooperativity of the cross-bridges—contrary to the general view that the cross-bridges act independently. Oplatka et al. propose a hydrodynamic theory for muscle contraction and for other primitive motile systems, which is fundamentally different from the current sliding filament hypotheses. Iwazumi's new field theory of muscle contraction is also fundamentally different from the general concept of cross-bridges, resting on the assumption that the cross-bridges are dipole field generators so that electrostatic forces are generated between the cross-bridges and the tips of dielectric thin filaments, and covers many phenomena which are not readily explained by the current sliding filament models, including the longitudinal and lateral stabilities of the filament-lattice structure.

As described above, this symposium contains many papers indicating various shortcomings of the current sliding filament/cross-bridge hypotheses. It underscores the necessity for careful reexamination of previous experiments which are generally regarded as supporting the current hypotheses. We believe that the papers and the controversial discussions following them will be useful for people interested in the cross-bridge mechanism, and hope that this volume will be an impetus for future progress in the study of muscle contraction.

H. Sugi
G. H. Pollack

Acknowledgements

The editors express their sincere thanks to Dr. Shoichi Okinaga, President of Teikyo University, for the generous financial support which enabled them to organize the symposium.

Our thanks are also due to Dr. Reiji Natori, President of Jikei University, for his encouragement and for his contributions towards the costs of the symposium, and to Drs. Martin J. Kushmerick and Michael J. Delay for preparing the typescript of the discussions.

In addition, we thank Drs. Hiroo Hashizume, Yasubumi Saeki, Kaoru Kometani, Kenichi Nishiyama, Takenori Yamada, Hidehiro Tanaka, and Tsukasa Tameyasu for their help in compiling the indices.

Participants

Symposium on the Current Problems of Sliding Filament Model and Muscle Mechanics held at the Hotel New Japan, Tokyo, 13-15 September, 1978

AMEMIYA, Y.
Engineering Research Institute, Faculty of Engineering, University of Tokyo, Bunkyo-ku, Tokyo 113, Japan,

ARATA, T.
Department of Biology, Faculty of Science, Osaka University, Toyonaka, Osaka 560, Japan

BARTELS, E. M.
The Open University, Oxford Research Unit, Foxcombe Hall, Berkeley Road, Boars Hill, Oxford, England

BLANGÉ, T.
Fysiologisch Laboratorium, Jan Swammerdam Instituut, Universiteit van Amsterdam, Eerste Constantijn Huygensstraat 20, Amsterdam, The Netherlands

BRUTSAERT, D. L.
Department of Physiology, University of Antwerp, 2020 Antwerp, Belgium

DAWSON, J.
Department of Physiology, University College London, Gower Street, London WC1E 6BT, England

DEWEY, M. M.
Department of Anatomical Sciences, School of Basic Health Sciences, Health Sciences Center, State University of New York at Stony Brook, Stony Brook, New York 11794, U.S.A.

DRABIKOWSKI, W.
Department of Biochemistry of Nervous System and Muscle, Nencki Institute of Experimental Biology, Polish Academy of Sciences, 3 Pasteur Street, Warszawa 22 Poland

DRAKE, A.
The Midhurst Medical Research Institute, Midhurst, West Sussex GU29 OBL, England

EBASHI, S.
Department of Pharmacology, Faculty of Medicine, University of Tokyo, Bunkyo-ku, Tokyo 113, Japan

EDMAN, K. A. P.
Department of Pharmacology, University of Lund, Sölvegatan 10, S-223 62 Lund, Sweden

ENDO, M.
Department of Pharmacology, Faculty of Medicine, Tohoku University, Seiryo-cho, Sendai-shi 980, Japan

FUJIME, S.
Mitsubishi-Kasei Institute of Life Sciences, Machida, Tokyo 194, Japan

GERGELY, J.
Boston Biomedical Research Institute, Boston, Massachusetts, U.S.A.

GILBERT, S. H.
Physilogie Générale des Muscles, Université Catholique de Louvain, Bruxelles, Belgium

GOTT, A. H.
Department of Physiology, Pharmacology and Biophysics, School of Medicine, Loma Linda Unversity, Loma Linda, California 92354, U.S.A.

HASHIZUME, H.
Engineering Research Institute, Faculty of Engineering, University of Tokyo, Bunkyo-ku, Tokyo 113, Japan

HILL, T. L.
National Institute of Arthritis, Metabolism and Digestive Diseases, National Institutes of Health, Bethesda, Maryland 20014 U.S.A.

HOH, J. F. Y.
Department of Physiology, The University of Sydney, Sydney, N.S.W. 2006, Australia
HUXLEY, H. E.
MRC Laboratory of Molecular Biology, University Medical School, Hills Road, Cambridge CB2 2QH, England
ISHIDE, N.
Department of Internal Medicine, School of Medicine, Tohoku University, Seiryo-cho, Sendai-shi 980, Japan
IWAZUMI, T.
Anesthesia Research Center, Department of Anesthesiology, School of Medicine, University of Washington, Seattle, Washington 98195, U.S.A.
KAMIYAMA, A.
Department of Physiology, School of Medicine, Teikyo University, Itabashi-ku, Tokyo 173, Japan
KAWAI, M.
Laboratory of Neurophysiology, Department of Neurology, College of Physicians and Surgeons, Columbia University, 630 West 168th Street, New York 10032, U.S.A.
KODAMA, T.
Department of Pharmacology, School of Medicine, Juntendo University, Bunkyo-ku, Tokyo 113, Japan
KOMETANI, K.
Faculty of Pharmaceutical Sciences, University of Tokyo, Bunkyo-ku, Tokyo 113, Japan
KRAMER, A. E. J. L.
Department of Urology, Academisch Ziekenhuis-Leiden, Rijnsburgerweg 10, Leiden, The Netherlands
KUSHMERICK, M. J.
Department of Physiology, Harvard Medical School, Boston, Massachusetts 02115, U.S.A.
MAÉDA, Y.
Department of Pharmacology, Faculty of Medicine, Tohoku University, Sendai-shi 980, Japan
MARUYAMA, K.
Department of Biology, Faculty of Science, Chiba University, Chiba-shi 280, Japan
MASHIMA, H.
Department of Physiology, School of Medicine, Juntendo University, Bunkyo-ku, Tokyo 113, Japan
MATSUBARA, S.
Department of Neuropsychiatry, Kanazawa Medical University, Ishikawa-ken 920-02, Japan
MATSUMOTO, Y.
Service de Chimie Physique II, Campus Plaine U.L.B., Boulevard de Triomphe, 1050 Bruxelles, Belgium
MATSUMURA, M.
Department of Physiology, Kawasaki Medical College, Kurashiki-shi, Okayama 701, Japan
MITSUI, T.
Department of Biophysical Engineering, Osaka University, Toyonaka, Osaka 560, Japan
MOHAN, R.
Physiologisches Institut, Lehrstuhl für Klinische Physiologie, Universität Düsseldorf, 4 Düsseldorf, Moorenstrasse 5, West Germany
MORIMOTO, S.
Department of Physiology, Jikei University Medical School, Minato-ku, Tokyo 105, Japan
NAMBA, K.
Department of Biophysical Engineering, Osaka University, Toyonaka, Osaka 560, Japan
NATORI, Reiji
President, Jikei University Medical School, Minato-ku, Tokyo 105, Japan
NATORI, Reibun
Department of Physiology, Jikei University Medical School, Minato-ku, Tokyo 105, Japan
NISHIYAMA, K.
Faculty of Pharmaceutical Sciences, University of Tokyo, Bunkyo-ku, Tokyo 113, Japan
NOBLE, M. I. M.
The Midhurst Medical Research Institute, Midhurst, West Sussex GU29 OBL, England

PARTICIPANTS

O'BRIEN, E. J.
Department of Biophysics, School of Biological Sciences, University of London King's College, 26-29 Drury Lane, London WC2B 5RL, England

OGAWA, Y.
Department of Pharmacology, School of Medicine, Juntendo University, Bunkyo-ku, Tokyo 113, Japan

OKINAGA, S.
President, Teikyo University, Itabashi-ku, Tokyo 173, Japan

OPLATKA, A.
Polymer Department, The Weizmann Institute of Science, Rehovot, Israel

OTSUKI, I.
Department of Pharmacology, Faculty of Medicine, University of Tokyo, Bunkyo-ku, Tokyo 113, Japan

POLLACK, G. H.
Anesthesia Research Center, Department of Anesthesiology, School of Medicine, University of Washington, Seattle, Washington 98195, U.S.A.

RALL, J. A.
Department of Physiology, The Ohio State University, 1645 Neil Avenue, Columbus, Ohio 43210, U.S.A.

RÜEGG, J. C.
II Physiologisches Institut, Universität Heidelberg, Im Neuenheimer Feld 326, 6900 Heidelberg 1, West Germany

SAEKI, Y.
Department of Physiology, School of Dentistry, Tsurumi University, Tsurumi-ku, Yokohama 230, Japan

SAGAWA, K.
Department of Medical Engineering, School of Medicine, The Johns Hopkins University, 720 Rutland Avenue, Baltimore, Maryland 21205, U.S.A.

SAKAI, T.
Department of Physiology, Jikei University Medical School, Minato-ku, Tokyo 105, Japan

SHIMIZU, H.
Faculty of Pharmaceutical Sciences, University of Tokyo, Bunkyo-ku, Tokyo 113, Japan

STEIGER, G. J.
Department of Physiology, School of Medicine, The Center for the Health Sciences, University of California at Los Angeles, Los Angeles, California 90024, U.S.A.

STIENEN, G. J. M.
Fysiologish Laboratorium, Jan Swammerdam Instituut, Universiteit van Amsterdam, Eerste Constantijn Huygensstraat 20 Amsterdam, The Netherlands

SUGI, H.
Department of Physiology, School of Medicine, Teikyo University, Itabashi-ku, Tokyo 173, Japan

SUZUKI, S.
Department of Physiology, School of Medicine, Teikyo University, Itabashi-ku, Tokyo 173, Japan

TAMEYASU, T.
Department of Physiology, School of Medicine, Teikyo University, Itabashi-ku, Tokyo 173, Japan

TANAKA, H.
Department of Physiology, School of Medicine, Teikyo University, Itabashi-ku, Tokyo 173, Japan

ter KEURS, H. E. D. J.
Laboratory for Physiology, State University of Leiden, Wassenaarseweg 62, Leiden, The Netherlands

TSUCHIYA, T.
Department of Physiology, School of Medicine, Teikyo University, Itabashi-ku, Tokyo 173, Japan

TREGEAR, R. T.
Department of Zoology, ARC Unit of Insect Physiology, University of Oxford, South Parks Road, Oxford OX1 3PS, England

UMAZUME, Y.
Department of Physiology, Jikei University Medical School, Minato-ku, Tokyo 105, Japan

WAKABAYASHI, K.
Department of Biophysical Engineering, Osaka University, Toyonaka, Osaka 560, Japan

WAKABAYASHI, T.
Department of Physics, Faculty of Science, University of Tokyo, Bunkyo-ku, Tokyo 113, Japan

WHITE, D. C. S.
Department of Biology, University of York, Heslington, York YO1 5DD, England

WILKIE, D. R.
Department of Physiology, University College London, Gower Street, London WC1E 6BT, England

WINEGRAD, S.
Department of Physiology, School of Medicine, University of Pennsylvania, Philadelphia, Pennsylvania 19104, U.S.A.

WOLEDGE, R. C.
Department of Physology, University College London, Gower Street, London WC1E 6BT, England

YAMADA, K.
Department of Physiology, School of Medicine, Juntendo University, Bunkyo-ku, Tokyo 113, Japan

YAMADA, T.
Faculty of Pharmaceutical Sciences, University of Tokyo, Bunkyo-ku, Tokyo 113, Japan

YAMAGUCHI, T.
Department of Biology, Division of Natural Sciences, College of Liberal Arts, International Christian University, Mitaka-shi, Tokyo 181, Japan

YAMAMOTO, T.
Department of Physiology, Faculty of Dentistry, Kyushu University, Fukuoka-shi 812, Japan

YANAGIDA, T.
Department of Biophysical Engineering, Osaka University, Toyonaka, Osaka 560, Japan

YOSHINO, S.
Mitsubishi-Kasei Institute of Life Sciences, Machida, Tokyo 194, Japan

Contents

Preface vii

Acknowledgements xii

Participants xiii

Introductory Remarks   S. Ebashi xxiii

I. SARCOMERE DYNAMICS

Structural Changes in Thick Filaments during Sarcomere Shortening in *Limulus* Striated Muscle
    M.M. Dewey, R.J.C. Levine, D. Colflesh, B. Walcott,
    L. Brann, A. Baldwin and P. Brink 3
    Discussion 20

Discrete Nature of Sarcomere Shortening in Striated Muscle
    G.H. Pollack, D.V. Vassallo, R.C. Jacobson, T. Iwazumi
    and M.J. Delay 23
    Discussion 35

Contraction in Overstretched Skinned Muscle Fibers
    Reiji Natori, Reibun Natori and Y. Umazume 41
    Discussion 49

Optical Diffraction Study on the Dynamics of Thin Filaments in Skinned Muscle Fibers
    S. Fujime, S. Yoshino and Y. Umazume 51
    Discussion 67

II. INSTANTANEOUS ELASTICITY AND STIFFNESS

Fast Response of Cardiac Muscle to Quick Length Changes
    M.J. Delay, D.V. Vassallo, T. Iwazumi and G.H. Pollack 71
    Discussion 81

The Origin of the Series Elasticity in Striated Muscle Fibers
    H. Sugi 85
    Discussion 100

Multiple Segment Analysis of Muscle Data
    A.H. Gott 103

Discussion 123

Muscle Stiffness in Relation to Tension Development of Skinned Striated Muscle Fibers
    J.C. Rüegg, K. Güth, H.J. Kuhn, J.W. Herzig, P.J. Griffiths and T. Yamamoto 125
Discussion 143

Effect of MgATP on Cross-Bridge Kinetics in Chemically Skinned Rabbit Psoas Fibers as Measured by Sinusoidal Analysis Technique
    M. Kawai 149
Discussion 166

Dynamic Stiffness of Heart Muscle in Twitch and Contracture
    K. Sagawa, Y. Saeki, L. Loeffler and K. Nakayama 171
Discussion 189

III. ISOTONIC AND ISOMETRIC TRANSIENTS

What Does Relaxed Insect Flight Muscle Tell Us about the Mechanism of Active Contraction?
    D.C.S. White, M.G.A. Wilson and J. Thorson 193
Discussion 209

Isometric Tension Transients in Skeletal Muscle before and after Inhibition of ATP Synthesis
    T. Blangé and G.J.M. Stienen 211
Discussion 222

Isotonic Velocity Transients and Enhancement of Mechanical Performance in Frog Skeletal Muscle Fibers after Quick Increases in Load
    T. Tsuchiya, H. Sugi and K. Kometani 225
Discussion 238

Load-Induced Length Transients in Mammalian Cardiac Muscle
    D.L. Brutsaert and P.R. Housmans 241
Discussion 256

Kinetic Analysis of Isometric Tension Transients in Cardiac Muscle
    G.J. Steiger 259
Discussion 271

IV. LENGTH-TENSION RELATION

The Length-Tension Relation in Skeletal Muscle: Revisited
    H.E.D.J. ter Keurs, T. Iwazumi and G.H. Pollack 277
Discussion 292

CONTENTS   xix

The Effect of Stretch on Contracting Skeletal Muscle Fibers
    K.A.P. Edman, G. Elzinga and M.I.M. Noble 297
    Discussion 309
Length-Tension Relation in Frog Skeletal Muscle Activated by Caffeine
and Rapid Colling
    T. Sakai 311
    Discussion 318
Contribution of Connectin to the Parallel Elastic Component in Muscle
    K. Maruyama and K. Yamamoto 319
    Discussion 328
Regulation of the Contractile Proteins in Cardiac Muscle
    W. Winegrad and G.B. McClellan 329
    Discussion 339
Discussion on the Length-Tension Relation 341

V. FORCE-VELOCITY RELATION

The Velocity of Shortening at Zero Load: Its Relation to Sarcomere
Length and Degree of Activation of Vertebrate Muscle Fibers
    K.A.P. Edman 347
    Discussion 356
Effect of "Viscosity" of the Medium on Mechanical Properties of Skinned
Skeletal Muscle Fibers
    M. Endo, T. Kitazawa, M. Iino and Y. Kakuta 365
    Discussion 374
Force-Velocity Relation in Tetanized Cardiac Muscle
    H. Mashima 377
    Discussion 387

VI. X-RAY DIFFRACTION STUDIES

Time Resolved X-ray Diffraction Studies on Muscle
    H.E. Huxley 391
    Discussion 401
The Use of Some Novel X-ray Diffraction Techniques to Study the Effect
of Nucleotides on Cross-Bridges in Insect Flight Muscle
    R.T. Tregear, J.R. Milch, R.S. Goody, K.C. Holmes and
    C.D. Rodger 407
    Discussion 421
X-ray Diffraction Studies on the Dynamic Properties of Cross-Bridges
in Skeletal Muscle
    Y. Amemiya, H. Sugi and H. Hashizume 425
    Discussion 441

The Structure of Thin Filament of Crab Striated Muscle in the Rigor State
    K. Namba, K. Wakabayashi and T. Mitsui 445

Arrangement of Troponin and Cross-Bridges around the Thin Filaments in Crab Leg Striated Muscle
    Y. Maéda 457
    Discussion (for the previous two papers) 469

VII. ENERGETICS

Problems of Muscle Energetics and Some Observations on the Period after a Stretch of Active Muscle
    R.C. Woledge and N.A. Curtin 473
    Discussion 479

An Explanation of the Shortening Heat Based on the Enthalpy Profile of the Myosin ATPase Reaction
    T. Kodama and K. Yamada 481
    Discussion 488

Myosin Isoenzymes in Skeletal and Cardiac Muscle
    J.F.Y. Hoh 489
    Discussion 498

Relation between Initial and Recovery Reactions in Skeletal Muscle
    M.J. Kushmerick 499
    Discussion 512

Studies of the Biochemistry of Contracting Muscle Using ^{31}P Nuclear Magnetic Resonance (^{31}P NMR)
    M.J. Dawson, D.R. Wilkie and D. Gardian 515
    Discussion 534

VIII. THEORIES OF CONTRACTION

Simplified Theory of the Huxley-Simmons T_0, T_1 and T_2 in Muscle Models with Two Attached States
    T.L. Hill and E. Eisenberg 541
    Discussion 560

Active Potential and Dynamic Cooperativity in the Chemo-Mechanical Conversion in Active Streaming and Muscle Contraction
    H. Shimizu, M. Yano, K. Nishiyama, K. Kometani, S. Chaen and T. Yamada 563
    Discussion 590

A Hydrodynamic Mechanism for Muscular Contraction
    R. Tirosh, N. Liron and A. Oplatka 593
    Discussion 608

A New Field Theory of Muscle Contraction
    T. Iwazumi 611
  Discussion 629

General Discussion 633

Concluding Remarks   H.E. Huxley 651

Contributors Index 657

Subject Index 659

Introductory Remarks

Setsuro EBASHI

Department of Pharmacology, Faculty of Medicine, and Department of Physics, Faculty of Science, University of Tokyo, Tokyo, Japan

It is my great privilege of having a chance to make introductory remarks on the occasion of the opening of this important Symposium.

The year 1954 was a very memorable one for muscle scientists. In an issue of *Nature* of that year, two brief notes, one by A.F. Huxley and R. Niedergerke and the other by J. Hanson and H.E. Huxley, were presented. At that time most muscle people had been attracted by *in vitro* studies on the actomyosin-ATP system as established by Albert Szent-Györgyi and his colleagues as early as 1941–2. These two papers which propose a novel idea made a strong impact on muscle scientists. We must not forget that the first paper by R. Natori, dealing with skinned muscle fibers, was also published in 1954.

H.E. Huxley then published a paper full of epoch-making electron micrographs in the *Journal of Biochemical and Biophysical Cytology* in 1957. This paper made the sliding concept more acceptable to all biological scientists. Moreover, the ingenious treatise of A.F. Huxley in *Progress in Biophysics and Biophysical Chemistry* in the same year has contributed a great deal to a deeper understanding of this concept. A. F. Huxley's paper helped in soothing the almost hostile opposition by some people, who had been insisting that this concept could not satisfy thermodynamic requirements. I personally was most impressed by H.E. Huxley's paper published in Scientific American in 1960 for the enlightment of "common people," in which the properties of the "crossbridge," *i.e.*, the subject of this Symposium, were adventurously described. It is my utmost pleasure to have him as the main speaker at this Symposium.

It is now almost half Jubilee of the sliding concept. It is no longer a hypothesis, but rather a fact. The remaining problem is to uncover details of this unique mechanism. In this connection many penetrating observations and sophisticated ideas have been presented towards the elucidation of the final mechanism, but we have not yet found the relevant route to the final goal. This may be the chief aim of the Symposium.

Towards this purpose, there are various approaches. One of them might be to examine the validity of observations and explanations by which the sliding concept has acquired support. It is not an easy task to be skeptical about evidence which seemingly supports the concept, but such skepticism is one of the things which may possibly emerge from this meeting.

I have heard that the chief qualification of a participant in this Symposium is that he should have dealt with living muscle, including Natori's fibers. Although I am a little reluctant to accept this principle in general, I completely agree with it in the present situation. Some young muscle scientists venture to ignore the real muscle contraction in our bodies and are concerned only with the *in vitro* reactions. One of them told me rather proudly, "I have never prepared myosin; it is a dirty system. I think we scientists should start from subfragment-1, but even it is . . . " I interrupted him, "What I can do for you right now is to teach you how to kill a rabbit mercifully."

Now I would like to ask the indulgence of the audience to allow me to give my personal opinion. As a person working on smooth muscle for the time being, I feel a little sorry that I can find virtually no paper dedicated to smooth muscle in the program. Skeletal muscle is certainly a beautiful creation. I understand that the people working on skeletal muscle wish to believe that every secret of contraction can be found in this muscle. Working with smooth muscle, however, I am more and more convinced that skeletal muscle has been highly specialized and differentiated in order to carry out its specific function: the rapid and precise contraction which can meet the incredible skill of athletes and artists. I am now inclined to think that the prototype of contraction is rather to be seen in smooth muscle. So, I would consider it a personal favor if smooth muscle were to be included as a formal topic in the second meeting of this Symposium, which I hope will be held in the near future.

Finally, I would like to express my hearty thanks to Professor H. Sugi, the organizer, President S. Okinaga, the sponsor, and the speakers and audience who have made it possible to have this meaningful Symposium. The planning of this Symposium as a satellite meeting has certainly contributed to the success of the Biophysics Congress in Kyoto. As one of the organizers of the Congress, too, my sincere thanks are also due to all of them.

I. SARCOMERE DYNAMICS

Structural Changes in Thick Filaments during Sarcomere Shortening in *Limulus* Striated Muscle

M. M. DEWEY, R. J. C. LEVINE, D. COLFLESH, B. WALCOTT, L. BRANN, A. BALDWIN and P. BRINK

Department of Anatomical Sciences, School of Basic Health Sciences, Health Sciences Center, State University of New York at Stony Brook, New York, U.S.A. and Department of Anatomy, The Medical College of Pennsylvania, Philadelphia, Pennsylvania, U.S.A.

ABSTRACT

Using phase optics, immunochemical staining for myosin and paramyosin (Levine, Dewey & de Villafranca, 1972) and electron microscopy of sectioned material (Dewey, Levine & Colflesh, 1973) we have confirmed de Villafranca's (1961, de Villafranca & Marschhaus, 1963) observation that during sarcomere; shortening the A-band shortens. As determined by electron microscopy, the A-band shortening has two phases: (1) from rest length to longer sarcomeres, the thick filaments shear past each other, and (2) from rest length to shorter sarcomeres, the thick filaments shorten. Further, we have isolated thick filaments from living, resting and K^+ or electrically stimulated and glycerinated muscle of various sarcomere lengths (Dewey, Walcott, Colflesh, Terry & Levine, 1977). Thick filaments isolated from fibers with short sarcomeres are short (3.4 μm), while thick filaments isolated from fibers with long sarcomeres are long (4.5 μm).

We will describe recent studies designed to elucidate the mechanism of thick filament shortening.

1. Determination of A-band length by Fourier analysis of optical diffraction patterns at various sarcomere lengths of living, fixed and glycerinated muscle fibers and ATP-induced shortening glycerinated fibers.

2. The ionic and ATP requirements for *in vivo* shortening of isolated long thick filaments.

3. Activation curves for isolated filament shortening at different concentrations of ATP and Ca^{++}.

---

[1] *In memorium:* George Warren de Villafranca, 1922–1977.
[2] This investigation was supported by grants from the National Institutes of Health, GM 20628 and HL 15835 to the Pennsylvania Muscle Institute. Rhea J. C. Levine is a recipient of a Research Career Development Award from the National Institutes of Neurological and Communicable Diseases and Stroke (70476).

4. Lengthening of shortened isolated thick filaments using chicken intestine alkaline phosphatase.

5. The changes in molecular packing during shortening of *Limulus* thick filaments determined by optical diffraction studies of isolated negatively stained thick filaments.

6. Using a similar approach we are studying thick filament shortening which occurs in lobster abdominal slow flexor muscle.

In 1963, de Villafranca & Marschhaus described shortening of the A-bands in glycerinated myofibrils of *Limulus polyphemus* striated muscle that had been induced to shorten by the addition of ATP. In addition, de Villafranca & Leitner (1967) also showed that this muscle contained paramyosin, a protein thought to be restricted to molluscs and annelids. These two surprising observations prompted us to examine this peculiar muscle from a number of different points of view.

Immunocytochemical localization of myosin and paramyosin

In these experiments we examined isolated fibrils of glycerinated *Limulus* muscle. We confirmed de Villafranca's observation that the A-band length changed with the sarcomere length (Dewey, Levine & Colflesh, 1973). We also demonstrated that both antimyosin and antiparamyosin bound exclusively to the A-band. We noted, however, that the particular staining pattern observed varied with the sarcomere length, and thus the A-band length in this muscle (Levine, Dewey & de Villafranca, 1972). At long sarcomere lengths (from 7 to 11 μm) antimyosin stained the entire A-band while antiparamyosin stained only the lateral margins of the A-band. In short sarcomeres (from 3 to 7 μm), however, the staining pattern was reversed: antimyosin now stained the lateral margins of the short A-bands and antiparamyosin stained throughout the A-band. These experiments convinced us that the A-band did change in length and further that molecular rearrangements involved in this phenomenon were probably responsible for the different staining patterns at different sarcomere lengths. In order to determine the basis of the change in A-band length, we undertook an ultrastructural analysis of the muscle fixed at different sarcomere lengths.

Ultrastructural analysis

Examination of longitudinally sectioned, glutaraldehyde/OsO$_4$-fixed muscle at different sarcomere lengths confirmed our and de Villafranca's observations with the light microscope. The overall A-band length varied linearly with the sarcomere length, and this A-band length change appeared to have two components. When the sarcomere length was increased from 7 to 11 μm, the A-band also increased in length, but the constituent thick filaments remained at a constant length of 4.9 μm. This A-band length increase appeared to be due to a skewing of the thick filaments relative to each other. As the sarcomere length decreased from 7 to 3 μm, the thick filaments were more aligned within the A-band, and the A-band shortening seemed to be due to a shortening of the thick filaments themselves (Fig. 1) (Dewey *et al.*, 1973). A number of other approaches were employed to study this phenomenon further.

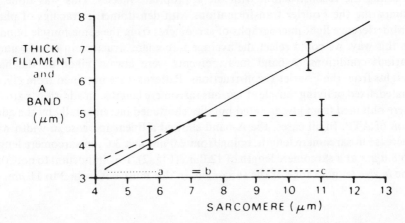

FIG. 1. Graph showing the relationship of both A-band and thick filament length to sarcomere length. Slopes were computed from measurements made on electron micrographs of longitudinally sectioned (sections cut normal to the fiber axis) freshly fixed muscle fibers. The thick filament lengths and overall A-band lengths (including misalignment) were measured independently on medium power (approximately 10,000–20,000 ×) electron micrographs of individual sarcomeres. Thick filament lengths were recorded only from filaments which could be traced entirely through the A-band and when at least ten filaments of the same length were measured per sarcomere. N in this case refers to sarcomeres having been analyzed, not individual filaments measured. Bars indicate standard errors of slopes. Confidence interval for all slopes: $P < 0.01$. Solid line, overall A-band length, $N = 232$; dash line, thick filament length, $N = 190$ (sarcomeres). a, b, and c along the abscissa are ranges of sarcomere lengths of medial telson levator muscles as determined by laser diffraction. Range includes variation among animals: a, telson maximally elevated; b, telson at 180° to carapace (straight out); c, telson maximally depressed.

Optical diffraction of muscle bundles

In *Limulus* muscle, as with other striated muscles, when a laser beam is passed through a single fiber or even a bundle of fibers, a diffraction pattern is obtained consisting of a series of layer lines. This pattern reflects the underlying pattern of refractive index differences in the muscle. One can, by measuring the spacing and relative intensities of the layer lines, create Fourier reconstructions of the refractive index along an average sarcomere. In order to do this, however, a high degree of order must be present. This can be judged by the number of layer lines and their width. In all our work (Dewey, Blasie, Levine & Colflesh, 1972) we have only used patterns in which the number of layer lines is equal to, or greater than the sarcomere length in microns less 1. Thus, in a fiber with a sarcomere length of 9 μm, there must be at least 8 layer lines on either side of the central spot. Since there are many possible Fouriers, the problem is one of selecting the reconstruction with the appropriate phases. This was done by comparing the Fourier transformations with densitometric tracings of phase or interference light micrographs of sarcomeres from the same muscle bundle. In this way, we could select the average sarcomere structure in muscle under various conditions. A-band measurements were always made as the half-heights from the Fourier reconstructions. Patterns were obtained from glycerinated, fixed or living muscle at various sarcomere lengths. In addition, patterns were obtained from glycerinated bundles shortened incrementally by the addition of ATP. In all cases, the A-band showed a linear increase in width with increase in sarcomere length, going from 2.0 μm at a 3.0 μm sarcomere length to 8.6 μm at a sarcomere length of 12.0μm (Fig. 2). It is important to note that the *in vivo* length range of the sarcomere in this muscle is from 3 to 11 μm. At

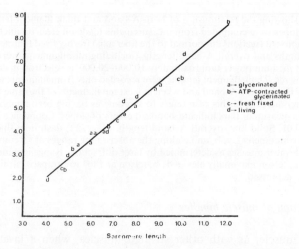

FIG. 2. The linear relation between sarcomere and A-band length determined in a number of different preparations by Fourier reconstruction from laser diffraction patterns.

7.1 μm and above, the reconstructions showed shoulders developing at the margins of the A-bands. This was interpreted as reflecting the skewing of the thick filaments. These experiments confirmed the length change of the A-band but did not provide any informaiton on the mechanism(s) underlying the change.

Isolation of thick filaments

Other than filament shortening, a possible mechanism for the observed A-band length change was the alignment and skewing of consistently short thick filaments. To examine this, we measured the lengths of negatively stained thick filaments which were isolated from muscle under various conditions (Dewey, Walcott, Colflesh, Terry & Levine, 1977). We found that filaments isolated from muscle briefly glycerinated at short sarcomere lengths were short (2.9 μm) while filaments isolated from long sarcomere muscle were significantly longer (4.4 μm) (Table 1). Similarly, the thick filaments were short when isolated from electrically stimulated or K^+-contracted fiber bundles, while they were long from unstimulated long sarcomere muscle bundles. These results confirmed the earlier measurements from longitudinal sections and showed that at least part of the A-band length change was due to a significant length change in the constituent thick filaments.

TABLE 1. Relationship between sarcomere, A-band and thick filament lengths under various conditions

	Sarcomere length (μm)	A-band width (μm)	Isolated thick filament length (μm)
Glycerinated	8.1 ± 1.2 (230)*	4.5 ± 1.9 (230)	4.4 ± 0.5 (95)
Glycerinated	6.4 ± 1.0 (208)	3.4 ± 0.6 (208)	2.9 ± 0.5 (90)
Living (EGTA)	7.7 ± 1.1 (273)	4.0 ± 0.6 (273)	4.0 ± 0.7 (127)
Living (stimulated) Electrically or High K^+	5.0**		2.9 ± 0.3 (100)

* Mean ± standard deviation (number of observations)
** Determined by laser diffraction

In a parallel series of experiments we examined the ability of the filaments to lengthen *in situ*. Small bundles of living intact fibers were electrically stimulated and allowed to shorten. The bundles were then soaked in either Ringer's or a relaxing solution containing ATP and EGTA. The bundles were then stretched and the thick filaments were isolated. The filaments isolated from the bundles soaked in Ringer's and then stretched were long, while those isolated from the relaxing solution treated-bundles were short as if the bundles had not been stretched. This result suggests that lengthening of the thick filaments *in situ* may depend on some thick and thin filament interaction.

Structural changes in the thick filament

We then sought to observe the structural modifications underlying the change in thick filament length. First, we measured thick filament diameters in cross-sections of muscle bundles of known sarcomere lengths (Dewey, Colflesh, Walcott & Levine, 1977). In all cases the thick filament profiles were elliptical and the long axes of the ellipses appeared to be randomly oriented in the fibril. From measurements of the diameters of both the long and short axes, we calculated the cross-sectional areas for many filaments at each sarcomere length. In all cases, filament diameter was measured in the A-I overlap region. The data show that as the sarcomere length increases from 7 to 9 μm, there is no change in the thick filament cross-sectional area. However, as the sarcomere length decreases below 7 μm, the thick filament cross-sectional area increased dramatically (Table 2). This agrees with our previous observation that thick filament shortening occurs only below 7 μm and further shows that as the filament shortens, it gets "fatter." As yet we have no evidence as to whether the change is uniform along the length of the filament or whether the change is incremental as the filament shortens. Measurements of filament diameters in the pseudo H zone showed no change as a function of sarcomere length.

TABLE 2. Thick filament dimensions at different sarcomere lengths

Sarcomere length (μm)	Thick filament diameter (nm)	Thick filament cross-sectional area (nm^2)	Number of observations
9.4	25.1 ± 2.5*	398 ± 6.1	746
7.4	24.6 ± 1.6	425 ± 2.1	111
6.4	26.6 ± 3.5	459 ± 7.7	1796
5.9	33.0 ± 3.2	673 ± 8.8	526
4.4	36.6 ± 4.6	841 ± 14.4	543

* Mean ± standard deviation

It would not be unreasonable to expect change in the surface organization of myosin as the filaments change length. Therefore, Millman (Millman, Warden, Colflesh & Dewey, 1974) obtained X-ray diffraction patterns from glycerinated bundles supplied by us at known sarcomere lengths. No changes were observed in either the 14.5 μm reflection or the strong layer lines at 38.0 and 59.0 μm in these bundles, which ranged from 4.5 to 9.0 μm in sarcomere length. Neither were Wray, Vibert & Cohen (1974) able to detect any change in the 14.5 nm reflection, although they did notice a loss of the sharpness of the pattern at longer sarcomere lengths which, as they suggest, may reflect the disorder due to skewing of the thick filaments. Therefore, any model for the filament shortening must take into account the constancy of the 14.5 nm period which is presumed to represent the myosin or underlying paramyosin period. This does not preclude, however, longer range shifts which might not be visible on even low angle X-ray.

Therefore, we (Levine & Dewey, in press) have examined the optical diffraction patterns that can be obtained from portions of negatively stained isolated filaments of different lengths. We have grouped the filaments into long (4.4 ± 0.5 μm), intermediate (3.7 ± 0.5 μm) and short (3.2 ± 0.3 μm) classes. All filaments produced 14.5 nm meridional reflections which is consistent with the X-ray results. However, the long filaments gave a layer line at 145 nm while the short filaments gave one at 220 nm. Intermediate filaments showed both layer lines (Fig. 3). These results suggest that the interbridge spacing remains constant, but that the helical repeat of the bridges along the filament may increase as the filaments become shorter and increases in diameter. Further experiments are needed to confirm this point.

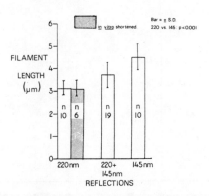

FIG. 3. Bar graph with ordinate = thick filament lengths and abscissa = layer line reflections observed using optical diffraction. Thick filaments isolated long showed a layer line at 145 nm. Short thick filaments gave a layer line at 220 nm. Intermediate thick filaments showed both 145 nm and 220 nm reflections. Note: Shaded bar represents diffraction of *in vitro* shortened fialments.

In vitro *thick filament shortening*

As noted earlier, thick filaments of different lengths can be isolated from muscle glycerinated at different sarcomere lengths. In all these experiments, the isolations were performed in a solution containing ATP and EGTA, so that the free Ca^{++} concentration was kept very low. Addition of Ca^{++} to a suspension of long thick filaments resulted in a significant decrease in thick filament length (Dewey, Baldwin, Colflesh, Walcott & Brink, 1977). To examine this further, we purified the thick filament preparations using electrophoretically pure DNase 1 (0.4 mg/ml) which binds in a mole:mole ratio with g actin (Hitchcock, Carlsson & Linberg, 1976), and phenylmethyl sulfonyl fluoride (1 %) and dimethyl sulfoxide (1 %) to reduce any contaminating protease activity. The preparations were then treated with DEAE cellulose which binds F-actin (Morimoto and Harrington, 1974). This whole procedure produced preparations which appeared relatively free of actin when examined with negative staining in the electron microscope or when run on SDS polyacrylamide gels. Using these preparations, we examined the ionic and energy requirements for filament shortening.

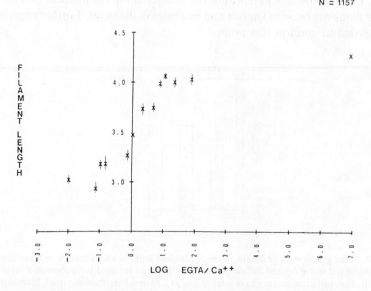

FIG. 4. *Limulus* telson muscle. Graph showing the effects of varying $EGTA/Ca^{++}$ ratio, incrementally changing free Ca^{++} in the presence of 5 mM ATP, 2mM $MgCl_2$, 0.1 M KCl, 1 mM DTT and 5 mM Tris, pH 7.0. The abscissa is the log ($ETGA/Ca^{++}$).

We determined that Ca^{++} was necessary for thick filament shortening and by varying the ratio of Ca^{++} to EGTA, obtained a curve of filament shortening (Fig. 4) related to Ca^{++} concentration. The curve is sigmoid and shows maximum shortening at an EGTA/Ca^{++} ratio of 1, which is an estimated pCa of 5.5. Intermediate Ca^{++} concentrations produced filaments of intermediate lengths

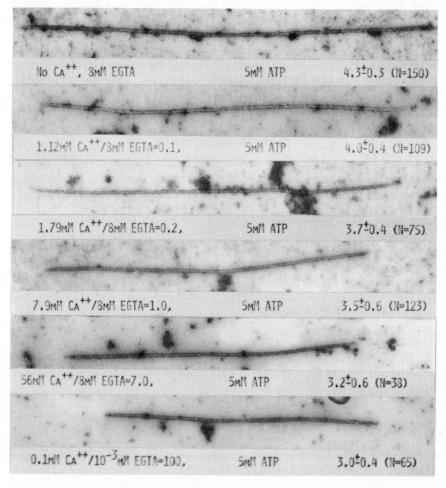

FIG. 5. *Limulus* telson muscle. Shortening of isolated thick filaments in the presence of increasing concentrations of free Ca^{++} (EGTA/Ca^{++}), ATP concentration constant at 5 mM, pH 7.0. Electron micrographs of negatively stained isolated thick filaments. Total Mag. = 25,000 ×.

(Fig. 5). This may mean that multiple Ca^{++} binding sites are involved and that Ca^{++} is not simply acting as a "trigger" in thick filament shortening.

Similarly, keeping the Ca^{++} concentration high and constant and varying the ATP concentration produces a curve of ATP activation (Fig. 6) showing that ATP is required for shortening as well. Again, intermediate ATP concentrations produce filaments of intermediate length (Fig. 7). More specifically, ATP hydrolysis is needed as shown by the substitution of a nonhydrolyzable analog of ATP (adenylyl imidodiphosphate, AMP-PNP) (Yount, Frye and O'Keefe, 1973) for ATP. Under these conditions, even with high free Ca^{++}, no thick filament shortening occurred (Fig. 8). These results show that both ATP and Ca^{++} at physiological levels are required for thick filament shortening (Fabiato and Fabiato, 1975; Kushmerick and Davies, 1969; Störer and Cornish-Bowden, 1976).

In vitro *thick filament lengthening*

A very important question involves the mechanism(s) of thick filament lengthening. It appears that ATP and Ca^{++} induce shortening but simple removal of these molecules and ions does not cause the filaments to lengthen.

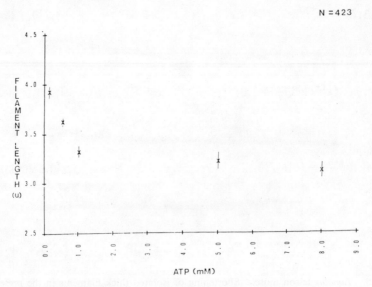

FIG. 6. *Limulus* telson muscle. Graph showing the effects of incremental changes in ATP concentration in the presence of an optimal Ca^{++} concentration ($Ca^{++}/EGTA = 0.7$), pH 7.0

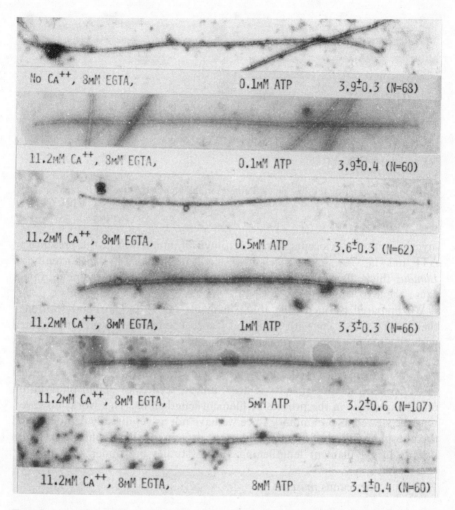

FIG. 7. *Limulus* telson muscle. Shortening of isolated thick filaments in the presence of an optimal Ca^{++} (Ca^{++}/EGTA = 0.7) concentration by increasing concentration of ATP, pH 7.0. Electron micrographs of negatively stained isolated thick filaments. Total Mag. = 25,000 ×.

Recent works on various vertebrate skeletal (Bremel & Weber, 1975; Weber & Oplatka, 1974; and Perrie, Smillie & Perry, 1972) and vertebrate and invertebrate smooth muscle (Lehman, 1977; Sobieszek & Bremel, 1975; and Bremel, Sobieszek and Small, 1977) have suggested that phosphorylation of

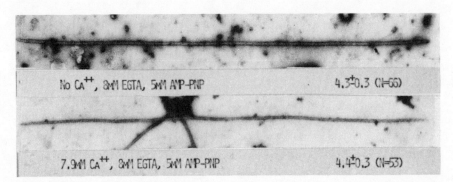

FIG. 8. *Limulus* telson muscle. Thick filaments in the presence of the ATP analog, adenyl imidodiphosphate (AMP-PNP), in the absence and in the presence of Ca^{++}, pH 7.0. Total Mag. = 25,000 ×.

myosin light chains occurs during actomyosin interaction in the presence of ATP and Ca^{++}. A similar mechanism might be involved in the shortening of *Limulus* thick filaments. If so, then dephosphorylating the filaments might cause them to lengthen.

We prepared purified long thick filaments (4.1 ± 0.5 μm) and shortened them to 3.0 ± 0.5 μm with ATP and Ca^{++}. The filaments were then treated with alkaline phosphatase (0.4 mg/ml, Sigma Chemical Corp.). The filaments when measured after this treatment had significantly relengthened (3.7 ± 0.4 μm, $p < 0.001$) them (Fig. 9) (Brann, Dewey, Baldwin, Walcott & Brink, submitted to *Nature*). From these results it does indeed seem that dephosphorylation may be involved in the process of filament lengthening. This work, however, must be viewed as preliminary since we have not as yet directly demonstrated the phosphorylation of any thick filament protein, or that dephosphorylation occurs during filament lengthening. These studies are under way.

Homarus americanus *muscle*

It could be argued that *Limulus* is a "living fossil" and that the phenomena we have described are unique to that animal. Therefore, we have begun to investigate muscles from other animals to determine the generality of the phenomenon. While other workers have observed A-band length changes in other muscles, staghorn beetle (Bowman, 1840) and barnacle adductor (Silverston, 1964), we have examined the slow and fast flexor muscles of the American lobster. In this system, the fast flexor muscle occupies the majority of the abdomen and is responsible for the rapid tail flip used by the animal during es-

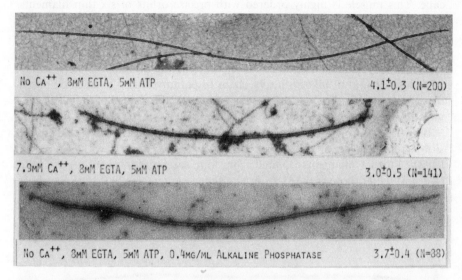

FIG. 9. *Limulus* telson muscle. Top: Thick filament isolated from a long sarcomere. Middle: Filament shortened in the presence of Ca^{++} and ATP. Bottom: Filament relengthened following incubation of 3 hr with alkaline phosphatase, pH 7.0. Electron micrograph of negatively stained isolated thick filaments. Total Mag. = 25,000 ×.

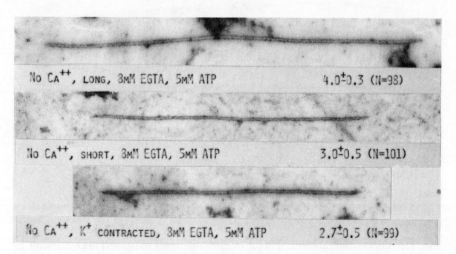

FIG. 10. Lobster slow abdominal flexor muscle. Thick filaments isolated under three conditions; long sarcomere, short sarcomere and high K$^+$ added to the bathing medium, pH 7.0. Electron micrographs of negatively stained isolated thick filaments. Total Mag. = 25,000 ×.

cape. This muscle is highly ordered with precise orbits of six thin filaments around each thick filament, and physiologically is a twitch-type of muscle. The slow flexor muscle, a thin ventral sheet, is responsible for maintenance of abdominal position and is probably not used during the tail flip. The slow muscle is very similar to *Limulus* muscle in structure although it has a somewhat higher ratio of myosin heavy chain to paramyosin. We have compared these two muscles, the slow and the fast, and looked for some of the phenomena observed in *Limulus*.

Both laser diffraction and measurements of fixed muscle show that in the slow abdominal flexor the A-band length increases linearly with sarcomere length. Further, filaments isolated from long sarcomeres (11.0 μm) are long (4.0 μm) while those isolated from shortened sarcomeres (4.0 μm) are short (2.7 μm) (Fig. 10). Preparations of long isolated thick filaments can be induced to shorten in the presence of ATP and Ca^{++} in concentrations similar to those

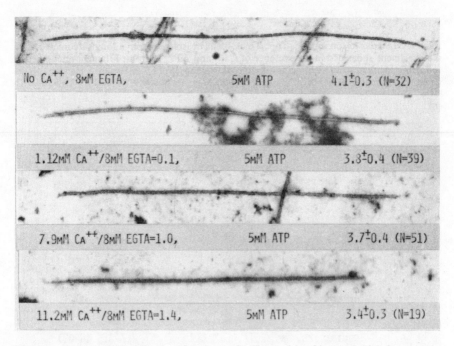

FIG. 11. Lobster slow abdominal flexor muscle. Shortening of isolated thick filaments by changing the free Ca^{++} concentration (5 mM ATP), pH 7.0. Electron micrographs of negatively stained isolated thick filaments. Total Mag. = 25,000 ×.

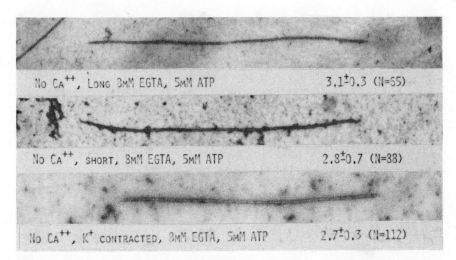

FIG. 12. Lobster fast abdominal flexor muscle. Thick filaments isolated under three conditions; long sarcomere, short sarcomere and high K^+ added to the bathing medium, pH 7.0. Electron micrographs of negatively stained isolated thick filaments. Total Mag. = 25,000 ×.

used in the *Limulus* experiments (Fig. 11). The fast muscle on the other hand appears to have filaments of constant length (sarcomeres ranging from 4.8 to 3.2 μm, thick filaments from 3.1 to 2.9 μm, $p < 0.05$) which do not change *in vivo* or when isolated and treated with high ATP and Ca^{++} (Fig. 12 and 13).

These preliminary experiments show that the phenomena seen in *Limulus* are more general and occur in some, but clearly not all, invertebrate striated muscles.

Our observations indeed do not run counter to the accepted sliding filament theory of muscle contraction. Instead it appears that at least several invertebrate muscles have incorporated an additional component into the contractile mechanism: thick filament shortening. This type of modification may relate to a specific functional requirement such as great extensibility of these muscles. It is interesting to note that Davidheiser, Delluva & Davies (1978) have found a 50% decrease in efficiency of *Limulus* muscle contraction when the starting muscle length is short when compared to a longer initial muscle length. The possibility exists that the decreased efficiency may result from a diversion of phosphate from actin-myosin activation to thick filament shortening. Many important questions regarding both the molecular and the regulatory mechanisms of thick filament length changes remain to be answered.

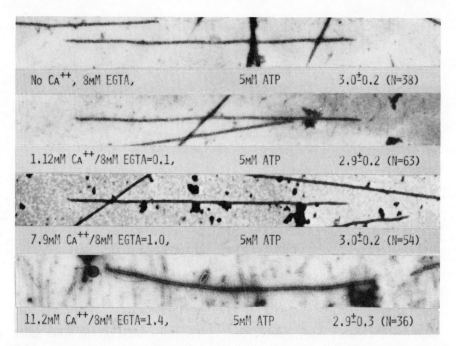

FIG. 13. Lobster fast abdominal flexor muscle. Incremental changes in Ca^{++} concentration (5 mM ATP), pH 7.0, do not produce shortening of isolated thick filaments. Electron micrographs of negatively stained isolated thick filaments. Total Mag. = 25,000 ×.

REFERENCES

Bowman, W. (1840). On the minute structure and movements of voluntary muscle. *Phil. Trans. R. Soc.* 451.

Bremel, R. D., Sobieszek, A. & Small, J. V. (1977). Regulation of actin myosin interaction in vertebrate smooth muscle. In *Biochemistry of Smooth Muscle.* ed. Stephens, N. L., pp. 533–549, Baltimore: Univ. Park Press.

Bremel, R. D. & Weber A. (1975). Calcium binding to rabbit skeletal myosin under physiological conditions. *Biochim. biophys. Acta* **376,** 366–374.

Davidheiser, S., Delluva, A. M. & Davies, R. E. (1978). Energy usage during A-band shortening in *Limulus* muscle. Presented at the ASBC/AAI Meetings, Atlanta, Georgia.

de Villafranca, G. and Leitner, V. (1967). Contractile proteins from horseshoe crab muscle. *J. gen. Physiol.* **50,** 2495–2496.

de Villafranca, G. & Marschhaus, C. (1963). Contraction of the A band. *J. Ultrastruct. Res.* **9,** 156–165.

Dewey, M. M., Baldwin, E., Colflesh, D. E., Walcott, B. & Brink, P. (1977). Activated *in vitro* shortening of isolated thick filaments of *Limulus* striated muscle. *J. cell Biol.* **75,** 322a.

Dewey, M. M., Blasie, J. K., Levine, R. J. C. & Colflesh, D. E. (1972). Changes in A-band structure during shortening of a paramyosin-containing striated muscle. *Biophys. Soc. Annu. Meet. Abstr.* **12**, 82a.

Dewey, M. M., Colflesh, D. E., Walcott, B. & Levine, R. J. C. (1977). Changes in length, diameter and volume of the thick filaments of *Limulus* striated muscle during sarcomere shortening. *Biophys. Soc. Annu. Meet. Abstr.* **17**, 160a.

Dewey, M. M., Levine, R. J. C. & Colflesh, D. E. (1973). Structure of *Limulus* striated muscle. The contractile apparatus at various sarcomere lengths. *J. cell Biol.* **75**, 366–375.

Dewey, M. M., Walcott, B., Colflesh, D. E., Terry, H. & Levine, R. J. C. (1977). Changes in thick filament length in *Limulus* striated muscle. *J. cell Biol.* **75**, 366–380.

Fabiato, A. & Fabiato, F. (1975). Effects of Mg^{++} on contraction activation of skinned cardiac cells. *J. Physiol.* **249**, 497–517.

Hitchcock, S. E., Carlsson, L. & Linberg, U. (1976). DNase I-induced depolymerization of actin filaments. In *Cell Motility*, pp. 545–560, eds. Goldman, Pollard and Rosenbaum. Cold Spring Harbor Conference on Cell Proliferation, Cold Spring Harbor, New York.

Kushmerick, M. J. & Davies, R. C. (1969). The chemical energetics of muscle contraction. II. The chemistry, efficiency and power of maximally working sartorius muscles. *Proc. R. Soc. B.* **174**, 315–353.

Lehman, W. (1977). Calcium ion dependent myosin from decapod crustacean muscles. *Biochem. J.* **163**, 291–296.

Levine, R. J. C. & Dewey, M. M. (in press). Changes in molecular packing during shortening of *Limulus* thick filaments. In *Motility in Cell Function*. John Marshall Symposium in Cell Biology, ed. Pepe, F. E., New York: Academic Press.

Levine, R. J. C., Dewey, M. M. & de Villafranca, G. (1972). Immunohistochemical localization of contractile proteins in *Limulus* striated muscle. *J. cell Biol.* **55**, 221–235.

Millman, B. M., Warden, W. J., Colflesh, D. E. & Dewey, M. M. (1974). X-ray diffraction from glycerol-extracted *Limulus* muscle. *Biophys. Soc. Annu. Meet. Abstr., Fed. Proc.* **33-5**, 622, 1333.

Morimoto, K. & Harrington, W. F. (1974). Evidence for structural changes in vertebrate thick filaments induced by calcium. *J. molec. Biol.* **88**, 693–708.

Perrie, W. T., Smillie, L. B. & Perry, S. V. (1972). A phosphorylated light chain component of myosin from skeletal muscle. *Biochem. J.* **128**, 105p.

Silverston, A. (1964). Mechanism of contraction in barnacle muscle. Masters Thesis, University of Oregon.

Sobieszek, A. & Bremel, R. D. (1975). Preparation and properties of vertebrate smooth muscle myofibrils. *Eur. J. Biochem.* **55**, 49–60.

Störer, A. C. & Cornish-Bowden, A. (1976). Concentration of Mg-ATP and other ions in solution. *Biochem. J.* **159**, 1–5.

Weber, M. M. & Oplatka, A. (1974). Physicochemical studies on the light chains of myosin. III. Evidence for a regulatory role of a rabbit myosin light chain. *Biochim. biophys. Res. Commun.* **57**, 823–830.

Wray, J. S., Vibert, P. J. & Cohen, C. (1974). Cross-bridge arrangements in *Limulus* muscle. *J. molec. Biol.* **88**, 343–348.

Yount, R. G., Frye, J. S. & O'Keefe, K. R. (1973). Inhibition of heavy meromyosin by purine disulfide analogs and adenosine triphosphate. Cold Spring Harb. Symp. quant. Biol. **37**, 113–118.

Discussion

Edman: Do you get any shortening of the thick filament during the isometric contraction?

Dewey: No. Once an isometric contraction is reached, the thick filaments will be graded in length, depending on whether the sarcomere length is 4 μm or 6 μm or 7 μm.

Edman: Oh, but that is set by resting length, isn't it?

Dewey: It is set by the length to which the muscle is allowed to shorten.

Edman: Does contractile activity alone cause any shortening?

Dewey: From our observations it doesn't cause any shortening above 7 μm, i.e., above rest length. There is another interesting aspect that we have looked at. If you take the muscle and shorten it either electrically or by raising extracellular K^+, then put it back into a physiological solution, then stretch it back, then isolate the thick filaments, they are long; on the other hand, if you take a similar bundle and shorten it by the same processes, and instead of transferring it to a physiological solution and stretch it, the thick filaments are shortened. Lengthening of the thick filament may require thick and thin filament interaction.

Sugi: Your presentation gives me the impression that the thick filament can shorten actively. Is that correct?

Dewey: Yes.

Sugi: Then, it seems to me to contradict the cross-bridge concept, because if the thick filament can shorten actively, then this may generate isometric force transmitted through the cross-bridge. So the force for generating tension may originate from the tendency of thick filaments to shorten, but not from cross-bridge motion.

Dewey: Considering the mechanical structure of this muscle, thick filament shortening must be tension generating. This is a different system for force generation. However, I do feel our work is consistent with the cross-bridge theory on the basis of the optical and X-ray diffraction of the muscle.

Pollack: In Fig. 4, I think there may be some confusion as to what you mean by steps. Do you mean as you increase the calcium concentration, the thick filament shortens progressively, or do you mean that as you increase the calcium concentration there are certain ranges in which you find no change of thick filament length, and as you increase it further, the thick filaments shorten further?

Dewey: That is what I meant. And then you reach a point, as you further increase the calcium, keeping ATP constant, that the filaments will shorten no further.

ter Keurs: If you stimulate *Limulus* muscles electrically, do you also see stepwise shortening of myosin?

Dewey: Yes. We do see that. It depends on how short you let the muscle go at any given length. You also see step-wise lengthening of myosin- and paramyosin-containing thick filaments.

ter Keurs: Is the phenomenon shown in Fig. 4 due to a fluctuation of the calcium ion concentration, or is it a property of the structure of the myosin?

Dewey: One can't say whether it is the myosin or paramyosin because we don't know which, if either of these, is responsible for the shortening. It could be a third protein. As we said, there is another protein associated with the contractile apparatus. And, yes, you would assume a variation in calcium concentration in the cell, *in vivo*. The thick filament shortens to a length characteristic of the isometric sarcomere length of the muscle, a sarcomere length below 7.0 µm. We've seen it in thick filaments isolated from living muscle.

Winegrad: Have you done the reverse of the phosphorylation study, *i.e.*, to take the long filaments and try to phosphorylate them with protein kinase?

Dewey: No. We are just in the process of doing that, and I can't give you of any data on it yet.

Huxley: In Fig. 2, you showed very clearly that the A filament length was related solely to the sarcomere length; that is, you had a number of different kinds of physiological states of the muscle, but the points all lay on the same curve.

Dewey: The A-band varies linearly with the sarcomere length. The thick filament remains constant in length, according to our measurements, from rest to long sarcomere lengths, but from rest to short sarcomere lengths, it shortens.

Huxley: Yes. But those two statements are true for a number of different types of muscles—relaxed muscle, contracting muscle, or glycerinated muscle, and so on; that is all right. I mean, it seems that the structure depends on the sarcomere length. It is uniquely related to the sarcomere length. But your other results showed that the A filament length was related to the state of activity, or to the calcium concentration; so, there seems to be a difference between those two. How do you reconcile them?

Dewey: The force generated by the shortening, if there is a force, is not overridden by the force generated by the contractile mechanism.

Huxley: Well, I am not familiar with the length-tension diagram of these muscles, but, supposing you have a sarcomere length of, say, 11 µm with a rather small amount of overlap, what sort of tension do you get there?

Dewey: You get 50% of maximum tension. We have obtained the length-tension curve (Walcott & Dewey, 1978).

Huxley: And then below rest length, supposing you go to, say, 6 μm, what sort of tension do you get there?

Dewey: I can't remember the data at 6 μm, but if you let me pick 4 μm, then you have again about 50% of maximum tension. It is a very broad length-tension curve.

Huxley: That would mean, then, that the tension being applied to the thick filaments was smaller at the long sarcomere lengths where there was very little overlap. And, therefore, you might think, if it was that force that was stopping the filaments from shortening, that they'd shorten more at the long length.

Dewey: Well, I agree, that is a puzzle. The biggest puzzle of all is that, in fact, you can get that much tension on the muscle when it should be all non-overlap. Even with the skewing it is still a puzzle. You have to somehow transmit tension through the sarcomere. I don't understand it. You do see bits of actin in the A-band, even at long sarcomere lengths.

REFERENCE

Walcott, B. & Dewey, M. M. (1978). Length-tension relation of *Limulus* muscle. *Biophys. J.* **21**, 55a.

Discrete Nature of Sarcomere Shortening in Striated Muscle

G. H. POLLACK, D. V. VASSALLO, R. C. JACOBSON, T. IWAZUMI and M. J. DELAY

Department of Anesthesiology and Division of Bioengineering, University of Washington, Seattle, Washington, U.S.A.

ABSTRACT

We measured sarcomere shortening in rat trabecular muscle and single fibers of skeletal muscle, using a light diffraction technique. Sarcomere shortening, measured in any sampled region along the specimen, was generally punctuated by a series of pauses, each of which persisted for up to tens of milliseconds (Pollack *et al.*, 1977). The same pattern of pauses and shortening was usually repeated over several succeeding contractions.

We present evidence ruling out a series of potential artifacts. Further, we present preliminary data showing that the pauses are associated with inflections in the rising tension. Thus we believe the phenomenon is genuine.

The existence of well-defined pauses is indicative of a high degree of synchrony of action among widely dispersed contractile elements: the sarcomere length computed from the light diffraction signal is derived from contributions from sarcomeres situated throughout the volume of tissue transsected by the laser beam, ca. $(100 \mu m)^3$. It is not possible to detect a pause in such a field unless either (i) a population of lengthening sarcomeres exactly cancels the effect of a population of shortening sarcomeres, giving a net velocity of zero; or (ii) all sarcomeres in the field pause synchronously. The latter possibility is supported by our data; the intensity profile of the first order remains invariant throughout the pause. Since the volume of tissue transsected by the beam contains approximately 10^9 myofilaments, the region of synchrony is extensive.

According to current views of contraction, the attachment of cross-bridges occurs stochastically. It remains to be seen whether the macroscopic and repeatable synchrony implicit in these observations can be accommodated within the cross-bridge theory.

Recent advances in photodiode technology have allowed us to make sarcomere length measurements with a time resolution of 200 μsec and sarcomere length resolution of 2-3 nm. These high resolution measurements turned up some unexpected features of the sarcomere shortening pattern.

The apparatus is based upon a light diffraction technique. An isolated specimen of muscle is transected by a laser beam. As there are regular striations along the specimen, the incident light is diffracted into a series of orders, the angles between which are uniquely related to the striation spacing in the volume of the fiber transected by the beam. On the basis of this principle the orders are directed onto a photodiode array whose output is used to measure the "instantaneous" sarcomere length.

In early investigations we first studied cardiac muscle (Pollack, Iwazumi, ter Keurs & Shibata). Papillary muscles from rat ventricles were held isometrically at 25–28° and were stimulated to contract. We observed "internal" shortening, *i.e.,* the shortening of sarcomeres allowed by the compliant regions adjacent to the clamps. After stimulation and latency, shortening began abruptly, ceased for several milliseconds, thenceforth resuming and ceasing in alternation. The number of such pauses which occurred during the shortening phase of contraction varied with the specimen and with the region for a given specimen. Frequently, several distinct pauses were followed by additional shortening in which the signal was too noisy to discern whether or not pauses were present. A representative example is shown in Fig. 1.

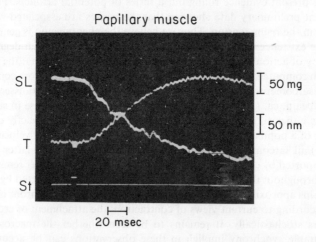

FIG. 1. Sarcomere shortening and tension development during isometric contraction. "St" denotes the stimulus pulse. As contraction proceeds the diffracted orders become progressively weaker and the signal to noise ratio diminishes. If pauses had been present late in contraction, they would, therefore, have been obscured.

We then examined ventricular trabeculae, which were somewhat thinner than the papillary muscles (typically 80 μm as opposed to 300 μm), as well as single fibers of frog skeletal muscle. The shortening patterns were qualitatively similar, *i.e.*, a series of pauses was observed between "bursts" of shortening. In both types of specimen the pauses generally showed up even more distinctly than in papillary muscles; *i.e.*, the transitions between pause and burst were sharper, occasionally being complete within several hundred microseconds, and the sarcomere shortening velocities during pauses were often indistinguishable from zero. It is possible that these differences arise out of the considerably sharper diffraction patterns in these two thinner preparations.

The pause durations were variable, depending on the order of the pause after stimulus, the type of specimen and the region observed. We found, in general, that the first pause tended to be the longest, lasting for 8.5 ± 0.8 msec in rat trabeculae at 26–28° ($n = 30$), while subsequent ones tended to be progressively shorter, but this rule was by no means inviolable. Moreover, when the diffraction pattern was sharp enough that the signal remained strong throughout contraction, long pauses (20–30 msec) were often noted at the peak of contraction as shown in Fig. 2. In the skeletal muscle specimens studied at comparable or slightly lower temperatures the pauses were shorter, rarely lasting more than 3 or 4 msec. Most often there were fewer pauses than in the cardiac specimens (Fig. 3), but regions that underwent substantial shortening often showed a series of pauses (see Fig. 6 below).

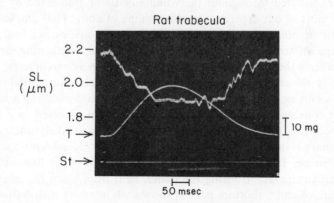

FIG. 2. Thin ventricular trabeculae (typically 80 μm) give crisper diffraction patterns than the papillary muscles. In this preparation pauses are seen throughout contraction, and are particularly lengthy at the time of peak tension.

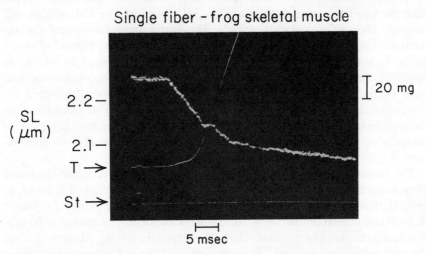

FIG. 3. The pauses seen in skeletal muscle fibers were generally the most distinct, though of relatively short duration.

The existence of well-defined pauses in the sarcomere shortening records is indicative of a high degree of synchrony of action among widely dispersed contractile elements, as is shown by the following: The sarcomere length computed from the light diffraction signal is derived from contributions from sarcomeres situated throughout the volume of tissue transected by the laser beam. In most of our experiments, this volume of about $(100 \ \mu m)^3$ contained roughly 10^9 myofilaments. It is not possible to detect a pause in a field containing so large an array unless either (i) a population of lengthening sarcomeres exactly cancels the effect of a population of shortening sarcomeres, giving a net velocity of zero, or (ii) all sarcomeres in the field pause synchronously. The first seems unlikely, for exact cancellation is too demanding a condition to be met consistently, and unambiguous pauses are observed in a majority of our records. Thus, possibility (ii) remains the most likely interpretation.

Preliminary results indicate that the sarcomeres do indeed remain stationary during a pause. If so, the intensity distribution across the first order peak should remain invariant during the pause, as it does when the muscle is at rest. Figures 4 and 5 (bottom panels) show such intensity distributions taken at various times during contraction in a single fiber of skeletal muscle. In Fig. 4 the scans of intensity distribution were obtained throughout the pause. The blanked regions on the sarcomere length trace (upper panel) denote the times of scan.

DISCRETE SARCOMERE SHORTENING

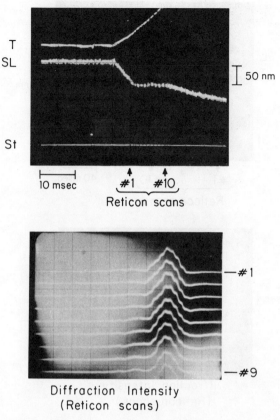

FIG. 4. Scans of intensity distribution of the first order peak made during a pause (lower). Blanked regions on sarcomere length trace (upper) denote the times of scan. The shape of the first order did not change substantially during the pause, indicating that the sarcomere length distribution remained fixed.

These records should be compared with those of Fig. 5. This is a subsequent, identical contraction, but the time scale on the upper trace is compressed, the interval between scans is increased, and we have shifted the display of the diffraction scans leftward. Note the progressive change in first order position and intensity distribution; by contrast, progressive changes during the pause (Fig. 4) are virtually absent. Thus, the sarcomere length distribution appears to remain stationary throughout the pause.

It is surprising that so vast an array of contractile elements can behave synchronously. According to current views, sarcomere shortening occurs as a result of the *asynchronous* cyclic interactions of myosin cross-bridges with

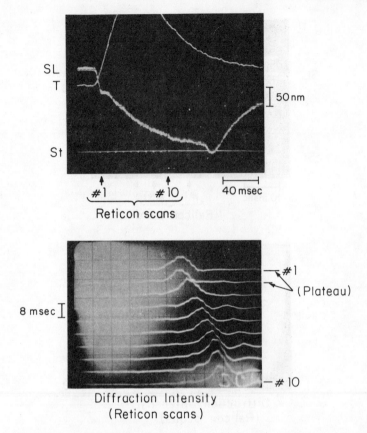

FIG. 5. Same fiber and similar contraction as in Fig. 4, except that scans were taken at wider intervals throughout the contraction. Note the progressive shift and change of shape of first order as shortening proceeds. The first two traces (labeled "plateau") were taken during the pause.

actin monomers of the thin filaments. *A priori,* one expects such a stochastic process to give rise to a large-scale motion which is smooth (A.F. Huxley, 1957). Thus, it is important to rule out the possibility that the phenomenon is artifactual.

The following are possible sources of artifact:
(i) the optical sensor;
(ii) slippage of the specimen from the clips;
(iii) ambient mechanical noise;
(iv) alteration of the angle of the diffracted beam by something other than a change of striation spacing;

(v) inaccuracies associated with the use of diffraction as a measure of sarcomere length.

The optical sensor used initially was a Reticon photodiode array, containing 128 elements in series. Although reasonably confident that the discreteness of the sensor did not give rise to an artifactually discontinuous shortening pattern, we implemented an entirely different type of optical sensor to test this possibility. A Schottky barrier photodiode (a diffused, non-discrete optical sensor) was combined with appropriate electronics to permit measurement of sarcomere length. Photokymography was used as a third independent method. When diffracted light was split between either of these two secondary devices and the Reticon array such that simultaneous records of sarcomere length against time were obtained, the records from these devices gave similar stepwise shortening patterns (Pollack et al., 1977). Consequently we were able to rule out the possibility of sensor artifact.

Slippage of the specimen from the clips could generate an artifactually discontinuous shortening pattern. If this occurred in several stages, it could conceivably give rise to a series of shortening bursts. The possibility of slippage seems unlikely since the pattern is repeatable over a large number of contractions (Fig. 6; also see Pollack et al., 1977). Furthermore, stepwise shortening can be seen in preparations of single, isolated cardiac muscle cells which have not been clamped (J.W. Krueger, personal communication).

Artifacts generated by ambient mechanical noise, created either by building vibrations or noise from the mechanical servo system are also ruled out by the repeatability of the signal over many contractions.

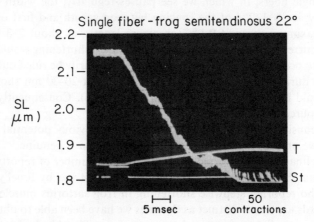

FIG. 6. Sarcomere length vs. time recorded in successive contractions on a storage oscilloscope. The pattern is repeatable.

The fourth possibility, that the angle of the diffracted beam might be altered by something other than a change of striation spacing, may be considered by tracing the beam of light from its source to the sensors. All elements are fixed rigidly in space except the specimen, the fluid flowing around it, and the air through which the beam travels. The repeatability of the phenomenon excludes random variations in local refractive indices of the air and fluid. Fluid surface vibrations are excluded since the objective used to collect the diffracted light is immersed in the fluid. The objective lens (Zeiss, 40 ×) and projection lens (E. Leitz) are common to all measurements. Aberrations of these lenses cannot be responsible for the pauses since we have been able to observe the pauses when other lenses were substituted. The remaining possibility along these lines is that the interface between the specimen and fluid suffers a series of undulations which bend the beam to and fro, giving an artifactually discontinuous pattern. We checked this by recording the position of the zeroth order, which should be similarly bent. Although the position of the zeroth order did exhibit minor random fluctuations in position during contraction, these were neither large enough nor consistent enough to account for the pauses.

Finally, a source of artifact may be the diffraction technique itself. Since the muscle does not act as a simple grating, but as a grating in three dimensions with perhaps several different sarcomere lengths, the sarcomere length computed from the diffraction pattern may show slight pauses because of interference between the different sarcomere length populations (Morgan, 1978). The maximum possible error in sarcomere length corresponds to about 5% of the width of the first order (Morgan, personal communication). For skeletal muscle single fibers, in which we see pauses regularly, the width of the first order is typically 1–2% of the spacing between zeroth and first orders. This gives a maximum error of 0.1% sarcomere length, or about 2–3 nm. If the pauses occurred over a time during which 2–3 nm of shortening would have been expected to occur, such an artifact could not necessarily be ruled out; however, the pause durations are often sufficiently long that 20–30 nm shortening, or more, would have occurred (*e.g.*, Figs. 1,3,4 and 6). Consequently, this is an unlikely source of artifact.

These considerations appear to rule out the obvious potential sources of artifact; we are inclined to believe the phenomenon is genuine.

This inclination is supported by the increasing number of reports of similar phenomena observed in other laboratories. The first was by Emel'yanov *et al.* (1966), who reported stepwise shortening in frog sartorius muscle; however, their records are not as distinct as the ones we have been able to obtain using a monochromatic light source and sensors of considerably higher resolution, and it is perhaps for this reason that their observations have gone essentially unnoticed. More recently, stepwise shortening has been observed by ter Keurs *et al.* (submitted) in rat ventricular trabeculae, and by Krueger & Wittenberg (submitted) in enzymatically isolated single heart cells. A more detailed study of the phenomenon is underway in the laboratory of W. Halpern, who has observed pauses regularly in single fibers of frog skeletal muscle (personal communication). The laser beam width used in Halpern's studies is 1 mm, considerably larger than the one we generally use; consequently his observations indicate that the region of synchrony can extend over at least 1 mm.

The ultimate proof of the existence of the phenomenon would come if one could discern its impact in another kind of signal. For example, if the tension waveform or the muscle length waveform showed pauses which were synchronous with those measured by diffraction, one would be assured that the signal could not be spurious.

Consider, for example, an isometric contraction during which sarcomeres are shortening and thereby stretching the elastic regions. This brings about a progressive increase of tension. Suppose *all* sarcomeres in the specimen suddenly pause. Tension would remain constant during this period since the series elastic elements would no longer be stretching. However, if the pause failed to occur synchronously throughout the length of the preparation, some sarcomeres would pause while others in series would shorten. The tension waveform would therefore not be expected to show a pause, but might possibly show an inflection.

Such inflections have been reported (Foulks & Perry, 1966). By careful examination of the first derivative of tension development in frog toe muscles contracting isometrically, Foulks and Perry found that the rise of tension was not smooth, but showed a series of definite inflections of slope. These inflections were repeatable, and exhibited a well-defined functional relation to sarcomere length and temperature. The authors concluded that, given the large number of contractile elements contributing to the rise of tension, such inflections were possible only if the contractile behavior among these elements were synchronous.

We have just initiated studies designed to better document this conclusion. Our studies are in single fibers rather than whole muscle, and we are examining the sarcomere shortening pattern as well as the tension waveform. Figure 7 shows a record we obtained in a preliminary attempt. The upper portion shows the behavior of the initial rise of tension, averaged over a large number of contractions. Averaging assures that any inflections occurring as a result of random noise are averaged out. There appears to be an inflection, and as shown in the lower record, the inflection occurs at the time of the pause. This pause was observed in several contiguous regions, indicating synchrony extending about 600 μm. We have very recently extended the natural frequency of our tension transducer and will be interested to see whether the tension inflections will show up more convincingly than in the preliminary result shown in Fig. 7.

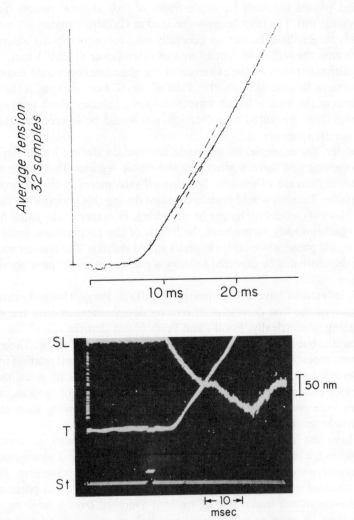

FIG. 7. Lower panel shows tension development and shortening observed in a single fiber of skeletal muscle. In the upper trace, the early phase of the tension record is shown, averaged over 32 successive contractions to remove the effects of random noise. The inflection on the tension record corresponds to the time at which the pause occurs.

The phenomenon of stepwise shortening has a number of potentially significant implications:

(i) *Contraction at the molecular level may occur in a succession of alternating bursts and pauses.* This follows directly from the records presented.

(ii) *The bursts and pauses may be synchronized over large regions.* The diffraction pattern gives a composite (but not necessarily the mean) value of sarcomere length sampled over the entire illuminated region. If the pauses occurring throughout this region did not take place synchronously, a standstill in sarcomere length could not have been detected. Since distinct pauses can be observed in regions as large as 1 mm, synchrony can extend over what, by molecular standards, must represent enormous dimensions.

(iii) *The pattern of bursts and pauses may be predictable.* Since the burst occurs synchronously over so vast a volume, it is logical to assume that the phenomenon is deterministic; that is, the local occurrence of the event should be predictable for a given set of environmental conditions. The repeatability of the pattern among successive contractions (Fig. 6) supports this view. These events would not, therefore, occur stochastically.

(iv) *At the myofilament level the size of a burst may be fixed.* This quantal hypothesis is speculative. Evidently the occurrence of a burst is related to the occurrence of some molecular event. Since the molecular structure is so regular, the simplest hypothesis is to presume that the magnitude of the event may be fixed. This will require careful testing. The records presented here are inadequate, whether they support the quantal hypothesis (*e.g.* Fig. 8) or not,

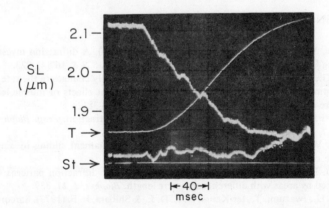

FIG. 8. Tension development and sarcomere shortening in a rat trabecular specimen. The third trace is related to first order intensity, and is of no relevance here. Note that the pauses occur at fairly regular intervals.

since they may reflect the contribution of more than one synchronous region. A narrower laser beam will be required, to assure that the sampled region lies entirely within a locally synchronous population.

Can the observations presented here be reconciled with the cross-bridge theory? Evidently the kind of synchronous behavior implied by our observations is not implicit in the theory; cross-bridge attachment is generally thought to be a stochastic event (Huxley, 1957). The theory, if it is to account for our observations, will require modification. Consider, for example, a series of shortening bursts and pauses occurring during a period of constant tension. One simple possibility is that the burst reflects the synchronous rotation of many S-1 heads which are attached to the thin filaments. Since tension is maintained after the burst, synchronous detachment of these heads would not be possible, unless another population of S-1 heads takes over to bear the tension. After a pause these heads would then rotate, giving the next burst, and so on.

Such an hypothesis raises many questions: Why should the bridges attach synchronously? Why should they rotate synchronously? Why should they detach synchronously? Why should at least two out-of-phase populations exist? How can tension remain constant as rotated bridges are replaced by non-rotated ones?

Evidently, if the basic model is to be retained, and if it is to incorporate the stepwise shortening phenomenon, it will require revision of a substantial nature. Alternatively, it is possible that a newer model may be required, in which synchronous, possibly quantal, behavior is an explicit feature.

REFERENCES

Emel'yanov, V. B., Efimov, V. N., & Frank, G. M. (1966). A diffraction investigation of "sliding" during muscle contraction. *Dokl. Akad. Nauk. S.S.R.* **167**(4), 923.

Foulks, J. G., & Perry, F. A. (1966). The time course of early changes in the rate of tension development in electrically-stimulated frog toe muscle: effects of muscle length, temperature and twitch-potentiators. *J. Physiol.* **185**, 355–381.

Huxley, A. F. (1957). Muscle structure and theories of contraction. *Prog. Biophys. biophys. Chem.* **7**, 257.

Krueger, J. W., & Wittenberg, B. A. Dynamics of myofilament sliding in single, intact cardiac cells. submitted.

Morgan, D. L. (1978). Prediction of some effects on light diffraction patterns of muscles produced by areas with different sarcomere length. *Biophys. J.* **21**, 889.

Pollack, G. H., Iwazumi, T., ter Keurs, H. E. D. J., & Shibata, E. F. (1977). Sarcomere shortening in striated muscle occurs in stepwise fashion, *Nature, Lond.* **268**, 757–759.

ter Keurs, H. E. D. J., Rijnsburger, W. H., Bloot, R., & Nagelsmit, M. J. (1979). Tension development and sarcomere length in rat cardiac trabeculae: evidence of length dependent activation. submitted.

Discussion

Edman: I am not really convinced that this reflects uniform behavior, because it is quite clear if you look at streak photographs of the diffraction pattern during the onset of contraction that the microstructure changes within the beam; it changes continuously, and I would think either that this reflected an interference phenomenon or that we really have non-uniform behavior of individual myofibrils. You actually measure the center of gravity of the first-order intensity profile, and the shortening steps may appear because there is a change of microstructure.

Pollack: What you are saying, I think, is if you look at the intensity profile in the first order, that under some conditions the intensity profile doesn't remain fixed but that the microstructure, or the shape of the intensity profile, has a tendency to change with time during contraction.

Edman: Yes.

Pollack: Well, we see changes in the intensity profile during contraction, too. But during a pause, as I showed in Fig. 4, there are no changes of intensity profile, and therefore presumably no movement of the microstructure.

Edman: But you can find it. This is more pronounced in different fibers, and I am sure that in some fiber regions you don't see these plateaus at all.

Pollack: We see these phenomena in perhaps 50 or 60% of regions that we look at. We find that in the fibers that are well mounted and have a uniform striation pattern, we see it more regularly than in fibers that are contracting non-uniformly and have a skewed distribution pattern.

Edman: I would think that this is more likely to be explained on the basis of non-uniform behavior of parallel myofibrils.

Pollack: If the behavior were non-uniform, then how would you be able to see pauses which persist as long as tens of milliseconds? You are sampling a region that is 200 μm in diameter or, in fact, can be as large as 1 millimeter in diameter in the measurements made by Bill Halpern at the University of Vermont, and the shortening velocity in so large a region remains zero during all this time. I should think that indicates uniformity, not non-uniformity.

Wilkie: I wonder whether you have checked another possibility? If the fiber sees not only a compliance in series with it, but also inertia, you get oscillations of a purely mechanical origin, as I showed about 30 years ago in experiments on flexing my own elbow joints (Wilkie, 1950). Now, when you scale down the quantities from a human being to a single fiber, I don't know what sorts of inertia would give these effects, but it might well be the inertia of the fiber, or of the fluid that is moved by it. Have you looked into the possibility that the purely mechanical setup of compliance and inertia could give oscillations? After all, oscillations could look like steps.

Pollack: No, we hadn't thought about that possibility, but one observation that we made perhaps speaks against it. The pauses that we see are almost exclusively of zero velocity, or close to zero velocity. If they were oscillations, then I would expect to see velocities that range all over the spectrum, as opposed to velocities that hover around zero. For that reason, and because they begin and end so abruptly, I would suspect that they are not oscillations but true pauses. And again, I refer also to Fig. 4 on the intensity profile of the first order; with time the order remains stationary. So, I think the pause is indeed not an oscillation.

Bartels: I am wondering why this stepwise behavior has never been seen in latency relaxation on experiments done by Haugen & Sten-Knudsen (1976) where they measure the sarcomere lengthening or shortening with a laser technique?

Pollack: Those experiments, as I understand, are restricted to the very beginning of the contraction in order to get extremely high resolutions. The first pause—depending on the preparation, and depending on the temperature—may occur perhaps some 10 or 15 msec after the beginning of the contraction. I wonder if that is being looked at in that particular laboratory?

Bartels: It is not. They are mostly interested in latency relaxation, perhaps only the first 10 msec or so of contraction.

Pollack: May I just make one comment on the period of latency relaxation? We have done some experiments with variation of temperature and we find in preliminary observations that, as the latency period decreases with temperature, the durations of the pause also change in a similar manner with temperature. It may be that latent period is very similar to a pause. In other words, the latent period may be the first pause before the first burst of shortening.

Winegrad: First, have you looked for these pauses during lengthening of an active muscle? Secondly, have you measured the frequency of these pauses? Is there any sort of constant frequency or relatively constant range of frequencies that you see?

Pollack: We haven't looked specifically at lengthening, although we tried one or two pilot experiments, and we did find an irregular type of lengthening. Regarding the frequency of pauses, perhaps the most exciting piece of information would be the value of this frequency, *i.e.*, the preferred value sarcomere length decrements between pauses; however, it is not so easy to tackle this question because of the following problem: The patterns of stepwise shortening don't occur uniformly throughout the preparations that we use. The pattern may be different for one region and another region 200 μm or so away. Because of that, if you use a laser beam that samples some arbitrary region, you may be getting contributions from one, two, three or more local regions of synchrony. If that is the case, then even if the shortening bursts were quantized, you might not pick that up. So, recently we have begun to restrict the laser beam diameter, going down to 30 or 40 μm, and we are just beginning to look at the data. It is interesting that with the laser beam of small diameter, the pauses are in fact rather long; often, most of the period of shortening seems to be taken up by pauses—perhaps up to 80%. But we haven't yet carried out the analysis of the preferred sarcomere length decrements between pauses.

Huxley: Were some of the records taken with the laser beam covering the whole diameter of the fiber?

Pollack: Yes, We have done it many ways. For example, with a 200 μm laser beam we virtually cover the full fiber width, but when we restrict the laser beam we look at a portion of the fiber width.

Huxley: So, when you say it is a pause, it is not a pause in the tension development by that region of the fiber; the tension is clearly increasing fairly smoothly, apart from that sort of slight pause in it that you see, but on the whole the tension is going up continuously. It is only the length that is moving in a stepwise way.

Pollack: Yes, it is true. I think what you are getting at is how possibly could the tension rise relatively smoothly if the sarcomere length is not changing? Is that the question?

Huxley: No. I think that if it is really a true phenomenon, one can think of explanations for it in terms of cross-bridges going on as the tension increases; you are getting more and more cross-bridges attached, but only after you get up to a certain number can you get a shortening of those—that is, if they start off attaching rather synchronously. But what bothers me is that this mechanism should be occurring at different times in different parts of the fiber, presumably unrelated to the moment of stimulus, and I wonder, how do you account for this asynchrony of the steps?

Pollack: By local variations in load or activation level. The pause durations appear, from preliminary evidence, to depend on both. For example, if one region of a fiber has a larger cross-section than another, its load is relatively lower and I would expect the pauses to be shorter. So the series of pauses in the two regions might go progressively out of phase with one another. That might explain why the steps are occurring at different times in different regions.

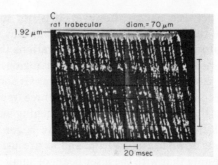

FIG. I–1. Time course of sarcomere length changes during a series of contractions (each displaced successively rightward), demonstrating the repeatability of the stepwise pattern. From Pollack et al., 1977. Calibration, 0.1 μm.

ter Keurs: I would like to make a remark concerning the question of Dr. Winegrad with respect to the frequencies at which the shortening steps occur in muscle. I'm now referring to steps of sarcomere shortening, not steps of thick filament shortening. During one of the experiments in cardiac muscle, we tried to question whether the phenomenon was a reproducible one, either in time or in space, and found that in a specific specimen the steps were varying in time as you will see in Fig. I-1 (Pollack, Iwazumi, ter Keurs & Shibata, 1977): In this figure, you see a shortening trace of sarcomeres in cardiac muscle as a function of time, and you see the result of about twenty contractions in that muscle. We shifted the sarcomere length axis along the abscissa, *i.e.,* along the time axis, because there occurred significant fluctuations of the duration of the shortening plateaus, so if you tried to superimpose them, you obtained a blurred picture. However, by shifting them along the time axis, you see that there are white dots on all traces which are occurring at about constant sarcomere length. So, with respect to the question of Dr. Winegrad, it seems to be a phenomenon linked in this case to the specific sarcomere length you are studying, and not to a specific frequency, because the plateaus were varying in time.

Mohan: Dr. Pollack, how significant are the first-order fringes so that you can measure the sarcomere length so accurately to see the steps?

Pollack: With a good diffraction pattern we estimate that the resolution is approximately two or three nanometers. The symmetry is not so great. What we find is that the first order intensity profile is rarely Gaussian. It usually has some substructure, as Dr. Edman was referring to, and what we do is to measure the median of the first order. We don't select the peak. The peak is susceptible to the influence of noise due to the substructural peaks, so by measuring the median sarcomere length as a function of time we observe the pauses. The median doesn't shift over a period of many milliseconds during a pause.

Winegrad: Have you looked at the sarcomere length at the ends of the single fiber at the same time that you are monitoring what you call "pause" but what one might also call a "give" as well, to see whether there are stepwise releases in the ends? As you know, the sarcomere length is quite different from what it is in the center. In cardiac muscle, the mechanical characteristics of the damaged ends are really totally unpredictable, and may even be nonuniform within a given area, and non-uniform over a period of time.

Pollack: No. I think it is an interesting study to do; and it is on our list. I wish Dr. Flitney were here, because if he were, he would perhaps have mentioned that if fibers are stretched, the amount of sarcomere lengthening is not proportional to the amount of stretch of the entire fiber. The "give" of sarcomeres seems to occur once the sarcomere is stretched beyond a certain point. Then it seems to jump to a longer length. I should mention that you can see stepwise shortening in the case where there is no load. If you can see it where there is no load, then the problem of "give" of the end sarcomeres, I should think, would be irrelevant, at least in terms of the genesis of the phenomenon.

White: Do you have a coil at each end in that experiment?

Pollack: No. We would dearly love to have a double servo system. In fact, Dr. Iwazumi has designed such a device to keep a specific region under the laser beam at all times, but we haven't built it.

White: Until you have such a system, you cannot really say that you are looking at the things you are suggesting. Presumably, you are getting differential sarcomere lengths along the length of the fiber which then run past your laser beam.

Pollack: Yes, we are indeed. However, the effect on the stepwise shortening pattern should be a blurring rather than an accentuation of the phenomenon. The way we do our experiments, to avoid such blurring, is to look microscopically at natural markers along the specimen to try to find a region that doesn't translate substantially during contraction. Most of our measurements are carried out on such regions. A double servo, however, would be wonderful.

Kramer: I want to extend Dr. Winegrad's remark on the possibility of the "give". It was shown at the Biophysics Congress in Kyoto that for *Physarum* strands, comparable things have been seen as you demonstrate here (Kamiya, 1978). The results there are interpreted as localized "give", somewhere in the strands, and I don't know if you've looked upon your measurements in that way, but I think it would be nice.

Pollack: You think that the burst of shortening occurs because there is "give" in a certain region of the specimen? If so, how does one then explain the presence of pauses in the case where the load is essentially zero? For example, in the case of John Krueger's results, where the specimen is shortening freely, what "gives" in such a case?

Kramer: It was also shown in *Physarum* strand that, in the case of a no-load or very little load situation, they see the give as a fast movement. They see a jumping movement even in the normal situation. So, I will not speculate on what is doing it.

Pollack: It is intriguing that a similar phenomenon is observed there.

Winegrad: Since we're talking about "give", Dr. Fabiato actually has a paper in press (Fabiato & Fabiato, 1978) showing "give" at the ends of his fragments of cardiac cells. This is not only a measurement of force, but it is also an optical picture of the sarcomere lengths, and there clearly is a "give".

Sugi: Concerning your experiment, I was rather surprised that the time course of shortening is rather linear over a considerable distance, though tension is increasing with time. In this connection, I wonder during the tension development whether the tension is evenly distributed in each myofibril or not?

Pollack: Your speculation is as good as mine, perhaps better. In many of our records we have noticed that the sarcomere length change during a burst of shortening remains linear, though the tension is increasing rather substantially during that time. At high loads we very often see shortening velocities which are the same as low loads; however, the pause durations are longer. It is possible that this can account for the force-velocity relation, i.e., with higher loads the pause durations are longer, giving you effectively a lower velocity. But, to return to your question, I can't really say anything about the load distribution among myofibrils.

REFERENCES

Fabiato, A. & Fabiato, F. (1978) Myofilament-generated tension oscillations during partial calcium activation and activation dependence of the sarcomere length-tension relation of skinned cardiac cells. *J. gen. Physiol.* **72**, 667–699.

Haugen, P. & Sten-Knudsen, O. (1976) Sarcomere lengthening and tension drop in the latent period of isolated frog skeletal muscle fibers. *J. gen. Physiol.* **68**, 247–265

Kamiya, N. (1978) Cyclic contraction of *Physarum* cytoplasm. *6th Internat. Biophys. Congr. Kyoto*, pp. 122

Pollack, G. H., Iwazumi, T., ter Keurs, H. E. D. J. & Shibata, E. F. (1977) Sarcomere shortening in striated muscle occurs in stepwise fashion. *Nature, Lond.* **268**, 757–759

Wilkie, D. R. (1950) The relation between force and velocity in human muscle. *J. Physiol.* **110**, 249–280

Contraction in Overstretched Skinned Muscle Fibers

Reiji NATORI, Reibun NATORI and Yoshiki UMAZUME

Department of Physiology, Jikei University School of Medicine, Tokyo, Japan

ABSTRACT

Skinned fibers of skeletal muscle can be extended in paraffin oil to nearly three times the length of the resting length without tearing them off. Connectin (Maruyama) plays a role in the prevention of tearing off due to stretching. On the basis of this high extensibility of skinned fibers, the properties of the thick and thin filaments in each sarcomere could be examined when the connection between thin and thick filaments due to formation of cross-bridges was absent. Furthermore, the properties of the thick and thin filaments *in situ* were compared with those of extracted actin and myosin.

INTRODUCTION

This report is concerned with some results of our recent experiments on the elastic structure and the properties of thick and thin filaments in skinned fibers of skeletal muscle.

Skinned fibers of skeletal muscle were stretched in paraffin oil to three times the length of the resting length without rupture and were restored to the resting length by withdrawing the stretching force.

The properties of the thick and thin filaments in the skinned fiber were examined when the overlap of the thick and thin filaments in each sarcomere was inhibited by stretching.

The aim of these experiments is partly the comparison of the properties of myosin and actin in living muscle fibers with those of extracted myosin and actin, and is partly an experimental trial for a unified explanation of the fundamental mechanism of various cell motilities including muscle contraction.

EXPERIMENTAL RESULTS

The skinned fibers of skeletal muscle of the Japanese toad and bullfrog were used as specimens in the experiments.

The sarcolemma of skeletal muscle fibers of the bullfrog were easier to remove in paraffin oil than those of the Japanese toad. However, the bright, thick fibers of the Japanese toad were suitable for obtaining skinned fibers able to respond to electrical stimulation with a twitch-like contraction, although no notable difference was observed with respect to $CaCl_2$-induced contractions of the skinned fibers of the bullfrog and Japanese toad.

(1) Elastic property of the skinned fibers of skeletal muscle

Skinned fibers could be extended to nearly three times the resting length, *i.e.*, the sarcomere was extended from 2.3–2.4 μm at resting length to 7–8 μm.

The single muscle fibers became ruptured in most cases when they were extended to twice the resting length. This might be caused by local overstretching due to partial rupture of the sarcolemma.

The stretching residual was observed in a skinned fiber immediately after withdrawal of the stretching force, but it tended to decrease gradually with the lapse of time. The stretching residual was small after the sarcomere had been stretched to less than 5 μm, but increased with an increasing extent of stretching. This might be due to rupture of the internal membrane, which will be discussed below.

When a skinned fiber was stretched to more than twice its resting length, after withdrawal of the stretching force local contraction was induced by application of $CaCl_2$ or by an electrophoretic local application of Ca^{++} through a micropipette. The tension of these Ca-induced contractions was less than that in normal skinned fibers. The tension varied greatly in different specimens, ranging from 200 to 500 g/cm^2.

T-tubules and the sarcoplasmic reticulum (the internal membrane system) may be considered to belong to the parallel elastic structure, and to play a dominant role in prevention of rupture against overstretching. However, since it is not likely that the structure of the internal membrane observed in electron micrographs is so strong as to be able to prevent some possible rupture at the portion of the triadic junction where an intensive tension may be induced during excessive stretching, it may be inferred that some supporting materials surround the internal membrane system.

In 1975, Maruyama isolated an elastic protein, which is different from collagen, by treating with alkali the stroma obtained during extraction of muscle protein (its Young modulus was 4×10^6 dyne/cm^2, close to that of

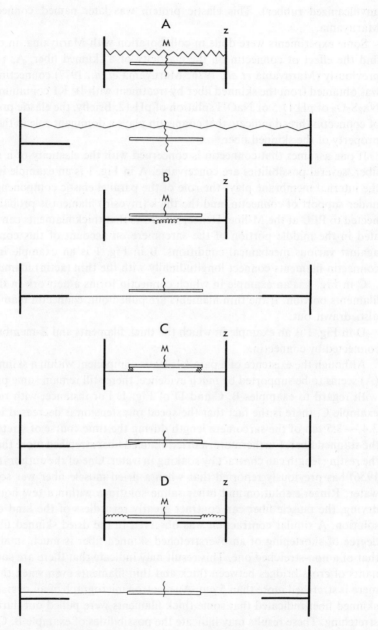

FIG. 1. Schematic illustration of some possible examples of elastic structure by connectin.

unvulcanized rubber). This elastic protein was later named connectin by Maruyama.

Some experiments were done in collaboration with Maruyama, in order to find the effect of connectin on the elasticity of a skinned fiber. As reported previously (Maruyama et al., 1976; Maruyama et al., 1977) connectin thread was obtained from the skinned fiber by treatment with 1M KI containing 0.1 M $Na_2S_2O_3$ of pH 11.5 or NaOH solution of pH 12. Briefly, the elastic properties of connectin thread indicate that connectin plays a dominant role in the elastic property of the skinned fiber.

If one assumes that connectin is concerned with the elasticity of a skinned fiber, several possibilities are conceivable. A in Fig. 1 is an example in which the internal membrane plays the role of the parallel elastic component (PEC) under support of connectin, and the thick (myosin) filament is probably connected to PEC at the M-line. Hence, it follows that thick filaments can be situated in the middle portion of the sarcomere on account of this connection against various mechanical conditions. B in Fig. 1 is an example in which connectin filaments connect longitudinally with the thin (actin) filaments.

C in Fig. 1 is an example in which connectin forms a network in the thick filaments portion. If the thin filaments are pulled out, connectin filaments are also drawn out.

D in Fig. 1 is an example in which the thick filaments and Z-membrane are connected by connectin.

Although the existence of a parallel elastic component within a skinned fiber (A) seems to be supported by much evidence, there still remain some problems with regard to examples B, C and D of Fig. 1. For instance, with regard to example C, there is the fact that the speed of extension is decreased in about $3.4 \sim 3.5$ μm of the sarcomere length during the time course of stretching of the skinned fiber. Furthermore, a dried skinned fiber stretched more than twice the resting length can contract by soaking in water. One of the authors (Natori, 1956) has previously reported that when a dried muscle fiber was soaked in water, Ringer's solution and other saline solutions within a few hours after drying, the muscle fiber can contract greatly regardless of the kind of saline solution. A similar contraction was observed in the dried skinned fiber. The degree of shortening of an overstretched skinned fiber is much smaller than that of a non-stretched one. This result may indicate that there are some remnants of cross-bridges between thick and thin filaments even when the sarcomere is stretched more than 5 μm. An electron micrograph of an overstretched skinned fiber indicated that some thick filaments were pulled out during over stretching. These results may indicate the possibilities of example B, C or D.

Hence, although inhibition of the overlap of thick and thin filaments was expected in an overstretched skinned fiber, such remnants of cross-bridges should be kept in mind for the quantitative analysis of the properties of overstretched skinned fibers.

(2) Properties of thin and thick filaments

One of the authors (Y. U.) found that, when a voltage of 30 to 50 volts/cm was applied to both ends of a stretched skinned fiber, the sarcomere length of which, measured by the diffraction pattern of a laser light beam, was more than 8 μm, the application of voltage induced a change in the intensity of light transmitted through the skinned fiber. This is called Umazume's effect.

In connection with this effect, the effect of voltage on the light intensity of each portion of the diffraction pattern of the laser beam and the birefringence of each portion of striation were examined. The application of voltage increased the intensity of diffraction due to actin filaments and decreased birefringence at the half I-band of one side of the Z-membrane and decreased birefringence at the half of the other side.

This effect was almost the same in a ghost skinned fiber in which myosin was extracted by immersion in KCl solution of a high concentration. These changes indicate the relation between the dipole moment of thin filaments and the change of electric field. It is inferred that the actin filaments diverge at one end and converge at the other depending on the direction of the electric field. Umazume and Fujime (1975) supposed that the dipole property of an actin filament is positive on the Z-membrane side and negative on the free end side. This dipole property of thin (actin) filaments is similar to that of extracted actin (Kobayashi *et al.*, 1964). For example, the curve of increase and decay of the first order of light diffraction, when the voltage was 50 V/cm and sarcomere length was 8.4 μm, was similar to that of extracted actin.

These results indicate that Umazume's effect is due to the dipole property of actin in living muscle, while this effect gives evidence for the applicability of the properties of extracted actin to the study of the properties of thin filaments in a living state.

When 1M KCl solution was applied to a portion of a skinned fiber which was immersed in paraffin oil, the striation of the sarcomere was observed to become gradually less clear, and the partial breakdown of the striation under phase-contrast microscope and the diffraction pattern of the sarcomere by laser light beam indicated a disorder of striation; a marked decrease in birefringence also was observed. Moreover, electron micrography of a skinned fiber treated with 1M KCl suggested the possibility of dispersion of myosin. When 1 M KCl solution was applied to an overstretched skinned fiber in which the thin and thick filaments were parted from each other, local contraction was observed by application of a small amount of CaCl₂ solution when withdrawing the stretching force. In this case the tension development was far smaller than that of a nontreated skinned fiber. However, when the concentration of KCl applied was rapidly decreased by addition of distilled water immediately after the application of CaCl₂, a very strong contraction was induced which relaxed with the lapse of time.

From these results, the author infers that tension development in the contraction process depends directly on the rate of aggregation of myosin.

CONCLUSION

A comment on the fundamental mechanism of muscular contraction

If the cross-bridge theory (Hanson & Huxley, 1955; Huxley, 1957; Huxley, 1969; Huxley & Simmons, 1971) is acceptable, it is possible that tension development is induced between one thin filament and the neighboring three thick filaments (Fig. 2).

Since, according to Huxley & Brown (1967), Haselgrove, Stewart and Huxley (1976), Matsubara & Yagi (1978), the change in X-ray diffraction pattern during contraction indicates that the head of heavy-meromyosin (HMM) approaches the thin filament during contraction, it is likely that con-

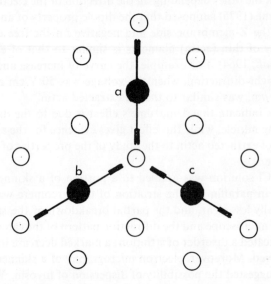

FIG. 2. Schematic illustration of cross-bridge. Black circle represents thick filament, and open circle thin filament.

tact between the head of HMM and the active portion (T) of actin in Fig. 3 is concerned with tension development. If it is assumed that the ratio of the 429 Å periodicity of cross-bridges to that of the active binding sites on actin filament is 9:8, the contact between the head of HMM and the active sites on actin filament (T) during isotonic contraction is shown schematically in Fig. 3. However, since actin filaments can be considered to increase in flexibility owing to Ca^{++} action, according to Fujime & Iwashita (1971) there is a possibility that the ratio changes to 9:7.5. If so, the probability of contact between the head of HMM and the active sites (T) on an actin filament is larger than that in the case of the ratio 9:8 (Fig. 3). If the scheme mentioned above is applicable in the explanation of the contraction process, the action of each cross-bridge may be linked synchronously. With respect to the mechanism of force due to this contact, the explanation proposed by Huxley et al. may be phenomenally acceptable. However, Pollack (1977) brought forward a revised opinion on the propriety of the cross-bridge theory. Besides, there still remains a substantial amount to be explained although there are several speculative explanations. Since actin filaments have a dipole property, as mentioned above, it may be considered that if the active portion of actin is assumed to be charged in the presence of Ca^{++} and the head of HMM be oppositely charged, an attraction due to coulombic force would occur between them, and coulombic force may be considered to play a dominant role in the sliding force in twitch and tetanus. Furthermore, the sliding mechanism may explain various types of contraction of muscle, but other possible mechanisms also may be considered. For instance, if the thin filaments are pliant at the knots of actin molecules, a force due to folding may be considered even though the grade of tension develop-

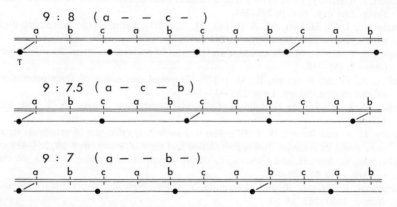

FIG. 3. Schematic illustration of sliding force. 9:8, 9:7.5 and 9:7 are the ratio of the unit length of myosin and that of actin respectively.

ment in the axial direction is smaller than that due to sliding. As reported previously, when a skinned fiber was stretched beforehand to over twice the resting length, after withdrawal of the stretching force contraction was induced by application of $CaCl_2$ solution. Electron micrography of these contracted skinned fibers indicated that in some instances thin filaments were bent and rolled at the border area of the A- and I-bands of the sarcomere. Moreover, the sarcomere could be shortened to 0.5 μm and in such a case a folding configuration of filaments was detected with electron micrography. Therefore, it may be considered that a type of movement due to folding of the thin filaments is possible in a certain arrangement of thin and thick filaments. As another mechanism, it may be possible to consider a jet-like force as reported previously by Oplatka (1973, 1974). Although a force due to this kind of water streaming (jet), even if present, may be considered to be small in the contraction process of a skeletal muscle fiber, there are some experimental results which should be considered. When 1M KCl solution or 1M KCl containing 0.75 μg/ml trypsin was applied to a skinned fiber, very slow contractions were induced by application of $CaCl_2$. In these cases myosin molecules are dispersed; hence, one would not be able to deny the possibility that the movement due to water streaming would play a role in a cell in which the myosin molecule was dispersed.

REFERENCES

Fujime, S. and Ishiwata, S.(1971). Dynamic study of F-actin by quasielastic scattering of laser light. *J. molec. Biol.* **62**, 251–265.

Hanson, J. & Huxley, H. E. (1955). The structural basis of contraction in striated muscle. *Symp. Soc. exp. Biol.* **9**, 228–264.

Haselgrove, J. C., Stewart, M. & Huxley, H. E. (1976). Cross-bridge movement during muscle contraction. *Nature, Lond.* **261** 606–608.

Huxley, A. F. (1957). Muscle structure and theories of contraction. *Prog. Biophys. biophys. Chem.* **7**, 255–318.

Huxley, A. F. and Simmons, R. M. (1971). Proposed mechanism of force generation in striated muscle. *Nature, Lond.* **233**, 533–538.

Huxley, H. E. (1969). The mechanism of muscular contraction. *Science, N. Y.* **164**, 1356–1366.

Huxley, H. E. and Brown, W. (1967). The low-angle X-ray diagram of vertebrate striated muscle and its behavior during contraction and rigor. *J. molec. Biol.* **30**, 384–434.

Kobayashi, S., Asai, H. and Oosawa, F. (1964). Electric birefringence of actin. *Biochem. biophys. Acta* **88**, 528–540.

Maruyama, K., Natori, R. and Nonomura, Y. (1976). New elastic protein from muscle. *Nature, Lond.* **262**, 58–59.

Maruyama, K., Matsubara, S., Natori, R., Nonomura, Y., Kimura, S. Ohashi, K., Murakami, F., Handa, S. and Eguchi, G. (1977). Connectin, an elastic protein of muscle. *J. Biochem., Tokyo* **82**, 317–337.

Matsubara, I. and Yagi, N. (1978). A time-resolved X-ray diffraction study of muscle during twitch. *J. Physiol.* **278**, 297–307.

Natori, R. (1956). Differences in physiological properties of myofibrils of small and large muscle fibers. *Jikeikai Med. J.* **3**, 36–42.

Oplatka, A. and Tirosh, R. (1973). Active streaming in actomyosin solutions. *Biochem. biophys. Acta* **305**, 684–688.

Oplatka, A., Gadasi, H., Tirosh, R., Lamed, Y., Muhlrad, A., and Liron, N. (1974). Demonstration of mechanochemical coupling in systems containing actin, ATP and non-aggregating active myosin derivatives. *J. Mechanochem. Cell Motility* **2**, 295–306.

Pollack, G. (1977). Sarcomere shortening in striated muscle occurs in stepwise fashion. *Nature, Lond.* **268**, 757–759.

Umazume, Y. and Fujime, S. (1975). Electro-optical property of extremely stretched skinned muscle fibers. *Biophys. J.* **15**, 163–180.

Discussion

Pollack: You mentioned that your model could somehow explain the kind of synchrony that appeared in the presentation I made. Could you elaborate on that somewhat?

Natori: In Fig. 3, the helical arrangement of the cross-bridges on the thick

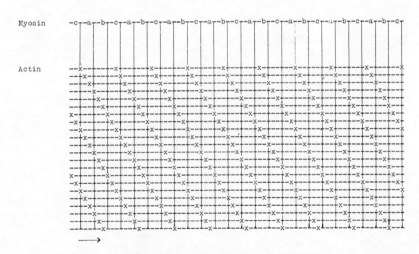

FIG. I–2. Diagram showing possible mode of interaction between the cross-bridges on myosin filament (a-b-c---) and the active sites (X) on actin filament (-----).

filament is represented as *a-b-c-* . . . , where the distance from *a* to adjacent *a* corresponds with the 429 Å periodicity. Meanwhile, the distance between the adjacent active binding sites on the thin filament (T) has a value smaller than 429 Å. Then, if it is assumed that a cross-bridge can attach to an active site on the thin filament only when they come to a fixed relative positions, the formation of cross-links might take place in a manner illustrated in Fig. I-2, as the thick and thin filaments slide past each other; the active site (X) on the thin filament (broken lines) can only interact with the cross-bridge when X meets with the vertical lines originating from the top line (*a-b-c-* - -) representing the thick filament. I am not, however, sure whether this serves to account for the stepwise sarcomere shortening which you observe.

Optical Diffraction Study on the Dynamics of Thin Filaments in Skinned Muscle Fibers

S. FUJIME,* S. YOSHINO ** and Y. UMAZUME†

* *Mitsubishi-Kasei Institute of Life Sciences, Tokyo, Japan*
** *Department of Physics, Faculty of Science, Nagoya University, Nagoya, Japan*
† *Department of Physiology, Jikei University School of Medicine, Tokyo, Japan*

ABSTRACT

When an electric field is applied along the fiber axis of noncontracting muscle, the intensities of all observable optical diffraction lines increase. This electro-optical effect was extensively studied both experimentally and theoretically, and it was confirmed that the effect comes from the interaction between electric dipoles of thin filaments and the applied field. From the present study, we elucidated that (1) the thin filament is a semiflexible rod, (2) the overtone (or the second) mode of the bending motion of thin filaments contributes to the electro-optical effect at higher frequencies of a sinusoidal field or shorter durations of a square field, (3) the induced moment has no appreciable effect, and (4) the estimated value of the flexural rigidity of thin filaments strongly depends on the concentrations of free calcium ions in the myofibrillar space and on the temperature.

A possible physiological implication of the thin filament flexibility will be discussed.

INTRODUCTION

The thin filament or F-actin (a main component of thin filaments) was proposed to be a semiflexible rod (Fujime, 1970), and the flexural rigidity of the thin filament has been measured *in vitro* (Fujime & Ishiwata, 1971; Ishiwata & Fujime, 1972) and *in vivo* (Umazume & Fujime, 1975; Yanagida & Oosawa, 1978). In *in vitro* experiments, however, the existence of aggregation and the exponential length distribution of actin filaments make the analysis very difficult. In *in vivo* experiments, on the other hand, absence of aggregation, the

**Predoctoral fellow of the Mitsubishi-Kasei Institute of Life Sciences

monodispersity of the thin filament length and the regularity of the muscle structure make it simple to treat the system.

Skinned fibers of the semitendinosus muscle of frogs can be stretched up to 8 μm or more in sarcomere length (L) and these extremely stretched fibers give quite sharp optical diffraction patterns (Fig. 1a). When an electric field is applied along the fiber axis, intensities of all observable diffraction lines increase (Fig. 1b; Umazume & Fujime, 1975). An analysis of the decay process of this excess intensity after turning off the field suggested that the flexural rigidity of thin filaments is 2–$3 \cdot 10^{-17}$ dyne cm². In that analysis, we assumed that this "electro-optical" effect was interpreted as due to the bending motion of the thin filaments as a result of the interaction between electric dipoles of the thin filaments and the external field. Very recently, we extended our previous theory to a more general form (Fujime & Yoshino, 1978), and made an extensive experimental study along the lines of the theory (Yoshino, Umazume, Natori, Fujime & Chiba, 1978). In this note, we briefly summarize our theoretical and experimental results on this effect.

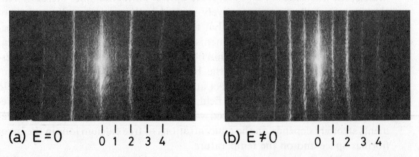

FIG. 1. Intensity modulation of diffraction lines by an external electric field E. Note the marked increase of odd-order line intensities. $L = 7.0$ μm and field strength = 80 V/cm.

THEORETICAL

Let us consider a one-dimensional model lattice (Fig. 2). If the optical path difference between the scattered rays from neighboring big particles (a's) is equal to the wavelength of incident light, i.e., $L \sin(\phi) = \lambda$, the total amplitude of the scattered rays from a's will be

$$A = a_1 + a_2 + \ldots + a_N = Na. \tag{1}$$

A similar relation will hold for the scattered rays from small particles (b's), i.e., $B = Nb$. Thus, the intensity of scattered rays from both a's and b's will

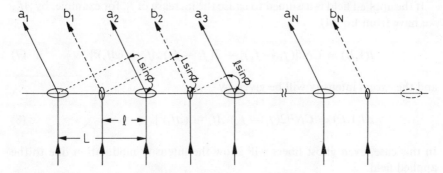

FIG. 2. One-dimensional model lattice illustrating diffraction.

be given by

$$I(1) = C|A + B\exp(i\delta)|^2 = CN^2[a^2 + b^2 + 2\,ab\cos(\delta)] \quad (2)$$

where

$$\delta = 2\pi(l/\lambda)\sin(\phi) = 2\pi(l/L) \quad \text{for} \quad L\sin(\phi) = \lambda \quad (3)$$

and C is a proportionality constant. Let us replace a- (b-) particles by A- (I-) bands. In skeletal muscle, l is equal to $L/2$. Thus, from Eq. (2), we have for the intensity of the first order line

$$I(1) = CN^2(f_I - f_A)^2 \quad (4)$$

where f_I and f_A are the *molecular* scattering factors of I- and A-bands, respectively, corresponding to the *atomic* scattering factors in X-ray diffraction.

If the applied field E (V/cm) is assumed to cause an electrophoretic shift of A- or I-bands by ξ, i.e., $l = L/2 + \xi$, we have from Eqs. (2–4)

$$I(1,E) = CN^2(f_I - f_A)^2 + CN^2 2f_I f_A[1 - \cos(2\pi\xi/L)] \quad (5)$$

and the excess intensity due to the field will be given by

$$\Delta I(1,E) = I(1,E) - I(1) = CN^2 2f_I f_A[1 - \cos(2\pi\xi/L)]. \quad (6)$$

For a small ξ-value, $\Delta I(1,E) > 0$, i.e., the intensity of the (first order) diffraction line will increase. However, if this effect is the main cause of the intensity modulation, ghost fibers will give *less* intensity modulation since $f_A \approx 0$.

If the applied field is assumed to cause the increase of f_I, for example, by Δf_I, we have from Eq. (4)

$$I(1,E) = CN^2[(f_I - f_A)^2 + 2(f_I - f_A)\,\Delta f_I + (\Delta f_I)^2] \tag{7}$$

and the excess intensity will be given by

$$\Delta I(1,E) = CN^2[2(f_I - f_A)\,\Delta f_I + (\Delta f_I)^2]. \tag{8}$$

In this case, even ghost fibers will show the intensity modulation due to the applied field.

In order to see which cause is the main one, a phoretic shift or a change in the f_I value, we measured the intensity increase $\Delta I_{skinned}$ of a skinned fiber at $L = 6.0\ \mu$m for a given strength and a duration of the applied field. Next, the same fiber was well treated with 1M KCl and then the ionic strength was reduced to 0.1M KCl. After the treatments, optical anisotropy of the A-bands observed by a polarizing microscope was very much reduced. Although not complete, we could expect that A-bands were dissolved to a great extent after the treatment, *i.e.*, $f_A \approx 0$. The treated fiber gave the intensity increase ΔI_{ghost} very close to $\Delta I_{skinned}$ for the same strength and duration of the applied field; the results for five preparations were $\Delta I_{ghost}/\Delta I_{skinned} = 1.1, 1.0, 0.9, 1.3$ and 0.9. This experiment clearly shows that the intensity modulation is mainly due to some changes in I-bands and, at the same time, excludes the possibility of phoretic shifts of A-band positions.

In the following we examine $\Delta f_I(t)$ in a more detailed manner. In extremely stretched fibers where there is no overlap between thin and thick filaments, packing of thin filaments becomes loose because of the repulsive force between them and because of their flexibility (Fig. 3). Since each thin filament has a dipole directing from the free end to the Z-line, when the electric field is applied, the packing of the thin filaments on one side of the Z-line becomes denser and that on the other side looser as shown in Figs. 3b and c. Moreover, the applied field may induce the overtone mode of the motion of thin filaments as shown in Fig. 3d. Even if there is a slight overlap between thin and thick filaments, the applied field may induce the lateral motion of thin filaments as shown in Fig. 3e.

Figure 4 shows the mathematical model for the theoretical treatment of the effect of the applied electric field. Using the notation in Fig. 4, we have the functional form of the refractive index of the I-band as ($|p(x)| \ll 1$)

FLEXIBILITY OF MUSCLE THIN FILAMENTS

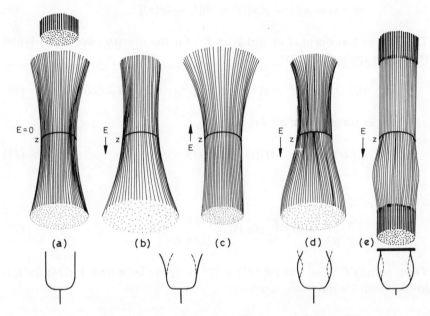

FIG. 3. An exaggerated illustration showing the effect of an applied field on the I-band.

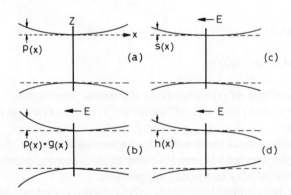

FIG. 4. Simplified model for the theoretical treatment [modified version of Fig. 11 in Umazume & Fujime (1975)]. (a) $E = 0$. $p(x)$ represents the bending of filaments due to a repulsive force between them. $p(x) = p(-x)$. (b) $E \neq 0$. $g(x)$ represents the effect of an applied field on electric dipoles of thin filaments, and is assumed to be a sum of $h(x, t)$ and $s(x, t)$. (c) An assumed form of $s(x, t) = s(-x, t)$, and (d) an assumed form of $h(x, t) = -h(-x, t)$.

$$n_1^{(a)}(x) = n_1^0[1 + p(x)]^{-2} \approx n_1^0[1 - 2p(x)]. \tag{9}$$

The Fourier transform of $n_1^{(a)}(x)$ gives $f_1^{(a)}$ for the overstretched fiber. When the field is applied, we have

$$n_1^{(b)}(x) = n_1^0[1 + p(x) + g(x,t)]^{-2} \approx n_1^{(a)}(x) - 2n_1^0 g(x,t). \tag{10}$$

The Fourier transform of $n_1^{(b)}(x)$ gives

$$f_1^{(b)}(t) = f_1^{(a)} - \Delta f_1'(t) - i\,\Delta f_1(t) \tag{11}$$

where

$$\begin{Bmatrix} \Delta f_1(t) \\ \Delta f_1'(t) \end{Bmatrix} = 4n_1^0 \int_0^a g(x,t) \begin{Bmatrix} \sin(2\pi x/L) \\ \cos(2\pi x/L) \end{Bmatrix} dx. \tag{12}$$

Then, putting $CN^2 = 1$ and $F^{(a)}(1) = f_1^{(a)} - f_A$, we have from Eq. (4) the following form for the excess intensity:

$$\Delta I(t) = \Delta I_1(t) + \Delta I_2(t) \tag{13}$$

where

$$\Delta I_1(t) = -2F^{(a)}(1)\,\Delta f_1'(t) + [\Delta f_1'(t)]^2 \tag{14}$$

$$\Delta I_2(t) = [\Delta f_1(t)]^2. \tag{15}$$

It is noted here that $\Delta I_1(t)$ comes from a possible "asymmetric" effect of the applied field. If $g(x,t)$ were an odd function of x, i.e., $s(x,t) = 0$ in Fig. 4, $\Delta f_1'(t)$ would vanish.

The same relations as those considered above also hold for the case where there is a slight overlap between thin and thick filaments, provided that appropriate forms of $p(x)$ and $g(x,t)$ are assumed.

To express the complete form of the excess intensity, we must apply a rather complicated and tedious mathematical treatment. Thus, we summarize here only the final forms of relevant expressions.

1) For an applied field in a square form: $E(t) = E$ for $-T \leq t \leq 0$ and $= 0$ otherwise:

$$\Delta I_1(t) = -8n_1^0 F^{(a)}(1)\eta \sum_k P_k \exp(-t/\tau_k) \tag{16}$$

$$\Delta I_2(t) = [4n_1^0 \sum_k R_k \exp(-t/\tau_k)]^2 \tag{17}$$

$$\begin{Bmatrix} P_k \\ R_k \end{Bmatrix} = E_k \tau_k [1 - \exp(-T/\tau_k)] \begin{Bmatrix} A'_k \\ A_k \end{Bmatrix} \tag{18}$$

where τ_k is the relaxation time of, and A_k (A'_k) is the sine (cosine) part of the form factor of, the kth mode of the bending motion of the thin filament, E_k, the kth component of the applied field, and η ($0 \leq \eta \leq 1$) is a parameter describing the "asymmetric" effect of the applied field.

2) For an applied field in a sinusoidal form, $E(t) = E \cdot \cos(\omega t)$:

$$\Delta I(t) = B_\omega \cos(\omega t - \phi_\omega) + \hat{B}_{2\omega} \cos(2\omega t - 2\hat{\phi}_{2\omega}) + B_{2\omega} \cos(2\omega t - 2\phi_{2\omega}) \tag{19}$$

where B_ω and $\hat{B}_{2\omega}$ contain a factor $F^{(a)}(1) \eta$ like Eq. (16). Since expressions for B's are not necessary in this study, we only quote the following:

$$\phi_\omega = \tan^{-1} \left| \frac{U_1 \sin(\phi_1) + U_2 \sin(\phi_2)}{U_1 \cos(\phi_1) + U_2 \cos(\phi_2)} \right| \tag{20}$$

$$\hat{\phi}_{2\omega} = \frac{1}{2} \tan^{-1} \left| \frac{U_1 \sin(2\phi_1) + U_2 \sin(2\phi_2)}{U_1 \cos(2\phi_1) + U_2 \cos(2\phi_2)} \right| \tag{21}$$

$$\phi_{2\omega} = \tan^{-1} \left| \frac{V_1 \sin(\phi_1) + V_2 \sin(\phi_2)}{V_1 \cos(\phi_1) + V_2 \cos(\phi_2)} \right| \tag{22}$$

$$\phi_k = \tan^{-1}(\omega \tau_k) \quad (k = 1, 2) \tag{23}$$

$$\begin{Bmatrix} U_k \\ V_k \end{Bmatrix} = E_k \tau_k \cos(\phi_k) \begin{Bmatrix} A'_k \\ A_k \end{Bmatrix}. \tag{24}$$

We qualitatively summarize here the theoretical results. The applied electric field induces the lateral motion of the rod; due to this motion, the intensity modulation of diffraction lines will result. As to the lateral motion of the rod, it is intuitive to imagine the vibration of a tuning fork, although this is a simple analogy. The fundamental ($k = 1$) and the lowest overtone ($k = 2$) modes of the vibration of the tuning fork are also shown in Fig. 3. If our model is valid, the contribution of the overtone mode will be detected. When there is a slight overlap between thin and thick filaments, one end of each thin filament is fixed at the Z-line and the other end is *hinged* or *fixed* at the overlap region. In such cases, the relaxation times of the lateral motion of the rod differ greatly from that in the case where there is no overlap (Table 1).

TABLE 1. Bending motion of a rod at various boundary conditions and the relaxation times τ_k for $k = 1$ and 2 modes.
(a) both ends free, (b) one end fixed and the other end free, (c) one end fixed and the other end hinged and (d) both ends fixed. $\tau_{(b,1st)}$ means τ_1 for model (b).

Model Mode	(a)	(b)	(c)	(d)
1st	⌣	Z⌢⎯M	Z⌢M	Z⌢M
$\tau_{(b,1st)}/\tau_1$	40	1	20	40
2nd	∿	Z∿M	Z∿M	Z∿M
$\tau_{(b,1st)}/\tau_2$	300	40	200	300

In the case of multi-exponential decay functions, such as Eqs. (16,17), we can get the initial decay rate by the cumulant expansion method (Koppel, 1972):

$$(1/\tau)_1 = [\sum_k (1/\tau_k) P_k]/[\sum_k P_k] \tag{25}$$

$$(1/\tau)_2 = [\sum_k (1/\tau_k) R_k]/[\sum_k R_k]. \tag{26}$$

The theoretical prediction of the initial decay rates is shown in Fig. 8.

In the case of the sinusoidal field, we can measure the phase retardation by use of a lock-in (or phase-sensitive) detector. When we use the detector in its ω-mode, the output is proportional to $B_\omega \cos(\theta - \phi_\omega)$ where θ is the phase angle of the detector. When we use the detector in its 2ω-mode (or its harmonics mode), the output is proportional to $B_{2\omega} \cos(\theta - 2\phi_{2\omega})$. In any case, we can determine the phase shift θ which gives maximum (or minimum) output. In this way, we can obtain the ϕ_ω versus ω or $\phi_{2\omega}$ versus ω relation. Theoretical dispersion relations are shown in Fig. 9. [Details of the theory will be found in Fujime & Yoshino (1978).]

METHODS

Mechanically skinned (Natori, 1954) and glycerinated single fibers of frog semitendinosus muscle were used. A muscle fiber on a glass slide covered with liquid paraffin oil was mounted on a sample holder, which was made of copper and was thermoregulated. The temperature near the muscle fiber was monitored with a thermister. The standard saline solution consists of 107 mM methanesulfonic acid potassium salt, 4 mM magnesium sulfate and 20 mM Tris maleate buffer, pH 6.8. If necessary, EGTA (ethylene glycol bis(β-amino ethylether)-N,N'-tetraacetic acid), $CaCl_2$ and/or sodium pyrophosphate were added to the above solution.

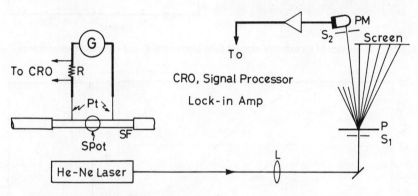

FIG. 5. Experimental set-up.
SF: skinned fiber, R: resister as a current monitor, Pt: platinum electrodes, G: square- and sine-wave generator with a current booster, L: lens, S :slits, and P: fiber position.

Figure 5 shows the schematic representation of the experimental set-up. The output of a photomultiplier tube was fed to a preamplifier, and then to a cathode ray oscilloscope, a signal processor in its averaging mode, a phase sensitive detector and so on. Field strength was 1–10 V/cm in the measurements of dispersion relations and 10–100 V/cm in the measurements of decay curves. (Details will be found in Yoshino et al., (1978).)

RESULTS

To see the outline of our measurements, we at first show the wave forms of the driving force (or an applied field) and the response in Fig. 6. The phase difference, $\phi_{2\omega}$, is given by the distance between the point of zero-crossing of the applied field and the minimum point of the output wave.

The dispersion relation was measured with a skinned fiber which had been well equilibrated with a solution containing 1 mM $CaCl_2$. The result (○ in Fig. 7a) gives the dispersion frequency ν_1 of 16 Hz or the relaxation time τ_1 of 10 ms.

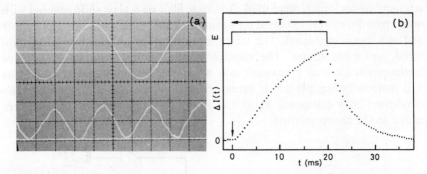

FIG. 6. Examples of an external electric field (upper trace) and response (lower trace).

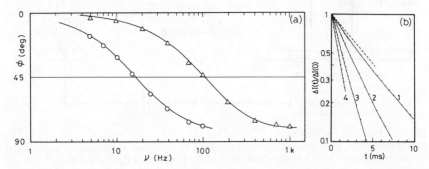

FIG. 7. Dispersion and decay curves showing the effect of Ca ions in the myofibrillar space. (a) The dispersion curves of: (o): skinned fiber at $L = 7.0\ \mu m$ was equilibrated with a saline solution containing 1 mM $CaCl_2$; and ($\triangle$): the above fiber was treated with a saline solution containing 2 mM EGTA to remove free Ca ions. Field strength ≤ 0.2 V/2 mm, 20°C. (b) The decay curves of: 1: as-prepared skinned fiber at $L = 8.0\ \mu m$; 2: a saline solution containing 10 mM EGTA (volume was several times that of the fiber) was poured onto the above fiber, 10 square pulses of strength of 30 V/cm and duration of 10 ms were applied, the saline solution around the fiber was removed and the response was measured; 3: further treatment as above was made; 4: further treatment was made; and - - - -: typical relaxation curve of swollen fibers. $E = 30$ V/cm, $T = 10$ ms and 20°C.

The fiber was then treated with a solution containing 2 mM EGTA to remove free Ca ions in the myofibrillar space, and the dispersion relation was measured at very low field strength ($E = 0.2$ V/2 mm $\equiv$ 1 V/cm). The result ($\triangle$ in Fig. 7a) gives ν_1 of 100 Hz or τ_1 of 1.6 ms. This change of ν_1 from 16 Hz to 100 Hz depending on the concentrations of free Ca ions was quite reversible.

It is supposed that the strong field induces the release of Ca ions from sarcoplasmic reticulum. A skinned fiber then was washed with a solution containing 10 mM EGTA under the influence of square pulses of 30 V/cm in strength and 10 ms in duration. The relaxation time τ_1 for the treated fiber became shorter

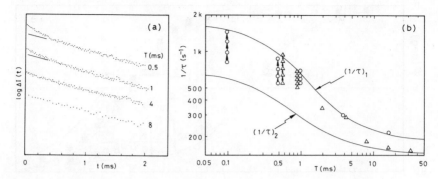

FIG. 8. Initial decay rates *vs.* durations of a square field.
(a) Examples of decay curves showing the two components at shorter durations of an applied field.
(b) Initial decay rate *vs.* duration. $L = 7.0$ μm, (———): theoretical curves based on Eqs. (25, 26), assuming $1/\tau_1 = 130$ and $1/\tau_2 = 5200$ s^{-1}. 20°C.

and shorter as the washing proceeded, and finally it was as short as 2 ms for the applied field strength of 30 V/cm (Fig. 7b). [Note that $\tau_1 = 2\tau_d$ where τ_d is the decay time.] The fiber was then equilibrated with 1 mM CaCl$_2$. The fiber then gave τ_1 of about 10 ms, which is the same as that of a fiber without the above treatment. Although not yet studied systematically, the midpoint of the change in the relaxation time or the dispersion frequency is supposed to occur at the physiological concentrations of free Ca ions.

Figure 8 shows some examples of the decay curves at different durations T. It is clearly seen that at shorter T values the decay curves consist of two components. Since the shortest binwidth of our signal processor was only 20 μs, the contribution from the overtone ($k = 2$) mode with $1/\tau_2 = 5000$ s^{-1} would be included in the first 10–15 bins because the contribution decays to the 1/e of the initial value within 200 μs. Thus, the cumulant expansion of the experimental data at shorter durations was made using, for example, first 50, 40, 30 and 20 points. Decreasing data points in the analysis gave increasing $(1/\tau)$ values (Fig. 8b). The change of the small $(1/\tau)$ values at longer durations to the big ones at shorter durations came solely from the decreasing contribution of the fundamental mode relative to the contribution of the overtone mode (see Eq. (18)). Experimental results in Fig. 8 suggest that at longer durations T (and hence lower field strengths) the main contribution to the response came from $\Delta I_2(t)$, whereas at shorter durations (and hence higher field strengths) the main contribution came from $\Delta I_1(t)$, probably due to the enhancement of ΔI_1 through the parameter η in Eq. (16). Actually, we can see in Fig. 4 of Umazume & Fujime (1975) that the initial increase of the excess intensity is proportional to time t at higher field strengths and to t^2 at lower field strengths. Anyway, the

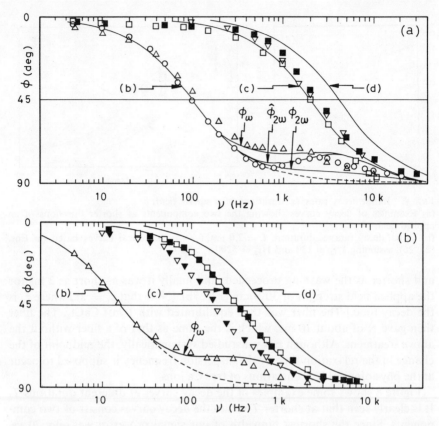

FIG. 9. Dispersion curves of glycerinated fibers at 20°C.
(a) In the absence of free Ca ions. (△): ϕ_ω and (○): $\phi_{2\omega}$ at $L = 4.0\ \mu$m (no overlap). (□, ■): in the absence of, and (▽): in the presence of, 4 mM Mg-pyrophosphat eat $L = 3.5\ \mu$m(slight overlap). (b) In the presence of 0.1 mM free Ca ions. (△): $\phi\omega$ at $L = 4.0\ \mu$m, (□, ■): $\phi\omega$ in the absence of, and (▽, ▼): ϕ_ω in the presence of, 4 mM Mg-pyrophosphate at $L=3.5\ \mu$m. (———): theoretical curves based on Eqs. (20–22); (b), (c) and (d) correspond to models b, c, and d in Table 1, respectively. (- - - -): $\phi = \tan^{-1}(\omega\tau_1)$.

fact that the experimental results lie between the theoretical curves of $(1/\tau)_1$ and $(1/\tau)_2$ supports our model.

To see the validity of our model from a different point of view, we show the results under different boundary conditions for the bending motion of thin filaments. When we use a *rigor* muscle fiber at, say, $L = 3.5\ \mu$m, one end of each thin filament is not free but will be *hinged* or *fixed* at the overlap region. Then, the dispersion frequency of such fibers is expected to be twenty or more times higher than that of the fibers at, say, $L = 4\ \mu$m, where there is no overlap between thin and thick filaments (see Table 1). Since, at $L < 6$ μm, the intensity of the first order line is very strong, we can also measure the dispersion relations in the ω-mode of the lock-in detector. Figure 9a shows the dispersion relations of glycerinated fibers at $L = 3.5$ μm(ω) and 4.0 μm (ω and 2ω) in the absence of free Ca ions. In the case of squares in Fig. 9a, the solution did not contain ATP, so that the fibers were not in the relaxed state. In the case of inverted triangles in Fig. 9a, there existed 4 mM Mg-pyrophosphate, so that the fiber was in the "relaxed" state. Figure 9b shows the dispersion relations of glycerinated fibers at $L = 3.5$ μm(ω) and 4.0 μm (ω) in the presence of 0.1 mM free Ca ions. In the case of squares in Fig. 9b, the fiber was in the "rigor" state. In the case of inverted triangles in Fig. 9b, on the other hand, there existed 4 mM Mg-pyrophosphate and 0.1 mM free Ca ions, so that the fiber was in the "contracting" state.

At the sarcomere length of 4.0 μm, the experimental data points largely deviate from the simple dispersion relation $\phi = \tan^{-1}(\omega \tau_1)$ (the dashed lines in Fig. 9). This deviation is strong evidence of the contribution from the overtone ($k = 2$) mode. At $L = 3.5$ μm, the experimental data points moved to the higher frequency region as the theory predicted. It should be noted that the theoretical dispersion curves in Fig. 9 were drawn by adjusting only one parameter τ_1 for model (b). Thus, we can see a surprising agreement between experimental and theoretical results.

DISCUSSION

As shown by dashed lines in Fig. 9, the contribution from the $k = 2$ mode is negligible at the frequency $\nu_1 \equiv \omega_1/2\pi$ where $\phi_{2\omega}$ (or ϕ_ω) $= \pi/4$ (the dispersion frequency of the $k = 1$ mode). Then we have (Umazume & Fujime, 1975; Fujime & Yoshino, 1978)

$$\omega_1 = 1/\tau_1 = \varepsilon (0.6\pi/a)^4/\zeta \qquad (27)$$

or

$$\varepsilon = [\zeta a^4/(0.6\pi)^4]\omega_1 = [k_B T a^3/D(0.6\pi)^4] \omega_1 \qquad (28)$$

where the Stokes-Einstein relation, $D = k_B T/\zeta a$, was assumed. Then we have the flexural rigidity ε for the length of a thin filament of $a = 1$ μm (for frog) and $D = 1.0 \cdot 10^{-8}$ cm²/s (at 20°C)

$$\varepsilon = 3.1 \cdot 10^{-19} (\eta_T/\eta_{293}) \omega_1 \qquad (29)$$

where ω_1 is in rad/s unit and η_T is the viscosity of water at $T°K$. The ε values at 20°C are $3.1 \cdot 10^{-17}$ dyne cm^2 in the presence of, and $1.9 \cdot 10^{-16}$ dyne cm^2 in the absence of free Ca ions in the myofibrillar space. As to the temperature dependence of ε values, the slopes of the straight lines in an Arrhenius plot of ε [manuscript in preparation] give about 5 kcal/mole in the presence of, and about 10 kcal/mole in the absence of free Ca ions. The present ε value in the presence of Ca ions is very close to those of previous results (Fujime, 1970; Fujime & Ishiwata, 1971; Ishiwata & Fujime, 1972; Umazume & Fujime, 1975; Takebayashi, Morita & Oosawa, 1977; Yanagida & Oosawa, 1978), but that in the absence of Ca ions is much bigger than a previous finding (Ishiwata & Fujime, 1972). The temperature dependence of about 5 kcal/mole in the presence of Ca ions is also very close to observations in earlier reports (Ishiwata, 1975; Takebayashi et al., 1977), whereas that of about 10 kcal/mole is very large.

The thin filament consists of F-actin, tropomyosin and troponin. It has been established that tropomyosin molecules associate with F-actin in such a manner that they settle in the grooves of an F-actin helix (Ebashi & Endo, 1968). Consequently the temperature dependence of the flexural rigidity of the thin filameat should originate at least from (1) loosening of actin-actin bonds, (2) loosening of actin-tropomyosin bonds and (3) the increase of the flexibility of tropomyosin rods. Loosening of actin-actin bonds will be responsible for the increase of the flexibility of F-actin alone with rising temperature. Loosening of actin-tropomyosin bonds is expected from the experimental fact that rabbit tropomyosin dissociates from rabbit F-actin *in vitro* above 37°C (Tanaka & Oosawa, 1971). Even below the dissociation temperature, bond loosening should occur. The increase of the flexibility of tropomyosin rods with rising temperature is also expected from the decrease of the heilx contents of tropomyosin rods (Tanaka & Oosawa, 1971). Thus, the observed change of the flexibility of thin filaments is a result of composite contributions from various origins. Of course, the flexibility change of the thin filament depending on free [Ca^{2+}] is closely related to the movements of tropomyosin molecules on the surface of F-actin (Haselgrove, 1972; Huxley, 1972). Anyway, the state of actin monomers in the flexible thin filament is supposed to be different from the state of actin monomers in the stiff thin filament. Actually it was found that F-actin, which was made stiff by chemical bridges between actin monomers, could activate the ATPase of myosin but could not show superprecipitation (Gadasi, Oplatka, Lamed, Hockberg & Low, 1974). Although speculative, the state of actin monomers is supposed to be closely related to the regulation of muscle contraction. This idea differs from the currently accepted blocking hypothesis (Wakabayashi, Huxley, Amos & Klug, 1975).

Finally, we would like to mention a correlation between instantaneous elasticity assumed in mechanical transients of muscle and the flexibility of the thin filament. For a macroscopic uniform rod, the flexural rigidity is equal to the product of the Young modulus Y and the moment of inertia I of the cross-section of the rod, i.e.,

$$\varepsilon = YI. \tag{30}$$

For a cylinder of the radius b, $I = \pi b^4/4$. If this relation is applied to a fibrous polymer of protein, the modulus Y obtained from ε is related to the microscopic elastic constant κ of the monomer-monomer bond. If, under an external force F, the monomer-monomer distance X is stretched by Δx,

$$F = \kappa \Delta x. \tag{31}$$

Therefore, the relation between Y and κ is given by

$$Y \equiv (F/A)/(\Delta x/X) = \kappa X/A \tag{32}$$

where A is the area of the cross-section, which is πb^2 for the cylinder. From Eqs. (30, 32), we have

$$\kappa \equiv YA/X = \varepsilon(A/XI) = 4\varepsilon/Xb^2. \tag{33}$$

If we assume appropriate values for X and b, we have κ of the order of 10^4 dyne/cm. This means that in activated muscle the thin filament having length 1 μm is stretched by 50–100 Å (Oosawa, Fujime, Ishiwata & Mihashi, 1972). This might have some correlation between the instantaneous elasticity and the flexibility of the thin filament (Ishiwata & Oosawa, 1974). F-actin is not a cylinder, but a two-stranded helical polymer. Very recently, various models of the F-actin structure were considered and it was concluded that the monomer-monomer bond strength lies between $1 \cdot 10^4$ and $3 \cdot 10^4$ dyne/cm (Oosawa, 1977). Although the instantaneous elasticity is assumed to reside in S-2 of the crossbridge (Huxley & Simmons, 1971), the muscle compliance of 80 Å per half sarcomere could be explained by the flexibility of the thin filament. It is most likely that the instantaneous elasticity resides in both S-2 and the thin filament.

REFERENCES

Ebashi, S. & Endo, M. (1968). Calcium ions and muscle contraction. *Prog. Biophys. molec. Biol.* **18**, 123–183.

Fujime, S. (1970). Quasielastic light scattering from solutions of macromolecules. II. Doppler broadening of light scattered from solutions of semiflexible polymers, F-actin. *J Physiol. Soc. Japan* **29**, 751–759.

Fujime, S. & Ishiwata, S. (1971). Dynamic study of F-actin by quasielastic light scattering of laser light. *J. molec. Biol.* **62**, 251–265.

Fujime, S. & Yoshino, S. (1978). Optical diffraction study of muscle fibers. I. A theoretical basis. *Biophys. Chem.* **8**, 305–315.

Gadasi, H., Oplatka, A., Ramed, R., Hockberg, A. & Low, W. (1974). Possible uncoupling of the mechanochemical process in the actomyosin system by covalent crosslinking of F-actin. *Biochim. biophys. Acta*, **333**, 161–168.

Haselgrove, J. C. (1972). X-ray evidence for a conformational change in the actin-containing filaments of vertebrate striated muscle. *Cold Spring Harb. Symp. quant. Biol.* **37**, 341–352.

Huxley, A. F. & Simmons, R. M. (1971). Proposed mechanism of force generation in striated muscle. *Nature, Lond.* **233**, 533–538.

Huxley, H. E. (1972). Structural changes in the actin- and myosin-containing filaments during contraction. *Cold Spring Harb. Symp. quant. Biol.* **37**, 361–376.

Ishiwata, S. (1975). PhD thesis, Nagoya University.

Ishiwata, S. & Fujime, S. (1972). Effect of calcium ions on the flexibility of reconstituted thin filament of muscle studied by quasielastic scattering of laser light. *J. molec. Biol.* **68**, 511–522.

Ishiwata, S. & Oosawa, F. (1974). A regulatory mechanism of muscle contraction based on the flexibility change of the thin filaments. *J. Mechanochem. Cell Motility* **3**, 9–17.

Koppel, D. R. (1972). Analysis of macromolecular polydispersity in intensity correlation spectroscopy: The method of cumulants. *J. chem. Phys.* **57**, 4814–4820.

Natori, R. (1954). Role of myofibrils, sarcoplasma and sarcolemma in muscle contraction. *Jikeikai Med. J.* **1**, 18–28.

Oosawa, F. (1977). Actin-actin bond strength and the conformational change of F-actin. *Biorheology* **14**, 11–19.

Oosawa, F., Fujime, S., Ishiwata, S. and Mihashi, K. (1972). Dynamic property of F-actin and thin filament. *Cold Spring Harb. Symp. quant. Biol.* **37**, 277–286.

Takebayashi, T., Morita, Y. & Oosawa, F. (1977). Electron microscopic investigation of the flexibility of F-actin. *Biochim. biophys. Acta* **492**, 375–363.

Tanaka, H. & Oosawa, F. (1971). The effect of temperature on the interaction between F-actin and tropomyosin. *Biochim. biophys. Acta* **253**, 274–283.

Umazume, Y. & Fujime, S. (1975) Electro-optical property of extremely stretched skinned muscle fibers. *Biophys. J.* **15**, 163–180.

Wakabayashi, T., Huxley, H. E., Amos, L. A. & Klug, A. (1975). Three-dimensional image reconstruction of actin-tropomyosin complex and actin-tropomyosin-troponin T-troponin I complex. *J. molec. Biol.* **93**, 477–497.

Yanagida, T. & Oosawa, F. (1978). Polarized fluorescence from ε-ADP incorporated into F-actin in myosin-free single fiber: Conformation of F-actin and its changes induced by heavy meromyosin. *J. molec. Biol.* **126**, 507–524.

Yoshino, S., Umazume, Y., Natori, R., Fujime, S. & Chiba, S. (1978). Optical diffraction study of muscle fibers. II. Electro-optical properties of muscle fibers. *Biophy. Chem.* **8**, 317–326.

Discussion

Kawai: What is the longitudinal compliance of the actin filament based on your measurement, when calcium is bound?

Fujime: We measured flexural rigidity. You must know the moment of inertia of the cross-section of the thin filaments to calculate Young's modulus. It is very difficult to calculate. Young's modulus of the thin filaments is probably about one-tenth of Young's modulus of aluminium wire.

Kawai: Is this comparable to cross-bridge compliance?

Fujime: Yes. About the same order of magnitude.

Kawai: Then can you say that, in the presence of calcium, the compliance lies half in the bridges and half in actin?

Fujime: Yes, maybe.

Pollack: I was wondering, if the frequencies involved are in the audible range, do you ever hear music from your muscles?

Fujime: No. I showed you an analogy of vibration with a tuning fork, but in the muscle the fibers are surrounded by a viscous medium. So, no vibration occurs. Overdamped motion occurs. It is only an analogy.

White: Your flexural rigidity would be concerned only with the bending of the I filament. You could only make inferences about the longitudinal compliance provided the material is isotropic, or if you know the degree of anisotropy. I would suggest that it is unknown, and so your experiment cannot say anything about the longitudinal stiffness of the filaments.

Fujime: My measurement was concerned with lateral motion, and not longitudinal, but we estimate the Young's modulus; this is very difficult because it is a two-stranded helical polymer, so I cannot say very conclusively.

II. INSTANTANEOUS ELASTICITY AND STIFFNESS

Fast Response of Cardiac Muscle to Quick Length Changes

M. J. DELAY, D. V. VASSALLO, T. IWAZUMI and G. H. POLLACK

Department of Anesthesiology and Division of Bioengineering, University of Washington, Seattle, Washington, U.S.A.

ABSTRACT

An apparatus has been developed which is capable of measuring the fast response to quick length changes. The output of a photodiode array, which captures laser light diffracted by the muscle, is used to compute sarcomere length every 200 μsec. A voice coil, driven by a fast amplifier, completes length changes in 250 μsec, independent of amplitude. Tension is measured with a transducer having a resonance frequency of 11 kHz in air.

Measurements were made on left and right ventricular trabeculae from rats. Only a contiguous region of the preparation having good sarcomere length uniformity was examined; these regions were typically 600 μm long, compared to the typical laser beam width of 100 μm. The length change was varied to cause a quick tension change, and the resulting change in sarcomere length was obtained by averaging over the uniform region.

In preliminary work it was found that between 6 and 8 nm shortening per half sarcomere is required to drop the tension to zero. This value is not far from the corresponding figure for skeletal muscle. Tension change is linear with sarcomere length change up to stretches of at least 12 nm per half sarcomere. The muscle length change needed to obtain 100% tension drop is variable, depending on the size of the damaged ends, but is typically 2% to 3%. Only a 0.6% sarcomere length change is required in the intact regions. The high compliance of the ends implies that restretching them to develop substantial post-release tension might take times of the order of 10 msec, thus masking the fast phases of tension recovery seen in skeletal muscle.

INTRODUCTION

It has been argued that a rapid component of skeletal muscle elasticity resides at the site of the cross-bridges (Huxley & Simmons, 1971; Ford, Huxley &

Simmons, 1977). This conclusion was reached by examination of the quick drop in tension and the fast component of subsequent tension redevelopment following the imposition of a quick release in muscle length during an isometric contraction. Similar experiments have been performed on cardiac preparations to determine the extent of structural similarity at the cross-bridge level between these two types of muscle. The results showed that significantly greater amounts of quick shortening were needed to drop the tension to zero in cardiac muscle (*e.g.*, Abbott & Mommaerts, 1959; Parmley & Sonnenblick, 1967; Steiger, 1977); the exact amount, however, varied with the experiment. It has been pointed out that part of the high compliance is located in the damaged ends where the preparation is clamped (Pollack & Krueger, 1975; Sugi & Kamiyama, 1977). The amount of this compliance would therefore depend on the extent of the damage, perhaps accounting for the considerable variation found in the results.

It was the aim of this study to examine the behavior of the rapid elasticity in cardiac muscle using the laser diffraction technique for measuring instantaneous sarcomere lengths in the intact regions in combination with an apparatus capable of generating quick length changes and measuring fast length responses.

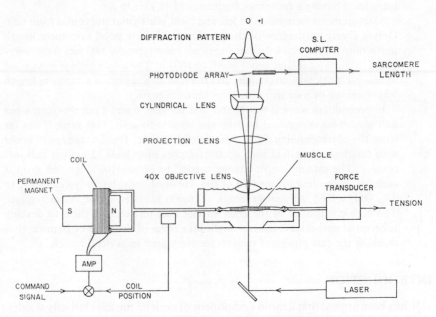

FIG. 1. Overall scheme of the apparatus.

METHODS

The experiments were performed with the apparatus diagrammed in Fig. 1. It consists of a sarcomere length measurement scheme based on the diffraction of a laser beam by the muscle, an electromagnetically driven coil to maintain isometric contractions and impose quick length changes, and a force transducer with wide-band frequency response.

Optical apparatus

The optical apparatus is described more fully elsewhere (Pollack et al., 1977; ter Keurs et al., 1978; Iwazumi & Pollack, in press). Briefly, the beam from a He-Ne laser (Spectra-Physics 120, 5 mW) is focussed to a diameter of 100 μm at the muscle specimen. Since the muscle is composed of alternating A- and I- bands with different refractive indices, and since the organization of contractile elements is regular, these bands act as a diffraction grating. The resulting diffraction pattern includes a number of intensity maxima whose distances in the far-field region from the zeroth order maximum are related to the sarcomere length. The diffraction pattern is collected by a Zeiss 40 × objective lens of numerical aperture 0.75 and passes through a phase centering telescope which is focussed on the back focal plane of the objective lens. Finally, a cylindrical lens condenses the diffraction orders onto their meridional line so as to increase their intensity and ameliorate the intensity fluctuations due to laser speckle. A photodiode array with 128 elements (Reticon 128EC) is positioned to capture one of the two first-order maxima over a range of distances from the zeroth order maximum corresponding to sarcomere lengths between about 1μm and 4μm. The intensity information as a function of position along the array is read out by a fast analog computer, which calculates the position of the median of the first order maximum and the corresponding sarcomere length. This information is updated every 200 μsec.

The system can resolve 3 nm and is accurate to 30 nm per sarcomere. Calibration is accomplished by replacing the muscle with a diffraction grating of known spacing, and adjusting the position of the photodiode array for the correct output length. Several gratings are used in the calibration process to assure correct behavior over a range of spacings.

Force transducer

The force transducer consists of a silicon beam with a diffused surface resistive element (Mikroelektronikk AE801) whose resistance varies linearly with the deflection of the beam from its resting position. A small glass stem is fastened with epoxy in a perpendicular position to the beam; the other end of the stem is attached to a small stainless steel clip which clamps the muscle preparation near one of its ends and transmits the muscle tension to the silicon beam. The glass stem extends into the muscle bath through an aperture small enough that the surface tension is sufficient to prevent leakage. The amount of viscous damping is adjusted by changing the separation between the two halves of a silicone oil-filled collar surrounding the stem. The transducer assembly has an effective mass of 3 mg, a compliance of about 2 μm/g, and a linear response to 10 g. In order to test the response of the complete transducer, the beam end was held away from its resting position and then quickly released, allowing the beam, stem and clip to oscillate freely; response frequencies are 11 kHz and about 3 kHz with the shaft in air and water, respectively, with approximately critical damping. The final force transducer output, after filtering and amplification, has a noise level corresponding to 0.5 mg. The base-line drift due to the effects of temperature is negligible in comparison.

Servo control

Servo control of either muscle length or tension is performed by a voice coil attached to the muscle; the coil is driven by a fast amplifier which accepts input from a control unit.

The coil assembly is shown pictorially in Fig. 2. It consists of a coil 5.4 cm in diameter and 3 mm in length, wound of aluminum wire having a rectangular cross-section of 1.25 mm $\times$ 0.15 mm. The coil is fastened to a balsa cylinder of identical diameter and length 1.2 cm, the other end of which is attached to a thin-wall aluminum shaft with appropriate balsa bracing; a small muscle attachment clamp protrudes from the far end of the shaft. The assembly is able to slide along the axis of a permanent magnet with field 20 kG.

Radial support is provided by a web of taut strings and by two bearing surfaces, one midway along the shaft and the other near the muscle clip. A slight amount of viscous damping is necessary to minimize overshoot, and is achieved by light silicone oil in the mid-shaft bearing. The mass of the movable assembly is 4.3 g. An infra-red light emitting diode is glued to the shaft; its light is focussed upon a position sensor (Schottky barrier photodiode) whose output is therefore a measure of the coil position.

The fast amplifier is of local design, and is able to provide up to 50 A of current with a rise time of about 1 μsec. The input to the amplifier is generated by a servo control unit, which can be programmed to control muscle length

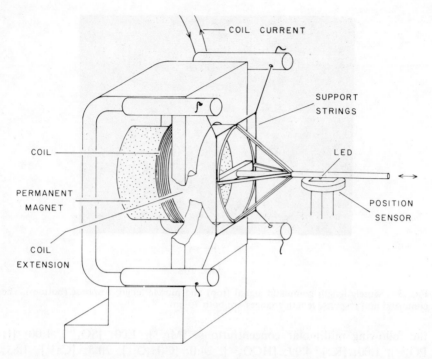

FIG. 2. Diagram of the voice coil assembly.

or tension. The command signal can include any combination of constant levels and length change (either ramped or with exponential behavior in time) with adjustable amplitudes and onset times. The command signal is then compared with the coil position signal, yielding a correction signal, which is compensated for by the frequency response of the coil before being sent to the amplifier.

Muscle length changes of up to 200 μm are made in a little over 250 μsec, with the response times being independent of amplitude. Longer changes are possible up to about 1 mm in about 400 μsec. Typical command and response signals are shown in Fig. 3. The tip of the shaft executes an almost pure translation during length changes, with less than 3 μm lateral pmovemet detectable.

Solutions

A physiological salt solution was used throughout the experiments with

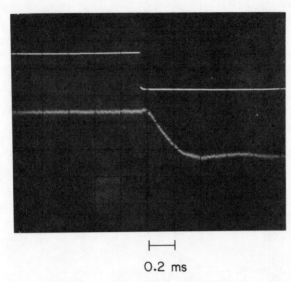

FIG. 3. Muscle length command signal (top) and muscle length response (bottom). The command and response length changes are both 70 μm.

the following millimolar concentrations: $[Mg^{2+}]$, 1.00; $[SO_4^{2-}]$, 1.00; $[HPO_4^{2-}]$, 1.00; $[K^+]$, 4.99; $[HCO_3^{2-}]$, 24.0; $[C_2H_3O_2^-]$, 20.5; $[Ca^{2+}]$, 1.83; $[Cl^-]$, 102.4; and $[Na^+]$, 145.5. Insulin was added (0.1 cc/l) along with 1.8 g/l of dextrose; the mixture was bubbled continuously with a gas of 95% O_2 and 5% CO_2. Experiments took place at solution temperatures between 27° and 29°C, with a pH of 7.4.

Muscle chamber and stimulation

The muscle chamber was made of Teflon, except for the bottom, where a glass slide was mounted to admit the laser light. Solution was passed into the chamber at a replacement rate of about 10 times per minute, sufficient for maintenance of the preparation, and was removed by an aspiration pump which was mechanically isolated from the chamber to prevent vibrational coupling to the force transducer. A thin glass coverslip was placed over the force transducer end of the chamber, and the fluid level was allowed to meet it, thereby preventing changes in the force transducer output due to fluctuations in surface tension. The stimulus electrodes were of platinum wire placed parallel to the muscle; stimulation was unipolar and took place at a rate of 12/min with a pulse length of 5–6 msec.

Experimental procedure

Ventricular trabeculae from young adult rats were used in the experiments because of the high quality and consistency of their diffraction patterns. The rat was anesthetized with ether and the heart was rapidly excised, placed in the dissection chamber, and infused continuously with solution. Either the right or the left ventricle was exposed to view, and an appropriate trabecula was selected and carefully removed. Useable preparations measured 2–3 mm in length, 60–300 μm in width and 50–100 μm in depth. The muscle was clamped in the apparatus, stimulation started, and the muscle allowed to equilibrate (for one hour or more) until it demonstrated the following: stable contractile tension, relatively uniform activation as confirmed by visual inspection, low resting tension (less than 5% of active tension at the sarcomere length studied), and a relatively noise-free diffraction pattern with a sharp first order and little "dithering" motion between contractions. Remounting was sometimes necessary to minimize nonuniform contraction and the consequent large amounts of lateral and longitudinal translation. Only those muscles which satisfied the above criteria were used in the experiments.

The first step in an experiment was to identify regions of the preparation with uniform sarcomere length so that there could be no translation of the illuminated region during the quick length change to one with quite different characteristics. For this purpose scans were made of the sarcomere length during contractions at 50 μm intervals along the muscle. A region was deemed acceptable if two intervals on either side demonstrated nearly identical traces of sarcomere length with time; hence, more than 200 μm of length change (about 10% of the resting length) could occur quickly with assurance that the sarcomere length change observed would not be an effect of nonuniformity. In practice this was easy to obtain since the uniform regions varied in length between 200 and 600 μm. It was typical that, for a sudden imposed change of sarcomere length of 10 nm, the observed change in sarcomere length was in error due to translation by no more than 10%.

Quick length changes were imposed at a time just approaching peak tension. The resting sarcomere length was 2.25 μm $\pm$ 0.05 μm. Each contraction having a quick length change was followed by 5 uninterrupted contractions. The muscle length was held constant during contractions, except for the quick length change, and was restored to its original value after the end of the contraction in an exponential manner with a time constant of about 2 sec. The amount of muscle length change was varied over the range -3% to $+10\%$ of the resting length, with all stretches being performed at the end of the session; if any signs of damage were noted, the experiment was terminated.

The changes in tension and sarcomere length (ΔT and ΔSL, respectively) observed as a result of a quick length change were measured over each uniform region at intervals of 100 μm. Each set ΔT and ΔSL was based on 16 measurements made over the muscle; ΔSL was averaged using an amplitude averager. Both T and SL were displayed on storage oscilloscopes and photographed for later analysis.

No compensation was made to the data for the mass of the force transducer or the muscle; the resultant error in ΔT is negligible. Each datum yielded values of ΔSL and $\Delta T/T$, where T is the tension just prior to the length change.

RESULTS

It was found that the quick length change does not precipitate the appearance of nonuniformity in sarcomere length over the illuminated region. This was shown by a series of recordings taken from the Reticon photodiode array showing the intensity of the diffracted light as a function of position along the array. The lack of broadening of the first order just after a release justifies the use of sarcomere length as a descriptive variable in this experiment.

Figure 4a shows the typical behavior of muscle length, tension and sarcomere length during a contraction with a quick release. The sarcomere length in Fig. 4b is shown averaged over a uniform region for a stretch. Occasionally this trace shows oscillations at a frequency above about 1 kHz immediately following the length transient; in these cases a line was drawn through the midpoints of the oscillations to the intersection with a line following the fast change in sarcomere length.

Figure 5 shows a summary graph of sarcomere length *vs.* relative change in tension. A typical error bar is shown, based on estimated resolutions and effects of translation. The line is the result of a least squares fit to the data with a straight line passing through the origin, assuming that the measurement error shown applies uniformly to each datum.

The sarcomere length change needed to just drop the tension to zero (ΔSL_0) is between 6 nm and 8 nm per half sarcomere, or 7.2 nm phs according to the fit line. The change in muscle length needed for the same effect was variable, but was typically about 2% to 3% of the resting length; in contrast, ΔSL_0 is between 0.5% and 0.7% in the intact regions.

The rapid elasticity has stiffness independent of the release magnitude up to stretches of about 10 nm per half sarcomere; beyond this point, there was considerably greater uncertainty in the measurement of tension at the end of the stretch.

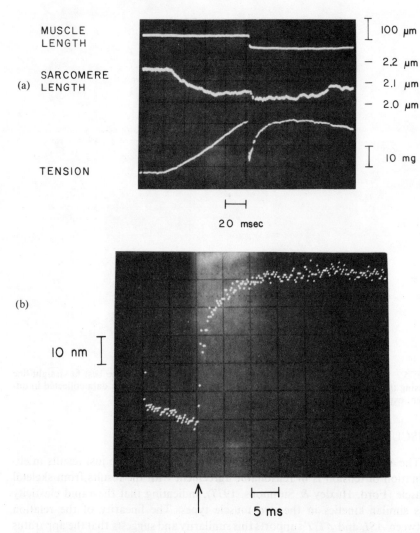

FIG. 4. (a) Behavior of muscle length, sarcomere length and tension during a contraction with a quick release. (b) Behavior of sarcomere length near the time of a large quick stretch, which occurs at the arrow. Inter-dot time difference is 0.2 msec. The two dots at the far left are for calibration purposes.

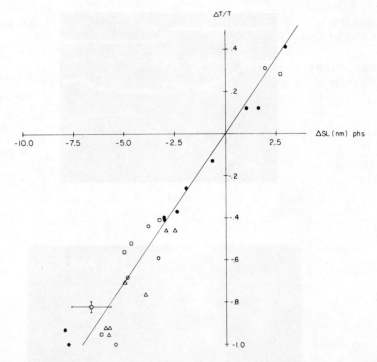

FIG. 5. Summary plot of ΔSL vs. $\Delta T/T$. A typical error bar and the best fit straight line passing through the origin are shown. The different symbols represent data collected in different experiments.

DISCUSSION

The amount of quick decrease of sarcomere length which just results in elimination of tension is in reasonable agreement with the results from skeletal muscle (Ford, Huxley & Simmons, 1977), indicating that the rapid elasticity has similar kinetics in the two muscle types. The linearity of the relation between ΔSL and $\Delta T/T$ supports this similarity and suggests that the apparatus has sufficient speeds of release and response to measure the elastic properties discussed.

The damaged ends, however, are perhaps 4 to 6 times more compliant than are the intact regions. Quick-release data on cardiac preparations obtained using muscle length as a descriptive variable, rather than sarcomere length, may therefore yield misleading interpretations. Furthermore, the development of substantial post-release tension, which requires restretching the ends, will occur at a speed given by the force-velocity relation for the intact regions as well as by the size and characteristics of the presumably passive end regions; in any case, this velocity will be less than V_{max}, the maximum shortening speed. Since the corresponding shortening time will be in the order of 10 msec, and because of the difficulty in evaluating the effects of the end compliance, it is unlikely that the fast phases of tension recovery seen in skeletal muscle could be easily resolved in the usual cardiac preparations.

Acknowledgements

This work was supported in part by grants NHLBI 18676 and AHA 75791. M. Delay was supported by NIH Postdoctoral Training Grant HLO 5147-19. D.V. Vassallo was supported by the CNPq of Brazil. G.H. Pollack is an Established Investigator of the American Heart Association. We acknowledge the indispensable assistance of Mr. R.W. Marvin, Mr. D. Hakala, Mr. S Moulton and Mr. E. Shibata.

REFERENCES

Abbott, B. C. & Mommaerts, W. F. H. M. (1959). Study of inotropic mechanisms in the papillary muscle preparation. *J. gen. Physiol.* **42**, 533–551.

Ford, L. E., Huxley, A. F. & Simmons, R. M. (1977). Tension responses to sudden length change in stimulated frog muscle fibers near slack length. *J. Physiol.* **269**, 441–515.

Huxley, A. F. & Simmons, R. M. (1971). Proposed mechanism of force generation in striated muscle. *Nature, Lond.* **233**, 533–538.

Iwazumi, T. & Pollack, G. H. (1979). On-line measurement of sarcomere length from diffraction patterns in muscle. *I. E. E. E. Trans. Biomed. Eng.* In press.

Krueger, J. W. & Pollack, G. H. (1975). Myocardial sarcomere dynamics during isometric contraction. *J. Physiol.* **251**, 627–643.

Parmley, W. W. & Sonnenblick, E. H. (1967). Series elasticity in heart muscle: the relation to contractile element velocity and proposed muscle models. *Circulation Res.* **20**, 112–123.

Pollack, G. H., Iwazumi, T., ter Keurs, H. E. D. J. & Shibata, E. F. (1977). Sarcomere shortening in striated muscle occurs in stepwise fashion. *Nature, Lond.* **268**, 757–759.

Steiger, G. J. (1977). Tension transients in extracted rabbit heart muscle preparation. *J. molec. Cell. Cardiol.* **9**, 671–685.

Sugi, H. & Kamiyama, A. (1977). The anatomical origin of the series elastic component in cardiac muscle. *Proc. Japan Acad.* **53**, Ser. B, No. 7, 297–301.

Ter Keurs, H. E. D. J., Iwazumi, T. & Pollack, G. H. (1970). The sarcomere length-tension relation in skeletal muscle. *J. gen. Physiol.* **72**, 565–592.

Discussion

Kawai: I think the reason why you do not see the fast transient phase is presumably related to the large length release. If you do releases of fraction of a percent, I think you can see it.

Pollack: We have done it. We have performed releases from the smallest that we could discern, to releases that dropped tension to zero, and we have not seen a distinct quick recovery phase. I am not making a major point of this because I am not sure if it is due to a problem of compliance at the ends, or if it really is a basic feature of mammalian cardiac sarcomeres.

Steiger: I did a couple of related experiments with glycerinated cardiac muscle (Steiger, 1977). We could drop to zero tension with a release of about 1.5% muscle length, and the speed of the muscle puller at that time was about 3 msec.

I would also like to make a comment dealing with experiments on living cardiac muscle during sustained contraction. With very small amplitude of length change, 0.1 to say 0.5%, we saw a fast recovery phase, and we also saw splitting up in two phases, following stretch, but the speed of the recovery was not as fast as in skeletal muscle as shown by Huxley and Simmons (1971) with frog muscle.

Pollack: Yes, I know those results. On some of the records that you see here, there is also a recovery which one might say is a "fast" recovery, but in the Huxley-Simmons results (Huxley & Simmons, 1971) the distinction between the very fast recovery and the slow recovery was so very much greater than we noticed in our results. I am not sure if such a large distinction is apparent in your results, either. Could you see the distinction so clearly?

Steiger: Yes, it is about an order of magnitude between the very fast one and the second one. Also, I will show in my upcoming presentation that the biochemical results of Marston and Taylor indicate that the whole actomyosin cycle is about an order of magnitude slower.

Sagawa: Dr. Pollack, some years ago you demonstrated an important nonlinearity. I remember that you obtained different coefficient values of elastic modulus for stretch and for release. However, in one slide you have shown this morning I did not see any difference. You have connected a straight line for both stretch and release. Is this because of the small percentage-wise stretch and release?

Pollack: There was an important distinction between the old and new results. The old result (Pollack, Huntsman & Verdugo, 1972) plotted a change of muscle length against change of tension. In the newer results we plotted change of sarcomere length against change of tension. The problem is that the damaged ends are unpredictable. There is a zone near the clips which we think behaves more or less like an elasticity; then there is a presumably healthy zone where we carry out our studies; and in between those two is a zone that one may call a "gray zone," which behaves in some intermediate manner that is difficult to predict. Perhaps it is this gray zone that causes the complexity of the results that are apparent in the old literature.

Mashima: I did the same experiments using about the same apparatus. I measured just the length of the whole muscle, and still I obtained similar results: the maximum extension of the series elastic component during contraction was about 3–4% of muscle length.

Pollack: What were the speeds of release?

Mashima: It is not so fast as yours. In heart muscle, the damaged cell soon deteriorates, but the neighboring cell recovers very soon, so that I do not think the damage is propagated.

Pollack: Well, the question is, what is the neighboring cell? When clamping occurs, certainly the cells adjacent to the clamp are destroyed and depolarized. Because of the low resistance junctions or nexuses, the depolarization will propagate to adjacent cells. To how many cells it will propagate depends on the length constant, but we can watch visually under the microscope to determine this. We see no striations near the clamp and then, gradually, there is a buildup of striations, and when you are, perhaps, four or five hundred microns from the clamp, the well-striated regions appear. That is the evidence that we have for the damage at the ends, plus the fact that in quick release most of the recoil occurs at the ends. Very little occurs in the central region.

REFERENCES

Huxley, A. F. & Simmons, R. M. (1971) Proposed mechanism of force generation in striated muscle. *Nature, Lond.* **233**, 533–338.

Pollack, G. H., Huntsman, L. L. & Verdugo, P. (1972) Cardiac muscle models. An overextention of series elasticity? *Circulation Res.* **31**, 569–579.

Steiger, G. J. (1977) Stretch activation and tension transients in cardiac, skeletal and insect flight muscle. In *Insect Flight Muscle*, pp. 221–268, ed. Tregear, R. T. Amsterdam: North Holland.

The Origin of the Series Elasticity in Striated Muscle Fibers

HARUO SUGI

Department of Physiology, School of Medicine, Teikyo University, Tokyo, Japan

ABSTRACT

High-speed cinematographic recordings of length changes at different parts of active striated muscles or muscle fibers were performed during the course of a quick release to give information about the origin of the series elasticity (SE) in various kinds of striated muscle. The highly compliant SE in dog papillary muscle (extension of SE with P_0, 6–9% of L_0) was shown to result mainly from the localized elastic extension and recoil of muscle tissue at both ends. The highly compliant SE in the superficial abdominal extensor muscle fibers of the crayfish (extension of SE with P_0, about 2% of L_0 at which the sarcomere length is about 9 μm) was found, on the other hand, to be distributed fairly uniformly along the entire fiber length, indicating that the compliant SE may largely originate from some structures in each sarcomere other than the cross-bridges. In frog semitendinosus muscle fibers, the cinematographic studies indicated that, during a quick release, the shortening was mostly localized at the fiber segment nearest the released fiber end until the total fiber shortened by 0.5–1%. This may imply that the tension changes resulting from too rapid length changes do not give correct information about the cross-bridges. When rabbit psoas muscle fibers in the rigor state were stretched by 10–30%, the resulting lengthening of sarcomeres was mostly taken up by the H-zone and the I-band while the amount of overlap between the filaments did not change appreciably, indicating the extensibility of the thick and thin filaments.

Although the mechanical behavior of active muscle can be explained by postulating an elastic component in series with a contractile component (Hill, 1938), the origin of the series elasticity (SE) in various types of striated muscle is not yet clear. In frog skeletal muscle, the extension of the SE with the maximum isometric tension (P_0) is about 1% of the slack length (L_0) (Jewell & Wilkie, 1958), and the force-extension curves of the SE are known to be scaled down in proportion to the isometric tension immediately before quick release (Huxley & Simmons, 1971a, 1973). These results have been taken as evidence that the SE is not separate from the contractile component, but largely resides in the cross-bridges. In other kinds of striated muscles, however, the extension of the SE with P_0, as measured by the minimum amount of quick release required to reduce the active tension from P_0 to zero, seems to be too large for the elastic extension of the cross-bridges (3–7% of L_0 in mammalian skeletal muscle, Close, 1972; 5–10% of L_0 in cardiac muscle, Sonnenblick, 1964; Bahler, Epstein & Sonnenblick, 1974). Large values of extension of the SE with P_0 have also been reported on chemically skinned or glycerinated frog muscle fibers (Julian, 1971; Wise, Rondinone & Briggs, 1973). Meanwhile, the dependence of the SE on the isometric tension (Huxley & Simmons, 1971a, 1973), can also be accounted for by assuming a passive SE having an exponential force-extension relation (Podolsky & Teichholz, 1970; Meiss & Sonnenblick, 1974; Hill, 1975).

As has been pointed out by Jewell & Wilkie (1958), the most straightforward way of locating the origin of the SE is to record the length changes at different parts along the length of a muscle or a muscle fiber during the actual course of a quick release. The following experiments were undertaken to examine the origin of the SE in various types of striated muscle by use of high-speed cinematography during the course of quick releases.

The series elasticity in cardiac muscle

The right ventricular papillary muscles were removed with the ventricular septa from dogs under pentobarbitol anesthesia. The ventricular septum was fixed securely to the bottom of an experimental chamber, and the septal artery was cannulated to perfuse the septum and the papillary muscle with oxygenated (95% O_2 − 5% CO_2) Tyrode solution containing the defibrinized blood of a sacrificed dog (Langer & Brady, 1963). The distal tendon of the papillary muscle was tied to an L-shaped arm extending from an electromagnetic vibrator (Ling, type 203), which was driven with rectangular current pulses to impose rapid length changes on the muscle. The tension transducer consisted of a pair of semiconductor strain gauges glued to the arm (resonance frequency, 500 Hz), while the displacement transducer was a light beam-photodiode system (Sugi, 1972). To avoid complications arising from the parallel elastic component, the initial muscle length was set at a point where the resting tension was less than 5% of the maximum isometric force. The muscle was stimulated to contract with 5 msec current pulses through a pair of Pt wire electrodes. At the peak of isometric tension the muscle length was quickly decreased to a variable extent in about 2 msec, and the coincident reduction of tension was observed with a dual-beam oscilloscope (inset in Fig. 1). All experiments were made at room temperature (24–26°C).

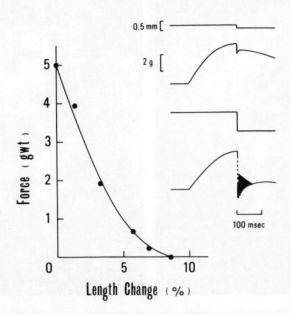

FIG. 1. Force-extension curve of the SE in dog papillary muscle. The tension immediately after a quick release is plotted against the amount of quick release expressed in percentage of the initial muscle length. Inset shows examples of length and tension changes by a quick release (Sugi & Kamiyama, 1977, by permission of *Proceedings of the Japan Academy*).

Figure 1 shows the force-extension curve of the SE in dog papillary muscle. In agreement with previous reports, the minimum amount of quick release required to reduce the peak tension to zero, *i.e.*, the extension of the SE with P_0, was 6–9% of the initial length (11–15 mm).

The anatomical origin of the above highly compliant SE was examined by putting a number of fine carbon particles on the muscle surface, and recording the length changes of the muscle segments divided by the particles during the development of isometric tension and during the quick release by means of a 16 mm high-speed cine-camera (500 frames/sec, Redlake Corp., Hicam) mounted on a Wild binocular microscope (Fig. 2). To avoid the distortion of the muscle image due to the disturbance of the surrounding fluid caused by the release, the muscle was exposed to air and illuminated by a Nikon xenon

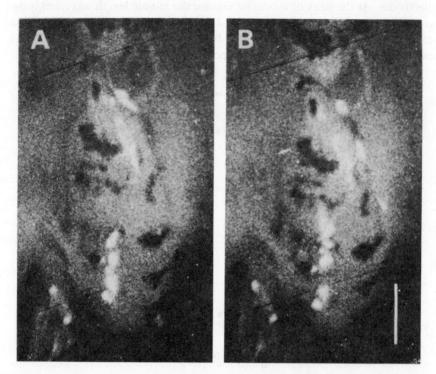

FIG. 2. Selected frames from a cinefilm of a papillary muscle at the peak of isometric tension (A) and immediately after a quick release (B) with which the tension fell to zero. The distal tendon end of the muscle is located at the upper part of each frame. Each dark mass on the muscle consisted of a number of fine carbon particles attached to the muscle surface. Calibration, 2.5 mm (Sugi & Kamiyama, 1977, by permission of *Proceedings of the Japan Academy*).

lamp during the cinematographic recording. Figure 3 shows a typical result of the length changes of the muscle segments during the development of isometric tension (left) and during a quick release (right) with which the peak tension fell to zero. The length of each fiber segment was not kept isometric during the tension development; some segments shortened by stretching others, though the total muscle length remained virtually constant (Krueger & Pollack, 1975). Such local tissue movements disappeared when the isometric tension reached the peak.

During a quick release, with which the peak tension was just reduced to zero, the shortening was mostly localized in the distal and the proximal end segments, while no appreciable shortening was observed in the other segments

SERIES ELASTICITY IN STRIATED MUSCLE

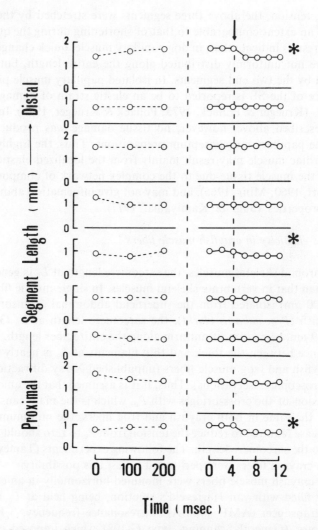

FIG. 3. Length changes of the muscle segments during the development of isometric tension (left) and by a quick release (right) with which the muscle length was reduced by about 9% so that the tension just fell to zero. The quick release was made at the time indicated by the arrow. Note that, during the quick release, the shortening is localized in the two end segments and in a segment at the middle part of the muscle. These three segments are indicated by asterisks (Sugi & Kamiyama, 1977, by permission of *Proceedings of the Japan Academy*).

except for a segment at the middle of the muscle where the shortening was sometimes also perceptible. It was noticed that, during the development of isometric tension, the above three segments were stretched by the other segments to an extent comparable to that of shortening during the quick release.

These results indicate that, in dog papillary muscle, quick changes in muscle length are not uniformly distributed along the entire length, but are mostly absorbed by the two end segments. In isolated papillary muscle preparations, the source of the SE is reported to be an elastic strain of damaged tissue at both ends (Krueger & Pollack, 1975; Pollack & Krueger, 1976). In the experiments described above, however, no tissue damage was produced at either end of the papillary muscle-septum preparations. Thus, the highly compliant SE in cardiac muscle may result mainly from the localized elastic strain and recoil of the muscle tissue due to the complex network of component muscle cells (Hort, 1960; Miur, 1965), and may not give information about the cross-bridge properties (Sugi & Kamiyama, 1977).

The series elasticity in crayfish muscle fibers

In arthropod skeletal muscles, the sarcomere length at L_0 is generally much longer than that in vertebrate skeletal muscles. In single muscle fibers (diameter 80–300 μm) isolated from the superficial abdominal extensor muscles of the crayfish (*Procambarus clarki*), the sarcomere length at L_0 (3.8–7.0 mm) is about 9 μm, being almost uniform along the entire fiber length. Meanwhile, the distance between the thick and thin filaments at L_0 is nearly the same in both crayfish and frog muscle fibers (unpublished X-ray diffraction and electron microscopic observations). Thus, if it is assumed that the amount of elastic extension of the cross-bridges with P_0, which is the main cause of the SE, is nearly the same in both crayfish and frog fibers, the minimum amount of quick release required to reduce the tension from P_0 to zero should be inversely related to the sarcomere length. The following experiments (Tameyasu & Sugi, 1979) on crayfish fibers were performed to test this possibility.

Single crayfish muscle fibers were mounted horizontally in an experimental chamber filled with van Harreveld's solution, being held at L_0 between the tension transducer (AME, type AE80; resonance frequency, 3 KHz) and a servo motor (General Scanning, type G-108) which imposed quick length changes on the fiber in about 0.5 msec. Since the fibers do not generate action potential, they were maximally stimulated by applying transverse alternating currents (4–6 V/cm, 1,000 Hz) through a pair of Pt plates covering two opposite walls of the chamber (Csapo & Wilkie, 1956). The resulting steady isometric tension (4–8 Kg/cm^2) was almost as large as the maximum potassium contracture tension, and was taken as P_0. The fiber length was quickly reduced at P_0 with the servo motor, and the length and tension changes were simul-

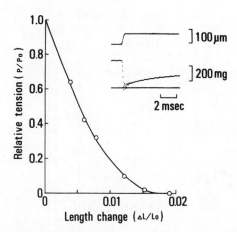

FIG. 4. Force-extension curve of the SE in single muscle fibers from the superficial abdominal extensor muscles of the crayfish. The tension (P) immediately after a quick decrease in length is plotted against the amount of decrease in fiber length (ΔL). P and ΔL are expressed relative to P_0 and L_0, respectively. Inset shows an example of length and tension changes by a quick decrease in fiber length. The bottom trace indicates the level of zero tension (Tameyasu & Sugi, 1979, by permission of *Experientia*).

taneously recorded with the oscilloscope (inset in Fig. 4). All experiments were made at room temperature (22–24°C).

Figure 4 shows a typical force-extension curve of the SE in crayfish muscle fibers. The minimum amount of quick release required to reduce the tension from P_0 to zero, *i.e.*, the extension of the SE with P_0, was about 2% of L_0, being definitely larger than the value in frog muscle fibers contrary to the expectation previously mentioned. Since it seems unlikely that each cross-bridge in crayfish fibers can be elongated by 2% of the length of a half sarcomere or about 900 Å, a value about one order of magnitude larger than that supposed for frog fibers, this result implies that the highly compliant SE in crayfish fibers cannot be explained as being due to the elastic extension of the cross-bridges. It should also be noted that, immediately after the completion of the fall of tension coincident with the quick decrease in length, the tension started to rise approximately exponentially without preceding rapid early tension recovery (inset in Fig. 4) which is known to occur in frog muscle fibers (Huxley & Simmons, 1971b, 1973).

The anatomical origin of the above highly compliant SE was examined by recording the length changes of the fiber segments in a single crayfish muscle fiber with a number of carbon particles attached to its surface during the course of a quick release. A 35 mm ultra high-speed cine-camera (Beckman & Whit-

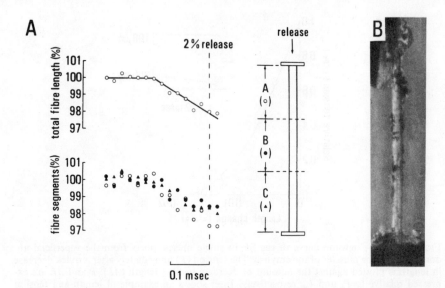

FIG. 5. Length changes of the fiber segments during the course of a quick release. In A, the length changes of the total fiber length and three fiber segments indicated by the inset are shown. Note that the shortening takes place fairly uniformly along the entire fiber length until the moment when the total fiber length is reduced by 2% (vertical interrupted line). B is a frame from a cinefilm of a tetanized muscle fiber during the course of a quick release. A number of carbon particles are attached to the fiber surface (Tameyasu & Sugi, 1979, by permission of *Experientia*).

ley, model 165) was used so that the length changes of the fiber segments could be recorded at 40,000–50,000 frames/sec (Fig. 5B). As shown in Fig. 5A, the time course of shortening of each fiber segment was fairly uniform until the total fiber length was shortened by 2%, *i.e.*, the amount of quick release required to reduce the tension from P_0 to zero (Fig. 4). This may indicate that the highly compliant SE in crayfish muscle fibers is fairly uniformly distributed along the entire fiber length. It seems possible that the SE originates mainly from extremely stretched sarcomeres distributed along the fiber length. This possibility was, however, excluded by microscopic observations (up to 400X) and cinematographic recording (Redlake Corp., Locam, at 500 frames/sec) of isometrically contracting crayfish fibers; though some fiber segments shortened by stretching others during the development of isometric tension as in the case of dog papillary muscles (Fig. 3), the resulting variation of sarcomere length along the fiber length was not so large.

Thus, the experiments on crayfish fibers strongly suggest that the highly compliant SE originates mainly from the extension of some structures in each sarcomere other than the cross-bridges, namely, the thick and thin filaments and the Z-disc. Experimental evidence for the extensibility of the thick and thin filaments in striated muscle fibers will be described later.

The series elasticity in frog muscle fibers

In 1971, Huxley & Simmons (1971a,b) studied the tension changes in single frog muscle fibers following quick length changes which were completed in 1 msec, and proposed a new sliding filament model for muscle contraction. According to them, each cross-bridge is composed of a myosin head and an elastic link extending from the thick filament, and the straight force-extension relation of the cross-bridge elasticity is truncated by the early rapid tension recovery due to the rotation of the myosin head (Huxley & Simmons, 1971b, 1973; Ford, Huxley & Simmons, 1977). As a matter of fact, the force-extension relation of the SE in frog muscle fibers is straighter the larger the velocity of quick release (Ford et al., 1977). If the SE resides mostly in the elastic link of the cross-bridges, the shortening is expected to take place uniformly along the entire fiber length. To ascertain whether this is actually the case, high-speed cinematographic recordings were also made on frog fibers during the course of a quick release (Sugi & Tameyasu, 1979).

Single muscle fibers (diameter, 50–120 μm) were isolated from the semitendinosus muscles of the frog (*Rana japonica*), a pair of stainless-steel wire connectors (0.1 mm in diameter and 2–3 mm in length, Civan & Podolsky, 1966; Sugi, 1972) being tied to both tendons with braided silk thread. The fiber was mounted horizontally by hooking the connectors to the tension transducer (AME, type AE80, resonance frequency, 3 KHz) and the lever of the displacement transducer (light beam-photodiode system, Civan & Podolsky, 1966; Sugi, 1972). The fiber was held at L_0, and maximally tetanized by applying transverse alternating currents (50 Hz) through a pair of Pt plates at both sides of the fiber. When the maximum isometric tension P_0 was developed, the fiber length was changed quickly by pushing the lever of the displacement transducer with a vibrator (Ling, type 203). The velocity of length changes (completed in 0.12–0.20 msec) was as large as or larger than that used by Ford *et al.* (1977). The length and tension changes were monitored with a dual-beam oscilloscope.

The length change of the whole fiber with a number of carbon particles attached to its surface was recorded during the course of a quick release or a quick stretch with the 35 mm ultra high-speed cine-camera at 40,000–50,000 frames/sec (Fig.6B). Attention was focused on the length changes of the fiber segments divided by the particles until the total fiber length shortened by 0.5–

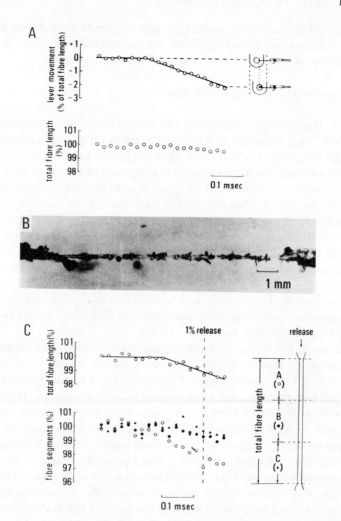

FIG. 6. Length changes in single muscle fibers from the semitendinosus muscles of the frog during the course of a quick decrease in length. A: a marked difference in the time course of lever motion and fiber shortening, indicating the detachment of the connector from the lever when the former is simply hooked to the latter as illustrated in the inset. B: a frame from a cinefilm of a tetanized muscle fiber during the course of a quick change in length. A number of carbon particles are attached to the fiber surface. C: time course of length changes of the fiber segments during the course of a quick decrease in length, the connector being clamped securely to the lever. The fiber was divided into three segments of nearly equal lengths (A, B and C in the inset). Note that the shortening is mostly localized in the segment A nearest the released fiber end (Sugi & Tameyasu, 1979, by permission of *Experientia*).

1%, since the tension is known to fall from P_0 to zero within the above range of quick decreases in length (Huxley & Simmons, 1971b, 1973; Ford et al., 1977). This was also confirmed in the present study, though the frequency response of the tension transducer was not high enough to follow the exact time course of the tension changes when the coincident length changes were complete in 0.12–0.20 msec. All experiments were made at room temperature (18–22°C).

At the early stage of the experiments, it was found that, if the wire connector was simply hooked to the lever imposing length changes on the fiber, the fiber did not shorten as quickly as the lever movement due to the detachment of the connector from the lever (Fig. 6A); since the diameter of the hole (0.8 mm) through which the connector was hooked to the lever was many times larger than the diameter of connector wire, the lever could move quickly in the direction of fiber shortening for a distance while the fiber shortened only with a much slower velocity (inset in Fig.6A). This may indicate that, even when the fiber generating the maximum isometric tension is suddenly allowed to shorten freely, the SE can only recoil with a velocity much slower than that of the lever movement used.

To prevent the above detachment of the connector from the lever, the connector was firmly clamped to the lever and similar experiments were repeated. Figure 6C is a typical example of the length changes of the fiber segments during a quick decrease in length. It can be seen that, until the total fiber length shortens by 0.5–1%, the shortening is not uniformly distributed along the entire fiber length, but is mostly confined to the fiber segment nearest the released end of the fiber. This seems to suggest that, the fall of tension during a quick release is not due to the elastic recoil of the cross-bridges taking place along the entire fiber length, but is due to a localized buckling of the fiber at its released end; the localized buckling may produce a mechanical impulse which is transmitted along the fiber to the opposite end to exert negative tension on the tension transducer. If this explanation is correct, the fact that the fall of tension during a quick release is straighter the faster the velocity of quick release (Huxley & Simmons, 1971b; Ford et al., 1977) might result from the fact that the larger the velocity of release is, the more marked is the localized buckling; the early rapid tension recovery following a quick release (Huxley & Simmons, 1971b; Ford et al., 1977) might be related to the disappearance of the buckling. Much more experimental work is needed to clarify the above point.

When, on the other hand, the fiber was stretched quickly, the lengthening of the fiber segments was observed to take place fairly uniformly along the entire

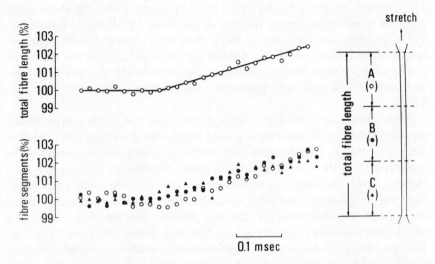

FIG. 7. Length changes of the fiber segments during the course of a quick increase in length. The fiber was also divided into three segments of nearly equal lengths (A, B and C in the inset). Note a fairly uniform lengthening along the entire fiber length (Sugi & Tameyasu, 1979, by permission of *Experientia*).

fiber length, as shown in Fig. 7. It remains to be investigated whether the quick stretch produces the elastic extension of the cross-bridges or produces the extension of other sarcomere structures such as the myofilaments.

In conclusion, these experiments on frog muscle fibers may be taken to indicate that the tension changes resulting from too rapid length changes do not necessarily give correct information about the cross-bridge properties, and the nature of the SE should be carefully examined from this point of view.

Extensibility of myofilaments

Though the thick and thin filaments in striated muscle have been assumed to be inextensible in the model of Huxley (1957), it seems possible that the myofilaments are extensible so that they are extended to a degree when a muscle is generating isometric tension to contribute to the SE. The contribution of the myofilament compliance to the SE should be much larger than that of the cross-bridge compliance in crayfish muscle fibers in which the SE is too compliant to be explained by the extension of the cross-bridges (Tameyasu & Sugi, 1979).

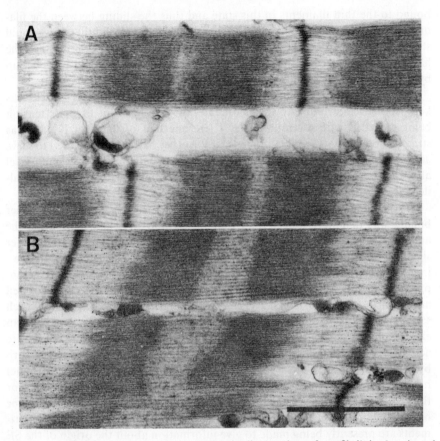

FIG. 8. Electron micrographs showing the longitudinal sections of myofibrils in glycerinated rabbit psoas muscle fibers fixed during the rigor state. The preparation was initially held isometric, and after the development of rigor tension one half was stretched by about 30% before fixation, while the other half was kept isometric. The sarcomere length in the stretched part (B) is much longer than the sarcomere length in the non-stretched part (A), while the amount of overlap between the thick and thin filaments is nearly the same between the two parts. Note that the elongation of the sarcomere is mostly taken up by the elongation of the H-zone and the I-band (Suzuki & Sugi, unpublished experiments). Calibration, 1 μm.

As stretching of resting or active muscle fibers primarily causes the sliding between the thick and thin filaments, it is not possible to measure the filament compliance by stretching muscle fibers unless the sliding between the filaments is inhibited. For this purpose, Dr. S. Suzuki and I recently made experiments on muscle fibers in the rigor state, since the cross-bridges are tightly bound to the thin filaments in this state and thus the sliding between the filaments cannot be readily produced by stretching. It was hoped that, if we could stretch each sarcomere without changing the amount of overlap between the filaments, then the resulting elongation of each sarcomere would be taken up by the elongation of the filaments in the non-overlap region.

A small fiber bundle obtained from glycerinated rabbit psoas muscle was kept isometric and put into the rigor state, the development of rigor tension being recorded with a strain gauge. Then, one half of the fiber bundle was stretched by 10–30%, while the length of the other half was not altered to serve as a control. When the tension in the stretched half reached a steady level after the completion of stretching, the whole fiber bundle was fixed in glutaraldehyde solution for electron microscopic observation.

It was found that the sarcomere length in the stretched half of the fiber bundle was definitely longer than that in the non-stretched half, while the amount of overlap between the filaments in the stretched half did not differ significantly from that in the non-stretched half (Fig.8). This may indicate that, in a state of rigor, the sliding between the filaments does not take place appreciably in spite of a marked elongation of sarcomeres as had been assumed at the start of the experiments. As can be seen in Fig.8, the elongation of each sarcomere is taken up by a marked elongation of the H-zone and also an elongation of the I-band, demonstrating that the thick and thin filaments are actually extensible. Since many myofibrils within the muscle fiber are torn by stretching, it is necessary to estimate carefully the proportion of intact myofibrils in the stretched fibers before calculating the filament compliance or the degree of extension of the filaments with P_0, which is related to the degree of contribution of the myofilaments to the SE in intact muscle fibers. Experiments are presently being made to give information about the origin of the SE in each sarcomere of striated muscle fibers.

Acknowledgement

I am grateful to my collaborators Dr. A. Kamiyama, Dr. T. Tameyasu and Dr. S. Suzuki for allowing me to include published and unpublished results from our recent experiments.

I am also indebted to Dr. H. Hashizume of the Engineering Research Institute, University of Tokyo, for providing facilities for the use of the ultra high-speed cinecamera, to Dr. T. Tsuno, Mr. Y. Nakamura and Miss S. Gomi for their technical assistance, and to Dr. Y. Hiramoto of the Tokyo Institute of Technology for suggesting the use of rigor state muscles for the examination of filament extensibility.

REFERENCES

Bahler, A. S., Epstein, F. & Sonnenblick, E. H. (1974). Series elasticity of *in vitro* mammalian cardiac muscle. *Am. J. Physiol.* **227**, 794–800.

Civan, M. M. & Podolsky, R. J. (1966). Contraction kinetics of striated muscle fibres following quick changes in load. *J. Physiol.* **184**, 511–534.

Close, R. I. (1972). Dynamic properties of mammalian skeletal muscles. *Physiol. Rev.* **52**, 129–197.

Csapo, A. & Wilkie, D. R. (1956). The dynamics of the effect of potassium on frog's muscle. *J. Physiol.* **134**, 497–514.

Ford, L. E., Huxley, A. F. & Simmons, R. M. (1977). Tension responses to sudden length change in stimulated frog muscle fibres near slack length. *J. Physiol.* **269**, 441–515.

Hill, A. V. (1938). The heat of shortening and the dynamic constants of muscle. *Proc. R. Soc.* **B 126**, 136–195.

Hill, T. L. (1975). Theoretical formalism for the sliding filament model of contraction of striated muscle Part II. *Prog. Biophys. molec. Biol.* 105–159.

Hort, W. (1960). Macroscopic and microscopic research on the myocardium of the left ventricle filled to varying degrees. *Arch. path. Anat. Physiol.* **333**, 523–564.

Huxley, A. F. (1957). Muscle structure and theories of contraction. *Prog. Biophys. biophys. Chem.* **7**, 255–318.

Huxley, A. F. & Simmons, R. M. (1971a). Mechanical properties of the cross-bridges of frog striated muscle. *J. Physiol.* **218**, 59–60P.

Huxley, A. F. & Simmons, R. M. (1971b). Proposed mechanism of force generation in striated muscle. *Nature, Lond.* **233**, 533–538.

Huxley, A. F. & Simmons, R. M. (1973). Mechanical transients and the origin of muscular force. *Cold Spring Harbor Symp. quant. Biol.* **37**, 669–680.

Jewell, B. R. & Wilkie, D. R. (1958). An analysis of the mechanical components in frog's sartorius muscle. *J. Physiol.* **143**, 515–540.

Julian, F. J. (1971). The effect of calcium on the force-velocity relation of briefly glycerinated frog muscle fibres. *J. Physiol.* **218**, 117–145.

Krueger, J. W. & Pollack, G. H. (1975). Myocardial sarcomere dynamics during isometric contraction. *J. Physiol.* **251**, 627–643.

Langer, G. A. & Brady, A. J. (1963). Calcium flux in the mammalian ventricular myocardium. *J. gen. Physiol.* **46**, 703–719.

Meiss, R. A. & Sonnenblick, E. H. (1974). Dynamic elasticity of cardiac muscle as measured by controlled length changes. *Am. J. Physiol.* **226**, 1370–1381.

Miur, A. W. (1965). Further observations on the cellular structure of carciac muscle. *J. Anat.* **99**, 27–46.

Podolsky, R. J. & Teichholz, L. E. (1970). The relation between calcium and contraction kinetics in skinned muscle fibres. *J. Physiol.* **211**, 19–35.

Pollack, G. H. & Krueger, J. W. (1976). Sarcomere dynamics in intact cardiac muscle. *Eur. J. Cardiol.* **4**, 53–65.

Sonnenblick, E. H. (1964). Series elastic and contractile elements in heart muscle: changes in muscle length. *Am. J. Physiol.* **207**, 1330–1338.

Sugi, H. (1972). Tension changes during and after stretch in frog muscle fibres. *J. Physiol.* **225**, 237–253.

Sugi, H. & Kamiyama, A. (1977). The anatomical origin of the series elastic component in cardiac muscle. *Proc. Japan Acad.* **53B**, 297–301.

Sugi, H. & Tameyasu, T. (1979). The origin of the instantaneous elasticity in single frog muscle fibres. *Experientia*, **35**, 227–228.

Tameyasu, T. & Sugi, H. (1979). The origin of the series elastic component in single crayfish muscle fibres. *Experientia*, **35**, 210–211.

Wise, R. M., Rondinone, J. F. & Briggs, F. N. (1973). Effect of pCa on series elastic component of glycerinated skeletal muscle. *Am. J. Physiol.* **224**, 576–579.

Discussion

Edman: Did you use a smaller release than 1%? Suppose you go down to 0.5%, or smaller, so that you do not get a complete drop in tension. What happens then?

Sugi: Our equipment has limitations because we cannot measure very small length changes precisely. I can try smaller releases, but it is extremely difficult to measure the resulting length change accurately.

Edman: The problem may be that you use relatively big movements so you have exceeded the maximum cross-bridge movement.

Sugi: As shown in Fig. 6A, the lever moves during release but the fiber does not move due to the detachment of the connector from the lever; this may give some indication. Anyway, due to technical limitations, I cannot apply very small releases to measure length changes of elementary parts precisely.

Pollack: Do you think that the quick recovery phase seen by Huxley & Simmons (1971) is due simply to the removal of the buckling?

Sugi: I expect so. During the quick recovery phase, non-uniformity may disappear. This is my expectation, but I have not yet done the experiment to ascertain this.

Pollack: Dr. Lydia Hill did some experiments two or three years ago (Hill, 1977), in which she did quick stretches. She measured the change of A-band width. Her contention was that there was no change in A-band width after quick stretch. I wonder if you could comment on the relevance of her results to yours.

Sugi: In those experiments, the muscle was not in a rigor state, so maybe sliding took place. Filament compliance may be masked by sliding.

Winegrad: Have you looked at transverse sections of these stretched rigor muscles to see how much misalignment there might be?

Sugi: A very important point. We have just started this kind of experiment about two months ago. We should do that.

Huxley: Perhaps I could return to really the essence of Dr. Edman's question again. In the regions of the fiber in which you apparently do not see any shortening, what is the maximum amount of shortening that could actually be taking place there, and you would still not see it? I think that in the Huxley-Simmons experiments, certainly in the more recent ones (Ford, Huxley & Simmons, 1977), the extent of shortening necessary to drop the tension to zero is really very small. I think it is about 0.5%. Can you say that such an amount of shortening is not taking place in these other regions of the fiber?

Sugi: It depends on the accuracy of measurement. This is perhaps a little smaller than 1%, but we cannot definitely measure 0.5%.

Huxley: Since you could not have detected a shortening of 0.5%, you can not claim that your results are inconsistent with those of Huxley and Simmons.

Sugi: I agree with you in that, to settle this problem completely, it is necessary to detect the non-uniformity with an accuracy of 0.5% or less.

Edman: I think the method to use here would be laser diffraction. I remember that Dr. Mason from Australia had a paper in *Nature* sometime ago (Barden & Mason, 1977) where they had visualized length changes which would correspond to those recorded by Huxley and Simmons. Have you done any?

Sugi: Not yet, but it would be interesting.

Kushmerick: I still want to return to the same point, and may ask the converse question, and that is the minimum amount of shortening that you can detect up to the region next to the lever. What percent was it?

Sugi: About 1%.

Rüegg: What do you think about measuring stiffness by stretching, and getting the curves by doing stretches, rather than releases? Is that all right?

Sugi: As far as this experiment is concerned, a quick decrease in length gives some non-uniformity, while a quick increase in length gives no non-uniformity. So, I think the two methods are qualitatively different, and the interpretations may not be simple.

Huxley: A comment on the experiments of stretching the rigor muscle. As you say that a number of fibrils ruptured during this process, presumably, the tension actually being applied to the remaining ones was a good deal larger, so, it seems to me that it does not really tell you anything about the elasticity.

Sugi: The purpose is to extrapolate the tension on the myofibrils back to the maximum isometric tension, *Po*. I am now estimating the percentage of fibrils remaining continuous, while we can measure total tension, so the only thing is to estimate the effective cross-sectional area. Then, we can draw some conclusions as to the elastic extension of the filaments with *Po*.

Huxley: Well, another way of getting at this is by looking for any change in the periodicity in the thin filaments during contraction, which we did by X-ray diffraction. We were not able to detect (up to an accuracy of about 1 or 2%) change in the 385 Å period in an isometric contraction, so I do not think under normal circumstances that thin filaments are extended.

Sugi: The point is that you do the experiment in the rigor state. Have you ever stretched a rigor muscle extensively, and observed periodicity changes?

Huxley: No. I mean, if you have a lot of rupture in the muscle, you have got no way of knowing what tension is being applied, but in isometric contraction you do know what tension is being applied to each part of the muscle.

Sugi: My interpretation is that filament compliance may be largely masked by sliding between the filaments without using rigor muscle.

REFERENCES

Barden, J. A. & Mason, P. (1977) Muscle cross-bridge stroke and activity revealed by optical diffraction. *Science, N. Y.* **199**, 1212–1213.

Ford, L. E, Huxley, A. F. & Simmons, R. M. (1977) Tension responses to sudden length changes in stimulated frog muscle fibres near slack length. *J. Physiol.* **269**, 441–515.

Hill, L. (1977) A-band length, striation spacing and tension change on stretch of active muscle. *J. Physiol.* **266**, 677–685.

Huxley, A. F. & Simmons, R. M. (1971) Proposed mechanism of force generation in striated muscle. *Nature, Lond.* **233**, 533–538.

Multiple Segment Analysis of Muscle Data

Allan H. GOTT

Volunteer Research Associate, Department of Physiology, School of Medicine, Loma Linda University, Loma Linda, California, U.S.A.

ABSTRACT

For fifty years, muscle physiologists have interpreted experimental results by using lumped models. A lumped model gathers all the ultrastructure of a particular type in a functional element. Two to four elements are configured to represent the experimental muscle unit. During the last two decades, investigators have used single isolated fibers, made very high-speed length changes at one end, and measured mechanical transients at the other end. Under such conditions, lumped models may mask important unequal dynamic states existing along the fiber. Most investigators use such mechanical transients to imply cross-bridge interaction with actin filaments, which is presumed to be the source of mechanical tension. The Multiple Segment Model uses charge-like mathamatics, accounts for unique ultrastructure properties, and describes mechanical transients of individual sarcomeres from one tendon to the other. These capabilities provide significantly finer detail than lumped models, and raise important questions regarding lumped models and presumed cross-bridge generation of mechanical tension. MSM results predict unique mechanical transients at the fixed and moving ends of a single fiber, which are not accounted for by cross-bridge theory. Ultra high-speed cine recording of a fiber during a quick length change offers force determination at all points along the fiber through the use of $F = ma$. Use of moving and fixed end transducers during high-speed quick length changes should confirm MSM predictions. These experimental results may provide new striated muscle function information, and raise interesting questions regarding presumed cross-bridge generation of mechanical tension.

INTRODUCTION

More than 50 years of experimental effort have been devoted to the study of striated muscle. A major result was the sliding filament model, which was developed during the last two decades of this period. A wide variety of investigations were undertaken throughout the entire interval, with the goal of obtaining data to establish and support a particular theory of striated muscle function.

During the later part of the five decades, a basic experimental procedure evolved for measuring and recording mechanical tension. A tension measuring device was attached to one end of some form of striated muscle. This end remained fixed throughout the experimental event. The other end of the muscle unit either remained fixed, or the position was varied, according to some predetermined protocol.

Analysis and interpretation of data obtained from such experiments was, of necessity, based on one of several lumped models. These methods required assumption of uniform (or mean) conditions throughout the muscle unit. This assumption was adequate during the early decades of experimental striated muscle function study. Typically, investigators would study a whole muscle (*e.g.*, frog sartorious).

Throughout this time period, another factor contributed to the retention of lumped models. The resolution of various optical instruments available for tissue study was not sufficient to resolve individual functional parts of striated muscle ultrastructure. Thus, the recorded tension time history was analyzed, along with other experimental conditions, according to whichever lumped model the investigator felt best represented the experimental epoch. Toward the end of these early decades of striated muscle function study, steady increases were made in the rates of muscle length change. Soon, it became increasingly difficult to explain experimental results (whether from controlled variation of load, or controlled variation of length), or the models failed altogether.

During the 1950's, the advent of the electron microscope provided a significant increase in resolution available for the study of striated muscle ultrastructure. This technology, together with cumulative experience in the preparation of single isolated fibers (*e.g.*, frog semitendinosus), led to a major new thrust in the experimental study of striated muscle function. This new effort shortly led to formulation of the sliding filament model (Huxley & Niedergerke, 1954; Huxley & Hanson, 1954).

Investigators became proficient in the preparation of a semitendinosus single isolated fiber (SIF). Study of SIF specimen ultrastructure by electron microscopy, and SIF macro-structural characteristics using medium-speed micro-cinematography, demonstrated large "uniform sarcomere length" regions between the ends of the SIF.

SIF experimental preparations were subject to intensive study. Movement of different points along an SIF axis were plotted during a fixed length (or isometric) contraction. Experiments were conducted both at "slack length," and at fiber lengths which precluded overlap of actin and myosin filaments comprising an individual sarcomere (Huxley & Peachey, 1961).

Of considerable significance, even though experimental data clearly demonstrated that points along an SIF axis moved at different rates, discussions and analyses of experimental data were still undertaken in terms of lumped models. As emphasized earlier, these methods necessarily presumed uniform or mean conditions between tendon ends. (Later in the same time period, other investigators applied x-ray diffraction techniques to the study of striated muscle ultrastructure (Elliot, Lowy & Worthington, 1963). However, this instrumentation provided statistical data on very large numbers of sarcomeres, which masked individual sarcomere function, and the data was still analyzed in terms of lumped models.)

At about the same time, mechanical function experiments were modified to minimize the effect of lumped model assumptions by maintaining length control over the uniform portion of the SIF. This servo technique presumed that the time histories of tension at the

(1) moving end tendon,
(2) moving end beginning of the uniform region (marked by a piece of gold foil),
(3) fixed or tension measuring end of the uniform region (also marked by a piece of gold foil), and
(4) fixed end tendon

were either equal, or, sufficiently close to a mean to allow interpretation of results in terms of a lumped model (Gordon et al., 1966a, b). These assumptions were necessary, since only lumped models were available for data analysis at that time. Such models allowed analysis only of fixed end tension histories in terms of effects generated by moving end displacement.

Some efforts were made to obtain data through the control of time-varying loads applied to the SIF moving end. However, difficulties in both experimental procedures and data analysis soon led to the almost exclusive utilization of controlled length variations applied to the moving end of the fiber. Experimental events were confined to shorter and shorter time periods, in an attempt to identify significant fine resolution time factors which might contribute to the high speed tension transients recorded at the fixed end. As in the early study of whole muscle, the increase in experimental rates yielded factors which caused lumped models to fail, unless progressively more complex assumptions were invoked.

Today, data for study of the sliding filament model is most often obtained by application of very high-speed controlled length changes to the moving end of an SIF, while a mechanical tension transient is recorded at the fixed end (Huxley & Simmons, 1971; Ford, Huxley, & Simmons, 1977). X-ray diffraction is still in use, and other efforts apply advanced optical instrumentation to direct measurement of the distances between Z-lines for a modest number of sarcomeres (Pollack, Iwazumi, ter Keurs & Shibata, 1977). The sarcomere set lies in a small region at a selected point along the axis of an SIF, while the SIF is subjected to an isometric or controlled length change event. Here again, analysis and interpretation of data must still rely upon lumped models, even though some experimental discussions recognize that an SIF has

(1) a unique ultrastructure configuration,
(2) unique rates of tension development, and
(3) unique rates of movement

at all points along its axis. Data analysis in terms of lumped models still requires increasingly complex assumptions, and relies upon the series elastic element (SE) concept as well as the stiffness concept, which requires mean properties assumptions. Finally, analytical methods only treat the narrow region in time immediately surrounding the experimental transient epoch, rather than from t_0 to the cessation of contractile activity.

CURRENT STATUS OF THE SLIDING FILAMENT MODEL

Evaluation of the sliding filament model is, in reality, concerned with two principal questions. The first is whether or not the actin and myosin filaments actually slide past each other. The second, which has been the subject of significant investigative effort during the last two decades, is concerned with identifying the basic mechanism which converts chemical energy to mechanical tension in striated muscle. Reports published shortly after the advent of the electron microscope appeared firmly to establish that actin and myosin filaments did indeed slide past each other during the contraction and relaxation processes (Huxley & Niedergerke, 1954; Huxley & Hanson, 1954). Other publications appearing not long afterwards were taken as evidence that overlap of actin and myosin filaments were necessary in order for the production of mechanical tension to take place (Gordon *et al.*, 1966a, b).

The postulate that interaction between myosin projections (cross-bridges) and sites along actin filaments was necessary for the production of mechanical tension was a major premise of these publications. Subsequent experimental effort has been devoted to the study of this postulate through two principal avenues: the analysis of the time history of fixed end mechanical tension during an experimental event, and the analysis of X-ray diffraction data to establish the spatial configuration of striated muscle ultrastructure during an experimental event.

The majority of investigators studying striated muscle function have accepted both the sliding filament model and the cross-bridge action postulate with few questions. These same investigators have also given little consideration to so-called "charge models," on the grounds that charge models do not account for significant aspects of striated muscle function. Even so, a broad evaluation of the sliding filament model and cross-bridge theory would note that an ideal theory of muscle function would account for the production of mechanical tension by both striated and smooth muscle, with little modification. Few, if any, publications have addressed this matter.

Classical cross-bridge theory is clearly not applicable to production of mechanical tension by smooth muscle. Additionally, publications in this area typically do not account for the radical difference in characteristics between tendon, sarcolemma, active, and viscous components of striated muscle. Publications which do address these matters, do so in qualitative, as opposed to quantitative, terms. Quantitative examination of these topics is necessary, if highly nonlinear properties of the various elements are to be adequately studied. Such studies should also include variations in ultrastructure cross-section area which are known to occur along the axis of a single isolated fiber. The acceptability of any model of striated muscle function would be significantly enhanced if it were to show distinct possibilities of accounting for the production of mechanical tension by smooth muscle.

Although (as stated above) charge models have not achieved general acceptance, some recent efforts have begun to achieve consistent results. In addition, methodical and detailed comparisons of cross-bridge and charge models have identified the growing complexity of assumptions required to support cross-bridge theory (Noble & Pollack, 1977).

Thus, a summary of the current status of the sliding filament model would seem to be as follows: Actin and myosin filament sliding is generally accepted. Cross-bridge action between myosin and actin filaments, as an *a priori* necessity for the production of mechanical tension, is strongly advocated by the majority of investigators. This type of muscle function does not appear to be directly applicable to the production of mechanical tension by smooth muscle. Even the widely accepted cross-bridge theory does not seem to quantitatively account for the different ultrastructure properties of the major components of striated muscle. A small but growing study of charge models has begun to yield interesting results.

The remainder of this paper is devoted to the exposition of a (multiple segment) charge model which does not require interaction between the myosin projections and actin filament sites for the production of mechanical tension. The multiple segment model also appears to quantitatively account for significantly different ultrastructure properties, as well as the different cross-sectional areas of these properties. Finally, the quantitative analytical approach utilized also seems capable of application to the smooth muscle question.

MULTIPLE SEGMENT MODEL BACKGROUND

The multiple segment model (MSM) is a charge model. Biophysical properties of the active, passive and viscous elements of the primitive element are directly determined by related three-dimensional ultrastructure. All assumptions are thus supported by identifiable ultrastructure (Gott, Stimson & Janz, 1971; Gott, 1978a). This methodology accounts for all material contained within an SIF extending from the end of one tendon, completely along the fiber axis, to the end of the other tendon. The influence of each structural element is modified by its own unique cross-sectional area.

In contrast to lumped model approaches, which mask the large number of dynamic states along an SIF, the multiple segment method divides the entire fiber into a number of segments. The forces resulting from each type of ultrastructure within a segment (as modified by related cross-sectional area) are summed, and entered into the expression $F^i = m^i a^i$ (Gott, 1978a). This expression forms the basis of a system of equations for relating both the component and net forces existing between individual SIF segments. Solution of the system of "stiff" equations by numerical integration (Gott, 1978b) will

(1) allow prediction of the forces from individual ultrastructure components within any segment at any point along the SIF axis from the moving tendon end to the fixed tendon end,

(2) account, for any given segment, for an entire mechanical tension time history from t_0, to the transient epoch, throughout the transient epoch, and to completion of the experimental event or t_{end}, and

(3) demonstrate an analytical reduction of the MSM system of equations to a lumped model form, and rigorously define the static and dynamic limits within which Tension $_{lumped\ model}$ = Tension $_{MSM}$.

These capabilities offer a powerful new method for analyzing striated muscle experimental data, while attempting to identify those factors which define the chemo-mechanical energy conversion process within the striated muscle sarcomere.

MULTIPLE SEGMENT MODEL FORMULATION OVERVIEW

The primitive model contains active, passive and viscous elements. The active element of interest is the Contractile Element (CE), which is formed from the instantaneous IT, ALT product. The IT has been analytically derived by expressing the Ca^{++} flow between the terminal cisternae, sarcoplasmic reticulum, myoplasm and myofilaments. Using rate constants derived from Ebashi (Ebashi & Endo, 1968; Gott et al., 1971), solution of the system of equations expressing the flow model yields a time history of calcium bound to the myofilaments which is congruent with an isometric twitch. This (IT) curve is considered to represent the time varying chemical capacity of the sarcomere to generate mechanical tension. The ALT has been analytically derived beginning with the expression for force between point charges in a vacuum, extending the expression to include line integrals representing actin and myosin filaments, and then summing all attractive and repulsive forces for filament positions defined by sarcomere ultrastructure. Displacement of sarcomere components at constant line integral charge density, while invoking the constant volume restriction, yields a series of net force values which closely approximate the ALT. This (ALT) curve is considered to represent the dimensionally varying capacity of the sarcomere to generate mechanical tension. Within the MSM Contractile Element (CE), sarcomere filaments are rigid and inextensible. This fact, together with the absence of cross-bridge function, defines that the MSM CE does not move until Z-line motion occurs.

(Note that extension of the charge force summation expression for straight line integrals to a charge force summation between 3-dimensional curves requires only the introduction of cosine terms. This modification would support a novel extension of the MSM algorithms from striated to smooth muscle.)

The IT instantaneous chemical energy input to the CE is modified by an acceleration-sensitive feedback term. The feedback requirement was identified by energy management analysis of output obtained from the MSM system of equations. While a detailed treatment of the analysis is beyond the scope of this publication, suffice it to say that the feedback curve has the classical shape of an ion exchange membrane. The plateau fulfills the function of allowing constant feedback along most of the length of an SIF, where individual segments have varying acceleration.

Instantaneous values for passive component inputs are directly and quantitatively specified by length tension curves for related ultrastructure.

Viscous components are still under investigation. Tendon viscosity is treated as being linear. Active material viscosity seems to demonstrate a remarkable nonlinearity. This nonlinearity is determined by the instantaneous concentration of calcium in transit through the myoplasm, before binding to the myofilaments. Although this postulation is still under investigation, it seems reasonable to assume an association between Ca^{++} concentration within the myoplasm and myoplasm viscosity because of the strength of the calcium ion.

Equations expressing these relationships are

$$p^i = F_{\text{tot}}^i = \{[(F_{\text{IT-norm}}(t) \cdot F_{\text{ALT-norm}}(\overline{SL})^i \cdot F_{\text{IT-max}}) \\ + (F_{\text{visc-act}}(l^i)] \cdot A_{\text{act-norm}}^i\} \\ + [F_{\text{PLT-slma}}(\overline{SL})^i \cdot A_{\text{slma-norm}}^i] \\ + \{[F_{\text{PLT-tend}}(\overline{SL})^i + (F_{\text{visc-tend}}(l^i)] \cdot A_{\text{tend-norm}}^i\} \quad (1)$$

which will depict the total force for any unique i th segment when properly initialized,

$$m^i \ddot{l}^i = p^i - p^{i+1} \quad (2)$$

which expresses the generalized inertial force balance between two adjacent segments, and

$$\dot{l}^i = l^i t \quad (3)$$

which defines the acceleration of the i th segment. Representative initializations of (1) are shown in Fig. 1.

A numerical integration solution of a system of equations formed from these basic elements yields a complete time-varying catalog of all component forces and displacement for each segment comprising a Multiple Segment Model. The comprehensive nature of this catalog represents a powerful tool for the methodical study of individual parametric variations. Additionally, provision of component forces and displacements for all segments of the SIF (besides the fixed end tendon) opens a complete new realm of experimental investigation not heretofore possible (Gott, 1978a).

MULTIPLE SEGMENT MODEL RESULTS

Early MSM investigations were based on the lumped model reduction (MSM-L). Later analytical investigations established the validity of the lumped form, and the limits within which the lumped model performed satisfactorily. Two different *in vitro* force velocity investigations were studied.

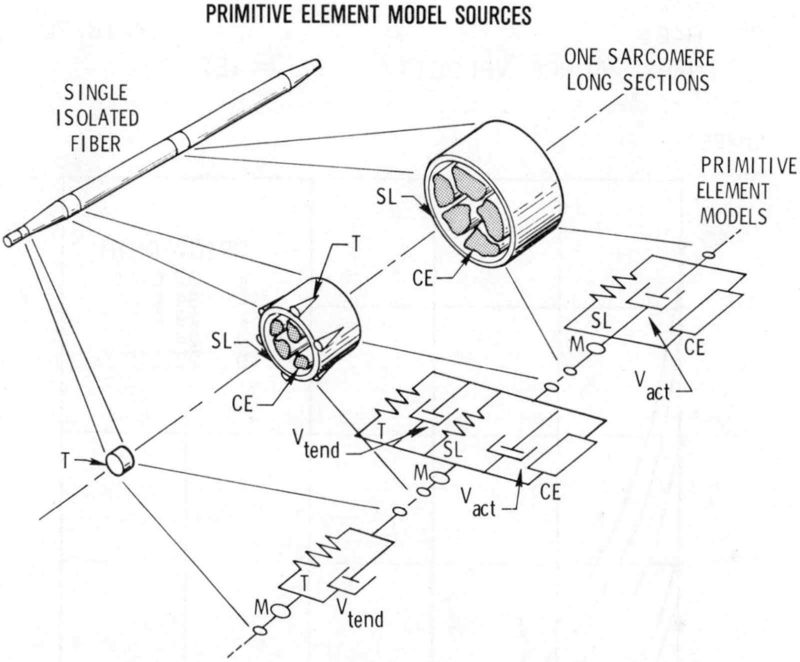

PRIMITIVE ELEMENT MODEL SOURCES

FIG. 1. MSM model formats. Various initializations of generalized MSM model, illustrating relationships of ultrastructure types to model elements.

The first analysis attempted to duplicate cat RV papillary muscle force velocity variation with temperature (Yeatman, Parmley & Sonnenblick, 1969). Figure 2 displays both laboratory data and MSM-L results. A remarkable correspondence was obtained, including the droop of high temperature curves at increasing afterload. This fingerprint reproduction of experimental results is best understood by referring to the published analytical form for force velocity prediction. The expression identifies the sharp peak of a high-temperature single-stimulus isometric twitch as the factor causing high temperature-high afterload force velocity droop (Gott, et al., 1971).

The second analysis studied a classical force velocity experiment based on the frog sartorius (Jewell & Wilkie, 1958). The original investigation measured tetanized tension and force velocity. Figure 3 demonstrates that when measured twitch data was used to drive the MSM-L model, a force velocity curve was produced which fell below the laboratory experimental force velocity

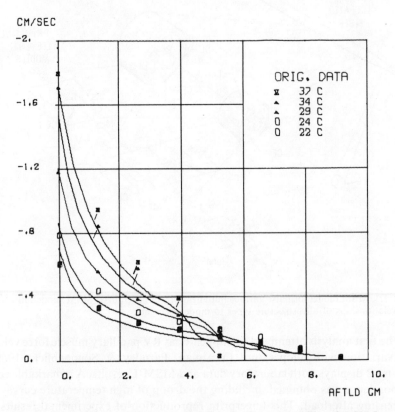

FIG. 2. Cat RV papillary muscle force velocity curve. Illustrates ability of MSM-L model to reproduce temperature-variant Force Velocity Curves when driven by experimental temperature variant IT's (Yeatman, *et al.*, 1969).

curve. When the recalculated twitch, based on an assumed series elastic (SE) element load extension curve was used, a force velocity curve was obtained which was congrueut with experimentally determined values (Gott *et al.*, 1971).

These successful reproductions of cardiac and skeletal force velocity curves seemed to support the initial MSM-L formulation. The MSM-L skeletal results provided an indication of energy supply feedback which was substantially

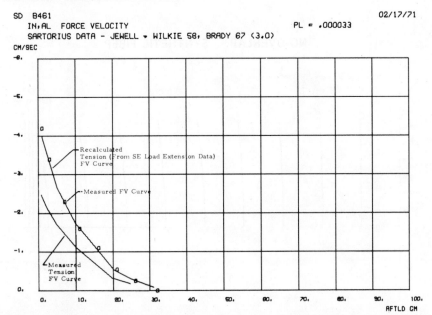

FIG. 3. Frog sartorius force velocity curve. MSL-L model performance in reproducing experimental Force Velocity Curves using measured and recalculated tetanized IT's (Jewell & Wilkie, 1958).

more clear than the feedback performance imbedded in the cardiac study.

Other quick length change (QLC) investigations revealed that the MSM-L model failed at non-physiological QLC rates. Analytical studies indicated a strong likelihood that different dynamic states existed along the axis of an SIF. The MSM-L model was then extended to the multiple segment form.

The first version was applied to the study of a classical single fiber experiment, where the fiber was stretched to a length which precluded the overlap of actin and myosin filaments (Huxley & Peachey, 1961). Tetanic stimulation of the fiber produced an isometric contraction, and results were interpreted to mean that contractile function occurred only in those regions near fiber ends which contained overlapping actin and myosin filaments. The first MSM configuration used in the study was limited to a "half fiber," and was based on the assumption that an isometric contraction would be symmetrical about the fiber midpoint. Algorithm initialization only accounted for variation in sarcomere length, and neglected variations in cross-sectional area for respective ultrastructure elements. Even so, Fig. 4 illustrates a remarkably close resemblance to published experimental data.

Shortly thereafter, experimental laboratory efforts began to study mechan-

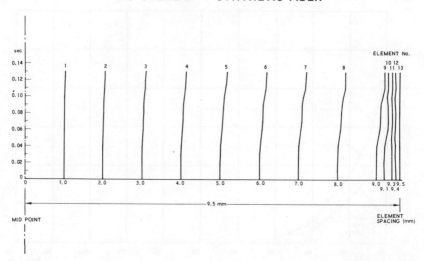

Fig. 4. Contractile activity, frog semitendinosus single isolated fiber stretched beyond overlap. MSM half fiber reproduction of experimental results (Huxley & Peachey, 1961) where MSM has no overlap and also excludes residual overlap.

ical transients produced by very high-speed QLC's applied to tetanized single isolated fibers. An important publication reported results which seemed to support a major advancement in the knowledge of cross-bridge function contribution to the production of mechanical tension (Huxley & Simmons, 1971). Of significant interest, large values of length change (both release and stretch, or QLC-R and QLC-S) produced mechanical tension transients demonstrating a notch or plateau.

Application of the first MSM half fiber configuration, which did not account for related ultrastructure cross-sectional areas, to these QLC studies was unsuccessful. Analysis of individual segment component forces and displacements clearly indicated the necessity for modification of individual ultrastructure element inputs by associated cross-sectional area. After this modification was accomplished, the MSM half fiber configuration demonstrated QLC-R performance illustrated in Fig. 5 and Fig. 6.

A detailed evaluation was made of the time-varying catalog of all component forces for each segment. This study defined the plateau or notch occurring during the mechanical transient (MT) recovery phase as resulting from the net sum of viscous (negative) and feedback-modified energy input (positive) re-

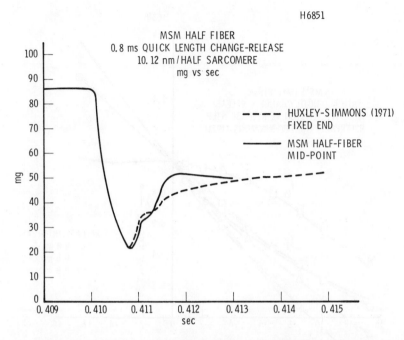

FIG. 5. Mechanical tension at midpoint of MSM half fiber (Quick Length Change-Release). Comparison of early MSM half fiber results demonstrating recovery phase notch and some tension overshoot before approaching stable recovery to experimental results (Huxley & Simmons, 1971).

maining approximately constant. It should be noted again that feedback magnitude for an individual segment is determined by instantaneous acceleration.

The MSM half fiber configuration was then applied to QLC-S investigations, with no success. Once again, evaluation of the component force catalog disclosed simulation details when the "half fiber" tendon end served as the moving end, and the midpoint served as the "fixed end." In summary, the sarcolemma and active elements (having midpoint cross-sectional areas) did not serve as an adequate "terminating pad."

The MSM half fiber configuration was then modified to the present full fiber form. It should be emphasized that each segment demonstrated a one-to-one relationship between ultrastructure elements and related cross-sectional areas. This is of particular importance in duplicating "nature's own terminating pads" formed by the tendon-sarcolemma-active material region between the primary contractile portion of the fiber and the tendons.

Application of the MSM full fiber configuration to a QLC-S event produced

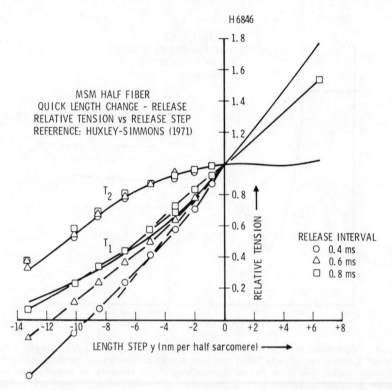

FIG. 6. T_1-T_2 variation with decrease in release interval for MSM half fiber (quick length change-release). Comparison of MSM half fiber performance at varying release intervals, predicting zero or negative T_2 values at large release magnitudes and small release times to experimental results (Huxley & Simmons, 1971).

the fixed end MT results illustrated in Fig. 7. The MSM results display a good fit with the published experimental mechanical transient. Here, the plateau results from an approximately constant net sum of viscous and feedback-modified energy input terms appearing in individual fiber segments near the fixed tendon end.

Another aspect of very high-speed QLC-S performance, not discussed in published laboratory experimental reports, relates to the influence of the QLC displacement time history. The highly nonlinear nature of MSM elements, and the acceleration-defined feedback curve, imply an extraordinary sensitivity of mechanical transient fingerprint details (for each and every segment) to the shape of the QLC input function. Tension magnitude and phase both have a nonlinear variation with QLC movement.

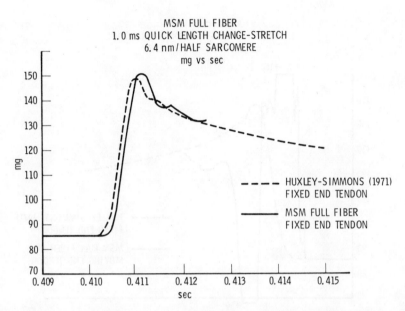

FIG. 7. MSM full fiber fixed end mechanical transient (quick length change-stretch). Comparison of MSM full fiber fixed end MT with experimental fixed end MT (Huxley & Simmons, 1971).

Thus far, only MSM-L and MSM fixed end outputs have been presented for comparison to experimental results. As stated earlier, the MSM approach yields a complete time-varying catalog of all component forces and displacements for each segment comprising a multiple segment model. This catalog may contain upwards of 12,000 component force and displacement values per millisecond of experimental time. Consequently, a detailed exposition is beyond the scope of this report.

However, a summary examination of the force components comprising the predicted moving end MT produces the result shown in Fig. 8. Clearly, the moving end MT is radically different from the conventionally recorded fixed end MT. It is instructive to examine the predicted MT's and displacements for segments representing moving and fixed ends of the "uniform sarcomere length" region. Mechanical transients are presented in Fig. 9, and displacements are presented in Fig. 10. The displacement curves appear qualitatively to resemble published data (Sugi & Tameyasu, 1979).

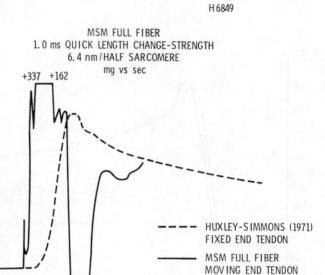

FIG. 8. MSM full fiber moving end mechanical transient (quick length change-stretch). Comparison of MSM full fiber moving end MT with experimental fixed end MT (Huxley & Simmons, 1971).

Both MT's and displacements are dissimilar for segments at opposite ends of the "uniform sarcomere length" (USL) region. Yet classical cross-bridge theory is based on lumped model analyses of data which presume that both ends of the uniform sarcomere length region may be treated as having equal performance. The different segment displacement time histories are of fundamental importance when evaluating these MSM results. The individual ultrastructure elements determine separate component forces, each of which are nonlinear with respect to velocity or acceleration. This being true, it is clear from $F^i = m^i a^i$ that differing rates of change of velocity (acceleration) and different velocities have nonlinear effects upon the total force.

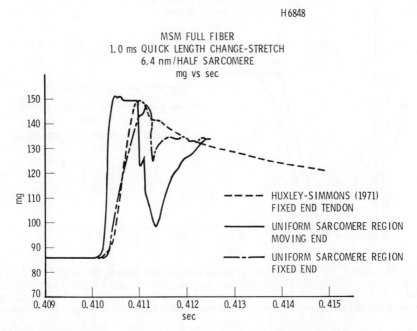

FIG. 9. MSM full fiber uniform sarcomere length region mechanical transients (quick length change-stretch). Comparison of MSM USL MT's with experimental fixed end MT (Huxley & Simmons, 1971).

SUMMARY OF STRIATED MUSCLE EXPERIMENTAL DATA ANALYSIS METHODS

Cross-bridge theory is studied by analyzing experimental data in terms of lumped models. Almost all data is obtained in terms of fixed end mechanical transients. Higher experimental rates require increasingly complex assumptions. Few of these assumptions are directly traceable to three-dimensional sarcomere ultrastructure for each unique sarcomere. Lumped models do not invoke specific passive length tension attributes of sarcolemma and tendon, nor do they require the constant sarcomere volume assumption. Three major experimental efforts (isotonic contraction of a single isolated fiber stretched beyond overlap (Huxley & Peachey, 1961), servo length controlled experimental determination of the ALT (Gordon et al., 1966a, b), and servo control of quick length changes (Ford et al., 1977)) all have differing analytical treatments of experimental data. Finally, neither lumped models nor cross-

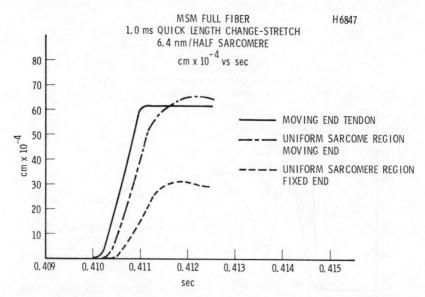

Fig. 10. MSM full fiber uniform sarcomere length region displacements (quick length change-stretch). MSM full fiber incremental displacements for moving end tendon and USL fixed and moving ends. Moving end USL overshoot of moving end tendon due to contraction of moving end region during rebound after termination of QLC-S movement.

bridge theory allow analytic derivation of the ALT or IT. The net result of these restrictions is to mask all static and dynamic performance characteristics of all points along the axis of an SIF, with the exception of the fixed end tendon. This masking of dynamic properties may conceal factors affecting the validity of SE and stiffness concepts.

Some charge models have been able to support analytic derivation of the IT and ALT (Ingels, 1967). Other more recent efforts had made a strong case for electrostatic force generation, as well as lateral and longitudinal stability (Iwazumi, 1970). Yet, neither of these efforts can support analytical methods which are capable of broad and comprehensive application to the analysis of experimental data from the preceding three experiments.

The Multiple Segment Model is based on charge theory. A single set of equations allows for analytic derivation of the IT and ALT, as well as a primitive model element based on three-dimensional active, passive and viscous ultrastructure attributes. This single set of equations seems to adequately represent, with no modification, any experimental data from isometric "stretched beyond overlap" experiments, up to and including very high-speed controlled length changes. Furthermore, the system of equations predicts and accounts for uniquely different high resolution tension and displacement histories for each point along the entire axis of a single isolated fiber. These MSM capabilities are not present within lumped models which support cross-bridge theory or in other charge models.

It is important to appreciate the significant difference between MSM data analysis capabilities and those provided by other methodologies. The resolution and sensitivity present within MSM equations implies a new requirement for refined experimental and data acquisition techniques. Traceable calibration is required throughout all instrumentation signal paths to a precision of $\pm$ 0.5%. QLC displacement time histories should be recorded to a precision of $\pm$ 0.5%, as should mechanical tension transient time histories. Methods for time correlation of all parameters are required. Publications should report all data to these same levels of resolution and accuracy. These steps are necessary if investigators are to be able to compare results between laboratories to a more quantitative level than typically utilized in current publications.

EXPERIMENTS TO CONFIRM MULTIPLE SEGMENT MODEL PREDICTIONS

Two realizable experiments would provide initial confirmation of Multiple Segment Model predictions. The first experiment would be based on high-speed cine recording of the full length of a single isolated fiber during a quick length change. The second experiment would depend upon instrumenting both ends of a single isolated fiber for tension recording during a quick length change.

High-speed cine recording of an SIF during QLC-R and QLC-S experiments, according to the techniques pioneered by Sugi (Sugi & Tameyasu, 1979), should establish the one-dimensional trajectories of all points along a single isolated fiber. The trajectories of the USL region fixed and moving ends would be of particular interest. A plot of the incremental displacements *vs.* time should confirm MSM prediction of these movements, which state that USL_{fixed} displacement $\neq USL_{move}$ displacement. MSM techniques predict that the phase and slope of these curves are not identical. Thus, it follows from $F = ma$ that the respective accelerations are not equal, and therefore Mechanical Transient $_{USL-fixed} \neq$ Mechanical Transient$_{USL-move}$. Confirmation of these predictions would substantiate major portions of MSM theory. Such confirmation would also raise interesting questions regarding important experiments relying upon servo-control of USL length during QLC-R and QLC-S experiments.

Development of a tension transducer capable of recording mechanical tension for the moving end tendon, without introducing artifacts, would allow comparison of both fixed and moving end mechanical transients, as recorded at the respective tendon ends. MSM predictions state that $MT_{tendon-fixed} \neq MT_{tendon-move}$. Furthermore, MSM predictions state that both $MT_{tendon-fixed}$ and $MT_{tendon-move}$ are highly sensitive to small variations in the overall time history of moving end tendon displacement (during both QLC-R and QLC-S experiments). Comparison of "fingerprint" variations of MSM predicted and experimentally recorded MT's should confirm these predictions and substantiate major portions of MSM theory. Confirmation of these predictions may raise other questions regarding major experiments using servo-control of USL length during QLC-R and QLC-S experiments.

CONCLUSIONS AND CURRENT PROBLEMS OF THE SLIDING FILAMENT MODEL

Sliding filaments are accepted. The majority of investigators prefer to explain the development of mechanical tension in terms of cross-bridge theory. The Multiple Segment Model offers comprehensive new analytical capabilities not heretofore available. The MSM predicts unique tension and displacement histories for all parts of a single isolated fiber in addition to the fixed end tendon, where conventional data is recorded. Realizable experiments may confirm these predictions, which seem to pose new questions regarding experiments (and results) providing fundamental support to theories of sliding filament cross-bridge function. Thus, it seems that the MSM offers a new alternative for the study of striated muscle experimental data, may supply additional new data for detailing the sliding filament model, and offers comprehensive analytical capabilities not available in other (lumped) cross-bridge or charge models.

REFERENCES

Ebashi, S. & Endo, M. (1968). Calcium ion and muscle contraction. *Prog. Biophys. molec. Biol.* **18**, 125–183.

Elliot, G. F., Lowy, J., & Worthington, C. R. (1963). An X-ray and light diffraction study of the filament lattice of striated muscle in the living state and in rigor. *J. molec. Biol.* **6**, 295–305.

Ford, L. E., Huxley, A. F. & Simmons, R. M. (1977). Tension responses to sudden length change in stimulated frog muscle fibers near slack length. *J. Physiol.* **269**, 441–515.

Gordon, A. M., Huxley, A. F. & Julian, F. J. (1966a). Tension development in highly stretched vertebrate muscle fibers. *J. Physiol.* **184**, 143–169.

Gordon, A. M., Huxley, A. F. & Julian, F. J. (1966b). The variation in isometric tension with sarcomere length in vertebrate muscle fibers. *J. Physiol.* **184**, 170–192.

Gott, A. H., Janz, R. F. & Stimson, M. J. (1971). Striated muscle function: theory, simulation and experimental observation. *Proc. San Diego Biomed. Symp.* **10**, 13–31.
Gott, A. H. (1978a). Striated muscle ultrastructure: Its functional contributions to mechanical tension. In preparation.
Gott, A. H. (1978b). FYBER, a Fortran IV computer program.
Huxley, A. F. & Niedergerke, R. (1954). Structural changes in muscle during contraction. *Nature, Lond.* **173**, 971–973.
Huxley, H. E. & Hanson, J. (1954). The cross-striations of muscle during contraction and stretch and their structural interpretation. *Nature, Lond.* **173**, 973–976.
Huxley, A. F. & Peachey, L. D. (1961). The maximum length for contraction in vertebrate striated muscle. *J. Physiol.* **156**, 150–165.
Huxley, A. F. & Simmons, R. M. (1971). Proposed mechanism of force generation in striated muscle. *Nature, Lond.* **223**, 533–538.
Ingels, N. B. (1967). An electrokinematic theory of muscle contraction. Ph.D. thesis, Stanford University.
Iwazumi, T. (1970). A new field theory of muscle contraction. Ph.D. thesis, Univ. of Pennsylvania.
Noble, M. I. M. & Pollack, G. H. (1977). Molecular mechanism of contraction. *Circulation Res.* **40**, 333–342.
Pollack, G. H., Iwazumi, T., terKeurs, H. E. D. W., & Shibata, E. F. (1977). Sarcomere shortening in striated muscle occurs in stepwise fashion. *Nature, Lond.* **268**, 757–759.
Jewell, B. R. & Wilkie, D. R. (1958). An analysis of the mechanical components in frog's striated muscle. *J. Physiol.* **143**, 515–540.
Sugi, H. & Tameyasu, T. (1979). The origin of the instantaneous elasticity in single frog muscle fibres. *Experientia*, **35**, 227–228.
Yeatman, L. A., Parmley, W. W., & Sonnenblick, E. H. (1969). Effects of temperature on series elasticity and contractile element motion in heart muscle. *Am. J. Physiol.* **217**, 1030–1034.

Discussion

Mohan: What kind of ion exchange mechanism do you envisage for the feedback system?

Gott: I think that the SR may act like an ion exchange membrane, and as it is mechanically stressed during a quick stretch or a quick release, as the wave propagates down the fiber, it would tend to change the rate at which calcium is either released or sequestered.

Muscle Stiffness in Relation to Tension Development of Skinned Striated Muscle Fibers*

J. C. RÜEGG, K. GÜTH, H. J. KUHN, J. W. HERZIG, P. J. GRIFFITHS and
T. YAMAMOTO

Department of Physiology II, University of Heidelberg, Heidelberg, F. R. G.

ABSTRACT

Muscle fiber stiffness was measured by recording tension changes during sudden length changes completed within 0.2, 0.5 or 0.8 msec. In mechanically or chemically skinned fibers of cardiac muscle, skeletal muscle (frog semitendinosus) or insect fibrillar muscle (*Lethocerus maximus* dorsal longitudinal muscle), stiffness appears to be Hookean in the range of $+0.5\%$ and -0.5% length change, *i.e.*, it does not depend on the direction and amplitude of the length change. Stiffness changes in proportion to tension when the latter is decreased by extension, reducing the actin-myosin overlap, or increased by Ca^{++} activation. Regardless of the size of isometric active tension, the linear portion of the force extension diagram (T_1-curve) extrapolates to the same abscissa intercept (about $-1\% L_0$). These findings are consistent with a contraction model where tension depends at any moment on the number of attached elastic myosin cross-bridges. In the model proposed by Huxley & Simmons, attachment of cross-bridges and force-generation by these attached bridges are distinct steps. Then, stiffness would be expected to alter in proportion to tension only as long as the ratio of force-generating and non-force generating attached cross-bridges is constant. However, this ratio is no longer constant under a variety of conditions. For instance, the ratio of stiffness to isometric tension is increased by lowering the temperature or the concentration of free Mg. An increase in force without concomitant change in stiffness was also found during the very fast tension recovery (quick phase) following a quick release. The increase in force at constant stiffness may be produced by a conformational change of attached cross-bridges (as suggested by the Huxley & Simmons theory).

* Supported by the Deutsche Forschungsgemeinschaft.

INTRODUCTION

It is widely accepted that tension is generated by actin-myosin cross-bridges. A. F. Huxley's theory (1957) proposes that the extent of tension development depends on the number of cross-bridges attached to actin at any one moment. According to Huxley & Simmons (1973), the relative number of attached cross-bridges may be estimated from measurements of immediate muscle stiffness; the latter is assumed to be Hookean and to be located to a large extent within the cross-bridges. This assumption is justified by two experimental findings: (1) stiffness is independent of the amplitude and direction of the length change used for stiffness measurements (Ford *et al.*, 1977); (2) stiffness depends on the extent of overlap between actin and myosin filaments and hence on the number of cross-bridges possible between filaments (Huxley & Simmons, 1973). In agreement with this cross-bridge hypothesis of tension generation it has been found that under a variety of conditions there is a strict proportionality between force generation and stiffness in living muscle. For instance, the ratio of stiffness to tension remains constant during the development of tetanic contraction in frog muscle (Huxley & Simmons, 1973), during the twitch of frog ventricular muscle, as well as during inotropic intervention in these muscles (Herzig, 1978); the same holds for contractions of vascular smooth muscle (Peterson, 1978) and of the anterior byssus-retractor muscle of *Mytilus* (Tameyasu & Sugi, 1976). It is of interest to know whether a tension-to-stiffness relationship similar to that in living muscle occurs in the case of isolated contractile structures such as glycerinated fibers or skinned fibers suspended in activating ATP-salt solution.

According to the new theory proposed by Huxley & Simmons (1973), attachment and force generation of cross-bridges are distinct steps in the cross-bridge cycle, and the stiffness-to-tension ratio will therefore remain constant only as long as the ratio of force generating and non-force-generating attached cross-bridge populations remains constant.

In muscle changes in these cross-bridge populations can be initiated by quick length changes leading to rapid tension transients. The Huxley-Simmons contraction model predicts a change in the stiffness-to-tension ratio during these rapid tension transients, and these will also be investigated in the case of skinned skeletal and glycerinated insect fibrillar muscle suspended in a ATP-salt solution.

METHODS

Mechanically skinned fibers of frog sartorius or semitendinosus were prepared according to a modified Natori method (Yamamoto & Herzig, 1978). Muscle fibers from the dorsal longitudinal muscle (DLM) of *Lethocerus maximus* were glycerinated and prepared according to Jewell & Rüegg (1966). A single fiber was mounted horizontally at L_0 ($\sim$ 4mm) between the length step generator and the tension transducer, using a fast-setting glue; the glycerol was washed out by immersing the fibers for a few minutes in a washing solution containing 50 mM KCl and 20 mM imidazole, then in an ATP relaxing solution containing 7–15 mM Mg-ATP and EGTA, pH 6.8 (for details see figure legends). The activating solution used in contraction experiments was similar to the relaxing solution except that EGTA was replaced by Ca^{++}-EGTA. Length changes measured by magnetic field plates were performed within 0.5 msec by a Ling dynamics vibrator (type 101) controlled by a velocity dependent feed-back system. Alternatively, step changes in length (measured by a photo-transistor) could be performed within 0.2–0.4 msec by means of a relay-type length step generator described by Güth & Kuhn (1978). The tension transducer (Aksjeselskapet AE 802) was mounted vertically on a firm metal arm, the position of which could be altered by means of a micrometer adjustment. The transducer stiffness was better than 1 mN per μm and its resonance frequency 8 to 14 kHz. It was possible to record both length and tension changes during a stretch or release performed within 0.25 msec and to plot tension and length in a form of a force-extension diagram on a digital oscillosocope. These force-extension curves may correspond to the T_1-curve described by Huxley & Simmons (1973); the slope of the linear portion of the curve was taken as a measure for immediate stiffness.

RESULTS

Linear relationship between tension and stiffness

As shown in Fig. 1a, there was a proportionality between stiffness and force during the tension rise following rapid calcium activation ("calcium jump," see Ashley & Moisescu, 1975) of a skinned skeletal muscle fiber. Skinned fibers of a frog semitendinosus muscle were relaxed in ATP-salt solution of very low calcium ion concentration buffered with 1 mM EGTA. After immersing the fiber in an activating ATP-salt solution containing 10^{-5} M Ca^{++} buffered with 20 mM calcium buffer (EGTA), the Ca^{++} ion concentration within the fiber was increased within less than 0.2 sec to the Ca^{++} ion concentration of the buffer. The tension rise, however was slower, peak tension being reached in > 1 sec (Fig. 1a). In some experiments the fiber was subjected to small amplitude oscillations of the muscle length (length change 0.05 % L_0, frequency $\sim$ 1 kHz),

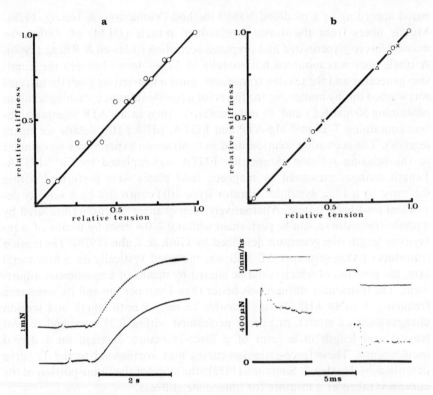

FIG. 1a. Above: relationship of stiffness and tension during contraction following a "calcium jump" in mechanically skinned frog sartorius muscle fiber (length 4 mm, sarcomere length 2.4 μm). Below: tension rise following a "calcium jump" with and without superimposed length oscillations. Stiffness was estimated from the band width of tension response to sinusoidal length oscillations (0.05% L_0, 1 kHz). Conditions: imidazole, 60 mM; ATP, 5 mM; Mg, 5 mM; Na, 20 mM; Cl, 41 mM; K, 36 mM; X^{2-}, 20 mM; creatine phosphate, 10 mM; creatine kinase, 20 units. ml^{-1}. pH 6.7, temperature, 10° C. pCa in relaxing solution 8, in activating solution 5.9. In the relaxing solution, X^{2-} represents 20 mM HDTA and 0.1 mM free EGTA, in the activating solution 20 mM total EGTA.

FIG. 1b. Above: dependence of immediate stiffness ($\Delta T/\Delta L$) upon isometric tension, altered by changes in free calcium concentration. Note that a distinct alteration in force is always accompanied by a corresponding proportional change in stiffness. Different symbols signify different experiments. Below: stiffness measurements by a quick stretch (left) or a quick release (right). The immediate elastic response (ΔT) occurring in phase with the length change (ΔL) is taken as a measure of stiffness ($\Delta T/\Delta L$). Single skinned semitendinosus fiber; conditions cf. Fig. 2.

and the corresponding tension changes were recorded on a digital oscilloscope. Stiffness was then estimated from the band width of these tension oscillations and it was shown to be linearly related to tension. A proportionality between stiffness and active force development was also observed when the degree of activation was varied by a change in the concentration of the activator Ca^{++} ions (Fig. 1b). If the sarcomere length was increased by stretching, there was a parallel decrease in force and stiffness due to the decrease in the overlap of actin and myosin filaments (Fig. 2). As in the case of similar experiments with surviving frog muscle fibers (Huxley & Simmons, 1973), these results were taken to mean that immediate stiffness depends on the number of cross-bridges attached to actin at any one moment.

In these experiments stiffness was determined by measuring the change in force due to, and in phase with, a rapid change in length (completed within 0.5 msec) (Fig. 3). Stiffness, *i.e.*, tension change per length change, is then given by the slope of the force-extension diagram, in which the tension change is plotted as a function of the length change (stretch or release). As shown in Fig. 3, the slope is fairly linear, indicating that the elastic elements responsible for immediate stiffness obey Hooke's law, *i.e.*, stiffness is independent of the size and direction of the length change. Though stiffness increases with increasing contractile force, the size of rapid length change required to abolish contractile tension in quick release experiments does not depend on this isometric force. A high and low contractile force can be abolished by a quick release corresponding to about 1 % length change (Fig. 3). These results are in agreement

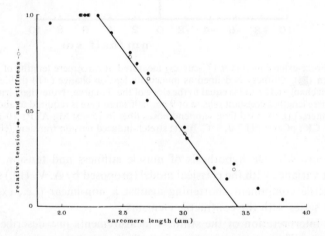

FIG. 2. Dependence of active tension (●) and stiffness (○) upon sarcomere length (skinned frog semitendinosus fiber). Note that tension and stiffness decrease proportionately with decreasing overlap. The deviation of sarcomere lengths > 3.2 μm is presumably due to the nonhomogeneity of sarcomere length occurring in highly stretched fibers. Sarcomere length was measured by laser diffraction. Conditions: 15 mM Mg-ATP; 10 mM CP; 30 units·ml⁻¹ CK; pCa 5; pH 7.0, 3°C.

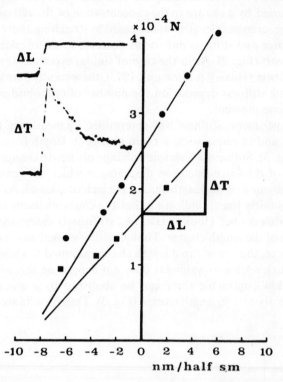

FIG. 3. Force-extension curves (T_1-curves) measured at sarcomere lengths of 2.2 μm (●) and 2.7 μm (■). Stiffness is defined as immediate tension change (ΔT) divided by immediate length change (ΔL) and is equal to the slope of the T_1-curve. Note that, irrespective of the sarcomere length, a constant release of 9 nm/half sarcomere is required to abolish tension (abscissa intercept). Skinned frog semitendinosus fiber in 15 mM Mg-ATP; 10 mM CP; 30 units·ml^{-1} CK; pCa 5; pH 7.0; 3°C. Inset: stretch-induced tension transient (cf. Fig. 1b)

with the cross-bridge hypothesis of muscle stiffness and tension. They are clearly at variance with the classical model (proposed by A. V. Hill) suggesting a contractile component shortening against a non-linear (*i.e.*, exponential) spring in series with the contractile elements.

In the interpretation of the stiffness measurements just described it is assumed that stiffness or the number of attached cross-bridges is not affected by the length step used for stiffness evaluation. In order to check the validity of this assumption, Güth & Kuhn (1976) devised a method for measuring stiffness *during* a quick change in length completed within about 0.2 msec. The parabolic and continuously accelerating length step (Güth & Kuhn, 1978) was imposed by a relay-type length step generator and tension changes were recorded during the length change of the digital oscilloscope and plotted against the length change. Provided that the length step was less than 0.5 %, the force extension curve of a glycerinated DLM fiber was fairly linear, *i.e.*, the slope stiffness remained constant, indicating Hookean elasticity (Güth *et al.*, 1978). At larger stretches however the force-extension curve was only linear during the first 0.5% of length change. At about 1% extension, a kink occurred in the force-extension diagram; during further extension from 1 to 2 %, the force no longer increased, behaving as though stiffness had suddenly dropped to zero. This phenomenon was interpreted as evidence for cross-bridge slippage involving rapid detachment of overstrained cross-bridges and reattachment of discharged cross-bridges (Fig. 4; c.f. Güth *et al.*, 1979, also Flitney & Hirst, 1978; Sugi, 1972). This experiment also shows that sudden length changes may be safely used to measure cross-bridge stiffness as long as they do not exceed $0.5\% L_0$. The method described can be used for determining the relative number of cross-bridges attached at any one moment within a fraction of a msec.

Stiffness during tension transients induced by quick changes in length

According to the model of Huxley & Simmons, force is generated when attached cross-bridges rotate from a vertical configuration into an angled con-

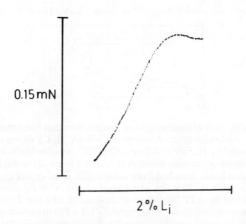

FIG. 4. Force-extension curve of single glycerinated DLM fiber (*Lethocerus maximus*) obtained during quick stretch ($1.5\% L_0$ within 0.4 msec). Conditions cf. Fig. 6.

figuration, thereby stretching an elastic component within the cross-bridge (Fig. 5). According to the model, cross-bridges are assumed to be elastic, and both attached cross-bridge populations therefore contribute to muscle stiffness, while only one kind of population (angled cross-bridges) contributes to active force. Huxley & Simmons (1971) assumed a rapid equilibrium between these two attached cross-bridge states which can be disturbed by a quick change in length. A quick release, for instance, is assumed to cause bridge rotation into a predominantly angled position; this cross-bridge rotation is assumed to be the cause of the observed rapid tension recovery (quick phase) following a quick release (cf. Fig. 6). Since it is assumed that no detachment or reattachment of cross-bridges occurs during the quick phase, the model predicts that

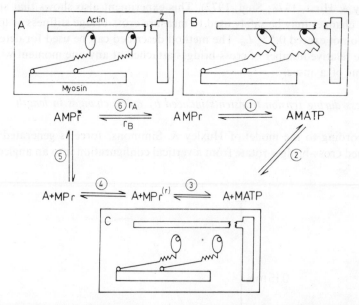

FIG. 5. Possible schematic representation of the link between biochemical and mechanical events during the cross-bridge cycle. *Mechanical cycle* (A, B, C) shows a contraction model similar to that proposed by Huxley and Simmons (1971): myosin heads attached to actin in a perpendicular, non-force-generating position in (A) and in an acute angled force-generating position in (B); myosin heads detached from actin in (C). *Biochemical cycle* shows 6 stages of the actin activated myosin ATPase. In step 1 product $Pr(-ADP.P_i)$ dissociation from actomyosin (AM) and reversible ATP binding to actomyosin (Lymn and Taylor) is involved. Step 3 represents the reversible hydrolysis of myosin-ATP (MATP) to myosin-product (Trentham). $MPr^{(r)}$ is the refractory myosin-product complex (Eisenberg and Kielly) the decomposition of which rate limits attachment of myosin to actin (C to A). In step 6 a conformational change (White & Taylor, 1976) of attached cross-bridges occurs; this step probably corresponds to cross-bridge rotation (A to B).

immediate stiffness remains constant during the release and the subsequent tension recovery. In order to verify this prediction of the model, stiffness was measured in glycerinated DLM-fibers at the beginning and at the end of the quick phase The stiffness at the beginning of the quick phase was estimated from the slope stiffness of the force-extension curve obtained during the conditioning length step (release) inducing the Huxley-Simmons force transient. After completion of a transient, *i.e.*, about 1.5 to 2.0 msec later, the released fiber was restretched (cf. Fig. 6) in order to measure stiffness at the end of the quick phase. This stiffness was again calculated from the slope of the force-extension curve (Fig. 7) during restretching. Figure 7 shows the linear force-extension curves obtained during the release and during the subsequent restretch. There is a parallel shift in the curves so that it is obvious that the slope stiffness is the same during release and during the subsequent restretch (Güth *et al.*, 1978; cf. also Ford *et al.*, 1977). Obviously the number of attached bridges is identical at the beginning and at the end of a quick phase.

The mirror image of the tension transient just described for release experi-

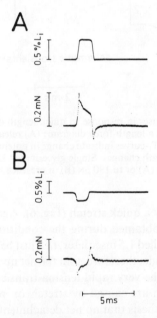

FIG. 6. Force transients induced by changes in length. Single glycerinated DLM fiber ($L_0 = 4$ mm) isometrically contracted to 190 μN (A) or to 150 μN (B) in solution containing 10 μM Ca^{++}; 7.5 mM Mg-ATP; 20 mM imidazole; 100 mM KCl; 10 mM NaN_3; 4 mM Ca EGTA; 15 mM HDTA; pH 6.7; 23° C.

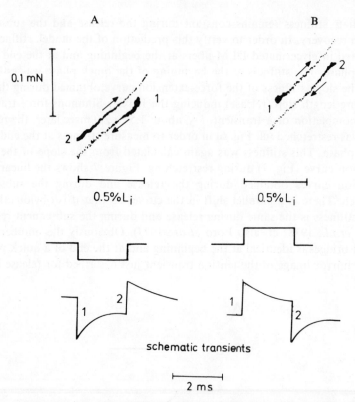

FIG. 7. Early tension responses in response to quick length changes (amplitude 0.4% L_i, duration 200 μsec) plotted in a length-force diagram: (A) release-stretch pulse; (B) stretch-release pulse. Parallel shifted T_1-curves indicate change in potential energy of the fiber during the quick phase following length changes. Single glycerinated DLM fiber (*Lethocerus maximus*). Isometric force: 210 μN (A) or to 190 μN (B) in presence of 10 μM Ca^{++} and Mg ATP. Conditions cf. Fig. 6.

ments may be induced by a quick stretch (Fig. 6). Again, the slope stiffness of a force-extension curve obtained during the conditioning stretch is identical to that of the test step applied 1.5 msec later, *i.e.*, just before the end of the quick phase (Fig. 7). We conclude, therefore, that the net number of attached bridges does not change during the very rapid tension transients described by Huxley & Simmons (1973) which are caused by stretch or release. The result is in agreement with the hypothesis that no net detachment or attachment of cross-bridges takes place during these rapid transients. These should therefore be ascribed to a configurational change of the cross-bridge as proposed by Huxley & Simmons. In other words, the transient reflects a change in the equilibrium of the force-generating and non-force-generating populations of attached cross-bridges.

The Huxley-Simmons model further predicts that the stretch or release induced shift in the equilibrium of force-generating and non-force-generating attached cross-bridges is rapidly reversed after restoring the initial length. In order to verify this prediction, Güth et al. (1979) checked whether the same mechanical state of the cross-bridges (same stiffness, same tension at same muscle length) can be restored by resetting the initial length after a quick length change. The experiments of Fig. 8 show the reversibility of tension after a quick stretch followed by a quick release to the initial length 7 msec later. The rapid reversibility to the initial state is, however, only observed as long as the stretches are smaller than 0.5 % of the muscle fiber length. At larger stretches, however, the effects are no longer reversible; after restoring the initial length by a quick release and the elastic tension-fall, tension still recovers but it does so only partly so that it remains considerably lower than before the stretch (Fig. 8). This suggests the occurrence of an irreversible process which according to Güth et al. (1979), may perhaps be identified with slippage of overstrained cross-bridges during large stretches. Indeed, a stretch exceeding 0.5 to 1 % length change induces a transient different to that observed after smaller length changes; there is an additional ultra-fast component of tension decay preceding the quick phase described above. These experiments may also be evidence for rapid detachment and reattachment of cross-bridges in the case

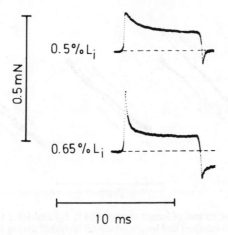

FIG. 8. Tension changes in response to stretch and release. Single glycerinated DLM fiber from *Lethocerus maximus,* isometric force 150 µN. Conditions cf. Fig. 6.

of large stretches and they are evidence against the occurrence of such processes during tension transients induced by small stretches or releases.

The next Figure (9) shows a tension transient following a quick release of an isometrically contracted insect fibrillar muscle fiber. Note that in this case there is again a rapid tension drop associated with the quick length change, which is followed by a quick tension recovery (quick phase), which is in turn followed by a delayed tension fall during several msec. The delayed tension fall however, is obviously associated with cross-bridge detachment. This is shown by a considerable decrease in stiffness; when the fibers were restretched, not immediately after the quick phase but several msec later, the slope of the force-extension curve was considerably smaller (Fig. 9B) than that observed during an early restretch, performed after the quick phase (cf. Fig. 9A). Another few

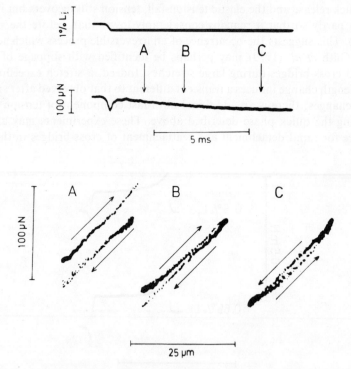

FIG. 9. Above: time course of length change (0.4% L_0) and force (L_i = 4mm).
Below: Plot of force (ordinate) and length (abscissa) recorded during the release (←) and restretch (→) performed at A, B or C; i.e., 2, 4 or 8 msec after initiation of the releases. Single DLM fiber; isometric force 150 μN.
Conditions cf. Fig. 6.

msec later the immediate stiffness increased again (Fig. 9C), suggesting cross-bridge reattachment; if the released fiber was restretched about 8 to 10 msec after the release, the slope of the force-extension curve was again similar to that during release. In summary, a quick release seems to cause first an elastic tension decrease due to a discharge of attached cross-bridges; this is followed by a quick tension recovery without a change in the net number of attached cross-bridges, and then by a delayed tension fall due to cross-bridge detachment, which is once more followed by cross-bridge reattachment.

Temperature effects

Figure 10 shows the effect of a temperature increase on the tension transient following a quick release. The rate of the tension recovery as well as the rate of

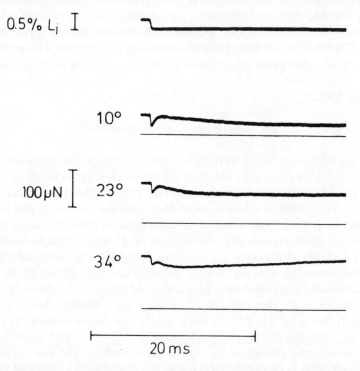

FIG. 10. Effect of temperature on early phases of tension changes in response to quick release (0.2% L_i) of contracted glycerinated single DLM fiber from *Lethocerus maximus*. Bottom line: zero tension.
Conditions cf. Fig. 6.

the delayed tension fall both increased with increased temperature. At very high rates of tension recovery, for instance at 35°C, the peak of the tension transient is truncated, so that the quick phase appears to be much less pronounced than at 10° C. According to the theory of Huxley & Simmons, the apparent rate constant of a quick phase may be related to the sum of the rate constant for the transformation (rotation) of vertically attached cross-bridges into the angled conformation and that for the reverse reaction. According to Kuhn's calculations these rate constants for rotation have temperature-coefficients of 3.5 and 1.5, respectively; at 15° C they are about equal (according to the calculations given by Kuhn et al., 1979). The ratio of the two kinds of cross-bridge populations in the attached state must change with temperature because of the different temperature coefficients; with increasing temperature the population of cross-bridges attached to actin in an angled, force-generating configuration must increase at the expense of vertically non-force-generating attached cross-bridges. Such a change must cause an increase in the ratio of tension-to-stiffness. Indeed, it could be shown experimentally that this is the case. The tension to stiffness ratio of contracting glycerinated *Lethocerus* DLM fibers is also decreased after lowering the concentration of magnesium ions.

DISCUSSION

"Quick phase" and chemomechanical energy conversion

It has been described that there is a parallel shift in the quasi-linear force-extension curve (T_1-curve) obtained during quick length changes just before and just after the quick tension recovery (quick phase) following a quick release. This result is taken to mean that immediate stiffness, and hence the number of cross-bridges attached to actin, remains constant during the force-generating quick phase—as predicted from the Huxley-Simmons model.

We conclude that during a quick phase following a release, elastic elements become extended, thereby storing potential mechanical energy. It appears that chemical energy must have been converted into mechanical energy without any change in the net number of attached cross-bridges during the quick phase. If this is so, it is difficult to escape the conclusion that energy transformation occurring during the quick phase is associated with some structural (conformational) change of the myosin cross-bridges. The kinetics of energy transformation is obviously reflected in the mechanical kinetics of the quick phase and it is of interest to relate it to the kinetics of one of the biochemial events during ATP-splitting.

It is usually assumed that one molecule of ATP is split during one cross-bridge cycle and that therefore the cycling frequency of cross-bridges corresponds to the molecular turnover number of the ATP splitting site. The turnover number of the ATPase activity of glycerinated fibers of the dorsal longitudinal muscle from *Lethocerus maximus* is of the order of 10 per sec, when the contractile system is fully activated (Breull et al., 1973). The rate of decomposition of the actin-myosin product complex probably corresponds to the ATP splitting rate (White & Taylor, 1976). This reaction is about 100 times slower than the conformational change of the cross-bridges as deduced from the analysis of the fast tension transients described in this paper. It would seem, therefore, that the decomposition of the actomyosin product complex is not directly linked with the rotational cross-bridge movement or the chemomechanical energy conversion but occurs later (cf. Julian et al., 1973):

$$AMPr^* \overset{(6)}{\rightleftharpoons} AMPr \to AM + Pr$$

$$Pr = ADP \cdot P_i$$

In *Lethocerus* flight muscle the equilibrium constant of reaction (6) is about 1 at 15° C under isometric conditions (Kuhn et al., 1979) so that there is no change in standard free energy between the two cross-bridge states. As discussed above, cross-bridges generate mechanical energy during cross-bridge rotation (reaction 6) which is stored in the elastic elements of the attached cross-bridges in state B (cf. Fig. 5). Cross-bridges in state B therefore store more mechanical potential energy than cross-bridges in state A. Since there is no change in standard free energy during the rotational process it follows that there must be a large drop in chemical free energy from state A to state B, *i.e.*, cross-bridges in state A must be relatively "energy rich." Evidence for an "energy rich" actin-myosin product complex stems from White & Taylor's work (1976) and from ATP-P_i exchange experiments suggesting a "high energy" actomyosin-ADP intermediate (Hotta & Fujita, 1971; Ulbrich & Rüegg, 1971, 1976, 1977). During contraction, cross-bridges move from state A to state B and perform a power stroke because the B state is continuously being removed from the equilibrium by the rate-limiting decomposition of the actomyosin product complex and the subsequent dissociation of actin and myosin by ATP.

The reversibility of energy transformation

We have just stated that the primary process of chemomechanical energy conversion associated with the A-B transformation of cross-bridges can be induced by a quick release causing a quick tension recovery (quick phase). This process can, however, be completely reversed by a quick stretch. Under these conditions the equilibrium of reaction (6) is disturbed, so that a certain fraction of cross-bridges in state B rotate backwards into state A, (where they have more chemical free energy at the expense of a decreased amount of mechanical potential energy). In other words, a quick stretch shifts the equilibrium so that the population of cross-bridges in a state rich in chemical energy increases at the expense of the cross-bridge population in state B which is rich in mechanical potential energy. This means, of course, that quick stretch causes a transformation of mechanical energy into chemical energy. Indirect evidence for such a reversal of a chemomechanical energy transformation was obtained before when it was found that a quick stretch facilitates the incorporation of inorganic phosphate ($^{32}P_i$) and the formation of $ATP^{32}P$ (Ulbrich & Rüegg, 1976).

The interaction of the ATP-analog AMP-PNP with actomyosin and the reversible formation of a ternary complex actin-myosin-AMP-PMP provides a model for the reversible transformation of chemical energy into mechanical energy by means of a conformational change in attached cross-bridges:

$$AM \cdot L \rightleftharpoons AM + L.$$

In this scheme the ligand L is the non-hydrolyzable ATP-analog AMP-PNP (Yount et al., 1971) which forms a ternary complex with actin and myosin that is in equilibrium with the dissociated complex. Half saturation occurs at about 0.1 mM AMP-PNP (Marston et al., 1976). At high concentration (e.g. 0.5 mM) a glycerinated fiber of insect fibrillar muscle (*Lethocerus maximus*) is partly relaxed and the cross-bridges are attached to actin at a predominantly vertical angle. After lowering the AMP-PNP concentration to below 10 μM the bridges rotate predominantly into the force-generating angled arrowhead configuration (e.g., Beinbrech et al., 1976) typical for rigor (cf. Reedy et al., 1965). This conformational change of the cross-bridge causes a tension increase of up to 80 μN per fiber (Kuhn, 1973) without any change in the net number of cross-bridges (Beinbrech et al., 1976). This force generation is associated with the generation of mechanical energy which is stored as mechanical potential energy in elastically strained cross-bridges. A work cycle can be performed: A fiber in rigor contraction is released by 0.5%, restretched in presence of 0.1 mM AMP-PNP, and allowed to generate force after reducing the AMP-PNP concentration to less than 0.01 mM (Kuhn, 1977). Under isotonic conditions and without load, the fibers shorten by about 0.3% after reducing the AMP-PNP concentration from 0.5 mM to 0. After stretching the fiber by about 0.3%

in the presence of AMP-PNP (concentration *e.g.*, 0.1 mM), cross-bridges in arrowhead configuration rotate backwards into a predominantly vertically attached position (Kuhn, 1978), a process which is associated with an increase of the amount of AMP-PNP bound to the cross-bridges (Kuhn, 1977). This is taken as evidence for the transformation of mechanical energy (supplied by stretching) into chemical free energy of binding. The quantitative aspects of this stretch-dependent AMP-PNP binding could be predicted from the theory of mechano-chemical systems (Kuhn *et al.*, 1960; Katchalsky *et al.*, 1960; Kuhn, 1973). The increased amount of ligand-binding by stretching is presumably due to an increased affinity between the AMP-PNP analog and its myosin binding site caused by stretching. By stretching fibers in a solution containing AMP-PNP of low concentration and releasing them in a solution of slightly higher concentration it was possible to transfer the AMP-PNP from one compartment to another against a concentration difference (Kuhn, 1977).

In conclusion, the evidence presented in this paper suggests that crossbridges are capable of transforming chemical energy into mechanical energy and *vice versa* by means of a conformational change while the bridges remain attached to actin. All experimental data can be easily interpreted in terms of a cross-bridge model where attachment of cross-bridges to actin and the generation of force and movement (Huxley & Simmons, 1973; cf. also Huxley, 1969) are distinct steps.

REFERENCES

Ashley, C. C. & Moisescu, D. G. (1975). The part played by Ca^{++} in the contraction of isolated bundles of myofibrils. In *Calcium Transport in Contraction and Secretion.* eds. Carafoli, *et al.*, pp. 517–525, Amsterdam, London, New York: North-Holland Publishing Company.

Beinbrech, G., Kuhn, H. J., Herzig, J. W. & Rüegg, J. C. (1976). Evidence for two attached myosin cross-bridge states of different potential energy. *Cytobiology* **12**, 385–396.

Breull, W., Steiger, G. & Rüegg, J. C. (1973). ATP splitting in relation to isometric tension-oscillation and cross-bridge cycling of insect fibrillar muscle. *J. Mechanochem. Cell Motility* **2**, 91–100.

Flitney, F. W. & Hirst, D. G. (1978). Cross-bridge detachment and sarcomere "give" during stretch of active frog's muscle. *J. Physiol.* **276**, 449–465.

Ford, L. E., Huxley, A. F. & Simmons, R. M. (1977). Tension responses to sudden length changes in stimulated frog muscle fibers near slack length. *J. Physiol.* **269**, 441–515.

Güth, K. & Kuhn, H. J. (1976). Cross-bridge elasticity in the presence and absence of ATP. *Pflügers Arch. ges. Physiol.* **362**, R25.

Güth, K. & Kuhn H. J. (1978). Stiffness and tension during and after sudden length changes of glycerinated rabbit psoas fibers. *Biophys. Stuct. Mechanism* **4**, 223–236.

Güth, K., Kuhn, H. J., Drexler, B., Berberich, W. & Rüegg, J. C. (1979). Stiffness and tension during and after sudden length changes of glycerinated single insect fibrillar muscle fibers. *Biophys. Struct. Mechanism.* in press.

Herzig, J. W. (1978). A cross-bridge model for inotropism as revealed by stiffness measurements in cardiac muscle. *Basic Res. Cardiol.* **73**, 273–286.

Hotta, K. & Fujita, Y. (1971). On the intermediate complex of actomyosin ATPase. *Physiol. Chem. & Physics* **3**, 196.

Huxley, A. F. (1957). Muscle structures and theories of contraction. *Prog. biophys. Chem.* **7**, 255.

Huxley, A. F. & Simmons, R. M. (1971). Proposed mechanism of force generation in striated muscle. *Nature, Lond.* **233**, 533–538.

Huxley, A. F. & Simmons, R. M. (1973). Mechanical transients and the origin of muscular force. *Cold Spring Harb. Symp. quant. Biol.* **37**, 669–680.

Huxley, H. E. (1969). The mechanism of muscular contraction. *Science, N. Y.* **164**, 1356–1366.

Jewell, B. R. & Rüegg, J. C. (1966). Oscillatory contraction of insect fibrillar miscle after glycerol extraction. *Proc. R. Soc. B* **164**, 428.

Julian, F. J., Sollins, K. P. & Sollins, M. R. (1973). A model for muscle contraction in which cross-bridge attachment and force generation are distinct. *Cold Spring Harb. Symp. quant. Biol.* **37**, 685–688.

Katchalsky, A., Lifson, S., Michaely, I. & Zwick, M. (1960). Elementary mechanochemical processes. In *Size and Shape Changes of Contractile Polymers.* ed. Wassermann, A., Oxford: Pergamon Press.

Kuhn, H. J. (1973). Transformation of chemical into mechanical energy by glycerol-extracted fibers of insect flight muscles in the absence of nucleoside triphosphate-hydrolysis. *Experientia* **29**, 1086–1088.

Kuhn, H. J. (1977). Reversible transformation of mechanical work into chemical free energy by stretch dependent binding of AMP-PNP in glycerinated fibrillar muscle fibers. In *Symposium on Insect Flight Muscle.* ed. Tregear, R. T., pp. 307–316, Oxford: North-Holland Publishing Company.

Kuhn, H. J. (1978). Tension transients in fibrillar muscle fibers as affected by stretch-dependent binding of AMP-PNP: A mechanochemical effect? *Biophys. Struct. Mechanism* **4**, 209–222.

Kuhn, H. J., Güth, K., Drexler, B., Berberich, W. & Rüegg, J. C. (1979). Investigation of the temperature dependence of the cross-bridge parameters for attachment, force generation and detachment as deduced from mechano-chemical studies in glycerinated single fibers from the dorsal longitudinal muscle of *Lethocerus maximus. Biophy. Struct. Mechanism.* in press.

Kuhn, W., Ramel, A., Walters, E. H., Ebner, G. & Kuhn, H. J. (1960). The production of mechanical energy from different forms of chemical energy with homogeneous and cross-striated high polymer systems. *Fortschr. Hochpolymer-Forschg.* **1**, 540–592.

Marston, S. B., Rodger, C. D. & Tregear, R. T. (1976). Changes in muscle cross-bridges when β-γ-imido-ATP binds to myosin. *J. molec. Biol.* **104**, 263–276.

Peterson, J. W. (1978). Relation of stiffness, energy metabolism, and isometric tension in a vascular smooth muscle. In *Mechanism of Vasodilatation.* eds. Vanhoutte, P. M. & Leusen, I., pp. 79–88, Basel: Karger.

Reedy, M. K., Holmes, K. C. & Tregear, R. T. (1965). Induced changes in orientation of the cross-bridges of glycerinated insect flight muscle. *Nature, Lond.* **207**, 1276–1280.

Sugi, H. (1972). Tension changes during and after stretch in frog muscle fibres. *J. Physiol.* **225**, 237–253.
Tameyasu, T. & Sugi, H. (1976). The series elastic component and the force-velocity relation in the anterior byssal retractor muscle of *Mytilus edulis* during active and catch contractions. *J. exp. Biol.* **64**, 497–510.
Ulbrich, M. & Rüegg, J. C. (1971). Stretch-induced formation of ATP-^{32}P in glycerinated fibers of insect flight muscle. *Experientia* **27**, 45.
Ulbrich, M. & Rüegg, J. C. (1976). Is the chemomechanical energy transformation reversible? *Pflügers Arch. ges. Physiol.* **363**, 219–222.
Ulbrich, M. & Rüegg, J. C. (1977). Mechanical factors affecting the ATP-phosphate exchange reaction of glycerinated insect fibrillar muscle. In *Symposium on Insect Flight Muscle.* ed. Tregear, R. T., pp. 317–333, Oxford: North-Holland Publishing Company.
White, H. E. & Taylor, E. W. (1976). Energetics and mechanism of actomyosin adenosine triphosphatase. *Biochemistry, N. Y.* **15**, 5818–5826.
Yamamoto, T. & Herzig, J. W. (1978). Series elastic properties of skinned muscle fibers in contraction and rigor. *Pflügers Arch. ges. Physiol.* **373**, 21–24.
Yount, R., Babcock, D., Ballantyne, W. & Ojala, D. (1971). Adenyl imidophosphate, an adenosine triphosphate analog containing a P-N-P linkage. *Biochemistry, N. Y.* **10**, 2484–2489.

Discussion

Pollack: I did not understand your interpretation of the results shown in Fig. 8. You said that by imposing a large enough stretch the transient becomes a bit different. You see, in the lower frame, a very sharp tension transient. The tension goes very much higher than in the upper panel and then it falls. I thought you said this high tension is due to the fact when you stretch far enough the cross-bridges are essentially ripped off.

Rüegg: Yes.

Pollack: If that is the case, why would you expect the tension to go so high? Would you not expect the tension to stop rising so sharply if bridges are being ripped off?

Rüegg: Well, the matter is complicated. If the amplitude of stretch is smaller than $0.5\% \ L_0$ (cf. Fig. 8 upper panel) the tension response in phase with the length change is almost proportional to the amplitude of stretch, and the tension decay following the stretch (quick phase) may be described by a single exponential (time constant 1 msec). If the stretch is larger than $0.5\% \ L_0$, the tension response increases nonlinearly with the length change (cf. also Fig. 4) and it is followed by an ultrafast tension decay (quick phase) completed within 0.3 msec (cf. Fig. 8 lower panel). This component superimposes on a slower component (time constant 1 msec). Because of the ultrafast quick phase the tension decay begins just after the steep portion of the s-shaped length change, hence giving rise to a sharp peak of the tension transient. Smaller stretches cause a tension increase followed by a slower tension decay which starts after the s-shaped length change (duration 0.3 msec) is completed, hence giving rise to a rounded "peak" of the tension transient.

Pollack: But were you not arguing before that the presence of that sharp tension transient indicated cross-bridge detachment?

Rüegg: Yes, but not only the sharp transient. There are two pieces of evidence suggesting rapid cross-bridge detachment during and after stretches greater than $0.5\% \ L_0$. The first evidence I mentioned already: the fact that the rate constant of the quick tension decay immediately after stretch is so much larger than that of the tension decay following smaller stretches implies that different processes may be involved in these two cases. The slower decay rate may be due to cross-bridge rotation as I discussed in my communication; the faster rate may be due to detachment. The other evidence is the shape of the force-extension diagram (cf. Fig. 4) obtained during large length changes (exceeding $1\% \ L_0$). In this case the tension increases during the first part of the stretch but does indeed stop rising during the later phase of extension, suggesting that cross-bridges may detach and reattach.

Tregear: May I take up the point? Were you saying there, Dr. Rüegg, that you thought there was additional attachment during that early blip, during that early phase, or that the number of cross-bridges remained constant during that early phase in the lower record?

Rüegg: If you plot tension versus length during stretch using larger stretches, above 0.5%, you very often get a linear curve up to about 0.5% but then, the curve gets steeper and steeper up to about 1% again. This is now very speculative, but if you want to know my views, we believe this means attachment of cross-bridges.

Tregear: I am very glad that you presented the results shown in Fig. 8 because it is critical to the interpretation, not only your interpretation, but everybody's interpretation of rapid stiffness measurements. Isn't it? I would have thought that another possible interpretation would be that the stiffness of the stretched bridge became greater as you stretched it more; not that there are more attached.

Rüegg: Yes, I quite agree. This would be possible.

Tregear: In which case one gets to the heart of the logic of saying, under what limitations of circumstance may one use stiffness as a measure of number of cross-bridges attached?

Sugi: Right. This is a very important question.

Rüegg: That was the difficulty of the experiment. That is why I said that

Güth and Kuhn were not content to measure stiffness in just the classical way, by measuring the peak tension during the stretch, and plotting it against length change, because they really wanted to know what happens.

Kushmerick: I would like to explicitly hear a discussion of how one can distinguish from this macroscopic stiffness measurement, which must mean n cross-bridges giving some stiffness per cross-bridge . . . how can you distinguish the changes in n from a change in stiffness per cross-bridge? It seems to me that there is no way in principle that one can distinguish these.

Rüegg: That is a hard question. Perhaps the experiments with overlap. The answer comes from experiments involving the extension of sarcomeres and finding a fairly linear decrease of tension with increasing sarcomere length and, just as the tension goes down, the stiffness goes down in parallel, *i.e.*, the scaling down of the T_1 curves, although we just heard that one has to be critical about that too. Well, that is the best I can answer. From this, I would say that in certain cases, stiffness may be a measure of the number of bridges attached, but I quite agree in other cases there could be a change in stiffness of a cross-bridge as well.

Sagawa: Dr. Rüegg said stiffness may reflect the number of cross-bridges, not the elastic property of the cross-bridges *per se*, so if there are a large number of cross-bridges bound, then the stiffness may also involve a number of end compliances, *i.e.*, the compliance of tendons and filaments and all these things in series with the bound cross-bridge section. I thought that could be one of the sources of confusion.

My question is related to the definition of stiffness. Would you mind redefining what you mean by stiffness?

Rüegg: Tension change per length change.

Sagawa: In a static sense or a dynamic sense?

Rüegg: In a dynamic sense, it was measured dynamically. A length change performed in 0.5 msec produces an increase in tension. If you take this tension increase versus length change you have the stiffness, but you can also take the differential of tension and length and take the ratio of those.

Sagawa: Decomposing the vector into real and imaginary axes?

Rüegg: Yes.

Sagawa: I would like to comment on that. If you really go to a very fast stretch or release, whichever is the case, at some point the mass of the system, whatever the system is, will have to come to play, as well as viscous properties.

Rüegg: The mass must be considered, of course.

Sagawa: If you go to extremes, the problem becomes very, very complicated. As a general comment, I think we should be careful. As Dr. Gott discussed, the use of models could tell us in which region of speed and magnitude of stretch we should start thinking about this; in other words, it could let us know the safe range for various purposes.

Rüegg: Yes. I am not a physicist but Dr. Güth is, and he did many calculations and control experiments with rubber strips. He was very worried about the mass coming in, and in a paper just published he goes into that (Güth & Kuhn, 1978). You can get around the problem by using, what many people avoided so far, parabolic length changes achieved by the movement of a relay continuously accelerating. He made controls using rubber strips and showed that you get reliable stiffness measurements after 100 μsec, but not before. He also calculated the transmission time, in his case for about a 3 mm length of fiber. It was about 80 μsec. It was much more than we used to think. So, for a 1 cm length of fiber, it was on the order of maybe 200 μsec, or so. Of course you have to be careful to avoid oscillations in the force transducer. It has 14 KHz resonance frequency. This is really a difficult technical problem.

White: I was just going to reply to Dr. Sagawa's question, to say that I think the transmission times that Dr. Podolsky's lab measured (Shoenberg, Wells & Podolsky, 1974) are relevant to this. These are comparable in speed or slightly faster than the ones Dr. Rüegg referred to, and I believe there was also discussion about this in the recent paper of Ford, Huxley & Simmons (1977), which talks about the effects fiber mass would have upon the properties. I think it shows that for stretches that are not faster than about 100 μsec, one can ignore fiber mass and also actually fiber viscosity, which in principle could come in.

ter Keurs: As a basis for your assumption that stiffness represents a measurement of the number of cross-bridges attached, you invoked the high correlation between stiffness and force as a function of sarcomere length. As this is a crucial observation, I would very much like to know how you derived the force sarcomere length relation in that experiment.

Rüegg: Since Dr. Yamamoto and Dr. Herzig did the experiment, could I pass on the question to Dr. Yamamoto?

Yamamoto: The difficulty of this kind of experiment is how to keep the sarcomere spacing constant. So, at the beginning of the experiment, we adjusted the sarcomere length to about 2.2 μm, and then we stretched the fiber in a calcium activated solution over a wide range of length change. We then returned the fiber to a relaxing solution. Then we measured sarcomere length again at, say, 2.5 μm or so. Then at that sarcomere length, we gave a stretch or a release to the fiber. The stretches covered a wide range, between −10 nm per half sarcomere length to about +5 nm per half sarcomere. Then, I compared the stiffness from the T_1 curve, but usually it is very difficult to determine the stiffness over the wide range of sarcomere lengths. So, for one preparation, we took two or three points of sarcomere spacing, measured the stiffness at each sarcomere spacing, and then compared the maximal stiffness from 2.2 μm to the other sarcomere spacings we tested. Usually, we took the stiffness from many preparations and plotted the stiffness on one figure.

Sugi: I think there are two basic assumptions implicit in the handling of the stiffness data. First, for small amplitude length oscillations, the resulting stiffness data is normally believed to be mostly due to cross-bridges. Next, if the amplitude is small, the cross-bridge is assumed to remain attached. Both of these two assumptions might not be correct.

Noble: I just wanted to reinforce your comment, Professor Sugi. The statement is that under certain circumstances, the ratio of the force change to the length change is proportional to the active tension. It seems to me that all you can say from that is that the force generator has this property, and your model for that force generator has to incorporate that property. Now I would have thought that many, many force generating models could be devised which had that property, and therefore, I think you are quite right to say that it is rather unjustified to extrapolate this always to cross-bridge mechanics.

Iwazumi: I have been working on skinned fibers for some time, and my initial experience was that when you put the skinned fiber in a contracting solution, regardless of whether it was mechanically or chemically skinned, you lose striations. Of course, in that state of contraction you can never tell what the sarcomere length is, and, in my opinion, the measurement of sarcomere length before or after contraction has nothing to do with the sarcomere length during contraction.

My experience is that during a contraction of a fiber which has no observable striations, the sarcomere length was all over the place, which means that you might have 4 μm sarcomere length and probably less than 1 μm sarcomere length. After you put the fiber in relaxing solutions, the sarcomeres more or less come back to their original length. The only way I could make it works was, as I showed in the Kyoto Biophysics meeting (Iwazumi & Pollack, 1978), with slow activation, in the sense that I could keep very good striations during contraction.

Rüegg: While, in many cases, the laser pattern was lost during contraction, there were quite a few cases where happily the pattern was maintained, even during contraction. And, more interestingly, the sarcomeres did not shorten during this isometric contraction. The mechanical behavior of these very good fibers was just the same as that of those that were not quite as good.

Kawai: If you release the length of the muscle, the stiffness also drops, and I think the cross-bridge is going slack in the rigor condition. I am wondering whether this is taking place in the activated muscle?

Rüegg: Are you saying that the stiffness per cross-bridge may really change as a function of stretch or release?

Kawai: That's right.

Rüegg: If the cross-bridges would have non-linear elasticity or go slack on release, the tension-length diagram of these elastic elements (T_1 curve) should be non-linear (as in the case of Fig. 4 or Fig. 9C). However, in Fig. 9C, this curve is linear during release and curved during restretch, while in Fig. 9A it is linear during stretch and release. Therefore, the cross-bridges cannot simply be regarded as non-linear springs which get stiffer as you stretch them. If this were so, why then should muscle fibers show non-linear behavior in one case (Fig. 9C) but not in the other (Fig. 9A)?

REFERENCES

Ford, L. E., Huxley, A. F. & Simmons, R. M. (1977) Tension responses to sudden length changes in stimulated frog muscle fibres near slack length. *J. Physiol.* **269**, 441–515.

Güth, K., Kuhn, H. J. (1978) Stiffness and tension during and after sudden length changes of glycerinated rabbit psoas muscle fibres. *Biophys. Struct. Mechanism* **4**, 223–236.

Iwazumi, T. & Pollack, G. H. (1978) Stretch responses of maximally activated skinned rabbit soleus fibers. *6th Int. Biophys. Congr. Kyoto*, 382.

Shoenberg, M., Wells, J. B. & Podolsky, R. J. (1974) Muscle compliance and the longitudinal transmission of mechanical impulses. *J. gen. Physiol.* **64**, 623–642.

Effect of MgATP on Cross-Bridge Kinetics in Chemically Skinned Rabbit Psoas Fibers as Measured by Sinusoidal Analysis Technique

Masataka KAWAI

Department of Neurology, College of Physicians and Surgeons of Columbia University, New York, N. Y., U.S.A.

ABSTRACT

Three exponential processes (rate constants: $2\pi a$, $2\pi b$, $2\pi c$ in the order of slow to fast) were determined as a function of MgATP (substrate for actomyosin ATPase) concentration in maximally activated, chemically skinned rabbit psoas preparations (1–5 fiber bundles) by the method called sinusoidal analysis. Experiments were carried out at 20°C in the presence of excess free ATP (5mM), creatine phosphate (20mM) and creatine phosphokinase (74 unit/ml), and at a constant ionic strength (225mM). The slow exponential lead (a; this corresponds to "phase 4" of Huxley, 1974) is insensitive to the substrate concentration in the range 1–20mM, but both b (middle frequency exponential lag; "phase 3") and c (fast lead; "phase 2") increase with increase in substrate concentration in the range 1–5mM, and saturate at 5–20mM. The present results show that MgATP binding and the subsequent dissociation of myosin from actin are rate-limiting for the fast processes ($2\pi b$, $2\pi c$) but not for the slow one ($2\pi a$). This substrate effect is the reverse of that expected from the models proposed by Huxley & Simmons (1971) or by Julian, Sollins & Sollins (1974). These models predict that the dissociation stage limits the slow processes corresponding to phases 3 and 4, and that the transition between multiple attachment sites limits the fast process corresponding to phase 2.

INTRODUCTION

The tension response to sinusoidal length changes can be resolved into 3 exponential processes in the frequency range 0.25–167Hz (corresponding to 1 msec time resolution in step analysis) in fully activated intact crayfish and frog (Kawai, Brandt & Orentlicher, 1977), and in skinned rabbit psoas (Kawai,

1978) muscle preparations. These processes are referred to as "(A)," a low frequency exponential lead, "(B)," a middle frequency lag (called "oscillatory work" by Pringle, 1967) and "(C)," a high frequency lead. They are correlated (see Discussion) with "phases (stages) 4, 3, 2", respectively, of tension transients that are observed after a stepwise alteration of muscle length (Huxley & Simmons, 1971; Heinl, Kuhn & Rüegg, 1974; Huxley, 1974; Abbott & Steiger, 1977; Ford, Huxley & Simmons, 1977). They can also be correlated with the length transients in response to step load changes as observed by Civan & Podolsky (1966) (see Huxley, 1974, for this correlation).

Processes (A), (B), (C) must reflect actively cycling cross-bridges since these processes are absent in the rigor or relaxed states where little cycling of the myosin cross-bridges takes place (Kawai et al., 1977). Skinned fibers allow one to monitor alterations in cross-bridge behavior with variation in the chemical environment and to interpret the physiological data in terms of known biochemistry. In this report, the dependence of the 3 processes on MgATP (substrate for actomyosin ATPase; Tonomura & Yoshimura, 1960) concentration is described in maximally activated, chemically skinned rabbit psoas muscle preparations. Rabbit psoas is used because much of the published biochemical work was carried out with extracts of this preparation.

In this study a method called sinusoidal analysis is employed to measure rate constants. With this technique, rate constants differing by many orders of magnitude can be resolved with high precision. Sinusoidal analysis is extensively used to study the kinetics of glycerinated insect muscles (Pringle, 1967; White & Thorson, 1973). It is a mechanical equivalent of spectrophotometry; energy transfer to and from the muscle is measured as a function of the frequency of the length oscillation. The tension response of muscle is not perfectly linear, but, as was shown by us earlier (Kawai, 1978), the non-linearity in the tension waveform is not as large as might be expected from results of step analysis methods, therefore the sinusoidal technique is suitable for the current experiments. At the present time our main emphasis is on the chemical dependence of the rate constants, rather than their length dependence, hence, the amplitude of the length oscillation is fixed so that results from different solution conditions can be directly compared.

In this paper we quantitatively investigate the relationships between the dissociation reaction and the experimentally measured (apparent) rate constants, and evaluate various models of muscle contraction. A brief account of the present results has been published (Kawai, 1978).

MATERIALS AND METHODS

Chemicals and solutions

H_4EGTA [ethyleneglycol-bis-(β-amino-ethyl ether) N,N'-tetra-acetic acid], Na_2H_2ATP, $CaCl_2 \cdot 2H_2O$, Na_2CP (creatine phosphate), CPK (CP kinase) were purchased from Sigma Chemical Company, and propionic acid, NaOH, $Mg(OH)_2$, Na_2SO_4, NaH_2PO_4, Na_2HPO_4, and imidazole from Fisher Scientific Company. Solution compositions are summarized in Table 1. All solutions were buffered to pH 7.00 ± 0.02 with 5mM imidazole and 7.14 mM phosphate. Ionic strength was adjusted to 225mM by addition or deletion of Na_2SO_4 when Na_2MgATP concentration was changed; hence, the concentration of monovalent cations (Na^+) remained unchanged. Activating solutions were made initially without Ca at pH 7.00. One ml of this solution lacking Ca was equilibrated with the muscle preparation before 50 μl of concentrated (50mM) $CaCl_2$ was injected to generate tension. The pH of the injecting solution was adjusted so that pH 7.00 ± 0.02 was attained after mixing.

TABLE 1. Solution compositions (total added concentrations in mM)

Application	$CaCl_2$	Na_2MgATP	Na_4ATP	P_i	Na_2SO_4	EGTA	CP	Imd
R, Relaxing	—	1.00	1.00	7.14	56.0	5	—	5
W, Washing	—	1.00	1.00	7.14	61.0	—	—	5
Activating								
1S	2.38	1.02	7.27	7.14	30.1	—	20	5
2.5S	2.38	2.55	7.24	7.14	28.7	—	20	5
5S	2.38	5.10	7.19	7.14	26.1	—	20	5
10S	2.38	10.2	7.09	7.14	21.0	—	20	5
15S	2.38	15.3	6.99	7.14	15.9	—	20	5
20S	2.38	20.4	6.89	7.14	10.8	—	20	5

Abbreviations: P_i = phosphate, Imd = imidazole. S indicates MgATP (chelated form) concentration in mM. All of the activating solutions have CPK (74 unit/ml), and their ionic concentrations are: Na^+ 140mM, Mg^{++} 0.02–0.4mM, pCa 4.04, CaATP 2.29mM, free ATP 5.0mM, ionic strength 225mM, pH 7.00 ± 0.02 (apparent association constants used for the above calculations are $10mM^{-1}$ for MgATP, and $5mM^{-1}$ for CaATP).

Preparation of rabbit psoas fibers and mounting

Strips of psoas muscle (2–4 mm in diameter and 30 mm in length) were excised from the parent muscle after tying them to small wooden sticks. They then were immersed immediately in the chemical skinning saline (in mM: 2 MgATP, 5 EGTA, 180 K propionate, 5 imidazole, pH 7.00). As with human biopsies (Wood, Zollman, Reuben & Brandt, 1975), the sarcolemma becomes permeable to ions of experimental interest in a few hours (Reuben, Wood & Eastwood, 1977), but the preparation was typically maintained in the same saline for at least 24 hours at 0–4°C before being used. In this way the selective permeability of the sarcolemma is destroyed, and Ca applied to the fiber does not reseal the surface membrane, in contrast to chemically skinned frog cardiac muscles (Winegrad, 1971). The preparations can be kept at 0–4°C for up to 2 weeks in the skinning saline. Alternatively, they can be maintained at −20°C for months in an isoionic glycerol relaxing saline (2 MgATP, 5 EGTA, 180 K propionate, 6 M glycerol, 5 imidazole, pH 7.00; Reuben et al., 1977). The chemical skinning and storage techniques were developed by Dr. D.S. Wood in this laboratory; the skinning solution is basically the relaxing saline.

One to 5 fiber segments (5 mm long) were dissected from the stock bundle at the time of an experiment and transferred to an experimental chamber filled with 1 ml of constantly stirred relaxing saline. Temperature was controlled to $20 \pm 0.5°C$. The muscle preparation was mounted between two plastic clamps, one of which was connected to a length driver (described in Kawai & Brandt, 1976) for length control, the other to a strain gauge (Bionix F-100) for tension detection (Fig. 1). The strain gauge was mounted on a Narishige micromanipulator to allow adjustment of muscle length (L_0) to approximately 10% above the slack length. Sarcomere lengths averaged 2.5–2.7μm, as measured by an optical diffraction technique (Cleworth & Edman, 1972; Kawai & Kuntz, 1973). The muscle registered little tension and stiffness at this length. After the length adjustment, the case of the strain gauge was secured to the experimental chamber to minimize extraneous vibrations. Diameters of the fibers were measured with an ocular micrometer attached to a dissecting microscope. This procedure slightly overestimates the cross-sectional area.

Calculation of complex stiffness (modulus)

"Complex stiffness" (defined in Eq. (1)) reflects physical properties of the muscle such as elasticity, viscosity, mass[1] and oscillatory work (negative viscosity). Since this quantity consists of both amplitude and phase at each frequency (f), the data are most conveniently represented as complex numbers.

As measurements of complex stiffness are complicated and time consuming, a mini computer (Nova 1220, Data General Corporation) is employed to control the experiments (Fig. 1). The experimenter triggers the computer, which in turn applies digitally-generated sinewaves to the length driver (40–400

---

[1] In the frequency range of the present experiments, the contribution of mass is insignificant; cf. Truong (1974); Schoenberg, Wells & Podolsky(1974).

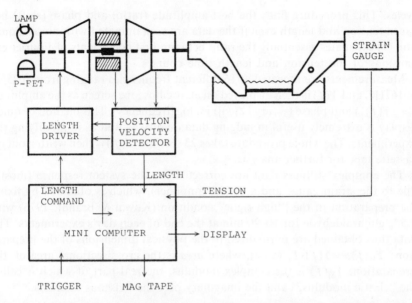

FIG. 1. A sketch of the experimental apparatus. The muscle preparation is held horizontally by two plastic clamps, one connected to a strain gauge for tension measurement, the other to a length driver. The performance of the length driver is calibrated against a photo-electric position detector at the left. The length driver is directly controlled by the mini computer, which also records length and tension time courses. Many interfaces are abbreviated for simplicity.

points/cycle) via a D/A (digital to analog) converter. The amplitude of the peak-to-peak length oscillation (l_p) is confined to 0.3–0.4% L_0, which corresponds to ± 20 to ± 26Å per half sarcomere. After the steady state has been achieved (0.25 sec), the time (t) courses of both length $L(t,f)$ and tension $P(t,f)$ are collected for the smallest number (n) integer cycles exceeding 0.4 sec. Data are collected via a multiplexed A/D converter every 50 μsec. During the collection period, data are reduced so that 8–40 points are stored for each cycle. The length driver is stopped for the next 0.25 sec, and in this period complex stiffness is calculated as follows:

$$Y(f) = \frac{\dfrac{f}{n} \cdot \displaystyle\int_{n\text{-cycles}} P(t,f) \exp(-2\pi f t i)\, dt}{\dfrac{f}{n} \cdot \displaystyle\int_{n\text{-cycles}} L(t,f) \exp(-2\pi f t i)\, dt} \tag{1}$$

where $i = \sqrt{-1}$, and the integrations (in practice, summations) are over n

cycles. This procedure finds the best amplitude (ratio) and phase (shift) between tension and length even if the data are contaminated with noise. Equation (1) calculates essentially the ratio between first coefficients of Fourier expansions of the tension and length time courses.

Measurements are repeated at 16 different frequencies ranging from 0.25Hz to 167Hz, and $Y(f)$ is then displayed on an oscilloscope screen as the amplitude $(= |Y(f)|)$ and phase $(= \arg[Y(f)])$ vs. log[frequency]. This method of quick display is extremely useful in judging data quality as well as in modifying the experiments. The whole procedure takes 23 sec, and $Y(f)$ is then written out on cassette tape for further analysis.

The complex stiffness data are corrected for the system response (mostly due to the strain gauge and a carrier amplifier), which is obtained by fixing the preparation in the "high rigor" condition (Kawai & Brandt, 1976) with 2.5% glutaraldehyde for 10–20 min at the end of each day's experiments. The data thus obtained are normalized to the physical dimensions of the preparation: $Y_M(f) = Y(f) \cdot L_0/(\text{area})$, where *area* is the cross-sectional area of the preparation. $Y_M(f)$ is the complex modulus, the real part of which is called the "elastic modulus," and the imaginary part the "viscous modulus."

RESULTS

In the experiments described below, the muscle preparation was soaked in a Ca-free solution, and it was then activated by an injection of $CaCl_2$ (Methods). As soon as the muscle developed a steady plateau tension, the computer was triggered and two sets of complex stiffness data were collected. Since the two complex stiffness functions do not differ with time in these rabbit preparations,[2] they were averaged before further treatment (except for the data in Fig. 3). The muscle preparation was subsequently relaxed with R (Table 2) and it was kept relaxed for at least 5 min before the next activation at a different MgATP concentration. The slow tension time course is reproduced in Fig. 2. This record shows that the isometric tension (P_0) decreases monotonically as the substrate concentration is raised from 1 to 20 mM (see also Fig. 4E below). This was not due to partial activation because an increase in free Ca concentration did not increase P_0. This decrease in P_0 is similar to observations made earlier in skinned crayfish muscle preparations by Orentlicher, Brandt & Reuben (1977).

Examples of complex modulus (stiffness) plots are displayed in Figs. 3A-C

---

[2] In intact crayfish and frog, we found that the magnitude of the oscillatory work changed with time while the muscle preparations were maximally activated (Kawai *et al.*, 1977).

FIG. 2. Slow pen trace of the tension time course after filtering through a second-order Butterworth (cut off at 1 Hz) filter. Each activation consists of the following sequence: the preparation is initially soaked in R, washed 2–3 times in W, exposed to an experimental saline at A, activated by Ca addition at C, and relaxed at R. Control records at 5S are repeated before and after the series. A slow drift contaminates the trace at the beginning. Calibrations are 100 dyn (tension) and 1 min (time).

at two different substrate concentrations (1S & 5S where S indicates MgATP concentration in mM). Figure 3A is a "Nyquist Plot" which facilitates identification of exponential rate processes. These processes are represented by hemicircles which center on the abscissa (Machin, 1964).

Figure 3B is a phase-frequency plot. This method of data plotting is convenient for examining changes in rate constants, since the plot has a minimum (~10Hz) near the characteristic frequency b, and a maximum (~50Hz) near c (see below). When the phase is positive (tension leads the length), the muscle is absorbing work from the oscillating driver. When the phase is negative (tension lags behind the length), the muscle is producing work on the driver. From Fig. 3B it is clear that the muscle was producing oscillatory work at 6–13Hz (1S) and at 6–30Hz (5S).

The length of the sarcomeres can be changed either by stretching elastic components such as the S-2 moieties of heavy meromyosin (Huxley, 1974), or by overall cross-bridge cycling. At high frequencies the stiffness registers all cross-bridges made at the moment of the length change; at the lower frequencies little stiffness is registered because cross-bridges have sufficient time to go through the chemical transitions required for a length adjustment. This is clearly demonstrated in Fig. 3C which is an amplitude-frequency plot. At intermediate frequencies, where the frequency of the imposed length oscillation approximates the intrinsic cross-bridge cycling rate, the stiffness response shows a complicated pattern (Fig. 3C).

Empirical description of the complex modulus

Three hemicircles are identifiable in Fig. 3A. This means that the length/

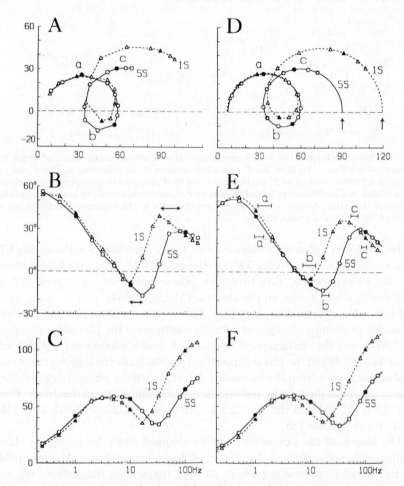

FIG. 3. Plots of complex modulus (stiffness) at two substrate concentrations (1S, 5S). *A & D*, Nyquist Plot. Viscous modulus (ordinate) *vs.* elastic modulus (abscissa), units in Mdyn/cm^2. *B & E*, Plot of phase (= arg $[Y_M(f)]$; ordinate) *vs.* frequency (abscissa in Hz). *C & F*, plot of dynamic modulus (= $|Y_M(f)|$; ordinate in Mdyn/cm^2) *vs.* frequency. *A-C*, original data. Symbols represent actual measurements and lines are drawn to connect them. Frequencies of measurements are 0.25, 0.5, *1*, 2, 3.13, 5, 7.14, *10*, 16.7, 25, 33, 50, 80, *100*, 133, 167 Hz (filled symbols are in *italics*). *D-F*, calculated data based on Eq. (2) with parameters shown in Table 2. In *E*, horizontal bars labeled *a, b, c*, represent locations of characteristic frequencies along the abscissa and their 95% confidence limits. The labels *a, b, c* over the bars belong to 1S and those under the bars to 5S. $1_p = 0.4\% L_0$.

tension relationship of activated muscle reflects at least 3 exponential processes (see Machin, 1964, for the Nyquist plots of single exponentials). A general transfer function containing 3 exponential terms can be written as:

$$Y_M(f) = H + \overset{(A)}{\frac{Afi}{a+fi}} - \overset{(B)}{\frac{Bfi}{b+fi}} + \overset{(C)}{\frac{Cfi}{c+fi}} \tag{2}$$

where (A), (B), (C) are three exponential processes with respective magnitudes A, B, C and characteristic frequencies a, b, c (in the order $a < b < c$). The apparent rate constants are therefore given by $2\pi a$, $2\pi b$ and $2\pi c$. These rate constants must be comparable to those obtained by step analysis methods (see Eq. (4) below). H is a constant that is usually small compared to other magnitude parameters. Process (B) has negative polarity in Eq. (2) because the corresponding arc goes through the 4th quadrant in the Nyquist plot (Fig. 3A) (the phase is negative; Fig. 3B).

The complex modulus data were fitted to Eq. (2). Fitting was performed by minimizing the sum of squared deviations of vectoral differences between experimental and predicted complex moduli. This is an expanded version of the least squares method applied to a non-linear equation. Figures 3D-F displays calculated data corresponding to Figs. 3A-C respectively. Comparison of these 2 sets of figures demonstrates adequacy of fit. a is usually found to be about 1 Hz, and here the viscous modulus becomes a maximum. b usually falls in the range of 8–20 Hz. The phase reaches a minimum in this vicinity. c is found in the range of 60–100 Hz, and in this range the phase becomes a maximum. Positions of these values are entered in Fig. 3 where appropriate.

Confidence limits of the fitted parameters

It is always important to assess the latitude of choice of the fitting parameters. This can be achieved by calculating the 95% confidence limit for each fitting parameter by use of Fisher's F-statistics. Table II lists the results of such calculations at 3 substrate concentrations. In Figs. 3E and 4A the confidence ranges for characteristic frequencies a, b, c are shown in bars. The 95% confidence limits are seen to be quite narrow. This is because 32 data points (2 for each frequency) were obtained over a range of frequencies approaching 3 decades.

Effect of substrate concentration on rate processes

Figures 3A, D show that process (A) was virtually unchanged when the sub-

TABLE 2. Fitted parameters, confidence limits, and P_0

Parameters	1S	5S	15S
H	7 ± 4 Mdyn/cm^2	7 ± 2 Mdyn/cm^2	7 ± 2 Mdyn/cm^2
A	65 ± 5 Mdyn/cm^2	61 ± 3 Mdyn/cm^2	48 ± 3 Mdyn/cm^2
B	53 ± 6 Mdyn/cm^2	59 ± 4 Mdyn/cm^2	42 ± 4 Mdyn/cm^2
C	100 ± 9 Mdyn/cm^2	84 ± 6 Mdyn/cm^2	54 ± 6 Mdyn/cm^2
H + A − B + C*	119 Mdyn/cm^2	93 Mdyn/cm^2	67 Mdyn/cm^2
$2\pi a$	8.4 ± 2.5 sec^{-1}	7.8 ± 1.4 sec^{-1}	6.9 ± 1.5 sec^{-1}
$2\pi b$	58 ± 16 sec^{-1}	116 ± 18 sec^{-1}	128 ± 25 sec^{-1}
$2\pi c$	396 ± 59 sec^{-1}	594 ± 62 sec^{-1}	670 ± 105 sec^{-1}
P_0	1.32 Mdyn/cm^2	1.15 Mdyn/cm^2	0.91 Mydn/cm^2

* Extrapolated (infinite frequency) elastic modulus. Confidence limits (approximately 95%), shown after ±, were obtained by Fisher's F-statistics. Degrees of freedom are 23 in all cases.

strate concentration was increased from 1 to 5mM, whereas Figs. 3B, E shows that characteristic frequencies b and c both increased with substrate. It may be seen from Fig. 3A that more oscillatory work was produced at the characteristic frequency b at 5mM than 1mM (see also Fig. 4E).[3] Figure 3D shows that the extrapolated (infinite frequency) stiffness (arrows) decreased at 5mM, presumably because of a smaller P_0 (Table 2).

Figure 4 summarizes the substrate effect. Figure 4A shows that the rate constant $2\pi a$ was unchanged, whereas both b and c increased over the substrate range 1–5mM and saturated at 5–20mM. All the magnitudes varied with substrate level, as shown in Fig. 4D. Magnitude B increased initially (this is a statistically significant increase), peaking at 2.5mM, and then decreased. Parameters A and C both decreased monotonically, as did the tension (Figs. 4D, E).

Figure 4B is a linear plot of the rate constants $2\pi b$ and $2\pi c$ vs. substrate concentration. Figure 4C is a double reciprocal (Lineweaver-Burk) plot of the same data. The continuous lines follow the Michaelis-Menten equation:

$$b = \frac{b_{\text{MAX}} S}{K_M + S}; \quad c = \frac{c_{\text{MAX}} S}{K_M + S} \tag{3}$$

As seen in Figs. 4B, C both rate constants follow Eq. (3); K_M values were 1.4 mM for b, and 0.8mM for c. Further studies are needed to determine whether this difference is statistically significant.

---

[3] Oscillatory power output is proportional to the negative viscous modulus times frequency:
Power $= -\frac{\pi f}{4} 1_p^2 \, \text{Im}[Y(f)] = \frac{(\text{area}) \pi f^2}{4 L_0 (b^2 + f^2)} 1_p^2 \, Bb = \frac{(\text{area}) \pi}{8 L_0} 1_p^2 \, Bb$ (at $f = b$).
The last term is plotted in Fig. 4E as a function of MgATP concentration.

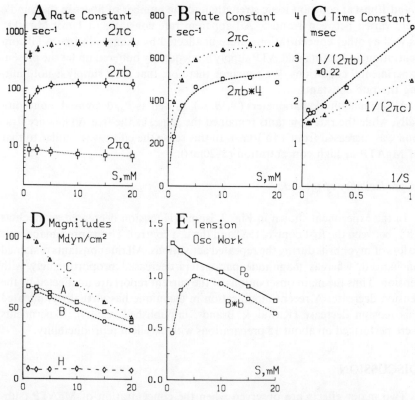

FIG. 4. Summary of the substrate effect. Symbols represent actual measurements, and, except for B and C, lines are drawn to connect them. A, apparent rate constants ($\square$, $\bigcirc$, $\triangle$) and 95% confidence ranges (vertical bars) in log scale; B, in linear scale, and C, double reciprocal plot; continuous lines are predicted data from Eq. (3). D, magnitude parameters. E, isometric tension (P_0) in Mdyn/cm², and oscillatory work production ($\propto B \cdot b$; normalized to the maximum) at characteristic frequency b. Abscissae in A, B, D, E are substrate concentrations in mM; in C, mM^{-1}. In B, C rate constant $2\pi b$ and its reciprocal are multiplied by certain factors (shown after *) for convenience in plotting. Figures 2–4 are based on data of 12/25/75, preparation 1 (a bundle of 4 fibers; 1 day old).

Tests for adequacy of ATP buffering

All experiments were performed with stirred baths and in the presence of 2.3mM CaATP, 5mM free ATP, and with an ATP regenerating system consisting of 20 CP and 74 unit/ml CPK. To test whether this is sufficient to buffer against ATP consumption and ADP build-up, the experiments were repeated as the free ATP (not bound to Ca or Mg) concentration was changed between

1 and 10mM at constant ionic strength, and the number of fibers in the bundle were varied between 1 and 5. All apparent rate constants at 1 mM MgATP, as well as other concentrations, were unaffected by changes in free ATP concentration, indicating that ATP supply is adequately buffered under the present experimental conditions. We conclude therefore that ATP supply is not limiting the rate constants.

All the magnitude parameters (A, B, C), as well as P_0, decreased monotonically, while the rate constants remained the same, as the free ATP concentration was increased from 1 to 10mM. In this sense the effect was similar to that of MgATP at high concentration (5–20mM).

Reproducibility of the data

In the experiment shown in Fig. 2, isometric tension decreased as much as 15% between the first control (5S) and the last control. This is presumably due to loss of myofibrils during the repeated activations. All rate constants remained unchanged, whereas magnitude parameters decreased proportionately with tension. Thus the main observations of the current report are unaffected by the tension decrease. A recent modification in technique has virtually eliminated this tension decrease (Kawai & Brandt, unpublished results). Experiments were performed on about 15 preparations with excellent reproducibility.

DISCUSSION

Two major effects are observed when the concentration of MgATP (substrate) is increased: (i) increase in apparent rate constants $2\pi b$ and $2\pi c$ at low concentrations (1–5mM), with no further change at 5–20mM (Figs. 4A, B); (ii) monotonic decrease in isometric tension (P_0) and magnitude parameters (A, B, C) in the range 5–20mM (Figs. 4D, E).

It is possible to interpret the first effect of MgATP in the context of biochemical schemes such as those of Lymn & Taylor (1971) or White & Taylor (1976). In their scheme, substrate (S) molecules bind to the myosin heads, which are in rigor-like linkages with actin (AM), and the substrate binding (AMS) promotes a rapid dissociation of myosin (MS) from actin (A):

$$AM + S \xrightarrow{k_1} AMS \xrightarrow{k_2} A + MS; \quad J = k_1 S \cdot (AM) = k_2(AMS)$$

where k_1 and k_2 are chemical rate constants, and J is the reaction rate in the steady state [(AMS) does not increase or decrease]. Both reactions can be reversible, but we consider a condition when S is being actively hydrolyzed, in the direction of the arrows.

In a mechanical scheme, the dissociation step is often written as (Huxley, 1957; Podolsky, Nolan & Zaveler, 1973; White & Thorson, 1973; Julian et al., 1974; Hill, 1974):

$$X \xrightarrow{g} \text{Detached State}; \quad J = gX$$

where X is an attached state (it can be an assembly of many chemical states), and g is the detachment rate constant. The mechanical scheme can have other attached states than X. If we assume states (AM) and (AMS) are mechanically indistinguishable, we can lump them together into the state X; then, from the above equations,

$$X = (\text{AM}) + (\text{AMS}) = \frac{J}{k_1 S} + \frac{J}{k_2} = \left(\frac{1}{k_1 S} + \frac{1}{k_2}\right) \cdot gX$$

$$\therefore g = \frac{k_2 S}{k_2/k_1 + S}.$$

Note that g is the effective rate constant for the composite attached state, (AM) & (AMS). Thus, the equation reduces to the Michaelis-Menten Eq. (3), with $K_M = k_2/k_1$. As shown in Figs. 4B, C, the observed rate constants $2\pi b$ and $2\pi c$ fit this equation very well, which supports the interpretation that MgATP binding is rate-limiting at low substrate concentrations ($S < K_M$), whereas the subsequent dissociation step is limiting at high concentrations ($S > K_M$). At K_M, observed to be in the vicinity of 10^{-3}M, both reactions equally limit the sequence ($k_1 S = k_2$). This is in good agreement with the data of Lymn and Taylor (1971), who found $k_1 = 10^6$ M^{-1} sec^{-1}, and $k_2 \geq 1000$ sec^{-1}. White & Taylor (1976) stated that k_2 is at least 500 sec^{-1} in comparable experimental conditions. Since the binding reaction is bimolecular, its rate must become comparable to the dissociation reaction somewhere in the mM range.

This hypothesis is at variance with the model proposed by Huxley & Simmons (1971). Based on the asymmetry of the tension response to step length changes, they postulated that the early tension transient (stage or phase 2), corresponding to our process (C), reflects transition(s) between multiple attachment sites. In their model, chemical energy is transduced into mechanical energy during phase 2. Julian et al. (1974) expanded this model by incorporating a full cycle of 3 states, two attached and one detached. A detailed examination of their model shows that the detachment step (and therefore MgATP concentration) affects the slower rate processes, presumably processes (A) and (B) in the current report. It also predicts that the transition between multiple attachment sites limits the fast process (C). This is contrary to our observation that substrate affects the faster processes (B) and (C), but not process (A).[4]

Possible objections to our conclusions are: (1) inadequate ATP buffering, (ii) effect may be specific to sinusoidal analysis, (iii) imperfect data fitting, (iv) inadequate system frequency response, and (v) the effect is specific to chemically skinned rabbit psoas. The evidence against (i) has already been presented in the Results section.

Concerning objection (ii), it can be shown rigorously that the tension time course $[\Delta P(t)]$ after step length change $[\Delta L]$ will be, from Eq. (2)

$$\frac{\Delta P(t)}{(\text{area})} = [\overset{(\text{phase 2})}{C \exp(-2\pi ct)} - \overset{(\text{phase 3})}{B \exp(-2\pi bt)} + \overset{(\text{phase 4})}{A \exp(-2\pi at)} + H] \cdot \frac{\Delta L}{L_0} \quad (4)$$

in the linear range. From Eqs. (2) and (4) the "phases" of step analysis and the "exponential processes" of sinusoidal analysis are readily correlated with one another. For this purpose the oscillatory work (B) serves as an unambiguous marker because of its unique polarity (negative sign in Eqs. (2) and (4)) and because of its universal presence in a variety of activated striated muscles (crayfish, frog, rabbit, insect, etc.). The negative polarity of (B) causes a delayed tension rise (decay) in step length increase (decrease), and thus it corresponds to phase 3 of Huxley (1974), Abbott & Steiger (1977), and Ford et al. (1977).[5] In sinusoidal experiments work is produced in this phase on the oscillating driver at the expense of ATP hydrolysis (Steiger & Rüegg, 1969).

Process (C) (phase lead) is the next faster process than (B). Hence, it corresponds to stage (phase) 2 of Huxley & Simmons (1971) or Huxley (1974). It is still to be determined whether this is better treated as the sum of two exponential leads (Abbott & Steiger, 1977), four closely spaced ones (Ford et al., 1977), or in some other way.[6] The slow exponential lead (A) resides between DC

---

[4] Slight decline in $2\pi a$ (if any; Fig. 4A) is opposite to that which the model predicts.
[5] Phase 3 exhibits a plateau when magnitude B is small, as shown by some of the step analysis data.
[6] If phase 2 splits into multiple exponential processes, (C) should correspond to the slower components. The same argument holds for "stage 2" of Huxley & Simmons (1971). Thus (C) unambiguously corresponds to the component on which Huxley & Simmons based their model.

(0 Hz) and the oscillatory work (B). Thus its identification is also unambiguous and it corresponds to phase 4.

The corresponding phases are entered above Eq. (4), and a computer simulation of the tension time course is depicted in Fig. 5 for the parameters given in Table 2. The above comparisons are useful when correlating data from a variety of muscle tissues under various experimental conditions. Even though the mapping of sinusoidal data to step data is not quantitatively exact, because of the nonlinearities associated with the exponential processes, their qualitative features, such as the polarity of the processes, and their relative order, cannot change.

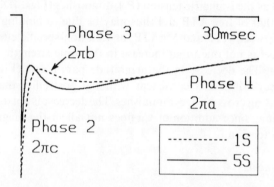

FIG. 5. Simulated data of tension time course after quick length release. Data are based on Eq. (4) with parameters shown in Table 2 at two substrate concentrations (1S, 5S). These parameters are the same as those used for the computations of Figs. 3D-F. Phases 2, 3, 4 of Huxley (1974) and rate constants $2\pi a$, $2\pi b$, $2\pi c$ of current report are labelled in the figure. Equivalent data of Figs. 3D-F.

The fitting equation is indeed an approximation. However, the substrate effect is evident in the data itself without reference to the fitting equation, as shown in Fig. 3B (arrows) by the shift of the minimum ($\sim$ 10Hz) and the maximum ($\sim$ 50Hz).

It is possible that other processes may exist in higher frequency ranges, as shown by the experiments of Abbott and Steiger (1977) or Ford et al. (1977), and that these additional processes might distort our measurements. The highest frequency used (167Hz) allows us to determine a time constant as low as 1 msec. According to Ford et al. (1977), phase 2 lasts at least 5 msec after length changes, therefore our system frequency response is adequate to resolve this phase. Errors associated with a finite bandwidth of measurement, as well as inadequacies in the fitting equation, are included in the confidence ranges.

Changes in rate constants in $2\pi b$ and $2\pi c$ produced by substrate concentration changes from 1S to 5S (or to 15S) far exceeds the confidence ranges of the rate constants (Table 2).

Concerning objection (v), we have obtained the complex stiffness data from a variety of striated muscle tissues: intact and skinned crayfish fibers, intact frog muscles, skinned frog fibers, as well as rabbit psoas. Their Nyquist plots are strikingly similar, suggesting commonalities in the underlying kinetic mechanisms.

The effect of MgATP at higher concentrations (5–20mM) is different from that at lower concentrations. Between 5 and 20mM, no change in the kinetic parameters (rate constants) is observed, but each magnitude parameter (A, B, C), as well as the isometric tension (P_0), diminishes (Figs. 4D, E). This effect is similar to that of free ATP and thus may be due to binding that does not distinguish between ATP and MgATP, or to a non-specific effect of polyanions. The effect is not due to an increase in the ionic strength, since this was kept constant. The decrease in the magnitude parameters (Fig. 4D) and infinite frequency stiffness is consistent with the possibility that the number of actively cycling cross-bridges diminishes. The decrease is not due to decreased actin activation, since addition of Ca does not affect the magnitude parameters nor the tension.

Acknowledgements

The author wishes to thank Drs. R. Cox, P.W. Brandt, M. Orentlicher and D.S. Wood for their stimulating discussions and critical reading of the manuscript, and to Professors H. Grundfest and J. P. Reuben for their constant encouragement. The present work is supported by grants from the National Science Foundation (PCM 78-08592), National Institute of Health (AM 21530), and grants to the H. Houston Merritt Clinical Research Center (NS 11766 and MDAA).

REFERENCES

Abbott, R. H. & Steiger, G. J. (1977). Temperature and amplitude dependence of tension transients in glycerinated skeletal and insect fibrillar muscle. *J. Physiol.* **266**, 13–42.

Civan, M. M. & Podolsky, R. J. (1966). Contraction kinetics of striated muscle fibers following quick changes in load. *J. Physiol.* **184**, 511–534.

Cleworth, E. R. & Edman, K. A. P. (1972). Changes in sarcomere length during isometric tension developed in frog skeletal muscle. *J. Physiol.* **227**, 1–17.

Ford, L. E., Huxley, A. F. & Simmons, R. M. (1977). Tension response to sudden length

change in stimulated frog muscle fibers near slack length. *J. Physiol.* **269**, 441–515.
Heinl, P., Kuhn, H. J. & Rüegg, J. C. (1974). Tension responses to quick length changes of glycerinated skeletal muscle fibers from the frog and tortoise. *J. Physiol.* **237**, 243–258.
Hill, T. L. (1974). Theoretical formalism for the sliding filament model of contraction of striated muscle, Part 1. *Prog. Biophys. molec. Biol.* **28**, 267–340.
Huxley, A. F. (1957). Muscle structure and theories of contraction. *Prog. Biophys. biophys. Chem.* **7**, 255–318.
Huxley, A. F. (1974). Muscular contraction. *J. Physiol.* **243**, 1–43.
Huxley, A. F. & Simmons, R. M. (1971). Proposed mechanism of force generation in striated muscle. *Nature, Lond.* **233**, 533–538.
Julian, F. J., Sollins, K. R. & Sollins, M. R. (1974). A model for the transient and steady-state mechanical behavior of contracting muscle. *Biophys. J.* **14**, 546–562.
Kawai, M. (1978). Head rotation or dissociation? A study of exponential rate processes in chemically skinned rabbit muscle fibers when MgATP concentration is changed. *Biophys. J.* **22**, 97–103.
Kawai, M. & Brandt, P. W. (1976). Two rigor states in skinned crayfish single muscle fibers. *J. gen. Physiol.* **68**, 267–280.
Kawai, M. & Kuntz, I. D. (1973). Optical diffraction studies of muscle fibers. *Biophys. J.* **13**, 857–876.
Kawai, M., Brandt, P. W. & Orentlicher, M. (1977). Dependence of energy transduction in intact skeletal muscles on the time in tension. *Biophys. J.* **18**, 161–172.
Lymn, R. W. & Taylor, E. W. (1971). The mechanism of adenosine triphosphate hydrolysis by actomyosin. *Biochemistry, N. Y.* **10**, 4617–4624.
Machin, K. E. (1964). Feedback theory and its application to biological systems. *Symp. Soc. exp. Biol.* **18**, 421–445.
Orentlicher, M., Brandt, P. W. & Reuben, J. P. (1977). Regulation of tension in skinned muscle fibers: effect of high concentrations of Mg-ATP. *Am. J. Physiol.* **233**(5), C127–C134.
Podolsky, R. J., Nolan, A. C. & Zaveler, S. A. (1969). Cross-bridge properties derived from muscle isotonic velocity transients. *Proc. Natl. Acad. Sci. USA* **64**:504–511.
Pringle, J. W. S. 1967. The contractile mechanism of insect fibrillar muscle. *Prog. Biophys. molec. Biol.* **17**, 1–60.
Reuben, J. P., Wood, D. S. & Eastwood, A. B. (1977). Adaptation of single fiber techniques for the study of human muscle. In *Pathogenesis of the Human Muscular Dystrophies*, ed. Rawland, L.P., pp. 259–269, 5th Int. Sci. Conf. MDA. Amsterdam: Excerpta Medica.
Schoenberg, M., Wells, J.B. & Podolsky, R.J. (1974). Muscle compliance and the longitudinal transmission of mechanical impulses. *J. gen. Physiol.*, **64**, 623–642.
Steiger, G. J. & Rüegg, J. C. (1969). Energetics and "efficiecny" in the isolated contractile machinery of an insect fibrillar muscle at various frequencies of oscillation. *Pflügers Arch. ges. Physiol.*, **307**, 1–21.
Tonomura, Y. & Yoshimura, J. (1960). Inhibition of myosin B-ATPase by excess substrate. *Archs. Biochem. Biophys.* **90**, 73–81.
Truong, X.T. (1974). Viscoelastic wave propagation and rheologic properties of skeletal muscle. *Am. J. Physiol.*, **226**, 256–264.
White, H. D. & Taylor, E. W. (1976). Energetics and mechanism of actomyosin ATPase. *Biochemistry, N. Y.* **15**, 5818–5826.
White, D. C. S. & Thorson, J. (1973). The kinetics of muscle contraction. *Prog. Biophys. molec. Biol.* **27**, 173–255.

Winegrad, S. (1971). Studies of cardiac muscle with high permeability to calcium produced by treatment with EDTA. *J. gen. Physiol.* **58**, 71–93.

Wood, D. S., Zollman, J. R., Reuben, J. P. & Brandt, P. W. (1975). Human skeletal muscle: properties of the "chemically skinned" fiber. *Science, N. Y.* **187**, 1075–1076.

Discussion

Tregear: About the experiment shown in Fig. 5, I just wanted to know what concentration of MgATP you had in that experiment.

Kawai: This particular experiment was carried out at 10 mM MgATP, but I did the same experiment at 1 mM MgATP, which is more important, because there the rate constants are limited. The rate constant is obviously slower at 1 mM MgATP, but it does not change at all with the change in free ATP concentration between 1 and 10 mM.

Tregear: So I suppose from that you can deduce that you are not seeing a competition between free ATP and MgATP for the site. Would you deduce that?

Kawai: I am not sure that I can jump to that conclusion, but I will think about it. The purpose of this experiment was to determine whether the free ATP supply was adequate.

Kushmerick: A biochemical question. If I remember rightly, sodium ions, at least in high concentrations, are inhibitory to actomyosin ATPase, and you necessarily are changing sodium in a decreasing direction as you increase the ATP.

Kawai: No. I am not changing the sodium concentrations at all. Because Na_2MgATP is replaced by Na_2SO_4. Both exist as divalent cations at pH 7, and therefore, you do not change sodium concentration or ionic strength.

Huxley: I am not very familiar with the use of these Nyquist plots, but what I am puzzled at is that the maximum frequency that you are making the measurements over is up to 100 Hz.

Kawai: 167 Hz.

Huxley: So the length change is taking place in about 6 msec.

Kawai: That is not correct. The comparable time scale has to be divided by 2π. For instance, if I detect 100 Hz, this corresponds to a rate constant of 2π times this, which is about 600 sec^{-1} so actually the correspondence of 167 Hz to the time scale is 1 msec, which is the limit of our measurement.

Huxley: But if the limit of your time resolution is then something on the order of 1 msec, is not that rather close to the Huxley-Simmons time constant to be comfortable?

Kawai: In their recent paper (Ford, Huxley & Simmons, 1977), their phase 2 lasts up to 5 msec, and then, between 5 and 10 msec, nothing is specified. I presume this period is a mixture of phases 2 and 3. If the time constants are comparable, I am sure 1 msec is good enough to evaluate at least the slower process in phase 2 (Phase 2 could be divided into many processes.) That is what I am talking about, and that is actually what Huxley and Simmons originally talked about.

Steiger: Does ADP have a similar effect to ATP?

Kawai: I have not done that experiment.

Steiger: How about phosphate?

Kawai: Phosphate changes the rate constant greatly at around 3 or 4 mM, and then saturates at higher concentrations.

Rüegg: An increase in rate constant?

Kawai: An increase in the rate constant of $2\pi b$ but it seems to me $2\pi c$ is not changing, and this is consistent with our conclusion that $2\pi c$ is reflecting dissociation and has nothing to do with the phosphate effect.

Steiger: Did you have any indication, when you lowered the temperature, that phase 2 was splitting into two components as we found (Abbott & Steiger, 1977)?

Kawai: If I lower the temperature, the Nyquist plot is a bit different from what I get at higher temperature. First of all, what we have to realize in the rabbit preparation is that if you lower the temperature you lower the tension, and then altogether lower the stiffness. If you lower the temperature by a factor of 2 in centigrade, say from 20 to 10°C, the tension is about half. In that case, if we start this kind of Nyquist plot; obviously the Nyquist plot shrinks as you lower the temperature. And not only that, you have some strange process coming in, such that the Nyquist plot is rather flat around this area, which I had a hard time to fit to any kind of transfer function. I am working on it presently, but it seems that this extra process cannot be fitted by an exponential rate constant, which may show up at lower temperature, but the effect is relatively less prominent as you increase the temperature, because this exponential process (C) is now higher at higher temperature. So, I think some different process is coming in, which is not necessarily an exponential process, but such as the one you see in the rigor preparation. Perhaps that is viscous interaction between the filaments.

Steiger: But if the rate constants would be too close, you could not resolve them.

Kawai: In some cases probably they are too close, and the processes are explained by one single rate constant, the precision of which is actually described by the confidence ranges, which indicate how good the fit is.

Rüegg: In connection with Dr. Kawai's studies, it may be interesting to refer to biochemical measurements of myosin detachment from actin. Eccleston, Geeves, Trentham, Bagshaw and Mrwa (1975) found a marked effect of Mg ATP concentration (in the millimolar range) on the rate of actomyosin dissociation, but at higher ATP concentrations a change in temperature (in the range 5 to 20° C) had little effect (Mrwa, personal communication).

Kawai: Oh, you did the same experiments, and the results generally agree?

Rüegg: An increase in Mg ATP concentration increases the rate of dissociation of the actomyosin complex, measured by the stop-flow technique.

Sagawa: Does the negative force delay phase become more prominent or less prominent with increasing temperature?

Kawai: The relative importance of that does not change with temperature.

Tregear: I just wanted to ask whether you could comment at all on the value of your apparent affinity of the actomyosin in the fibers for substrate. It is about 1 mM, isn't it?

Kawai: Yes.

Tregear: That is quite a large concentration compared to other people's ways of measuring the same thing. Well, we have measured it, for instance, using glycerinated fibers, and found values in the region of 20 μM (Marston, 1973).

Kawai: The difference seems to be pretty big, I agree.

Woledge: In your experiment you changed the substrate MgATP concentration in a system where phosphocreatine was constant and creatine kinase was present. I think the action of this ATP regenerating system would be to hold constant the ratio of ATP to ADP. This means that when you change the substrate concentration, you are also changing the product concentration. I think that rather complicates the interpretation of your results. I would like to suggest that it would be better to do these experiments in such a way that the product concentration did not change.

Kawai: What kind of concentration of ADP do you expect in the usual conditions?

Woledge: Well, of course, the concentration of ADP will be very low because of the rather high equilibrium constant. What really matters is that the change in substrate concentration is just the same as the change in the product concentration.

Kawai: Well, the next thing I am planning to do is the effect of ADP concentration, but I do not suppose the micromolar or lower concentration of ADP has any effect.

REFERENCES

Abbott, R. H. & Steiger, G. J. (1977) Temperature and amplitude dependence of tension transients in glycerinated skeletal and insect fibrillar muscle. *J. Physiol.* **266**, 13–42.

Eccleston, J. F., Geeves, M. A., Trentham, D. R., Bagshaw, C. R. & Mrwa, U. (1975). The binding and cleavage of ATP in the myosin and actomyosin ATPase mechanism. In **26**. *Colloquium Ges. biol. Chem. Mosbach/Baden, Molecular Basis of Motility*, eds. Heilmeyer, L., Rüegg, J. C. & Wieland, T. Berlin: Springer.

Ford, L. E., Huxley, A. F. & Simmons, R. M. (1977) Tension responses to sudden length changes in stimulated frog muscle fibres near slack length. *J. Physiol.* **260**, 441–515.

Marson, S. (1973) The nucleotide complexes of myosin in glycerol-extracted muscle fibers. *Biochim. biophy. Acta* **305**, 397–412

Dynamic Stiffness of Heart Muscle in Twitch and Contracture*

Kiichi SAGAWA, Yasutake SAEKI, Louis LOEFFLER III and Kiyoshi NAKAYAMA

Department of Biomedical Engineering, School of Medicine, The Johns Hopkins University, Baltimore, Maryland, U.S.A.

ABSTRACT

We analyzed the mechanical properties of cat papillary muscles at rest, in Ba^{2+} contracture and in isometric twitch by a frequency response method. The muscle length was perturbed sinusoidally with an amplitude less than 0.3% of L_{max} over a frequency range from 0.1 to 0.6 Hz to determine the dynamic stiffness, $|F(\omega)|/|L(\omega)|$, in which $|F(\omega)|$ = amplitude of the force response wave, $|L(\omega)|$ = amplitude of sinusoidal length wave, and ω = frequency, and the phase shift of $F(\omega)$ relative to $L(\omega)$.

All the data supported the hypothesis that the passive (resting) and the active properties are independent of each other, and led to a parallel combination model of passive and active branches.

In Ba^{2+} contracture, the well-known delayed force response was noted between 0.2 to 1 Hz at 22°C. At 36°C it occurred between 2 and 8 Hz with an attenuated intensity. The overall frequency response curve was approximated using a first-order visco-elastic model, which consists of two elastic elements and one viscous element coupled in two alternative configurations.

In twitching muscle, we used an "envelope technique" to gain the response in a low frequency range. The resultant model parameters at the time for the peak isometric force agreed well with those in muscles in contracture with 1 mM Ba^{2+}, although we should be very careful in the interpretation of the results owing to the limitations of the technique.

INTRODUCTION

Stiffness is a loosely defined term which expresses resistance to deformation. Buchthal and Rosenfalck (1960) used "dynamic stiffness" to mean the ratio of

---

* This research was supported in part by US PHS Grants HL 14903 and HL 18912.

the amplitude of the sinusoidal force wave, $F(\omega)$, to the amplitude of the input sinusoidal length wave, $L(\omega)$. Since muscle is a visco-elastic system, dynamic stiffness of muscle thus defined is a frequency-dependent complex quantity: $|X| \equiv |F(\omega)|/|L(\omega)|$. In this paper, too, we use stiffness in this sense. Based on the vibration theory, Buchthal and Rosenfalck (1960) decomposed the sinusoidal force vector into the elastic component in phase with the length sinusoid and the viscous component which is $+90°$ out of phase. They called these components elastic and viscous stiffness, respectively. In rheology these are termed dynamic modulus and dynamic loss (Alfey and Gurnee, 1950). Templeton et al. (1973, 1974, 1976) followed Buchthal's decomposition of stiffness in their analyses of time-dependent change in stiffness of the ventricle and papillary muscle. Although both groups of investigators were fully aware of the frequency dependence of the stiffness, they analyzed the stiffness observed at 100 Hz (Buchthal & Rosenfalck, 1960) or at 30 and 10 Hz (Templeton et al., 1973), instead of presenting a frequency spectrum of stiffness over a reasonably wide range. We decomposed the dynamic stiffness of heart muscle differently by studying the frequency spectrum extending from 0.1 to 30 Hz.

From the viewpoint of systems analysis, the frequency spectrum of the response of a system to sinusoidal perturbations in an appropriately chosen frequency range should yield a complete representation of the dynamic characteristics of the system (Milsum, 1966). How extensive the frequency range should be depends on the system and the purpose of investigation. From the features of the amplitude ratio (stiffness) curve and the phase shift between the input length waves and the output force waves as a function of frequency, one can estimate the order of complexity of a model which adequately represents the observed dynamic properties of muscle. Although a model reduced by the frequency analysis is phenomenological, it still aids in gaining some insight into the basic visco-elasticity of muscle by placing a restraint on the choice of models.

For the frequency response method to be valid the black box system must be linear and time-invariant. In the studies reported here the linearity requirement was met by using a small amplitude of sinusoidal length change around several mean lengths as the input (graduated linearization). The steady-state requirement can be met in skeletal muscle by studying tetanic contraction. Although heart muscle can be tetanized in a solution with high concentrations of Ca^{2+} and caffeine (Forman, Ford & Sonnenblick, 1972), the resultant tetanus is not adequately stable for a long enough period (10 to 30 minutes) as that needed for the frequency response analysis. Therefore, we used Ba^{2+} contracture in Study I (Saeki, Sagawa & Suga, 1978) after the method reported by Steiger, Brady & Tan (1978). We assumed that the state of the contractile machine in muscle in Ba^{2+} contracture is analogous to the end systolic state of heart muscle. In Study II (Loeffler and Sagawa, 1975) we attempted to determine the small signal frequency response of isometrically twitching papillary muscle by using an envelope technique. Although this technique is a compromise with a fundamental problem as pointed out by Hunter & Noordergraaf (1976), the frequency response estimated by the technique appears consistent with that in Ba^{2+} contracture. The results from both studies seem to supplement each other and point to a model which consists of a parallel combination of the passive and active branches, the latter comprising two elastic elements and one viscous element.

METHODS

Principle of the method

Bode plot for system identification: In systems engineering, the frequency response is shown in a diagram which plots the amplitude ratio on a logarithmically scaled ordinate and the phase shift on a linearly scaled ordinate against the frequency which is plotted on a logarithmically scaled abscissa. The diagram is called the Bode plot. Illustration of the use of the Bode plot by examples of simple physical systems will prepare the reader for choosing the simplest model from a given data set on the input-output relation of a system (Fig. 1). It is extremely important to remember that the system property represented by the Bode plot critically depends on the choice of input and output variables. In all of the examples in Fig. 1, the input is length change $L(\omega)$ and the output force change $F(\omega)$.

In the case of a perfectly elastic spring coil with a stiffness K (Fig. 1A), the dynamic stiffness is frequency independent and the phase shift is zero at any frequency. It is expressed in the Bode plot by the horizontal stiffness line in the middle panel and the phase line in the bottom panel. In the case of a purely viscous element with a coefficient of viscosity C (Fig. 1B), its dynamic stiffness is frequency dependent, approaching zero at infinitesimally small frequency and infinity at an infinitely large frequency. Thanks to the bi-logarithmic scales used in the Bode plot, the frequency spectrum of the stiffness of a viscous system can be represented by a straight line with a slope 10 fold per decade increase in frequency. Its phase curve is a horizontal line representing a 90° lead at any frequency. If a spring coil K and a dash pot C are coupled in parallel (Fig. 1C), the stiffness is described:

$$\frac{F(\omega)}{L(\omega)} = K\left(1 + j\frac{C}{K}\omega\right). \tag{1}$$

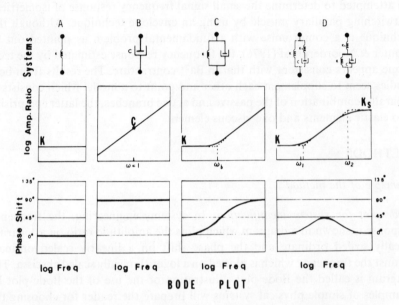

FIG. 1. Bode plots of various physical systems with viscosity and elasticity.

When ω is very small, the system behaves like a pure spring coil as shown in the left side half of the Bode plot. In the frequency region where ω approaches K/C, the property of the viscous element becomes manifest and the stiffness curve rises with ω. The phase angle increases to $+45°$ at $\omega = K/C$ and asymptotically approaches $+90°$ as ω increases toward infinity. The combination of elastic and viscous element properties is well indicated in the Bode plot. The frequency ω_b, at which the inflection of the stiffness curve occurs, is called a break frequency. It allows estimation of C because $\omega_b = K/C$ and K can be evaluated as the stiffness in the ultralow frequency region.

If a second spring coil, K_s, is added to the system in Fig. 1C, in either one of the two configurations shown at the top of Fig. 1D, a similar form of plot results. The stiffness curve for both systems consists of two plateaus, one in the lower frequency region, another higher plateau in the higher freqfiency region, and an s-shaped transitional part over the middle frequency region. The stiffness for the system in the right side of Fig. 1D is:

$$\frac{|F(\omega)|}{|L(\omega)|} = \left\{ \left(\frac{K_s K}{K + K_s}\right)^2 \left[\frac{1 + \left(\frac{C}{K}\right)^2 \omega^2}{1 + \left(\frac{C}{K + K_s}\right)^2 \omega^2}\right] \right\}^{1/2}. \tag{2}$$

The lower and upper break frequencies ω_1 and ω_2 of the stiffness curve are related to the elastic and viscous element parameters as follows:

$$\omega_1 = K/C \quad \text{and} \quad \omega_2 = (K + K_s)/C.$$

The phase angle shows a lead between ω_1 and ω_2. The stiffness for the system in the left side of Fig. 1D is:

$$\frac{|F(\omega)|}{|L(\omega)|} = \left\{ K^2 \left[\frac{1 + \left(\frac{K_s + K}{KK_s}\right)^2 C^2 \omega^2}{1 + \left(\frac{C}{K_s}\right)^2 \omega^2}\right] \right\}^{1/2} \tag{4}$$

and the break frequencies of the stiffness curve are given by:

$$\omega_1 = K \cdot K_s / C (K + K_s) \quad \text{and} \quad \omega_2 = K_s/C. \tag{5}$$

Reversing the deductive process to the inductive process, one can infer that if a visco-elastic system gives a Bode plot like that of Fig. 1D, it should have, at least, two elastic components and one viscous component which are coupled in either of the configurations. They are the simplest, if not the only, models compatible with the real system behavior in the chosen range of frequency. From the break frequencies and plateau values of stiffness, the system parameters K_1, K_s and C can be estimated. This is exactly the approach used in our studies. We chose a frequency range from 0.1 to 30 Hz because the fundamental frequency of the isometric contraction curve of heart muscle at 20°C was about 1/2 Hz and there are insignificant harmonics above 10 Hz. The preliminary findings also assured us that this range was sufficient, if not perfect, for our immediate purpose. To gain additional information to assess the cross-bridge dynamics, however, transient force responses to small step changes in length were studied in muscles resting in the normal Ringer solution and in contracture in the Ba^{2+}-Tyrode solution.

Experimental methods

Since details of the methods are described elsewhere (Loeffler & Sagawa,

1975; Saeki, Sagawa & Suga, 1978) only a brief account of the preparation and apparatus will be given here.

a) *Preparations*: For both Study I and Study II, excised cat papillary muscles with an average cross-sectional area of about 0.5 mm² and length of about 6.8 mm were used. The length-diameter ratio at L_{max} was always greater than 4.3.

After setting the muscle in a length-force control system, it was made to contract isometrically with a resting force of 0.5 to 1.0 g and a frequency of 12 min^{-1} for several hours in Tyrode's solution at 25°C. To cause Ba^{2+} contracture in Study I, the bath solution was switched to Ca^{2+}-free Tyrode's solution after obtaining the control data on the isometric force-length relation and L_{max}. When twitch force became unobservable in 5 to 20 minutes, the solution was switched to a solution which contained Ba^{2+} at 1 mM. A very stable contracture occurred and continued for hours.

b) *Apparatus:* The apparatus used in Study I for muscle length control and force measurement was essentially the same as that reported by Pinto *et al.* (1975). The overall compliance of the length control system was 1 $\mu m \cdot g^{-1}$ and the frequency response was flat to 100 Hz for a peak-to-peak amplitude of 100 μm. The apparatus used for Study II (which actually was performed earlier Study I) had an overall compliance of 6.5 $\mu m \cdot g^{-1}$, which was corrected at the time of data analysis.

RESULTS

Dynamic stiffness of muscle in Ba^{2+} contracture (Study I)

a) *Frequency response analysis:* The analysis was first applied to muscle resting in the normal Tyrode solution. The amplitude of the sinusoidal length perturbation, $|L(\omega)|$, was 18μm peak-to-peak, or 0.2% of L_{max} of this muscle, as shown in the bottom panel of Fig. 2. The frequency was increased from 0.1 to 32 Hz. The middle panel shows the total muscle force responding to the length perturbations, whereas the top panel indicates the sinusoidal force response after a 10-fold amplification. The amplitude of the force response, $|F(\omega)|$, and therefore the dynamic stiffness of the relaxed muscle, increased only very slightly with increase in frequency. By contrast, when the same muscle was in a state of contracture (Fig. 3), $|F(\omega)|$ first slightly decreased with increase in frequency to 1 Hz, but then markedly increased with a further increase in frequency.

Figure 4 shows a representative Bode plot of the dynamic stiffness (top) and the phase shift (bottom) obtained from one muscle. The solid circles represent the data from the resting muscle. The stiffness increased only by 2.7

DYNAMIC STIFFNESS OF HEART MUSCLE 177

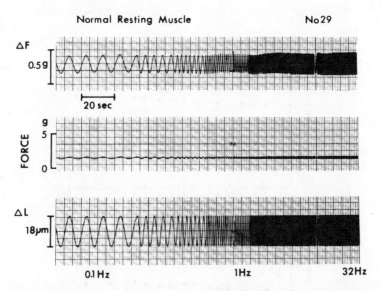

FIG. 2. Frequency response in resting muscle. The bottom channel shows given sinusoidal length changes from 0.1 to 32 Hz. The middle channel shows the total muscle force and the top channel indicates the sinusoidal force response after a 10-fold amplification.

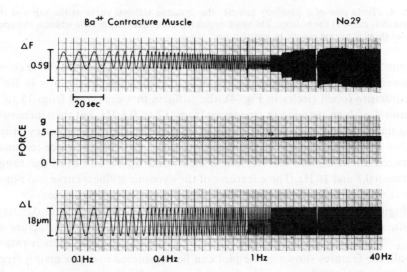

FIG. 3. Frequency response in muscle in contracture. For explanation, see the legend for Fig. 2.

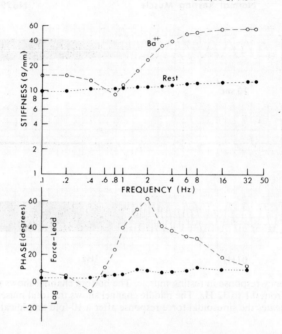

FIG. 4. Bode plot of a papillary muscle: the dynamic stiffness curve at the top and the phase shift curve at the bottom. The solid circles represent a resting muscle whereas the open circles show the same muscle in contracture.

g·mm^{-1} with the increase in frequency from 0.1 to 32 Hz and the phase curve showed a barely measureable degree of lead. When the muscle was in Ba^{2+} contracture (open circles in Fig. 4), the stiffness first decreased from 15 to 9 g·mm^{-1} with an increase in frequency from 0.2 to 0.8 Hz, but then increased and finally plateaued at a level of about 50 g·mm^{-1} over the frequency range above 8 Hz. The phase relation showed a mild lag ($-10°$) of force response between about 0.3 and 0.5 Hz, but a clear lead up to 60° over the range between 0.8 and 16 Hz. These features of the dynamic stiffness curve and phase curve were common to all of the Bode plots obtained from 8 muscles.

Figure 5 plots the mean values and the standard errors of total dynamic stiffness determined at 13 frequencies in 8 muscles in Ba^{2+} contracture at L_{max}. As the frequency-dependent change of resting muscle stiffness is rather small, the features shown in the plot can be considered to derive mainly from the active muscle elements. The features are that (1) the stiffness is relatively

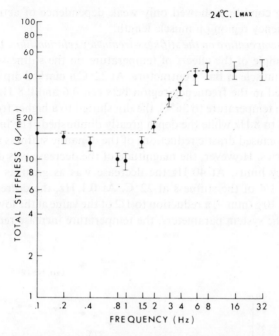

FIG. 5. Stiffness-frequency relationship averaged from 8 muscles in contracture. The vertical bars represent the standard errors of the mean. The dotted line indicates the theoretical stiffness curve expected for such first-order systems as shown in Fig. 1D.

independent of frequency in the high (> 8 Hz) and low (< 0.2 Hz) frequency regions, and (2) the stiffness decreases and then increases with increase in frequency in the intermediate range of frequency. The dotted line indicates a theoretical stiffness curve expected for the first-order system depicted in Fig. 1D with break frequencies at about 1.5 and 7 Hz. The fit of the mean data to the dotted line is excellent, except in the frequency region between 0.4 and 1.5 Hz in which a mild decrease in stiffness and a small phase lag occurred. The values of K, K_s and C for both configurations in Fig. 1D were calculated and are listed in Table 1.

b) *Stiffness-frequency relation at different muscle lengths:* Although the absolute value of the total dynamic stiffness depended on muscle length, the shape of the stiffness curves was independent of the muscle length showing only parallel vertical shifts at lengths between 0.96 L_{max} and L_{max}. This was also true of the stiffness-frequency relation in resting muscle. The difference between the lower plateau of the stiffness curve in contracture and the stiffness in the resting condition showed only weak dependence of active stiffness in the low frequency region on muscle length.

c) *Effects of temperature on the stiffness-frequency relationship:* Figure 6 shows a typical example of the effects of temperature on the stiffness-frequency relationship of muscle in Ba^{2+} contracture. At 22°C a marked dip in the stiffness curve occurred in the frequency region between 0.6 and 0.8 Hz. With an increase in bath temperature to 36°C, the dip shifted to a higher frequency range from about 4 to 8 Hz while the depth greatly diminished. The increase in temperature also caused drastic reductions of the dynamic stiffness at the low and high frequencies. However, the magnitude of the decrease was different at the two frequency limits. At 40 Hz the decrease was as great as 60 g·mm^{-1}, a reduction to 1/4 of the stiffness at 22°C. At 0.1 Hz, the decrease amounted only to about 10 g·mm^{-1}, a reduction to 1/2 of the value at the low temperature. In terms of the system parameters, the temperature increase reduced C most,

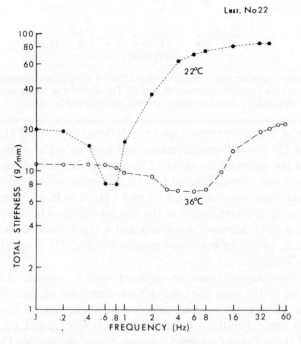

FIG. 6. Effects of temperature on the stiffness-frequency relationship in a muscle in contracture.

K_s next, and K least. In resting muscle, too, the stiffness curve shifted vertically downward (roughly by 4 g·mm^{-1}) with the increase in temperature.

e) *Stiffness-force relationship at different lengths*: Figure 7 shows a plot of dynamic stiffness at 32 Hz as a function of mean muscle force in a resting muscle and the same muscle in Ba^{2+} contracture. There is a significant difference in the slope of the stiffness - muscle force relation curve between the resting condition and the contracture state. The broken line was drawn through the three points which were obtained from the resting muscle at three different lengths (L_{max}, 0.98 L_{max} and 0.96 L_{max}). The solid lines were drawn through three pairs of stiffness data in the resting and contracture state at the three muscle lengths. The relation curve between the resting stiffness and resting force is significantly steeper than those between the active stiffness and active force.

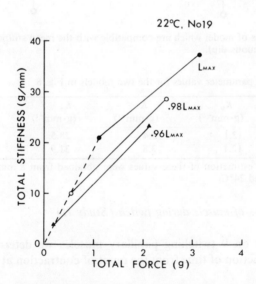

FIG. 7. Total stiffness (at 32 Hz)—total muscle force relationship. The broken line was drawn through the three data points obtained from a resting muscle at three different lengths. The solid lines connect these points with three active stiffness data points at identical lengths.

In Fig. 8 two alternative models are shown which are compatible with the major features of the frequency response of heart muscle in Ba^{2+} contracture.

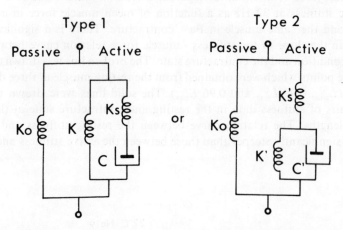

FIG. 8. Two types of model which are compatible with the mean stiffness curve in Fig. 5 (except the conspicuous dip).

TABLE 1. System parameter values for the two models in Fig. 8

Parameters	K_0 (g·mm^{-1})	K (g·mm^{-1})	K_s (g·mm^{-1})	C (g·sec·mm^{-1})
Type 1 Model	12.1	3.4	28.5	0.9
Type 2 Model	12.1	3.8	31.9	1.1

The data used for estimation of these values were obtained from 8 muscles in Ba^{2+} contracture at L_{max} and 24°C.

Dynamic stiffness of muscle during twitch (Study II)

In this study on 9 twitching papillary muscles we determined dynamic stiffness as a function of time from the onset of contraction at various muscle lengths.

a) *Envelope technique*: At a given frequency of length change, stiffness of twitching muscle changes with time from the onset of contraction. To know the stiffness at a particular instant in time, the peak-to-peak force excursion resulting from the sinusoidal length variation must be measured. When the input frequencies of length change were high ($>$ 10 Hz) compared with the frequency of muscle contraction, the envelope of maximum and minimum force excursions was not difficult to estimate. At lower frequencies ($<$ 10 Hz), however, the stiffness changed significantly during one cycle of length change and the maximum and minimum force excursions could not be recorded over the short interval within which the stiffness could be regarded as time-independent. To overcome this problem, we input each one of 16 perturbation frequencies (0.1 to 35 Hz) repeatedly with 0, 72, 144, 216, 288° phase shift with respect to the stimulation time. By superimposing on a memory scope these force responses to the set of 5 length perturbations with an identical frequency but with shifted phases, we could trace a relatively well-defined envelope of peak-to-peak excursions in the low frequency regions.

b) *Bode plot as a function of time*: Figure 9 is an example of the Bode plot of total muscle stiffness obtained from a twitching muscle at a length of 0.85 L_{max}. The data measured at times 0.2, 0.4 and 0.6 sec after stimulation are superimposed in this figure to show that, although the absolute magnitude of the stiffness varied with time, the basic features of the stiffness curve remained the same; namely, there were always two plateau, a lower plateau over the low frequency region and a higher plateaus over the high frequency region. In this example, the curve showed a relatively complex course over the intermediate frequency region. A simpler transition with a slope of 10-fold increase in stiffness per decade increase in frequency was noted in other muscles.

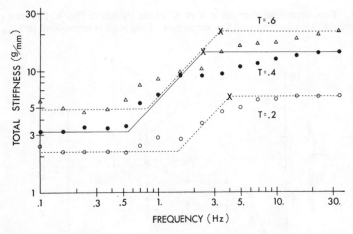

FIG. 9. Dynamic stiffness of twitching heart muscle at times 20, 40 and 60% T_{max}.

c) *Time and length dependence of K (elasticity coefficient)*: Assuming that the models shown in Fig. 8 are the simplest valid models for the stiffness curve obtained at various points in time, we calculated the values for the K element in the model. The mean of their time courses, obtained from 8 muscles all contracting at $L = 0.88\ L_{max}$, are shown in Fig. 10. Compared with the time course of active force F_A, which is also plotted, the K value tends to level off after 60% of the peak tension time is passed and stays at the plateau until 170% of the peak time.

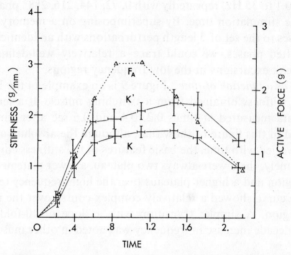

FIG. 10. Time-dependent changes in K or K' of the models in Fig. 8. The time course of active force (F_A) is superimposed for comparison. Time scale is normalized to T_{max}.

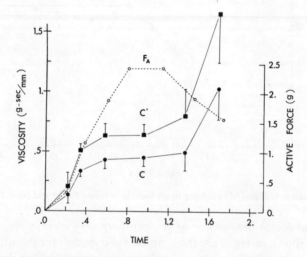

FIG. 11. Time-dependent changes in C or C' of the model in Fig. 8. Time scale is normalized to T_{max}.

d) *Time dependence of C (coefficient of viscosity)*: The same trend was found in the time course of the calculated C value. As shown in Fig. 11, however, C indicated a further increase after 160% T_{max}. This secondary rise became obscure at a muscle length greater than 0.98 L_{max}.

e) *Active force dependence of K_s (elasticity coefficient of a series element)*: The calculated values of K_s also appeared to be time-dependent. However, when the K_s value was plotted against simultaneous values of active muscle force, there was an extremely strong linear correlation between the two (Fig. 12). The linear regression coefficient of this correlation was independent of muscle length. Thus, we judged that the time-varying K_s is only apparently (or indirectly) dependent on time and is primarily dependent on active force. In other words, K_s represents a force-dependent, non-linear, passive elastic element, as indicated in many earlier studies (*e.g.*, Parmley & Sonnenblick, 1967).

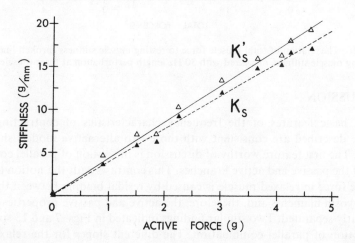

FIG. 12. Force dependence of K_s or K'_s of the models in Fig. 8.

f) *Stiffness-force-length relation*: When the stiffness at 30 Hz was plotted against the total muscle force during relaxation and contraction at three lengths of a muscle, the graph shown in Fig. 13 resulted. The relation of passive and active muscle stiffness to muscle force strongly resembles that indicated in Fig. 7. Namely, the slope of the stiffness-force correlation line in resting muscle is steeper than that in twitching muscle.

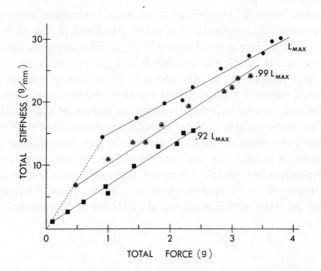

FIG. 13. The relation of total muscle force to resting muscle stiffness (broken line) and to twitching muscle stiffness measured with 30 Hz length perturbation at three muscle lengths.

DISCUSSION

The basic features of the frequency characteristics of contracting heart muscle described are consistent with those two alternative models shown in Fig. 8. The first feature worthy of discussion is the notion of parallel combination of the passive and active branches. This contrasts with the notion that the passive force in relaxed muscle is caused by residal binding between the actin and myosin filaments and, therefore, the active and passive properties cannot be clearly separated. Two similar findings indicated in Figs. 7 and 13 speak for the notion of parallel combination. The different slopes for the relaxed and contracted states demonstrate the presence of a qualitative, rather than merely quantitative, difference between the two states. In reference to the cross-bridge binding model, the implication of these findings is that almost all of the cross-bridges must be detached in relaxed muscle and nothing in the contractile machine contributes to the passive force.

The second feature of our model is that it does not have a "contractile element" as such. In the model the contractile machine is represented by three mechanical elements, K, K_s and C, among which K and C are active time-varying elements whereas K_s is a non-linear, passive, elastic element.

Force generation and shortening explained by the three-stage theory of cross-bridge reaction is expressed in this model by waxing and waning of the elasticity coefficient, K, and the viscosity, C. If the length of the model were not held constant as in the experiment, the model would shorten owing to the systolic increase in stiffness. Since K (the low frequency stiffness) deals only with changes in force and length, it must be integrated with respect to muscle length (over the length range in which active force generation occurs) in order to obtain an absolute active muscle force F_i at length L_i and time t_i:

$$F_i(L_i, t_i) = \int_{L_0(t_i)}^{L_i} K(x, t_i)\, dx \quad [x \text{ is a dummy variable}]$$

To perform this integration $K(L, t_i)$ must be known over the entire range of the effective length, L_0 to L_i.

We can only speculate about the correspondence of those components of dynamic stiffness to the parameters of cross-bridge kinetics. By definition, $K(L, t)$ represents the slope of the force-length relation curve (below L_{max}). Therefore, it is related to the *possible* increment or decrement of cross-bridge bindings per unit elongation or shortening of muscle. This interpretation of $K(L, t)$ explains its earlier increase than the isometric force (and the active stiffness) during twitch and the plateau of the K value over the period between 0.5 and 1.6 T_{max} (Fig. 10). The interpretation is also compatible with the independence of K from length between 0.80 and 0.94 L_{max} and its decay with elongation above 0.94 L_{max} in the left side version of the model in Fig. 8. The K' value calculated for the right side version was found to be length-dependent over the entire range of length, though it decayed more strongly at lengths above 0.94 L_{max}. C is considered to be closely related to the reciprocal of the rates at which the cross-bridges bind and unbind to thin filaments and/or they rotate backward or forward on the thin filament. Although C is expressed as a single parameter in the model, it probably represents a complex function of various rate parameters including those two reaction rates. On the other hand, if the length change used in our studies (0.1 $\sim$ 1 nm per half sarcomere or less) can be accommodated by the muscle without detachment and reattachment of bound myosin heads (Flitney & Hirst, 1978), C is related mainly to the rate of myosin head rotation on actin filaments.

The linear relation of K_s to F_a is probably because both F_a and K_s are a linear function of the number of bound cross-bridges and stretched myofilaments. The real non-linearity of the filaments cannot be denied, however.

The model is valid only if we disregard the peculiar dip in the stiffness curve and the slight phase lag of force wave behind the length wave over a certain frequency range in Study I. This peculiar response was shifted to a higher frequency range and greatly attenuated at the normal body temperature.

Therefore, we consider that the significance of the phenomenon itself for cardiac pumping function is doubtful. As a clue to speculation on the operating mechanism of the contractile machine, however, the phenomenon can be quite important (*e.g.*, Thorson & White, 1969; Abbot, 1972). In reference to the transient force response to step length change in heart muscle (Steiger, 1977; Steiger, Brady & Tan, 1978; our unpublished observation), it seems reasonable to consider that the decrease in dynamic stiffness occurs with an increase in frequency because the rate of delayed force development becomes too slow relative to the speed of stretch (Abbot, 1972) to develop its full potential. We can also speculate that, when the frequency is further increased, even the relatively high rate of detachment and back rotation of the cross-bridge becomes significant relative to the perturbation frequency and, with the series elastance K_s which comes from the elastic modulus of the myofilaments, begins to cause a frequency-dependent increase in stiffness. A more rigorous discussion is possible only with an analytical model analysis which we have not yet performed.

Figures 2–13 were reproduced from Saeki *et al.* (1978) and Loeffler & Sagawa (1975) with the permission of the American Heart Association.

REFERENCES

Abbott, R. H. (1972). An interpretation of the effects of fiber length and calcium on the mechanical properties of insect flight muscle. *Cold Spring Harbor Symp. quant. Biol.* **37**, 377–378.

Alfrey T, Jr. & Gurnee E. F. (1956). Dynamic of viscoelastic behavior. In *Rheology*, Vol. 1, ed. Eirich R., pp. 387–429, New York: Academic Press.

Buchthal, F. & Rosenfalck, P. (1960). Dynamic elasticity in the initial phase of an isotonic twitch. *Acta physiol. Scand.* **49**, 198–210.

Flitney, F. W. & Hirst, D. G. (1978). Crossbridge detachment and sarcomere 'give' during stretch of active frog's muscle. *J. Physiol.* **276**, 449–465.

Forman. R., Ford, L. E. & Sonnenblick, E. H. (1972). Effect of muscle length on the force-velocity relationship of tetanized cardiac muscle. *Circulation Res.* **31**, 195–206.

Hunter, W. & Noordergraaf, A. (1976). Can impedance characterize the heart? *J. Appl. Physiol.* **40**, 250–252.

Loeffler, L., III & Sagawa, K. (1975). A one-dimensional viscoelastic model of cat heart muscle studied by small length perturbation during isometric contraction. *Circulation Res.* **36**: 498–512.

Milsum, J. H. (1966). *Biological Control Systems Analysis.* pp. 142–149, New York: McGraw-Hill.

Parmley, W.W. & Sonnenblick, E.H. (1967). Series elasticity in heart muscle. *Circulation Res.* **20**, 112–123.

Pinto, J. G., Price, J. M., Fung, Y. C. & Mead, E. H. (1975). A device for testing mechanical properties of biological materials—the "Biodyne." *J. Appl. Physiol.* **39**, 863–867.

Saeki, Y., Sagawa, K. & Suga, H. (1978). Dynamic stiffness of cat heart muscle in Ba^{2+}-induced contracture. *Circulation Res.* **42**, 324–333.
Steiger, G. J. (1977). Tension transients in extracted heart muscle preparation of rabbit. *J. molec. Cell. Cardiol.* **9**, 671–685.
Steiger, G. J., Brady, A. J. & Tan, S. T. (1978). Intrinsic regulatory properties of contractility in the myocardium. *Circulation Res.* **42**, 339–350.
Templeton, G. H., Adcock, R., Willerson, J. T., Nardizzi, L, Wildenthal, K. & Mitchell, J. H. (1976). Relationships between resting tension and mechanical properties of papillary muscle. *Am. J. Physiol.* **231**, 1679–1685.
Templeton, G. H., Donald, T. C., III, Mitchell J. H. & Hefner L. L. (1973). Dynamic stiffness of papillary muscle during contraction and relaxation. *Am. J. Physiol.* **224**, 692–698.
Templeton, G. H., Wildenthal, K., Willerson, J. T. & Reardon, W. C. (1974). Influence of temperature on the mechanical properties of cardiac muscle. *Circulation Res.* **34**, 624–634.
Thorson, J. & White, E. C. S. (1969). Distributed representations for actin-myosin interaction in the oscillatory contraction of muscle. *Biophys. J.* **9**, 360–370.

Discussion

Kawai: It would be preferable to plot data in terms of a Nyquist plot or a root locus plot, because in this way, you can identify the rate processes, and also, you can judge the quality of the data very well because semicircles are very easy to judge by eye.

Sagawa: I think it is a matter of choice. You are probably more familiar with Nyquist plots, whereas I am more accustomed to Bode plots. Furthermore, I have my own opinion about calling your plot a Nyquist plot because, as you know, Nyquist is the one who established the Nyquist criterion for judging stability of negative feedback systems, and when you plot the open loop performance of a negative feedback system, 180° phase shift always exists at a very low frequency, and therefore, the tradition is that you regard 180° phase shift of the output relative to the input as 0°, and from there you start plotting either the advance or the delay. Now, if muscle people do not know this, and they keep calling that a Nyquist diagram, and then they talk to engineers who deal with negative feedback system, serious confusion can happen. The engineers hear Nyquist diagrams, and think, "Ah, ha! . . . there is already a 180° phase shift," although this is what you call 0°.

I checked with my colleague, Dr. Nakayama, about whether engineers usually call the kind of thing you showed Nyquist diagrams. He said yes, so you are right, but I do not like that engineering practice either.

Kawai: I should agree with you as far as the use of terminology.

Steiger: I also did experiments at 35°–36° with rabbit papillary muscle. My experience was the following: If you raise the temperature, for the first couple of minutes the loop opens up, so we get a bigger work output, but only for a very limited time. Then if the tension rose, even the isometric tension rose, and we got a similar phenomenon as we observed earlier in insect flight muscle; so it might be that the oxygen supply to the muscle was not enough to keep it at a normal, functional level at high temperatures.

Sagawa: That is a good point. Dr. Saeki, did you get this data mostly shortly after, or some time after the temperature change?

Saeki: After a high temperature of 36° is imposed, as Dr. Steiger said, the steady state of barium contracture tends to decrease quicker than at lower temperature; so I tried three or four preparations at a relatively early stage of the barium contracture, up to 30 minutes.

Sagawa: Yes. The entire frequency analysis itself takes 30 minutes at least, so it is impossible to do it in a short, early phase. The muscle size was very small. We carefully chose very small muscle but we cannot deny the oxygen delivery factor.

Steiger: I could maintain it for a longer time if I used less barium for a contracture, *e.g.*, 0.5 mM. When we had only half maximum activation, the loop was open for a longer time.

ter Keurs: It has been known that the region close to clamps in cardiac muscles is damaged. I do not see a separation in your model between the clamped, damaged area and the central, presumably healthy area. Do you have evidence that you can exclude the effect of the damaged areas?

Sagawa: No, I have no evidence that we can, and the reason is we do not have a nice system to look at the belly of the muscle, by either diffraction or other markers, and that is why I dropped by your laboratory and tried to learn your system, as well as Dr. Pollack's. I have no problem in accepting that there must be damaged ends, and the effect is pretty large, but to check ourselves about this problem we tried to measure compliance and other properties in this *in vivo* muscle. This is a big muscle in the dog's right ventricle. The tendon was cut and was attached to both the length control system and tension system. The root of this papillary muscle was attached and perfused by blood. But somehow you have to fix the root of the muscle; if you stretch it or fix it too rigidly then ischemia must result. So you have same problem in a different form. However, the result we have attained quantitatively was not so different from that for excised papillary muscle.

III. ISOTONIC AND ISOMETRIC TRANSIENTS

What Does Relaxed Insect Flight Muscle Tell Us about the Mechanism of Active Contraction?

D. C. S. WHITE*, M. G. A. WILSON* and JOHN THORSON**

* Department of Biology, University of York, Heslington, York, England
** Department of Zoology, University of Oxford, Oxford, England

ABSTRACT

A characteristic feature of insect flight muscle is that the fibers are very stiff, even in the relaxed state. In order to understand the contribution from the cross-bridges to the active tension it is essential to know whether the mechanical properties of the structures responsible for the relaxed stiffness contribute to the performance of the active muscle, and if so, whether their contribution is the same as that found in the relaxed muscle.

Several lines of evidence suggest that the structures responsible for this relaxed stiffness do exert the same effect in active and relaxed muscle:

a. Comparison of the stiffness of the muscle at different tensions in various types of muscle;

b. Investigations of the effect of calcium on the properties of (i) rigor muscle (ii) AMP.PNP-treated muscle and (iii) the relaxed muscle immediately after the application of calcium ions.

It follows that any analysis of the mechanical properties of the active muscle must take proper account of this contribution from the structures responsible for the stiffness of relaxed muscle. One important consequence of this is that, at high calcium ion concentrations, the dynamics of the active muscle are no longer adequately explained in terms of an exponential delay, even for small amplitude perturbations, as has been suggested in many papers over the last ten years; a transfer function more akin to a finite delay is required. Control of activity of the cross-bridges via A-filament strain (rather than sarcomere strain) provides a possible answer.

INTRODUCTION

It has been known for many years that one characteristic property of insect flight muscle is that the relaxed muscle is very much stiffer than relaxed vertebrate skeletal muscle (Machin & Pringle, 1959). There is a good deal of evidence to suggest that the A-filaments of insect flight muscle are linked to the Z-line by what have been termed C-filaments. This evidence is reviewed by White & Thorson (1973) and Pringle (1978).

Vertebrate skeletal muscle does exhibit a small visco-elasticity in the relaxed state; Hill (1968) and Ford, Huxley & Simmons (1977) provide evidence that this visco-elasticity does not contribute to the mechanical response of the active muscle. The visco-elasticity of the relaxed vertebrate muscle is thought to be a property of the cross-bridges.

It is essential to know whether the high resting stiffness of the insect muscle contributes to the response of the active muscle, and if so whether its contribution is the same as that in the relaxed muscle. The reason this knowledge is essential concerns the phenomenon of stretch activation which is fundamental to the normal function of insect fibrillar flight muscle; our entire view of the dynamics of the active process depends critically on the stance taken upon what the C-filaments are doing during activation. Donaldson & White (1977) measured the stiffness of active insect muscle at different tensions, and demonstrated that the muscle behaved as though the stiffness of the relaxed muscle was still present. However, there are reports that the stiffness of the relaxed muscle decreases when calcium ions are added under circumstances when the muscle is not activated (in the presence of an ATP analog, AMP.PNP; Chaplain & Frommelt, 1968).

In this paper we investigate this problem further and find no evidence for a decrease in the contribution of the properties responsible for the high stiffness of the relaxed muscle to those of the active muscle, and we discuss the implications of this result for models of the behavior of the active muscle.

MATERIALS AND METHODS

We have used the following muscles in this work: rabbit psoas muscle, the dorsal longitudinal flight muscles of the giant water bug *Lethocerus cordofanus* and of the bumblebee *Bombus terrestris*, and the dorso-ventral flight muscles from the dragonflies *Aeshna grandis* and *Aeshna cyanea*. The *Lethocerus* and *Bombus* flight muscles are fibrillar (their control in the animal is asynchronous); the dragonfly muscle is non-fibrillar (the control in the animal is synchronous) (Pringle, 1957). The muscles were all glycerol-extracted in a solution containing 50% glycerol, 20mM phosphate buffer, 3mM sodium azide, 2mM dithiothreitol, pH 7.0, and were used within a few days of being extracted. Some of the experiments on the *Bombus* muscle were performed after washing in glycerol solution for only 3–5 min.

The solutions used in the experiments are detailed in Table 1. All the experiments were performed at 15°C.

The experiments were performed on bundles of 1 to 3 fibers. The fibers were mounted between two hooks, 3 mm apart, using cellulose acetate dissolved in acetone as a glue. One hook was connected to a strain gauge (Akers, type 803) with a resonant frequency of about 2.2 kHz (measured when loaded with an activated fiber) and critically damped with a damping vane and oil. The other hook was connected to the moving arm of an electromagnetic vibrator (Ling Dynamics, type 101) whose movement was controlled by a servoloop, enabling length changes with rise times of the order of about 0.8 msec to be applied to the muscle. Sinusoidal analysis was performed with a Solatron Resolved Component Indicator (type VP 253.3). Photographic records were made of the oscilloscope recordings of the step responses, and the photographs analyzed with a digitiser (Hewlett-Packard model 9864A connected to a model 9830 calculator).

RESULTS

1. Measurements of stiffness during development of activation

Jewell & Rüegg (1966) first demonstrated that, following a step change of length, *Lethocerus* muscle showed large delayed tension changes. White & Donaldson (1977) showed, by measuring the stiffness of the muscle during these tension changes (by applying a second, small, test step of length and measuring the magnitude of the tension response), that the stiffness was linearly related to the tension, and that if the line was extrapolated back to zero tension, the intercept on the stiffness axis was equal to the stiffness of the relaxed muscle. Thus, the structure responsible for the stiffness of the relaxed muscle acts as though it were in parallel with the cross-bridges, and contributes to the response of the active muscle.

In such experiments the bridges which are attached at the instant of applying the initial length change will be distorted by the relative movement of the A- and I-filaments. If the lifetime of the attachment of these bridges is comparable to the interval between the conditioning step and the test step, then some of these distorted bridges will remain attached and distorted at the time of the test step. Such distorted bridges would affect the measured stiffness vs. tension differently than if they were not distorted. A situation in which distortion cannot build up is the slow rise of tension following introduction of Ca^{2+} under isometric conditions, and the question is thus whether similar stiffness-tension plots are obtained when one makes such measurements. We have measured the relationship between tension and stiffness during activation for the four different muscles used in this study.

The fibers were placed in relaxing solution. The insect flight muscles all showed marked stiffness in the relaxing solution at lengths about equal to their normal body length. The rabbit muscle showed negligible stiffness. The insect muscles were stretched 1% above the length at which the relaxed tension is just zero; the rabbit muscle was used at sarcomere lengths between 2.2 and 2.4 μm. The stiffness of the relaxed muscle was measured by applying a step length change of either 0.2% or 1.0% of the fiber length. The fiber was then transferred to the activating solution, and the stiffness measured during the rise of tension. Figure 1 shows the time course of such an experiment using rabbit muscle, together with sample records of the stiffness measurements from both rabbit and bee muscle.

The stiffness of the different muscles, as a function of their tensions, are shown in Fig. 2. Also plotted in Fig. 2, for the same *Lethocerus* muscle, are measurements made in a similar way to those described by Donaldson & White (1977), *i.e.*, during the delayed rise in tension following a step change of length, for which sample records are shown in Fig. 3.

From these results it is apparent that the stiffness-tension lines of the insect muscles all cross the stiffness axis at a value approximately equal to the stiffness of the relaxed muscle. (Note that the stiffness of the relaxed muscle also increases with stretch of the muscle and therefore with relaxed tension. The relaxed stiffnesses plotted on the graphs represent the stiffness measured at 1% stretch). The relationship for the rabbit muscle passes close to the origin, as found by Rüegg, Kuhn, Herzig & Dickhaus (1975), as we would expect because of the presumed absence of C-filaments.

If calcium ions have an especially strong effect upon the relaxed stiffness, it is possible that this would show up as a transient change of stiffness after the fibers are transferred to the activating solution, but before the tension starts to rise. We have looked carefully at this region in *Lethocerus* and bumblebee by applying alternating step increases and decreases of length of 0.2% at intervals of 300 msec, but found no detectable change of stiffness during this period.

If the structure responsible for the stiffness of the relaxed muscle is indeed a connection between the A-filament and the Z-line, then, in insect muscles below rest length, this connection is likely to be slack. Experiments on insect muscle at this length would, therefore, be likely to show a stiffness-tension relationship different from that found when the fiber is slightly extended, possibly resembling the vertebrate case. We are grateful to Mr. R. Cuminetti of Oxford for suggesting this experiment. The above experiments during activa-

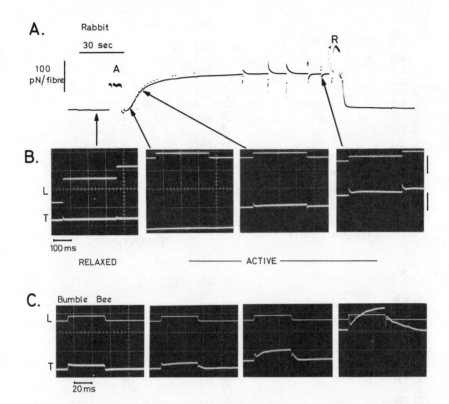

FIG. 1. A. Pen recorder trace of tension *vs.* time for an experiment on rabbit muscle. The fibers are in relaxing solution at the beginning of the trace. At 'A' they are transferred to activating solution. Step length changes of 0.2% of the muscle length, lasting 500 msec are applied to the fiber during the rise of tension, and give rise to the blips on the trace. At 'R' the fibers are returned to the relaxing solution.
B. Oscilloscope records of length (L) and tension (T) of four of the step responses made in the experiment shown in Fig. 1A. Calibrations: the side of one CRO square represents: Length, 0.83%; Tension, 15 μN/fiber for recording on left, 60 μN/fiber for the others; Time, 100msec.
C. Recordings made from a similar experiment on bumblebee muscle. Calibrations: Length, 0.83%; Tension, 30 μN/fiber; Time, 20 msec.

tion were therefore repeated on *Lethocerus* at 3% below the slack length.

The results are shown in Fig. 4. Open symbols represent measurements made during development of activation at a muscle length 1% above the length at which the tension is just zero (rest length); closed symbols represent measurements made at a length 3% below rest length, at which length the relaxed

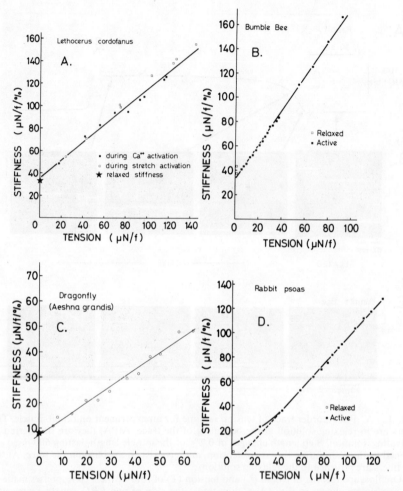

FIG. 2. Stiffness vs. tension during the development of activation for the four different muscles used in this study. The experiment was performed as described for Fig. 1. The stiffness was measured from the tension change occurring during the application of the change of length, i.e., from the magnitude of the initial spike of tension.

fibers are visibly slack, buckling slightly. The open squares denote measurements made before, and the open circles measurements made after, the two runs made at 3% below the slack length. In relaxing solution the fibers became buckled again after each activation at this length. The stiffness of the relaxed

Lethocerus

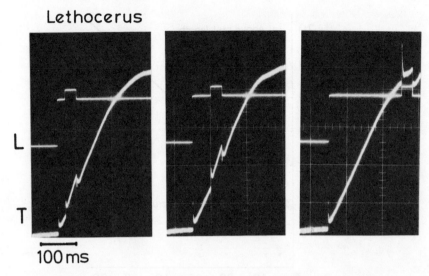

FIG. 3. Stiffness vs. tension during the development of tension after applying an initial conditioning step on *Lethocerus*. A 1% conditioning step is followed a short time later by a 0.2% test step. The stiffness is measured from the magnitude of the spike of tension obtained in response to the test step. Calibrations: Length, 0.83%; Tension, 30 μN/fiber; Time, 100 msec.

fibers returned to within 10% of the original value when they were re-extended after the runs at the slack length. The line on this figure is the regression line through the open symbols.

It can be seen that the stiffness-tension relationship of the active fibers at muscle lengths below rest length is very close to that when they are slightly extended, except at tensions below about 10 μN/fiber.

2. Effect of calcium ions on rigor muscle

If calcium has any marked effect upon the stiffness of the contractile proteins, then it is conceivable that this would show up in measurements on the stiffness of rigor muscle in the presence and absence of calcium. This experiment would be unlikely to detect even large changes in the properties of the C-filaments, but would presumably detect changes in A- or I-filament stiffness.

This effect was tested by first allowing fibers of *Lethocerus* to develop tension by transferring them from relaxing solution to rigor solution (as described by White, 1970). Their stiffness was then measured by applying a step length decrease of 0.1% in solutions containing, or not containing,

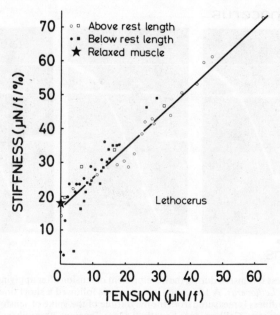

FIG. 4. Stiffness vs. tension during the development of activation in *Lethocerus*. Open symbols—fiber at 1% above the rest length; closed symbols—fiber at 3% below the rest length. A full description of this experiment is given in the text.

calcium ions. The pCa values of the two solutions, calculated using the dissociation constants of White & Thorson (1972), were 4.5 and about 10. The results are shown in Fig. 5. The recordings were made in the order indicated. The stiffness in the pCa 4.5 rigor was about 13% greater than that in the normal rigor solution.

3. *Effect of calcium ions on muscle in AMP.PNP solutions*

Chaplain & Frommelt (1968) reported that the stiffness of relaxed *Lethocerus* muscle is virtually unchanged when the fiber is placed in a relaxing solution in which the ATP is replaced by 6.5mM AMP.PNP, but that when 5.37×10^{-8} M Ca^{2+} is added to the AMP.PNP relaxing solution the stiffness is reduced by about 20%. This result is grossly incompatible with the findings of Marston, Rodger & Tregear (1976) and Marston, Tregear, Rodger & Clarke (1979) who report (a) that the stiffness of the AMP.PNP "relaxed" muscle is similar to that of the rigor muscle if measured at the same tension (this is many times greater than the stiffness of the relaxed muscle), and (b) that if the magnesium

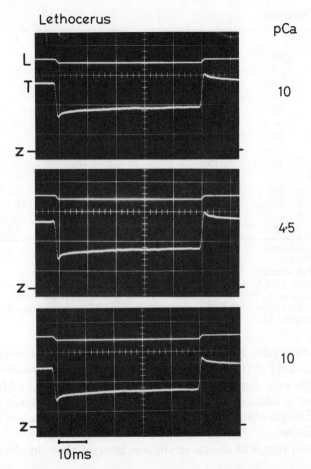

FIG. 5. Oscilloscope recordings of length and tension *vs.* time for an experiment on *Lethocerus*, in the presence and absence of calcium ions. The experiments were made in the order shown. Z indicates zero tension, L denotes length and T denotes tension. Calibrations: Length, 0.83%; Tension, 60 μN/fiber; Time, 10 msec.

in the solutions is replaced by calcium, then the tension change obtained with Ca-AMP.PNP is less than that obtained with Mg-AMP.PNP. The maximum concentration of AMP.PNP used by Marston *et al.* was 1mM.

Since Chaplain & Frommelt used a higher concentration of AMP.PNP, and because the total replacement of magnesium by calcium is not the same as adding calcium to solutions containing millimolar concentrations of magnesium, Maxine Clarke of Oxford kindly repeated Chaplain & Frommelt's experi-

TABLE 1. Solutions

	1	2	3	4	5	6
$MgCl_2$	10	10	5	5	5	5
EGTA	5	5	5	5	4	4
$CaCl_2$	—	5	—	5	—	4
KCl	—	—	50	50	50	50
Histidine	20	20	20	20	—	—
PIPES	—	—	—	—	10	10
ATP	10	10	—	—	—	—
AMP.PNP	—	—	—	—	6.5	6.5
CrP	10	10	—	—	—	—
PK(mg/ml)	2	2	—	—	—	—
HK(mg/ml)	—	—	—	—	0.2	0.2
Glucose	—	—	—	—	10	10
$AP_5A(\mu M)$	—	—	—	—	50	50
pH	7.0	7.0	7.0	7.0	7.1	7.1

1: Relaxing solution; 2: Activating solution;
3: Rigor solution; 4: Ca-rigor solution;
5: AMP.PNP solution; 6: Ca-AMP.PNP solution
Abbreviations: EGTA: Ethylene glycol bis-(β aminoethyl ether)
-N,N'-tetracetic acid; PIPES: piperadine-N,N'-bis (ethanesulphonic acid);
AMP.PNP: β,γ-imido-ATP; CrP: creatine phosphate; PK: phosphokinase;
HK: hexokinase; AP_5A: P1,P5 diadenosine pentaphosphate.

ments with us. Table 2 shows the results of measuring the stiffness of *Lethocerus* fibers in AMP.PNP solutions (see Table 1) with and without calcium. The experiments were done in the same way as those described by Marston et al. (1979), but in additon the stiffness of the fibers was measured by applying sinusoidal length changes and measuring the dynamic stiffness at three different frequencies.

There was very little change in stiffness brought about by the addition of calcium.

We also repeated the experiment which tests the effects of calcium on rigor muscle on the fibers used for the above experiment with AMP.PNP. On average, an increase of about 15% in stiffness of the rigor muscle, in the presence of calcium, was found.

4. Dynamic properties of insect fibrillar flight muscle

The above results indicate that the C-filaments are important in the active case, which underlines in turn that data ought to be collected carefully on the same fiber for the relaxed and the active states. Moreover, special attention must be paid to reproducibility.

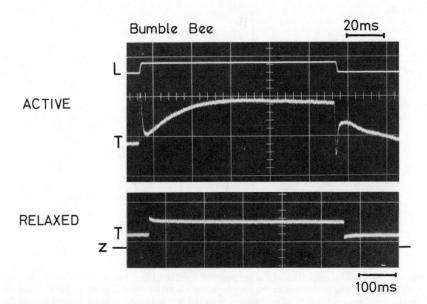

FIG. 6. Step responses of bumblebee muscle. The top tension trace was obtained with the fibers in relaxing solution, the lower one with the fibers in relaxing solution. The traces have been positioned so that "Z" denotes zero tension for both responses. The length step was 0.2% of the muscle length. The recording shown for the relaxed muscle was made immediately before that for the active muscle. A similar recording made immediately afterwards had a prestep tension of about 5 μN/fiber and a stiffness about 82% that of the recording shown. Calibrations: Length, 0.83%; Tension, 30 μN/fiber; Time, 20ms for the active muscle, 100 msec for the relaxed muscle.

Most publications in this area have not been comprehensive in this sense. We therefore present a complete set of small-signal behavior of the fully active insect fibrillar flight muscle, which at least offer a fair comparison of the relaxed and active responses. Note in particular that the "difference step" dips below zero (Fig.8B - see the discussion).

Figures 6 and 7 show such responses, made on bumblebee muscle that had been immersed in glycerol-extraction solution for about 5 minutes only. The initial tension before the application of the 0.2% step in the relaxing solution was 9 μN/fiber, and in the activating solution was 95 μN/fiber.

DISCUSSION

The stiffness-tension plots of Fig. 2 are most easily interpreted in terms of the structures responsible for the relaxed stiffness of the insect flight muscle

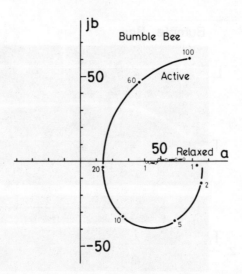

FIG. 7. Sinusoidal responses of the preparation used to obtain Fig. 6, plotted in the form of a Nyquist plot. The amplitude of the length changes used was 0.08% peak-to-peak. The units on the axes are μN/fiber/%. The frequencies used are indicated against the points on the active muscle response.

TABLE 2. Effect of calcium on stiffness of *Lethocerus* in AMP.PNP

f(Hz)	AMP.PNP	Ca-AMP.PNP
1	190	192
5	199	199
60	225	227

A 0.13% amplitude length change was applied to the fibers at the frequencies indicated and the tension changes measured on a Solatron JM1600 transfer function analyser. The stiffnesses are given in μN/fiber/%.

acting in parallel with the cross-bridges, and exerting the same effect in active and relaxed muscle. That such a structure is the C-filament, which links the A-filament to the Z-line, is obviously an attractive hypothesis.

The results of Fig. 4, showing similar stiffness-tension plots for both stretched and 3% slack starting conditions, are more difficult to explain. Conceivably the C-filament material acts as though it were compressible, rather than slack, when the muscle is shortened below resting length and then activated. Had the C-filaments become slack they would have contributed nothing to the mechanical properties when the muscle was held at 3% below the resting length, and the tension-stiffness relationship ought then to have been like that of the vertebrate muscle and passed through the origin.

The events underlying the slack case in Fig. 4 might arise in the following way. When the relaxed muscle is shortened to 3% below resting length, the fibers became buckled The tension in the A-filaments and the C-filaments will presumably be about zero. Upon activation the cross-bridges act to move the A-filaments relative to the I-filaments. The A-filaments move toward the Z-line, causing the fiber to straighten, and ultimately thereby compressing the C-filaments. (This kind of compression is an intra-sarcomeric phenomenon; it does not imply that the sarcomere exerts negative tension.) Such a compressive effect, even though it opposes the tension-producing property of cross-bridges, could provide the required stiffness offset in the observed stiffness-tension plot for the slack case. That is, if compression can be realized then it does not matter to a first-order approximation whether a slight extension of the muscle results in a greater pull on the end of the A-filament or a lesser effective push.

The measurements on the effect of AMP.PNP upon rigor muscle are similar to those reported by Marston, Rodger & Tregear (1976), and the effect of adding calcium ions, at the concentrations used to maximally activate the muscle in ATP solutions, was minimal. This result is in serious conflict with those of Chaplain & Frommelt (1968). Their low stiffness, equivalent to that of relaxed muscle, could perhaps have been due to ATP or ADP contamination in their solutions (the muscle contains sufficient myokinase to convert ADP to ATP to allow relaxation with ADP), but we have no explanation for their effects with calcium.

Had calcium had a marked effect upon the stiffness of the A- and I-filaments this might have shown up in rigor. We did observe a small but significant *increase* in stiffness in rigor solutions containing physiological concentrations of calcium by comparison with those with no added calcium.

Figures 6 and 7 illustrate the performance of the passive and active muscle measured on the same fibers. The difference plots obtained by removing the passive stiffness as though it were in parallel with the structure responsible for activity are illustrated in Figs. 8 and 9. The curves of Fig. 6 have been replotted in Fig. 8A, and the difference between them plotted in Fig. 8B. Note that the prestep tensions of the active and relaxed muscle are very different (95 and 9 μN/fiber, respectively). The position of the photographs in Fig. 6 shows these relative absolute tensions approximately. In Fig. 8A the zero of tension represents the prestep tension of each trace, and Fig. 8B shows the difference between the traces of Fig. 8A. In particular, the absolute values of both the relaxed and active stiffnesses and tensions are fundamental to any further analysis.

In conclusion, we have tried all the variants of the known procedures for identifying the effects of activation on filament properties which have occurred

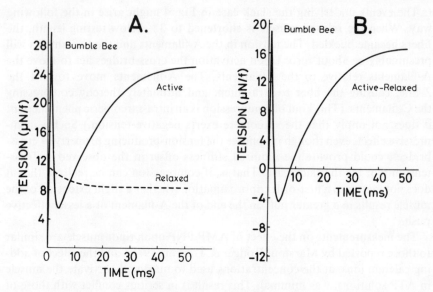

FIG. 8. A. The active and relaxed responses of Fig. 6 plotted on the same time axis. Zero tension represents the prestep tension for each response.
B. The difference between the active and relaxed responses drawn in part A.

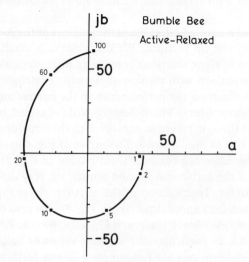

FIG. 9. The Nyquist plot obtained by subtracting the data obtained on the relaxed muscle from those obtained on the active muscle, using the results shown in Fig. 7.

to us and our colleagues; we find no evidence of any reduction of filament stiffness in any test. Wherever changes are noticed, they represent modest increases in stiffness of the sarcomere as a whole.

A further possible strategy might be to look for a protein similar to that suspected to represent the C-filaments (Bullard, Bell & Luke, 1977)—one which can be studied apart from the cross-bridge involvement present in fibrillar muscle. One structural protein, at least, is known to have considerable sensitivity to Ca^{2+} in the range of calcium ion concentrations required to activate and relax muscle. This is spasmin (Amos, Routledge, Weis-Fogh & Yew, 1976) found in the spasmoneme of certain ciliated protozoa. Increased Ca^{2+} in this range in fact produces contraction.

Ten years ago, Chaplain & Frommelt (1968) made the point that the C-filament stiffness must be much reduced during activation if one is to describe the mechanical dynamics via an exponential delay; they also remarked that if the C-filaments were not so altered, the parallel active component is then like a finite delay at low frequencies, as Machin & Pringle (1960) once considered. Chaplain & Frommelt in fact suggested that the C-filament stiffness decreased by just such an amount with Ca^{2+} that an exponential delay would fit their data. Abbott's simplification (1973a,b) of our early proposals (Thorson & White, 1969) have been applied by Abbott (1977, see also Pringle, 1978) to compare high Ca^{2+} small-signal data with the rate constants of cross-bridge attachment and detachment; they reach specific conclusions about the number of required states in the cross-bridge cycle. In the most straightforward interpretation of a 2-component model (passive and active element in parallel) his (Abbott, 1977) analysis, like Chaplain & Frommelt's, depends implicitly but clearly upon a drastic reduction in stiffness of the parallel elastic component upon activation (by as much as an order of magnitude; Abbott, 1973a, Fig. 3; c.f. $A' = A - B$ in Appendix, Abbott, 1973b).

However, if C-filaments retain most of their stiffness during activation—a situation compatible with the evidence reviewed here—then the usual 2-component model, with the active component dominated by a single exponential delay, cannot be made even to approximate the low frequency small-signal data; thus in that event there is no known basis for comparison of the dynamic data with the rate constants of cross-bridge theory.

Acknowledgements

We would like to thank Professor J.D. Currey for numerous helpful discussions.

REFERENCES

Abbott, R. H. (1973a). The effects of fiber length and calcium ion concentration on the dynamic response of glycerol-extracted insect flight muscle. *J. Physiol.* **231**, 195–208.

Abbott, R. H. (1973b). An interpretation of the effects of fiber length and calcium on the mechanical properties of insect flight muscle. *Cold Spring Harb. Symp. quant. Biol.* **37**, 647–654.

Abbott, R. H. (1977). The relationship between biochemical kinetics and mechanical properties. In *Insect Flight Muscle*, ed. Tregear, R. T., pp. 269–276, Amsterdam, London, New York: North-Holland.

Amos, W. B., Routledge, L. M., Weis-Fogh, T. & Yew, F. F. (1976). The spasmoneme and calcium-dependent contraction in connection with specific calcium-binding proteins. *Symp. Soc. exp. Biol.* **30**, 273–302.

Bullard, B., Bell, J. L. & Luke, B. M. (1977). Immunological investigation of proteins associated with thick filaments of insect flight muscle. In *Insect Flight Muscle*, ed. Tregear, R. T., pp. 41–52, Amsterdam, London, New York: North-Holland.

Chaplain, R. A. & Frommelt, B. (1968). On the contractile mechanism of insect flight muscle. I. The dynamics and energetics of the linearised system. *Kybernetik,* **5**, 1–17.

Donaldson, M. M. K. & White, D. C. S. (1977). The elasticity of insect flight muscle. *J. Physiol.* **271**, 22P–23P.

Ford, L. E., Huxley, A. F. & Simmons, R. M. (1977). Tension responses to sudden length change in stimulated frog muscle fibers near slack length. *J. Physiol.* **269**, 441–515.

Hill, D. K. (1968). Tension due to interaction between the sliding filaments in resting striated muscle: the effect of stimulation. *J. Physiol.* **199**, 637–684.

Jewell, B. R. & Ruegg, J. C. (1966). Oscillatory contraction of insect fibrillar muscle after glycerol extraction. *Proc. R. Soc. B.* **164**, 428–459.

Machin, K. E. & Pringle, J. W. S. (1959). The physiology of insect fibrillar muscle. II. Mechanical properties of a beetle flight muscle. *Proc. R. Soc. B.* **151**, 204–225.

Machin, K. E. & Pringle, J. W. S. (1960). The physiology of insect fibrillar muscle. III. The effect of sinusoidal changes of length on a beetle flight muscle. *Proc. R. Soc. B.* **152**, 311–330.

Marston, S. B., Rodger, C. D. & Tregear, R. T. (1976). Changes in muscle cross-bridges when β-imido-ATP binds to myosin, *J. molec. Biol.* **104**, 263–276.

Marston, S. B., Tregear, R. T., Rodger, C. D. & Clarke, M. L. (1978). Coupling between the enzymatic sites of myosin and the mechanical output of muscle. In press.

Pringle, J. W. S. (1957). *Insect Flight*. Cambridge University Press, Cambridge.

Pringle, J. W. S. (1978). Stretch activation of muscle: function and mechanism. *Proc. R. Soc. B.* **201**, 107–130.

Rüegg, J. C., Kuhn, H. J., Herzig, J. W. & Dickhaus, H. (1975). Effect of calcium ions on force generation and elastic properties of briefly glycerinated muscle fibers. In *Calcium Transport in Contraction and Secretion*, eds. Carafoli, E. *et al.*, Amsterdam, London, New York: North-Holland.

Thorson, J. & White, D. C. S. (1969). Distributed representations for actin-myosin interaction in the oscillatory contraction of muscle. *Biophys. J.* **9**, 360–390.

White, D. C. S. (1970). Rigor contraction and the effect of various phosphate compounds on glycerinated insect flight and vertebrate muscle. *J. Physiol.* **208**, 583–605.

White, D. C. S. (1973). Links between mechanical and biochemical kinetics of muscle. *Cold Spring Harb. Symp. quant. Biol.* **37**, 201–213.

White, D. C. S. & Thorson, J. (1972). Phosphate starvation and the nonlinear dynamics of insect fibrillar flight muscle. *J. gen. Physiol.* **60**, 307–336.

White, D. C. S. & Thorson, J. (1973). The kinetics of muscle contraction. *Progr. Biophys. molec. Biol.* **27**, 175–225.

Discussion

Kawai: How do you visualize the presence of resting stiffness, which is rather high in insect muscle?

White: In terms of the connections between the A filaments and the Z lines.

Kawai: The C filaments?

White: Often called C filaments, yes.

Kawai: This would not change when you activate the preparation with calcium. Is that right?

White: Well, in principle it could change for two reasons: One reason is that the calcium ions might affect that material directly, as indeed calcium ions are known to affect some proteins.

Kawai: What is the evidence?

White: I am saying we do not have any evidence. I would agree that our evidence is not firm on that because we cannot find a way to test the filament stiffness of the relaxed muscle in the presence of calcium without activating it. There is no known analog of ATP that we know to keep the muscle relaxed, and enables us to do that experiment. But, so far as we have been able to test it, we cannot see any reduction. The second reason why you might get a reduction in stiffness with calcium would be that the effect of the cross-bridges is in principle to extend the A filament. If the A filament shows compliance, then you might expect the C filaments to be unloaded and, thereby, to exert lower stiffness. So there are two reasons why you might expect a reduction in stiffness from the relaxed muscle. But as I say, our experiments can see no sign of any such reduction.

Kawai: I wonder why the stiffness at some frequencies is lower on activation than in the relaxed condition?

White: That is exactly the point of this talk; it is really to say that, because the stiffness of the relaxed muscle does not change, this effect is absolutely vital, and that none of the theories that has been published so far during the 1970s in any way gives anything like an explanation for it; they all do not. You are then asking, how do I explain it? John Thorson and I are doing some work at the moment, which is based entirely upon the assumptions that we made in our 1969 paper (Thorson & White, 1969), and making those assumptions we can get the kind of effect.

Tregear: Your argument is clear. How far are your models restricted by the further fact, which you will know of, though I think other people will not know, that in the more recent measurements at Oxford, the peak of the response in the active flight muscle, *i.e.*, the absolute value of the stiffness of the activated muscle, drops very nearly to zero! So, presumably in your arguments, you will have had to be subtracting two almost equal and opposite quantities; the relaxed and the activated have to be almost equal and opposite.

White: No. That is not right. I think, if I understand you right, Richard, what you are saying is that some of the recent work from Oxford has shown that the active Nyquist plot goes very close to the origin, but it makes no difference to my argument. It strengthens my argument rather than weakens it.

REFERENCE

Thorson, J. & White, D.C.S. (1969). Distributed representations for actin-myosin interaction in the oscillatory contraction of muscle. *Biophys. J.* **9,** 360–390

Isometric Tension Transients in Skeletal Muscle before and after Inhibition of ATP Synthesis

T. BLANGÉ and G. J. M. STIENEN

Department of Physiology, University of Amsterdam, Amsterdam, The Netherlands

ABSTRACT

The dynamic properties of the contractile mechanism of skeletal muscle were investigated in electrically stimulated frog sartorius muscle at 0°C by analysis of the tension responses to quick ramp-shaped displacements.

1. The T_1 and T_2 curves were determined from the responses as a function of the displacement at different rates of displacement.

2. The slow recovery phase in the tension response is dominated by one exponential, which is practically independent of the amplitude of the shortening.

3. Incubation of the muscles with the metabolic inhibitors IAA and FAA or IAA and FDNB caused, after exhaustion but before the muscle passed into the rigor state, a change in the fast recovery phase and the plateau phase which depends on the amplitude of the shortening.

4. Stiffness was determined during the tension recovery by application of a second test length change.

5. The measurements at lowered ATP concentrations as well as the stiffness measurements suggest detachment of cross-bridges during the plateau phase in the tension response. It is put forward that possibly detachment is preceded by a transfer of the cross-bridge to a mechanically distinct state.

INTRODUCTION

Biochemical and ultrastructural data about the contractile mechanism obtained, give rise to the assumption that it might be possible to identify the processes which find expression in the transient force response (Lymn & Taylor, 1971; Huxley, 1974; Abbott, 1977).

In this paper some experiments will be reported concerning the tension response of the electrically stimulated frog sartorius muscle. First, attention is given to the early part of the responses during and shortly after a quick shortening in relation to the size and speed of the displacement. Subsequently, the effect of metabolic inhibition and exhaustion of the muscle is investigated.

METHODS

Preparation

The sartorius muscle of the frog, *Rana esculenta*, was dissected with part of the os pubis attached. This permitted a firm mounting in the measuring apparatus. The ties were consolidated with fast drying glue. The free part of the tibial tendon was kept as short as possible. The bathing solution contained NaCl: 115 mM, KCl: 2.5 mM, $CaCl_2$: 1.8 mM and a sodium phosphate buffer (pH 7.3). In later experiments d-tubocurarine was added to the solution in a concentration of 10 mg/l. Supramaximal field stimulation was applied via two silver electrodes, one on either side of the muscle. The temperature of the preparation was kept at about 0°C during the experiment, unless otherwise noted.

Equipment

The distal end of the muscle was fixed to a strain gauge force transducer with a resonance frequency of 7,000 Hz (Blangé, Karemaker & Kramer, 1972). The proximal end was fixed to a coil moving on an air bearing in the field of a permanent magnet. The displacement of the coil was measured by means of an interferometer device (Stienen, Blangé & Schnerr, in press). The displacement signal was electronically filtered and fed back to the coil. The system permitted application of displacements to the end of the muscle of 60 μm or about 0.002 of the resting length with a rise time of 0.15 msec. The available power was the limiting factor and therefore greater shortenings were given the shape of a ramp, which then had an acceleration and a deceleration phase of 0.15 msec each.

Metabolic inhibition and exhaustion

Two methods of metabolic inhibition were used with the purpose of inhibiting ATP-synthesis. The first one was incubation for 60 min in 1 mM iodoacetic acid (IAA) and 1 mM fluoro-acetic acid (FAA). IAA is known to interrupt the glycolysis and FAA inhibits the Krebs cycle. Measurements were done during tetanic stimulation at about resting length (L_0) until rigor tension showed up.

Sometimes tetanic stimulation was interchanged with a series of twitches in order to speed up exhaustion. The second method of inhibition consisted of incubation with 0.38 mM 1-fluoro- 2,4-dinitro-benzene (FDNB) for 40 min and 1 mM IAA for 60 min. In this case only twitches were applied for the exhaustion and the measurements.

Sarcomere length

The sarcomere length was measured by microscope directly after termination of the experiment. During the experiment it was controlled in the period between measurements by inspection of the diffraction pattern made by means of a He-Ne laser. All experiments were done at a sarcomere length of about 2.3 μm.

RESULTS

In Fig. 1 some typical responses to ramp-shaped shortenings and lengthenings are shown. If a quick ramp-shaped shortening is applied to the frog sartorius muscle during an isometric tetanus, the tension response shows a decrease during the shortening followed by the fast recovery, plateau phase and slow recovery. According to the nomenclature applied by Ford, Huxley & Simmons (1977), the minimum in tension is called T_1, while T_2 is found as the intersection of the extrapolated plateau phase, in which the rate of change of the tension is low or almost zero, with the initial fall in tension. In Fig. 2 T_1 and T_2 are plotted as a function of the displacement. The T_1 values depend on the rate of displacement. T_2 does not depend on this velocity in the range under consideration. In Fig. 2 for completeness some T_1 points from lengthenings are given, where they represent the first maximum in tension.

The general shape of the curves is as can be expected from the work of Julian & Sollins (1975) and of Ford, Huxley & Simmons (1977). The T_2 curve is convex, so that its slope is increased at larger shortenings. The T_1 curves start steeply at small shortenings and their slope decreases with increasing shortening with the result that the T_1 and T_2 curves approach each other. At greater shortenings (roughly above 6 nm per half-sarcomere in our measurements) the slope of the T_1 curve does not decrease further or even increases again, depending on the velocity of shortening.

In the range of shortenings and shortening velocities under consideration the responses are non-linear, as noticed by Huxley & Simmons (1971) for fiber preparations. This is also shown in Fig. 3. Three responses to shortenings of 4.4, 8.8 and 13.2 nm per half-sarcomere are superimposed after scaling in pro-

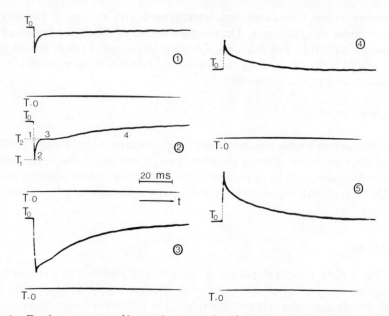

FIG. 1. Tension responses of isometrically tetanized frog sartorius muscle at rest length to ramp-shaped changes in length. Left column: responses to shortenings of 4.3 (1), 8.5 (2) and 17 nm/half-sarcomere (3); right column: responses to lengthenings of 4.3 (4) and 8.5 nm/half-sarcomere (5). Displacement rate per half-sarcomere: 8.5 nm/msec. The tension response to a shortening can be divided into a decrease in tension during the ramp (1) followed by the fast recovery (2), the plateau phase (3) and the slow recovery (4).
Experimental conditions: muscle rest length $(L_0) = 24.5$ mm, sarcomere length $(s_0) = 2.2$ μm, isometric tetanic tension $(T_0) = 2.2 \cdot 10^5 \text{N/m}^2$, temperature $= 2°C$.

portion to the respective shortenings. The duration of the shortenings was 1 msec.

During the first few tenths of a millisecond the responses coincide within the accuracy of the measurements. Then they diverge so that the minima in the responses to the smaller shortenings are lower. After the shortening the responses intersect in approaching the plateau phase and are reversed in order with respect to their magnitude. Furthermore, the responses at the onset in the range wherein they coincide show a rapid change in tension. After this rapid change the rate of decrease in tension is less. In the case of the 100 and 200 μm shortening the response follows approximately a straight line during the phase of constant shortening velocity. The responses suggest too the presence of a rapid change in tension corresponding to the deceleration phase in the displacement although certainly less clearly than at the onset.

If we now turn our attention to the time course of the slow recovery, its ap-

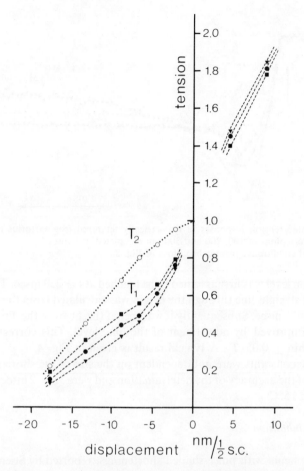

FIG. 2. T_1 and T_2 curves at different displacement rates of 4.4 (■), 8.8 (●) and 13.2 (▼) nm/msec (per half-sarcomere). The T_1 values (filled symbols) denote the extreme in tension reached during the ramp, normalized to T_0. The T_2 values (circles) denote the plateau in the responses relative to T_0. The T_1 values clearly depend on the displacement rate, in contradistinction to the T_2 values which superimpose at the velocity of the ramps used. At small displacements it is not possible to apply the desired displacement rates, in view of the duration of the acceleration and deceleration phase in the displacement.
Experimental conditions: $L_0 = 27$ mm, $s_0 = 2.4$ μm, $T_0 = 2.1 \cdot 10^5$ N/m², temperature = 2°C.

pearance suggests an exponential course. This was investigated by making a semilogarithmic plot (Fig. 4A). The plot was started ($t = 0$) when the plateau phase was considered to be over at about 25 msec after the shortening. The ulti-

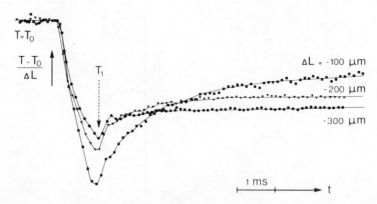

FIG. 3. Scaled tension responses of isometrically tetanized frog sartorius muscle superimposed. Shortenings of 100, 200 and 300 μm completed in 1 msec. Experimental conditions: identical to those of Fig. 2.

mate tension level was first assumed to be reached at $t = 250$ msec. The fit of the plot with a straight line through the origin was calculated from the points during the first 75 msec. Subsequently it was seen if the fit over the first 150 msec could be improved by adjustment of the final level. This correction always stayed within $\pm 0.05 \cdot T_0$. A typical result is given in Fig. 4.

The time constants were not dependent on the size of the shortening within the limits of the accuracy of the determination and were 37 ± 2 msec at 0°C and 8.5 msec at 15°C.

Metabolic inhibition

Measurements with pulse-shaped shortenings, reported by Stienen, Blangé & Schnerr (in press), in which the muscle after a shortening was set back to its original length showed a loss of force, which suggested detachment of cross-bridges during the fast recovery or plateau phase. Since it is generally assumed that ATP is needed for detachment of cross-bridges, it was tried by lowering the ATP concentration in the muscle if the time course of the fast recovery or the plateau phase could be affected. This experiment was performed by incubating the muscle with FDNB and IAA or with IAA and FAA. In the first case only twitches were used and the shortenings were applied at the apex of the twitch. In the second case the response early in the plateau of the tetanus was studied, but sets of twitches without shortenings were sometimes used after incubation to bring the muscle to the desired condition in a shorter time.

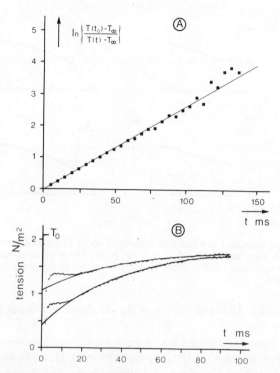

FIG. 4. A. Semilogarithmic plot of the slow recovery after a shortening of 17.6 nm/half-sarcomere. T_∞ denotes the ultimate tension level approached. The starting point (t_0) of the slow recovery at 25 msec after the displacement is taken as the origin of the horizontal axis. B. Comparison of the tension responses to shortenings of 8.8 and 13.2 nm/half-sarcomere and the calculated exponential fit of the slow recovery. Time constant: 38 msec. Experimental conditions: identical to those of Fig. 2.

The response to a certain shortening was determined before incubation. Directly after incubation a control measurement was done. During the subsequent contractions the tetanic or twitch tension decreased. When the isometric tension was reduced to about 60 or 70% it was found that at shortenings of 0.01 L_0 or more the bump caused by the fast recovery and the plateau phase had been decreased clearly or even abolished, a situation which was gradually reached during the repeated stimulations. A few contractions later rigor tension started to develop and the experiments were then terminated. A typical result in the stage before rigor tension was observed is shown in Fig. 5. For easy comparison a record taken directly after incubation and a typical one after exhaustion are superimposed after scaling with respect to the force just

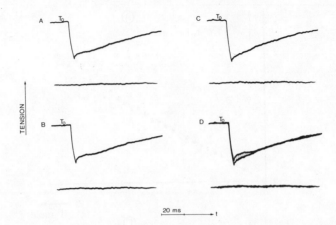

FIG. 5. Tension responses to a quick shortening of about 12 nm/half-sarcomere (A) before incubation with IAA and FAA, $T_0 = 2.2 \cdot 10^5$ N/m²; (B) after incubation, $T_0 = 2.2 \cdot 10^5$ N/m²; (C) after subsequent stimulation, $T_0 = 1.3 \cdot 10^5$ N/m². In (D) the scaled traces B and C are superimposed.

before shortening. The time course of the slow recovery is not affected at this stage.

When the same experiment was done with a small shortening of less than 0.005 of the resting length no change was observed. It can be noted that in this case the plateau phase is difficult to distinguish, but this does not apply to the fast recovery.

In view of the suggestion of detachment of cross-bridges during the early phases of the recovery from a shortening, it is interesting to measure the stiffness during the recovery in the unpoisoned muscle. In Fig. 6A the minimum in tension T_1 is shown as a function of the displacement ΔL. In this case the displacements are all completed in 1 msec. As already has been noted, $T_1 (\Delta L)$ is non-linear in the case of shortenings; it is practically proportional to the size of the lengthening until 200 μm. During the tension recovery after a shortening a second test length change (ΔL_t) can be applied. In Fig. 6B, the difference between the tension at the onset and the extreme at the end of the second length change, $T_1(\Delta L_t)$ is shown as a function of ΔL_t. $T_1(\Delta L_t)$ also appears to be roughly proportional to the size of the lengthening until 200 μm. This feature always appeared in the $T_1 (\Delta L_t)$ curves, except when the test length change was applied during the fast recovery phase. Owing to this proportionality the value of $T_1(\Delta L_t)$ over ΔL_t is taken as an estimate of the complex stiffness. In Fig. 6C this stiffness, measured with lengthenings of 200 μm relative to the stiffness determined from a similar length change at the isometric tension

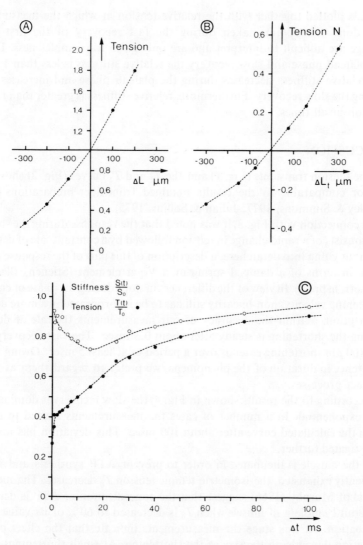

FIG. 6. Stiffness after shortening determined by a test displacement (ΔL_t) after an initial displacement (ΔL_i).
(A) Length-tension plot at isometric tetanic tension. The displacements are completed in 1 msec. The points in the plot are the corresponding T_1 values selected from Fig. 2.
(B) Length-tension plot at 15 msec after an initial shortening of 12.3 nm/half-sarcomere. The test displacements are completed in 1 msec.
(C) Time course of the stiffness normalized to the isometric stiffness (○) compared to the relative tension (●). $\Delta L_i = 17.2$ nm/half-sarcomere, $\Delta L_t = 8.6$ nm/half-sarcomere. Experimental conditions B and C: $T_0 = 2.0 \cdot 10^5$ N/m², $s_0 = 2.5$ μm, $L_0 = 30$ mm, temperature = 2°C.

level, is plotted together with the relative tension at which the measurement was done. The points taken during the fast recovery of the first length change are difficult to interpret and are only given for completeness. During the plateau phase and slow recovery the relative stiffness is less than 1, while the relative stiffness decreases during the plateau phase and increases again during the slow recovery. Furthermore, relative stiffness is greater than relative tension at all times.

DISCUSSION

The tension transients (Fig. 1) and the T_1 and T_2 curves (Fig. 2) show a behavior comparable to the results obtained from fiber preparations (Ford, Huxley & Simmons, 1977; Julian & Sollins, 1975).

In connection with Fig. 3, it was noted that the response during the shortening consists of a rapid change in tension followed by a constant rate of decrease. Without going into detail here, a description of this part of the response can be given in terms of a damped spring or a Voigt element (Stienen, Blangé & Schnerr, in press). In view of the differences in slope during the phase of constant shortening velocity, non-linearity still has to be incorporated in a more detailed description. Within the accuracy of our measurements the rate of decrease during the shortening is steady after about 0.3 msec. The fast recovery after the 100 μm shortening extends over a period of at least 5 msec. Owing to this difference in duration of the phenomena, we prefer to regard them as due to distinct processes.

According to the results shown in Fig. 4, the slow recovery is dominated by one exponential. In a number of cases the measurements started to deviate from the calculated curve after about 100 msec. This deviation has not been investigated further.

If the muscle is incubated in order to prevent ATP synthesis and is subsequently exhausted, the isometric tetanic tension T_0 decreases. The measurements of Murphy (1966) suggest that the concentration of ATP is decreased to about 1 μmole/g of muscle when T_0 is decreased to 60% of its value before incubation. At this stage the measurements indicate that the effect of ATP depletion depends on the size of the shortening. At small shortenings (about $0.005 \cdot L_0$) no changes were observed, at shortenings of about $0.01 \cdot L_0$ the plateau phase and fast recovery were abolished.

Kawai (1978) derived, from experiments on chemically skinned rabbit muscle fibers with small sinusoidal length changes, that at low ATP concentrations two components, which in his view after a quick length change give rise to a fast recovery and plateau phase, are slowed down. This could be in agreement with our findings that the change in the ATP concentration affects the first part of the recovery, *i.e.*, the fast recovery and the plateau phase. Differences in the preparations used as well as differences in the experimental set-up, however, prevent a detailed comparison.

The experiments with a second (lengthening) displacement yield evidence that in terms of cross-bridges detachment takes place during the plateau phase. It can be expected that detachment of cross-bridges causes a decrease in stiffness. From Fig. 6C it can be seen that, during the slow recovery the relative force is smaller than the relative stiffness. This suggests that, in approaching the final isometric cross-bridge strain distribution, in majority cross-bridges with high strain have to be replenished.

These results and the measurements at lowered ATP concentrations can be interpreted in terms of mechanically distinct states of the cross-bridges. It can then be assumed that as a result of the action of ATP the cross-bridge first passes to another state, in which it exerts more force. In the tension response of the unpoisoned muscle this would contribute to the fast recovery. The cross-bridge can be supposed to detach subsequently. This process would correspond with the plateau phase. The fact that we did not observe changes in the exhausted muscle in the case of small shortenings, suggests that in the fast recovery other processes too are involved.

Acknowledgements

We are very much in debt to Prof. J.Th.F. Boeles and Prof. L.H. van der Tweel for encouragement and valuable discussions, and to Miss M. Zethof for general assistance. This study has been aided in part by a grant from the Netherlands Organization for the Advancement of Pure Research (ZWO).

REFERENCES

Abbott, R. H. (1977). In *Insect Flight Muscle*, ed. Tregear, R. T., pp. 269–273, Amsterdam, London, New York: North-Holland.

Blangé, T., Karemaker, J. M. & Kramer, A. E. J. L. (1972). Elasticity as an expression of cross-bridge activity in rat muscle. *Pflügers Arch. ges. Physiol.* 336, 277–288.

Ford, L. E., Huxley, A. F. & Simmons, R. M. (1977). Tension responses to sudden length change in stimulated frog muscle fibers near slack length. *J. Physiol.* 269, 441–515.

Huxley, A. F. (1974). Muscular contraction, review lecture. *J. Physiol.* 243, 1–43.

Julian, F. J., Sollins, K. R. & Sollins, M. R. (1974). A model for the transient and steady-state mechanical behaviour of contracting muscle. *Biophys. J.* 14, 546–562.

Julian, F. J. & Sollins, M. R. (1975). Variation of muscle stiffness with force at increasing speeds of shortening. *J. gen. Physiol.* 66, 287–302.

Kawai, M. (1978). Head rotation or dissociation? *Biophys. J.* 22, 97–103.

Lymn, R. W. & Taylor, E. W. (1971). Mechanism of adenosine triphosphate hydrolysis by actomyosin. *Biochemistry, N. Y.* **10**, 4617–4624.

Murphy R. A. (1966). Correlations of ATP content with mechanical properties of metabolically inhibited muscle. *Am. J. Physiol.* **211**, 1082–1088.

Stienen, G. J. M., Blangé, T. & Schnerr, M. (1979). Tension responses of frog sartorius muscle to quick ramp-shaped shortenings and some effects of metabolic inhibition. *Pflügers Arch. ges. Physiol.* **376**, 97–104.

Discussion

Pollack: In Fig. 2, you showed that the T_1 and T_2 curves intersected the abscissa at somewhere between 20 to 25 nm per half sarcomere. This would seem, I think, to be 3 or 4 times the value given by Ford, Huxley, & Simmons (1977). Can you account for these differences?

Blangé: The T_2 curve is right. The T_1 curve was taken with fixed velocity. So, that is different from the work of Ford, Huxley & Simmons (1977). They only simulated responses for this type from their measurements and those do not intersect the axis at a low value.

Pollack: Weren't your releases done at approximately the same speed, or even faster than those of Huxley and Simmons?

Blangé: They were done with a velocity of 1 % of muscle length per msec. So, it is perhaps not too far off, but it is less.

Winegrad: If I interpreted your data correctly on the effects of low ATP, what they indicate is that ATP is necessary for the quick recovery phase, phase 2 in the Ford, Huxley, Simmons model. That is the phase associated with the change of position of the cross-bridge. That would suggest, then, I think, that more than one ATP is split during a cycle of a cross-bridge. Is there any other way that you can explain it?

Blangé: Well, I forgot to stress one important point—that we could get this kind of change in the response at moderate and large shortenings. But if we make a rather small shortening of 100 μm or so, then, as I indicated already, the plateau phase was not well developed, and the fast recovery was not clearly affected by metabolic inhibition. I did not mean that I was going to explain all of the fast recovery in this way. There is an ATP effect involved in the first recovery, but on the other hand, it does not show up clearly at small shortenings, which means that the first recovery, in the case of a small shortening, still contains the two phenomena as I indicated in Fig. 3 (see *Discussion* with respect to Fig. 3).

Winegrad: But isn't the short release more meaningful for determining the properties of the cross-bridge than the larger releases that you are referring to?

Blangé: I would not know why.

Winegrad: Because in the former case, you presumably are not getting detachment, while in the latter case, you probably are.

Blangé: Well, if you are referring to the behavior of the cross-bridge itself, I think that detachment is part of the behavior of the cross-bridge. So, you have to take account of that, too; but if you want to talk only about different states of the cross-bridge, then it is more relevant to consider small shortenings, and I cannot say much about that.

Rüegg: May I make a comment which may give some answer to your question? We did experiments with skinned fibers in the presence and absence of ATP (Yamamoto & Herzig, 1978). When you wash out the ATP you get a rigor-like state rather similar to the one you described after poisoning the muscle, and, interestingly, after a quick release in rigor, the quick-phase is almost completely missing. If you interpret this in terms of the Huxley-Simmons model, you would say there is no rotation of cross-bridges under these conditions.

Blangé: Do you have a clear plateau phase in that case?

Rüegg: No. The tension just stays down. In contrast to your experiment, there was very little tension recovery. We take this to mean that the ADP bound to myosin is necessary to allow the transiton from the two attached cross-bridge states, rather than thinking in terms of attachment-detachment with the requirement for ATP. I think, in summary, the quick-phase requires ATP, not because of attachment-detachment processes, but because it requires the ADP to be bound to the actin quarternary complex.

Winegrad: Have you done the experiment where you replace ATP with ADP, and then do you see a difference?

Rüegg: Yes, we did. The results are ambiguous. I cannot talk about it yet, but there is not too much change.

Tregear: That is precisely the experiment that Dr. Blangé has done, because he effectively replaced ATP with ADP.

Rüegg: Yes, what you showed is not rigor but something happening just before rigor.

Blangé: Yes, just before rigor.

Rüegg: It could be the ATP depletion, but it could also be the accumulation of ADP. After seeing your results, I think it would be very interesting to see what ADP does to the skinned fiber.

Blangé: I agree.

Kushmerick: I think you would like to think that it is due to either a decrease in ATP or an increase in ADP. But there are of course other products, *i.e.*, inorganic phosphate, creatine and possibly protons. So, one does not know exactly which product to consider. It seems that there are important sulfhydryls on myosin with which iodoacetate might have reacted, so conceivably that is also influencing the results.

Blangé: There might be all those possibilities, yes, that is certainly true. On the other hand, we have tried other ways to accomplish the same thing. I did not mention that we also did incubation with FDNB and IAA. Then we had to use twitches, but nevertheless the result was about the same.

Sugi: Dr. Blangé, several years ago you once challenged Huxley and Simmons by publishing your paper in *Nature* (Blangé, Karemaker & Kramer, 1972), stating that the Huxley-Simmons phenomenon might be due to the conbined effect of mass and damping, and I think your concept was very interesting. Why have you changed your mind? Why are you interpreting the data in a Huxley-Simmons manner?

Blangé: It is not so much that I changed my mind. I think things like that are going on, and that is why I made that clear distinction between small shortenings and larger shortenings. I think, again, at small shortenings we have already two phenomena going on, and at large shortenings we can influence the fast recovery by metabolic inhibition. But at small shortenings certainly there is that fast phenomenon of a few tenths of a msec, followed by a slightly slower recovery which takes a few msec, and I think this kind of mass and damping are certainly involved there.

REFERENCES

Blangé, T., Karemaker, J. M. & Kramer, A. E. J. L. (1972) Tension transients after quick release in rat and frog skeletal muscles. *Nature, Lond.* **237**, 281–282.

Ford, L. E., Huxley, A. F. & Simmons, R. M. (1977) Tension responses to sudden length changes in stimulated frog muscle fibres near slack length. *J. Physiol.* **269**, 441–515.

Yamamoto, T. & Herzig, J. W. (1978) Series elastic properties of skinned muscle fibres in contraction and rigor. *Pflügers Arch. ges. Physiol.* **373**, 21–24.

Isotonic Velocity Transients and Enhancement of Mechanical Performance in Frog Skeletal Muscle Fibers after Quick Increases in Load

Teizo TSUCHIYA,* Haruo SUGI* and Kaoru KOMETANI**

* Department of Physiology, School of Medicine, Teikyo University, Tokyo, Japan
** Faculty of Pharmaceutical Sciences, University of Tokyo, Tokyo, Japan

ABSTRACT

When the load on tetanized frog muscle fibers was quickly increased from P_0 to 1.05–1.3 P_0, they exhibited marked oscillatory length changes. The period of oscillation increased with increasing magnitude of step changes in load, and decreased with increasing temperature, as has been the case in the isotonic shortening velocity transients. If the load was increased from P_0 to more than 1.4 P_0, the fibers started to lengthen without oscillatory changes.

The stiffness of the fibers during isotonic lengthening was examined by applying small load steps and measuring the resulting length changes. The stiffness increased with increasing isotonic load from P_0 to about 1.3 P_0, and then started to decrease with further increase in isotonic load, decreasing sharply when the load exceeded 1.6 P_0. This suggests a decrease in the number of attached cross-bridges when the fibers are lengthened rapidly under a load above 1.6 P_0.

By applying quick decreases in load during the oscillatory length changes or the isotonic lengthening following quick increases in load, it was found that the fibers could shorten against a load equal to or slightly larger than P_0, indicating an increased load-bearing capacity after quick increases in load. Unexpectedly, the increased load-bearing capacity was observed even in the fibers during rapid isotonic lengthening under a load of more than 1.6 P_0 with a markedly reduced stiffness.

Although Hill (1938) assumed that the shortening velocity of the contractile component in muscle is uniquely and instantaneously determined by a given load, Podolsky (1960) showed that, when the load on muscle was suddenly reduced from the full isometric force to a lower value, the resulting steady isotonic shortening was preceded by non-steady motions which were attributed to the response of the contractile component. Civan & Podolsky (1966) examined in detail the isotonic velocity transients after quick decreases in load, and presented evidence that the transients reflect the kinetics of turnover of the cross-bridges. Similar isotonic velocity transients have also been observed by Armstrong, Huxley & Julian (1966). In spite of these contributions, however, no systematic work has hitherto been performed on the isotonic velocity transients after quick increases in load. The present experiments deal with some features of the isotonic velocity transients in isotonically lengthening frog muscle fibers in order to give information about the kinetic properties of the cross-bridges when they were displaced in a direction opposite to shortening. The opportunity was also taken to examine the stiffness of the muscle fibers during isotonic lengthening under various loads.

METHODS

All experiments were made with single muscle fibers or small bundles consisting of two to four muscle fibers isolated from the semitendinosus muscles of the frog (*Rana japonica*). A pair of stainless steel wire connectors (0.1 mm in diameter and 2–3 mm in length, Civan & Podolsky, 1966; Sugi, 1972) were tied to both tendons with braided silk thread, the length of tendinous material between the connectors being less than 0.5 mm. The preparation was mounted horizontally at its slack length (0.8–1.0 cm excluding tendons) by hooking the connectors to the tension and displacement transducers. Pre-cooled Ringer solution (2–3° C) was continuously circulated through the chamber, unless otherwise stated.

Tension changes were recorded with a capacitance-gauge type transducer (natural frequency of oscillation, about 3 KHz). Length changes were recorded with a light beam-photodiode system (Civan & Podolsky, 1966; Sugi, 1972). Its moving element was a magnesium lever pivoted on the bearings (equivalent mass, about 3 mg). The preparation was hooked to the long arm of the lever at a point 1.8 cm from the pivot, while the short arm was connected to the loading spring at a point 1 mm from the pivot (Civan & Podolsky, 1966).

The preparation was first kept isometric by fixing the lever in position, and was tetanized maximally with 2 msec rectangular current pulses at 15–30 Hz given through a multi-electrode assembly. When the full isometric tension P_0 was developed, the load on the preparation was quickly changed from P_0 to a new value by withdrawing the stops restricting the lever movement, the amount of isotonic load being varied by changing the length of the loading spring. When the load on the preparation was varied in two steps, the length of the loading spring was quickly changed with an electromagnetic device. The inertial oscillations of the lever following quick changes in load were damped to some extent by use of a dashpot device filled with silicon oil. Thus, the damped oscillations following the quick changes in load almost disappeared within 3–6 msec, so that the isotonic velocity transients could be studied in detail.

All experiments were performed within the range of fiber lengths where the resting tension was negligible (0.9–1.2 cm, sarcomere length 1.9–2.7 μm), to avoid complications arising from the development of resting tension. The stray compliance of the whole recording system including ties was estimated by substituting a length of stainless steel wire for the fiber, and was found to be about 8.5 μm/g. Since the maximum isometric tension of the preparation ranged from 50 to 300 mg, the stray compliance was not more than 0.1 % of the fiber length even under a load of 2 P_0. No corrections for this compliance were made in the measurement of the instantaneous elasticity.

RESULTS AND DISCUSSION

Isotonic velocity transients following step increases in load

Figure 1 shows the records of early length changes of the preparation when the load was quickly increased from P_0 to various values. Following the step increases in load from P_0 to 1.05–1.30 P_0, the preparation exhibited marked oscillatory length changes (Armstrong *et al.*, 1966). Immediately after the initial rapid lengthening coincident with the load step, the preparation showed a transient isotonic lengthening followed by a transient isotonic shortening until it started to lengthen with a nearly constant velocity. After the initial reversal in the direction of isotonic movement, there was a phase of distinct retardation in the lengthening velocity, so that the whole time course of length changes gave an appearance of heavily damped oscillation. The period of the oscillatory length changes increased with increasing magnitude of step increases in load. If the load was increased from P_0 to more than 1.4 P_0, the oscillatory length changes were no longer observable; following large step increases in load, the preparation started to lengthen, with velocities slowly decreasing with time, and there was no distinct phase of isotonic lengthening with a constant velocity.

Meanwhile, the time course of length changes following step decreases in load was very similar to that observed by Civan & Podolsky (1966), Armstrong *et al.* (1966), and Huxley & Simmons (1973), as shown in Fig. 2. The early length changes were also oscillatory, though the reversal in the direction of movement was only barely perceptible following extremely small load steps. In Fig. 3, the half period of oscillatory length changes following step changes in load is plotted against the magnitude of isotonic load attained after the load step. It can be seen that the half period increases with increasing isotonic load, irrespective

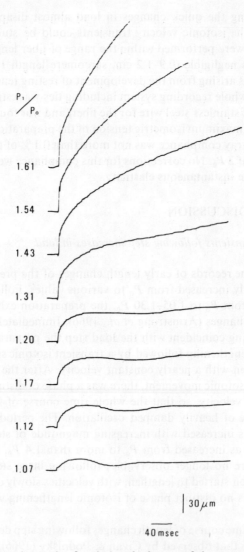

FIG. 1. Experimental records of early length changes of the preparation following step increases in load from P_0 to P_1. The values of P_1 relative to P_0 are shown on the left of each record.

of whether the load is less than P_0 (*i.e.*, after a step decrease in load) or more than P_0 (*i.e.*, after a step increase in load).

As has been the case in the isotonic velocity transients following step de-

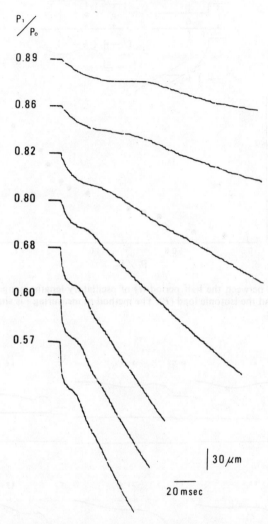

FIG. 2. Experimental records of early length changes of the preparation following step decreases in load from P_0 to P_1. The values of P_1 are also shown on the left of each record.

creases in load (Civan & Podolsky, 1966), the period of the oscillatory length changes following step increases in load decreased as the temperature was increased (Fig. 4), and its temperature coefficient appeared to be the same as that of the period of the isotonic shortening velocity transients (null time, Civan &

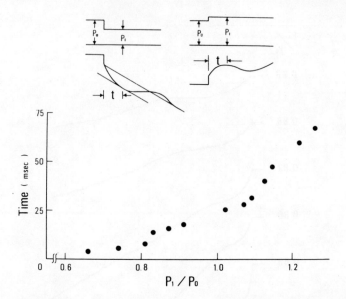

FIG. 3. Relation between the half period (t) of oscillatory length changes following step changes in load and the isotonic load (P). The method of measuring t is shown in the upper part of the figure.

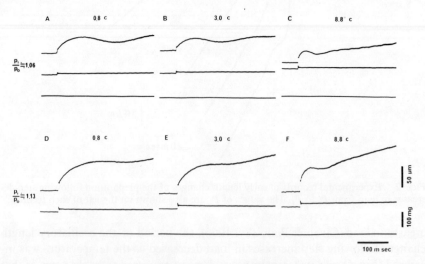

FIG. 4. Effect of temperature on the oscillatory length changes following step changes in load from P_0 to about 1.06 P_0 (A-C) and from P_0 to about 1.13 P_0 (D-F).

Podolsky, 1966). This may be taken to imply that both types of velocity transients originate from the kinetic properties of the cross-bridges following step changes in load.

Computer simulation of the oscillatory length changes following step increases in load

In order to throw some light on the underlying mechanism of the marked oscillatory length changes following step increases in load, attempts have been made to simulate the above isotonic velocity transients on a digital computer using the two-state contraction model (*e.g.*, Huxley, 1957; Podolsky & Nolan, 1973), in which appropriate values were chosen for the rate constants for making and breaking the cross-bridges (f and g, respectively) depending on the distance (x) between the site on the actin filament and the equilibrium position at which an attached cross-bridge exerts zero force. The digital computer (Hitachi) was operated by one of us (K.K.).

After a number of trials and errors, we reached a model (Fig. 5), in which f has positive values only over a range of the first 120 Å on the right of the ori-

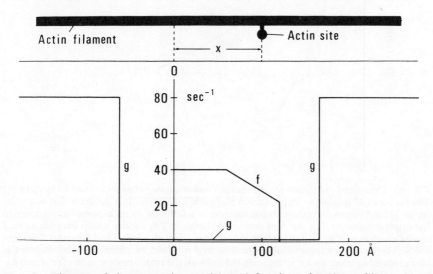

FIG. 5. Diagram of the contraction model used for simulating the oscillatory length changes following step changes in load. In the upper part, x denotes the distance between the site on the actin filament and the equilibrium position at which an attached cross-bridge exerts zero force. Dependence of f and g on x is illustrated in the lower part. For further explanation see text.

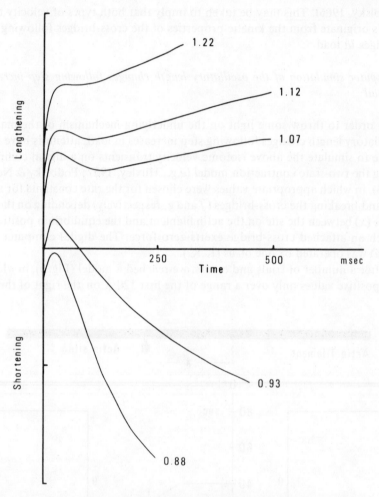

FIG. 6. Computed oscillatory length changes following step changes in load from P_0 to P_1. The values of P_1 relative to P_0 are shown alongside each curve. It is assumed that a step increase in load causes an instantaneous increase in g followed by an exponential recovery to the original value, so that the time course of change in g following a step increase in load can be given as $g = g_0 + Ae^{-bt}$, where A and b are taken as 10 sec^{-1} and 50 msec^{-1}, respectively. Meanwhile, an instantaneous increase in g followed by an exponential recovery is also assumed to occur by a step decrease in load with the same absolute values for A and b.

gin, while g has two different values everywhere: g has a small value in the vicinity of the origin at both sides, including the region where f has finite values, whereas it has a large value beyond -63 Å on the left and beyond 165 Å on the right of the origin. The shape of the g function on the left of the origin is analogous to that in the Podolsky-Nolan model, and has been shown to be effective in simulating the isotonic velocity transients following step decreases in length (Podolsky & Nolan, 1973). In the case of the isotonic velocity transients following step increases in load, however, the g function on the right of the origin was not sufficient for simulating the marked oscillatory length changes in which the direction of movement was reversed by turns (see Figs. 1 and 4). For this purpose, two additional assumptions were made: (1) that the value of g increased instantaneously to a higher value at the moment when a step increase in load was applied, and then (2) that the value of g returns to its initial value exponentially with time.

As shown in Fig. 6, it was possible to simulate the oscillatory length changes following step increases in load to some extent by the model described above. For step decreases in load, on the other hand, it is at present tentatively assumed that the value of g also increased to a higher value at the moment of application of the load step, and that the magnitude of the quick change in the g function was the same in both step increases and decreases in load. The responses of the model are apparently not satisfactory in the case of isotonic shortening transients, and further trials are being made to obtain a better fit to the actual responses.

Relation between the stiffness of the series elasticity and the isotonic load

The stiffness of the series elasticity of the muscle fibers was measured under various isotonic loads from the relative magnitude of tension and length changes, assuming that the initial rapid length changes coincident with the load steps are the responses of the series elasticity (Jewell & Wilkie, 1958). At 50–100 msec after the beginning of isotonic lengthening or isotonic shortening, small lengthening or shortening steps (less than 0.1 P_0) were applied for the measurement of the stiffness.

A typical result is illustrated in Fig. 7A. In agreement with Julian & Sollins (1975), the stiffness of the fibers during the isotonic shortening decreased with decreasing isotonic load, approaching a finite value as the isotonic load tended to zero. The stiffness of the fibers during the isotonic lengthening, on the other hand, increased with increasing load above P_0, reaching a maximum at 1.2–1.3 P_0. With further increase in the isotonic load above 1.5 P_0, the stiffness began to decrease sharply, and this appeared to be related to the tendency of the fibers to lengthen rapidly when the load was increased from about 1.5 P_0 towards 1.8–2.0 P_0 ("give," Katz, 1939), as shown in Fig. 7B.

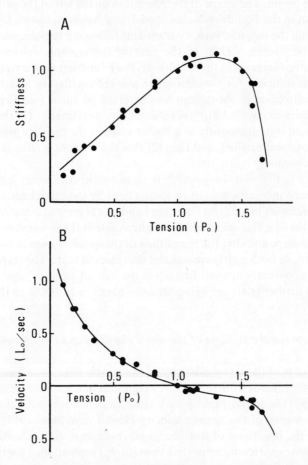

FIG. 7. Dependence of the stiffness of the series elasticity on the isotonic load (A) and the force-velocity relation (B) obtained on one and the same preparation. In B, the velocity was measured at the steady phase of isotonic movements. Note that the steep decrease in the stiffness with loads above 1.5 P_0 in A is related to the tendency of the fibers to lengthen rapidly in B.

Enhancement of the mechanical performance after step increases in load

The change in the mechanical performance of the muscle fibers after step increases in load was examined by applying step decreases in load after a period (50–100 msec) of isotonic lengthening to allow the fibers to shorten isotonically under various loads. Figure 8 shows an example of the experiments in which the load on the lengthening fibers was quickly decreased from about 1.4 P_0 to various values. It can be seen that the fibers show no marked lengthening under a load of about 1.1 P_0 (Fig. 8A), and can shorten for a considerable distance against a load as large as P_0 (Fig. 8B).

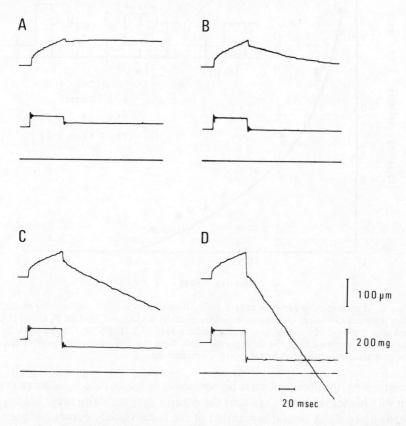

FIG. 8. Records of length changes in the fibers when the load was first quickly increased from P_0 to about 1.4 P_0 and then quickly reduced to various values. Note that the fibers can shorten against a load of P_0 (B) after a period of isotonic lengthening, indicating an increase in the load-bearing capacity.

Figure 9 shows force-velocity curves obtained by the application of step decreases in load on the isometrically contracting preparation (open circles), on the isotonically lengthening preparation under a load of 1.32–1.38 P_0 (filled circles) and also on the isotonically lengthening preparation under a high load of 1.61–1.67 P_0 (dotted circles). It will be seen that, after a period of isotonic

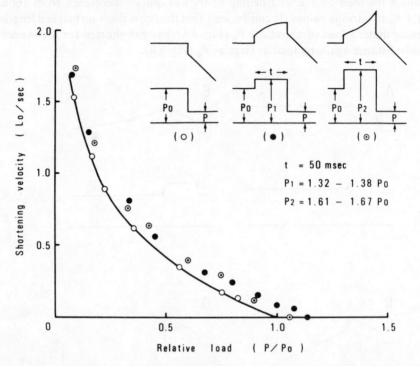

FIG. 9. Force-velocity curves obtained during the ordinary isometric tetanus (open circles), after a period (50 msec) of isotonic lengthening under a load of 1.31–1.38 P_0 (filled circles) and after a period of isotonic lengthening under a load of 1.61–1.67 P_0 (dotted circles). The curves were obtained from the same preparation. Note that the mechanical performance of the preparation is enhanced by the isotonic lengthening.

lengthening, the fibers can exert higher isometric forces than P_0, and can shorten with higher velocities than does the control over the entire range of isotonic loads up to P_0. A similar separation of the force-velocity curves has been reported to take place when the frog muscle fibers are made to shorten isotonically after stretch (Edman, Elzinga & Noble, 1978).

The stiffness of the series elasticity is generally taken as a measure of the number of attached cross-bridges in the preparation (*e.g.*, Huxley & Simmons, 1973; Podolsky & Nolan, 1973). Thus, the result that the stiffness in the isotonically lengthening fibers under a load of 1.3–1.4 P_0 is definitely larger than that in the isometrically contracting fibers (Fig. 7A) suggests the possibility that the above enhancement of the mechanical performance caused by the isotonic lengthening under a load of 1.3–1.4 P_0 (Fig. 8) results from the recruitment of active cross-bridges. This explanantion can not, however, be applied to the enhancement of the mechanical performance caused by the isotonic lengthening under a load of 1.6–1.7 P_0, since the fibers are lengthened rapidly under a load of more than 1.5 P_0 (Fig. 7B), being accompanied by a markedly reduced stiffness (Fig. 7A). In the sliding filament model of A.F. Huxley (1957), the reduced stiffness in both rapidly shortening and rapidly lengthening fibers (Fig. 7A) should result from the decrease in the number of attached cross-bridges when the thick and thin filaments are made to slide past each other rapidly. On this basis, the enhancement of the mechanical performance during isotonic lengthening under a load of more than 1.5 P_0 seems to be puzzling, throwing some doubt on the meaning of the stiffness. Much more experimental work is needed to clarify the nature of the muscle stifiness for the complete understanding of the mechanism of muscle contraction.

Acknowledgement

We wish to thank Miss S. Gomi for excellent technical assistance and Professor H. Shimizu for discussions.

REFERENCES

Armstrong, C. M., Huxley, A. F. & Julian, F. J. (1966). Oscillatory responses in frog skeletal muscle fibres. *J. Physiol.* **186**, 26–27P.
Civan, M. M. & Podolsky, R. J. (1966). Contraction kinetics of striated muscle fibres following quick changes in load. *J. Physiol.* **184**, 511–534.
Edman, K. A. P., Elzinga, G. & Noble, M. I. M. (1978). Enhancement of mechanical performance by stretch during tetanic contractions of vertebrate skeletal muscle fibres. *J. Physiol.* **281**, 139–155.
Hill, A. V. (1938). The heat of shortening and the dynamic constants of muscle. *Proc. R. Soc.* B **126**, 136–195.
Huxley, A. F. (1957). Muscle structure and theories of contraction. *Prog. Biophys. biophys. Chem.* **7**, 255–318.
Huxley, A. F. & Simmons, R. M. (1973). Mechanical transients and the origin of muscular force. *Cold Spring Harb. Symp. quant. Biol.* **37**, 669–680.
Jewell, B. R. & Wilkie, D. R. (1958). An analysis of the mechanical components in frog's striated muscle. *J. Physiol.* **143**, 515–540.
Julian, F. J. & Sollins, M. R. (1975). Variation of muscle stiffness with force at increasing speeds of shortening. *J. gen. Physiol.* **66**, 287–302.
Katz, B. (1939). The relation between force and speed in muscular contraction. *J. Physiol.* **96**, 45–64.
Podolsky, R. J. (1960). The kinetics of muscular contraction: the approach to the steady state. *Nature, Lond.* **188**, 666–668.
Podolsky, R. J. & Nolan, A. C. (1973). Muscle contraction transients, cross-bridge kinetics, and the Fenn effect. *Cold Spring Harb. Symp. quant. Biol.* **37**, 661–668.

Sugi, H. (1972). Tension changes during and after stretch in frog muscle fibres. *J. Physiol.* **225**, 237–253.

Discussion

Rüegg: I was very much interested in these findings in which you showed that after loading the muscle with a load of about 1.8 P_0 you got this rapid lengthening, and yet after removing the load there was a very rapid shortening. The question is, why can it shorten so quickly, despite the fact that, apparently based on the load-stiffness relation, there are few cross-bridges present? Perhaps the muscle lengthens so quickly not because cross-bridges detach, but because they slip, *i.e.*, they detach and reattach, so at the moment you unload, there again it can shorten. Would you agree?

Sugi: Yes. I think the number of attached cross-bridges may not decrease during a rapid lengthening. Then, the fundamental question arises, "What is the meaning of stiffness?"

Edman: I think your measurements of stiffness and the force-velocity relationship refer to different times. I think that when you make your measurements of force-velocity curves, you take these measurements after you have given the system time to recover after your initial stretch, so it is somewhat of a different situation, isn't it?

Sugi: Yes, it is complicated. It may be time-dependent.

Edman: What I mean is that you have given the bridges a chance to take another position than when you are stretching. Actually, Dr. Noble is going to show a rather similar result later, and it seems as if these long stretches recruit some parallel elastic element.

Sugi: Have you any definite idea about this?

Edman: No, not whether this element is changing.

Pollack: A tentative interpretation is that, as you stretch the fibers, the cross-bridges become detached and then perhaps reattach again so that the total number remains constant; wouldn't you expect that as the cross-bridge stretches to a point just before detachment, it is generating a maximum amount of force, and as soon as it detaches it should recoil, so that when it attaches once again, the amount of force that it generates should be less than at the time just before it detached?

Sugi: I think that some factors may cancel out one another to keep the force constant.

Brutsaert: May I add to this that in cardiac muscle the force does fall?

Huxley: But surely in this experiment you are applying a constant force.

Hill: In simulating the oscillatory length changes following a load step, you introduced a rate constant, with a term that had an exponential dependence on the time (see Figs. 10 and 11). I was wondering what the rationale was for that?

Sugi: Probably Dr. Kometani may have some excuse for this, because he devised this trick.

Kometani: When the load step is applied, the attached cross-bridge may be distorted instantaneously, and such distorted cross-bridges may be assumed to detach more rapidly than the normal ones in the isometric condition; then the detached and distorted cross-bridges start to reattach and thus the population of distorted cross-bridges decays exponentially.

Sugi: Is that all right?

Hill: There is an exponential decay, but is a rate constant that is also exponentially depending on time necessary? This is rather a fine point; perhaps we should pass it on.

Sugi: Anyway, I only mean that we could succeed in simulating the length oscillations by assuming a rate constant decaying exponentially with time.

Hill: It seemed rather empirical.

White: I would like to carry on with Dr. Hill's point. It seems to me that another way to do that would be to say that the rate constant for detachment would depend on the degree of distortion of the bridge, and to put it in terms of length changes and not in terms of time. I mean that is the way that Podolsky has treated it (Podolsky & Nolan, 1973), and that A. F. Huxley treated it in his '57 paper (Huxley, 1957).

Sugi (added after the discussion): In our simulation of the oscillatory length changes following step increases in load, the instantaneous increase in the rate constant for detachment of the cross-bridges (Fig. 11) may be taken to imply that a large proportion of cross-bridges detach after distortion at the same time, and then start to reattach more or less synchronously. This suggests that the length oscillations result from synchronized turnover of the cross-bridges induced by the load step.

Pollack: Isn't the phenomenon you showed just the opposite of the phenomenon that Dr. Edman has discussed, *i.e.*, the phenomenon of deactivation induced by shortening? Edman has shown that the ability to develop tension and to shorten is reduced after the muscle has shortened (Edman, 1975); in your case the muscle has lengthened, and you find that the ability to generate force and velocity has increased.

Sugi: I think it is somewhat puzzling, because Dr. Edman mostly focused his attention on the amount of tension redeveloped, but not on the shortening ability. The two phenomena may be related to each other, but I have at present no definite idea about this.

REFERENCES

Edman, K. A. P. (1975) Mechanical deactivation induced by active shortening in isolated muscle fibres of the frog. *J. Physiol.* **246**, 255–275.

Huxley, A. F. (1957) Muscle structure and theories of contraction. *Prog. Biophys. biophys. Chem.* **7**, 255–318.

Podolsky, R. J. & Nolan, A. C. (1973) Muscle contraction transients, cross-bridge kinetics, and the Fenn effect. *Cold Spring Harb. Symp. quant. Biol.* **37**, 661–668.

Load-Induced Length Transients in Mammalian Cardiac Muscle

Dirk L. BRUTSAERT and Philippe R. HOUSMANS

Department of Physiology, University of Antwerp, Antwerp, Belgium

ABSTRACT

Abrupt alterations in load (load clamps) during isotonic afterloaded twitch and tetanic contractions of mammalian cardiac muscle induced a viscoelastic lengthening of the muscle. The step change in load itself caused an initial rapid lengthening (phase 1), reflecting the extension of undamped series compliance. Subsequently, the clamped load was borne for a considerable time due to delayed isotonic lengthening. This isotonic yielding typically proceeded in a slow, viscous-like (phase 2) and a much faster (phase 3) lengthening, and was eventually followed by an abrupt fall of force during the subsequent isometric relaxation. The overall length change of phase 2 was limited to about 1% of muscle length at l_{max}, regardless of the time and magnitude of the load step. Accordingly, after the first two-thirds of isotonic shortening, small to moderate load clamps consistently elicited this three-step yielding as the first part of a prematurely load-induced relaxation. Moreover, clamped loads which were made to exceed isometric twitch and tetanic force could be sustained at a supra-isometric level at the expense of such three-phasic length transients, and were again followed by premature relaxation. A well-functioning sarcoplasmic reticulum appeared to be an essential requirement for this phenomenon to become apparent.

Our present working hypothesis is that, in the presence of a sufficiently lowered myoplasmic calcium level, an abrupt increase of load acutely disrupts a delicate balance that exists between force potential at a given time during shortening and the prevailing load, resulting in a premature relaxation. The isotonic three-phasic lengthening may then possibly reflect back rotation of attached cross-bridges mainly during phase 2, and eventual detachment predominantly during phase 3; the initial part of spontaneous isotonic relaxation may be governed by the same mechanism. This hypothesis was further supported by experiments in which isotonic relaxation could be postponed by appropriate unloading of the muscle at the onset of relaxation.

The study of length or force responses to abrupt or gradual alterations of load (Podolsky & Nolan, 1973; Sugi & Tsuchiya, 1975; Housmans & Brutsaert, 1976; Brutsaert & Housmans, 1977; Tsuchiya & Sugi, 1979) or length (Rack & Westbury, 1974; Julian & Sollins, 1975; Ford, Huxley & Simmons, 1977; Edman, Elzinga & Noble, 1978; Flitney & Hirst, 1978a, b; Amemiya, Sugi & Hashizume, 1979; Blangé & Stienen, 1979; Delay, Vassallo, Iwazumi & Pollack, 1979; Rüegg, Güth, Kuhn, Herzig, Griffiths & Yamamoto, 1979; Steiger, 1979) has led to various interpretations and models relating to mechanisms underlying contractile behavior of skeletal and cardiac muscle (Huxley & Simmons, 1973; Julian, Sollins & Sollins, 1973; Brady, 1974; Huxley, 1974; Julian & Moss, 1976; Noble & Pollack, 1977).

In isolated cardiac muscle preparations, analysis of force transients in response to length steps is complicated by the presence of high series compliance. The need to eliminate the contribution of high series compliance has led to experiments in which either sarcomere length was controlled (Pollack & Krueger, 1976), or load steps were applied and load-induced length responses studied at constant load in skeletal (Tsuchiya & Sugi, 1979) and cardiac muscle preparations (Housmans & Brutsaert, 1976; Brutsaert & Housmans, 1977).

We have studied changes in length that occur in response to step changes in load in mammalian ventricular cardiac muscle (cat papillary muscle), during contraction and relaxation of both twitch and tetanic contractions. The interpretation of our experimental data has led to a unifying view of the contractile mechanism of cardiac muscle in terms of cross-bridge interactions between actin and myosin myofilaments.

I. Analysis of responses in length to step increase in load (load clamps): three-step yielding

We have previously (Housmans & Brutsaert, 1976) shown, in twitch and tetanic contractions of cat papillary muscle, that an abrupt increase in load induced lengthening of the muscle, followed by a premature isometric relaxation. A typical example of this phenomenon during tetanus is illustrated in Fig. 1. This lengthening typically proceeded in three phases: an initial rapid (phase 1) extension accompanied the rise in load itself and has been ascribed to extension of undamped series compliance. At the clamped load, the subsequent delayed isotonic lengthening proceeded in a slow viscous-like phase at first (phase 2), and ihen more quickly (phase 3). This three-step yielding clearly differs from the oscillatory responses seen when either length or load is changed in skeletal muscle (Armstrong, Huxley & Julian, 1966) and in strontium-

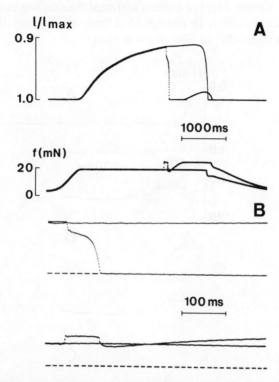

FIG. 1. Length responses to sudden increases in load in cat papillary muscle during tetanus. Panels A and B have the same length and force gain.

Panel A (slow sweep): An abrupt increase of load was imposed during steady shortening of an afterloaded tetanic contraction. Shortening was abruptly interrupted and rapid yielding of the muscle was provoked, followed by a transient fall in isometric force, despite the imposed clamped load being maintained at the clamped level throughout the remaining part of the tetanus. Subsequent recovery of force development towards the level of the higher clamped load, and of isotonic shortening at the clamped load was markedly delayed. Control afterloaded tetanus was superimposed. Muscles were tetanized by imposing trains of high intensity electrical stimuli in a bathing solution containing 10 mM calcium chloride and 10 mM caffeine (Henderson, Forman, Brutsaert, & Sonnenblick, 1971). Peak shortening of the control tetanus coincides with the time at which the train of electrical stimulation ended.

Panel B (fast sweep): Same load clamp as in panel A (sweep was triggered shortly before the load clamp). The load step again induced three-step yielding of the muscle. An initial fast extension (phase 1) accompanied the load step and was followed by a much slower viscous-like (phase 2) and a final faster (phase 3) lengthening. After an abrupt fall in force, a slow recovery of force (phase 4) proceeded towards the level of the clamped load. The train of electrical stimulation was continued throughout the entire sweep.

Muscle characteristics: length at l_{max}, 5.25 mm; mean cross-sectional area 0.62 mm^2; ratio of resting to total developed tension (R/T) 8.9%; temperature 19°C.

(Henderson & Cattell, 1976) or barium-activated mammalian cardiac muscle; and it could always clearly be distinguished from occasionally observed slow oscillatory responses due to alterations in load.

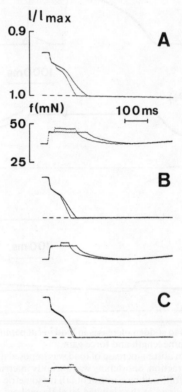

FIG. 2. Analysis of responses of length to sequential (loading) load clamps in cat papillary muscle during tetanus. Format is the same as in Fig. 1B.

Panel A: An abrupt increase in load during steady shortening of an afterloaded tetanic contraction initiated a three-step yielding response. A second loading "test" load clamp of small amplitude was applied during isotonic phase 2 lengthening. The step increase in load resulted in a small initial fast extension (phase 1′) at the time of the step itself, a phase 2-like slower lengthening (phase 2′), and a faster phase 3-like yielding (phase 3′). The summed length change of phase 2 (before the "test" load clamp) and of phase 2′ (after the "test" load clamp) did not exceed a value of about 1 % of muscle length at l_{max} and, for small amplitudes of "test" clamp, was equal to the overall length change of phase 2 of the control.

Panels B and C: An additional small load was imposed after the completion of phase 2 (Panel B) and during phase 3 (Panel C) of the isotonic three-step yielding induced by the first load clamp. Phase 2′ progressively shortened (Panels A and B) with later times of the additional clamp, followed by an eventual complete fusion of phase 1′ and 3′ (Panel C).

Muscle characteristics: length at l_{max} 9.0 mm; mean cross-sectional area 1.24 mm^2; R/T 9.3%; temperature 19°C.

Analysis of phases 2 and 3 of this three-step yielding at various times during shortening, for different magnitudes of load clamp (Fig. 6 B), and for various loading conditions before the clamp, revealed that the total extent of length change during phase 2 did not exceed a particular value, which for twitch contractions averaged $0.80 \pm 0.01\%$ (mean $\pm$ S.E.M.; $n = 14$) of l_{max} and for tetanic contractions $1.01 \pm 0.04\%$ (mean $\pm$ S.E.M.; $n = 17$). This average value of displacement during viscous phase 2 corresponds to the extent of motion ascribed to cross-bridges in skeletal muscle (Huxley & Simmons, 1973; Podolsky & Nolan, 1973; Barden & Mason, 1978; Flitney & Hirst, 1978a).

To test this phase 2 length change further, "test" clamps of very small amplitude in either direction (loading and unloading) were imposed about halfway through phase 2 of the control clamp (Figs. 2, 3, 4). The length responses to such load clamps were of a similar appearance to the three-step yielding: that is, after an initial elastic phase 1', a delayed lengthening with a slow phase 2' and a fast phase 3' were apparent. Moreover, the summed length changes during the interrupted phase 2 of the control clamped load and the subsequent

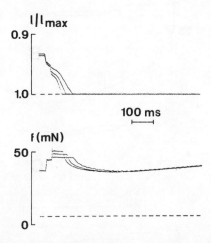

FIG. 3. Analysis of responses of length to successive loading clamps in cat papillary muscle during tetanus. Format is the same as in Fig. 1B. During steady shortening of an afterloaded tetanic contraction, two successive load clamps were imposed. The first load clamp was of the same amplitude for the three superimposed contractions. "Test" load clamps of three different amplitudes were imposed during phase 2 of the isotonic lengthening induced by the first load clamp. The "test" clamps caused an elastic phase 1' extension, a slow phase 2' and faster phase 3' lengthening. For the three superimposed contractions shown, the summed length changes of phase 2 (before the additional load clamps) and of phase 2' (induced by the additional load clamps) did not exceed 1% of muscle length at l_{max}.
Muscle characteristics: see Fig. 2; temperature 19°C.

phase 2' of the "test" load clamp again did not exceed the value given above, no matter at which instant during phase 2 a "test" clamp of small amplitude was superimposed. This was also the case when similar additional load clamps of various magnitudes were imposed during the original phase 2 (Fig. 3). For somewhat later "test" clamps, e.g., after the completion of phase 2 (Fig. 2B) or during phase 3 (Fig. 2 C) of the first clamp, no phase 2' was identified, leaving only fused phase 1' and 3'. Unloading test clamps of small amplitude (Fig. 4), or even of the same amplitude as the control load clamp, did not result in an immediate recovery of force or shortening, but, depending on the time of the

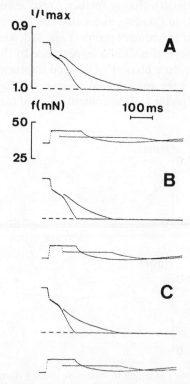

FIG. 4. Analysis of responses of length to sequential (unloading) load clamps in cat papillary muscle during tetanus. Same format as in Fig. 2.

Panel A: An additional (unloading) "test" clamp of a small amplitude was applied during isotonic phase 2 lengthening of a three-step yielding response which had been initiated shortly before by an abrupt increase in load during steady shortening of an afterloaded tetanus. The step decrease in load resulted in an initial rapid shortening (phase 1'), due to elastic recoil; thereafter, further lengthening proceeded with a slower phase 2' and a faster phase 3'.

Panels B and C: With later times of the "test" unloading clamp, phase 2' shortened (Panel B) or completely disappeared (Panel C), leaving only a phase 3' yielding followed by a slow recovery. Muscle characteristics: see Fig. 2; temperature 19°C.

"test" clamp during phase 2 or 3 of the first clamp, consistently showed delayed phase 2' and 3' lengthening (Fig. 4A).

Accordingly, once phase 2 lengthening had been initiated by a load clamp, phase 2 viscous lengthening always proceeded through its full extent, even when the muscle was unloaded again back to its original value (Housmans & Brutsaert, in preparation).

This phenomenon of three-step yielding has previously been interpreted in terms of the Huxley-Simmons model of cross-bridge interaction between actin and myosin filaments (Huxley & Simmons, 1973). It was then postulated (Housmans & Brutsaert, 1976) that phase 2 isotonic lengthening following a load clamp probably results from back rotation of cross-bridges over their full extent, with eventual detachment and sliding of filaments to their original position during phase 3. Early detachment of cross-bridges already during phase 1 and also during phase 2 could then explain the continued yielding and the eventual delayed recovery of shortening following reversal of the load clamp, leaving an insufficient number of attached cross-bridges to sustain the original load. Our results are in close agreement with those recently reported for skeletal muscle, where tension responses during ramp and hold stretches showed a discontinuity after an average sarcomere displacement of 1.2% of muscle length (Flitney & Hirst, 1978a). The conclusion is that sarcomeres must accommodate a particular extent of lengthening against a high resistance before "giving."

During cardiac muscle tetanus, the load-induced three-step yielding and in particular phase 3 lengthening, was followed by a slow recovery (phase 4) despite the continued state of tetanization. This differs from previous findings in tetanized skeletal muscle, where an increment of load was sustained throughout the stimulation period by a slow viscoelastic extension, not followed by any kind of fast phase 3 yielding (Katz, 1939). This particular behavior of mammalian cardiac muscle could result from an insufficiently fast reattachment of forcibly broken cross-bridges, despite the tetanized state. One possibility is that, even during tetanus, cat papillary muscle would still be submaximally activated. An alternative interpretation could reside in some "refractory period" of the cross-bridges, due to plastic configurational deformation of myosin heads after forcible detachment from the actin filaments. Similar delayed recovery of tension after stretch was also recently described in skeletal muscle (Flitney & Hirst, 1978b).

II. Analysis of load clamps at supra-isometric levels: supra-isometric force potential

During phases 2 and 3 delayed lengthening of three-step yielding, muscle could sustain an augmented clamped load for a considerable time (Figs. 5A & 6A). Even when the load clamp was augmented to levels beyond the isometric force level, both in twitch (Figs. 5B, C) and tetanic contractions (Fig. 6B), delayed isotonic lengthening allowed for sustainment of these supra-isometric loads. A limiting force level, or "maximal supra-isometric force potential" (Brutsaert & Housmans, 1977) was reached when delayed lengthening had completely fused within elastic phase 1. In line with the above hypo-

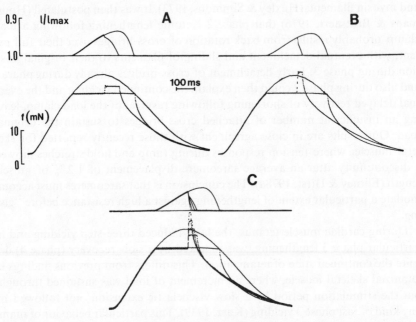

FIG. 5. Supra-isometric force during isotonic twitch contractions of cat papillary muscle. All panels have the same force and length gain. Load clamps of various magnitudes were applied during isotonic shortening of heavily afterloaded twitch contractions. Control afterloaded isotonic and isometric twitches were superimposed in each panel.

Panel A: Following a load clamp of a moderate amplitude, the augmented clamped load was sustained for some time at the expense of a slow extension of the muscle.

Panels B and C: Even when total load was clamped to beyond peak isometric force, these loads were held at supra-isometric levels for some time, at the expense of the concomitant delayed lengthening. Load clamps of increasing magnitude progressively accelerated the delayed lengthening (panel C). Analysis of this delayed lengthening showed that a slower phase 2 could be distinguished from a faster phase 3 only for load clamps of a moderate amplitude (panel A); it was always followed by an abrupt fall of force, resulting in a premature isometric relaxation.

Muscle characteristics: length at l_{max} 9.00 mm; mean cross-sectional area 0.44 mm²; R/T 12.0%; temperature 29°C; frepuency of stimulation 0.2 Hz. Reproduced by permission of *The Journal of Physiology* (Brutsaert & Housmans, 1977).

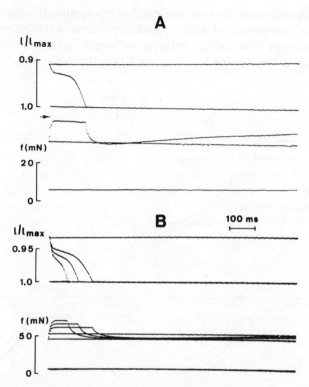

FIG. 6. Supra-isometric force during isotonic tetanus of cat papillary muscle. Same format as in Fig. 1 B.

Panel A: A load clamp of moderate amplitude (not exceeding the level of isometric tetanic force, see arrow at the left side of the panel) induced three-step yielding, again followed by a fast fall in force (see also Fig. 1 B).

Panel B: Three superimposed load clamps to supra-isometric force values again induced a similar three-step yielding response. Isometric control tetanus is also shown. In both panels, control isotonic afterloaded tetanus was superimposed.

Muscle characteristics: Panel A: length at l_{max} 4.50 mm; mean cross-sectional area 0.69 mm^2; R/T 11.0%. Panel B: length at l_{max} 5.00 mm; mean cross-sectional area 1.24 mm^2; R/T 8.0%. Temperature 19°C. Reproduced by permission of *The Journal of Physiology* (Brutsaert & Housmans, 1977).

thesis of back rotation and detachment of cross-bridges, this maximal supra-isometric force potential would then reflect the maximal extra force borne by cross-bridges that still remained attached immediately after the load step, at the expense of delayed phase 2 and 3 yielding. For twitch (Fig. 5C) and tetanic contractions, maximal supra-isometric force potential was about 15 to 20% higher than isometric force. This experimentally measured force potential actually underestimated the true maximal force potential; this may in part be explained by detachment of some cross-bridges already during phase 1, and by the limitations inherent in obtaining sufficiently large amounts of load-induced muscle lengthening in afterloaded twitch contractions.

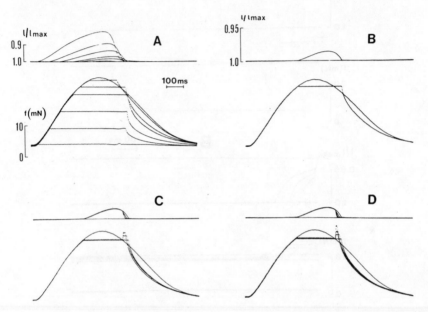

FIG. 7. Supra-isometric force during isotonic relaxation of heavily afterloaded twitch contractions. All panels have same force gain. Length gain in panel A is the same for panels C and D but is different for panel A.

Panel A: A series of afterloaded twitch contractions against various loads, along with a preloaded isotonic and an isometric twitch. Peak shortening always occurred after peak isometric force.

Panel B: A heavily afterloaded twitch contraction and control isometric twitch are superimposed. The end of the isometric force plateau extends to beyond the force of the isometric twitch; hence, load is sustained at supra-isometric levels during isotonic relaxation.

Panel C: Same contractions as in panel B, with two superimposed load-clamped contractions. During the supra-isometric force plateau of the after-loaded twitch, a small further increase in load was achieved by means of load clamp; after a first elastic extension, the clamped supra-isometric load was held for some time at the expense of delayed lengthening.

Panel D: Same contractions as in panel C, with an additional load-clamped contraction. Load clamps of increasing magnitudes progressively abbreviated the supra-isometric force plateau both by increasing the elastic extension at the time of the step and by accelerating the delayed lengthening. Isotonic lengthening during relaxation in either control (panels A and B) or load-clamped (panels C and D) was always followed by an abrupt fall of isometric force, thus undershooting the force levels of the isometric control contraction. Muscle characteristics: see Fig. 5; temperature 29°C; frequency of stimulation 0.2 Hz. Reproduced by permission of *The Journal of Physiology* (Brutsaert & Housmans, 1977).

Supra-isometric forces were also observed during isotonic relaxation of control (Figs. 7 A, B) and load-clamped (Figs. 7 C, D) heavily afterloaded twitch contractions. During isotonic lengthening of the relaxing afterloaded contraction, the load could still be carried at a time when the force in an isometric twitch had fallen considerably (Figs. 7 A, B). Isotonic lengthening during relaxation was always followed by an abrupt fall of subsequent isometric force, thus undershooting the force levels of the isometric control contraction. The apparent ability of a muscle to sustain a load for longer times under isotonic conditions was investigated further by applying loading steps of various magnitudes during the time when isotonic force was higher than that in a control isometric twitch (Figs. 7 C, D). Again, a maximal force response was obtained for load clamps where delayed lengthening was just separable from phase 1. In all isotonic contractions, whether load-clamped or not, isotonic lengthening during isotonic relaxation was always followed by an abrupt fall of isometric force undershooting the fall of force of the isometric control contraction.

From these observations, it thus appears that load-induced three-step yielding, supra-isometric force and supra-isometric force levels during the isotonic lengthening phase of relaxation in heavily afterloaded twitch contractions are three closely linked phenomena; it is therefore appropriate to postulate that all three have a common underlying mechanism and that the initial part of spontaneous isotonic relaxation may also be governed by the same mechanism.

III. Load-sensitive relaxation of mammalian cardiac muscle

The previous discussion leads us to consider the decay of the load-bearing capacity of cardiac muscle during spontaneous isotonic relaxation, and to answer the question whether load *per se* is relevant to the onset of spontaneous isotonic relaxation. In ventricular cardiac muscle preparations obtained from various mammalian species (cat, rat, rabbit, pig), relaxation was indeed sensitive to the loading conditions (Fig. 8) (Brutsaert, De Clerck, Goethals & Housmans, 1978). On the other hand, relaxation of single, partially skinned, Brij-58 pretreated ventricular cardiac cells of these same species appeared insensitive to load, and instead controlled only by the decay in activation (Brutsaert, Claes & De Clerck, 1978); these membrane-deprived cellular preparations resembled intact frog ventricular myocardium in this respect. From these results and from known differences in excitation-contraction and excitation-relaxation coupling mechanisms, we have proposed that relaxation of cardiac muscle is governed by the interaction of an activation-controlled and a load-controlled decay mechanism.

Load sensitivity of relaxation of mammalian cardiac muscle would therefore

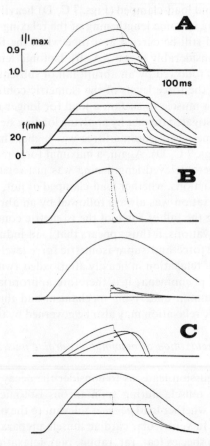

FIG. 8. Load sensitivity of isotonic relaxation of mammalian ventricular cardiac muscle during twitch. All panels have the same length and force gain.

Panel A: A series of afterloaded isotonic contractions against various loads, along with an isometric twitch contraction. Isotonic contractions were always of a shorter overall duration than the isometric contraction; however, the onset of isotonic relaxation at peak shortening was always later than the time to peak force of the superimposed isometric twitch: this implies that the shorter duration of isotonic contractions is due predominantly to an abbreviation of relaxation in isotonic conditions. Moreover, the isometric relaxation phases of the various afterloaded twitches are well separated in time.

Panel B: An isotonic preloaded, an isometric, and three load-clamped contractions were superimposed. At the time of peak shortening of the preloaded isotonic twitch, small step increases of load immediately abbreviated the contraction by inducing a premature isotonic and subsequent isometric relaxation. The relaxation traces of the load-clamped and of the control isometric twitch are well dissociated in time; again, this illustrates the load-sensitive nature of relaxation in mammalian heart muscle.

require the presence of a well-functioning calcium sequestering system, so that load, but not the decay in activation, would become rate-limiting in relaxation. The load-sensitive nature of relaxation could then be the mechanical expression of the relation between the force-generating potential of the muscle at any time and the prevailing load. With very low levels of myoplasmic calcium, achieved by the sequestering membrane systems, progressively fewer force-generating sites would be formed, so that the prevailing load would put a correspondingly increasing strain on cross-bridges. Due to detachment of cross-bridges, with minimal or no reformation of new force-generating bonds, load would then progressively outweigh the declining force potential, and cause cross-bridges to rotate backwards and to detach, with subsequent sliding of myofilaments back to their original position.

By contrast, when total load is appropriately reduced just before the onset of spontaneous isotonic relaxation, the imbalance between the number of prevailing cross-bridges and load is being temporarily readjusted, so that the onset of isotonic lengthening is proportionally delayed (Fig. 9). Successive steps of delayed lengthening by successive appropriate unloading steps would then allow the muscle to sustain these smaller loads at supra-isometric levels throughout isometric relaxation; this illustrates again the close link between isotonic lengthening and supra-isometric forces under conditions where load and number of cross-bridges are critically matched.

In conclusion, from these observations it thus seems plausible that load-induced viscous behavior of the contractile machinery could explain all observed phenomena. In the currently accepted Huxley & Simmons cross-bridge model (Huxley & Simmons, 1973), this viscous behavior could be attributed mainly to back rotation and detachment of cross-bridges with subsequent sliding of myofilaments back to their original position.

In support for this working hypothesis is the fact that in some experimental conditions where relaxation was insensitive to load, the load-induced three-step yielding could equally not be identified. Examples of such conditions are frog ventricular myocardium, and mammalian ventricular myocardium with the insults of caffeine and/or hypoxia (Le Carpentier, Chuck, Housmans, De Clerk & Brutsaert, 1979). Moreover, when relaxation of frog ventricular myocardium was reversed from load-insensitive to load-sensitive, as *e.g.*, in

---

Panel C: A loading and an unloading step of the same magnitude were imposed at the same time during the contraction. The control afterloaded isotonic contraction to which the muscle was clamped and the control isometric contraction are also shown. The duration of contraction and the course of relaxation appear greatly affected by the previous loading history. Muscle characteristics: length at l_{max} 6.50 mm; mean cross-sectional area 1.10 mm^2; R/T 11.0%; temperature 19°C; frequency of stimulation 0.2 Hz.

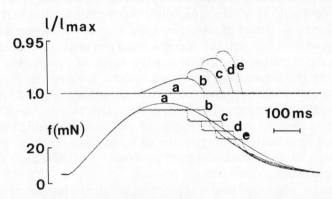

FIG. 9. Supra-isometric force during isotonic relaxation after appropriate unloading. Starting from a control afterloaded isotonic twitch contraction (b), the muscle was stepwise unloaded with load clamps of appropriate magnitude at the onset of isotonic relaxation. The step reduction in load was accompanied by an initial fast shortening due to elastic recoil; the smaller clamped load allowed some further shortening and therefore delayed isotonic relaxation. Three stepwise unloaded contractions with one (c), two (d) and three (e) successive unloading clamps were superimposed along with the isometric contraction (a). This procedure postponed the onset of isotonic relaxation further; the clamped loads (c,d,e) were therefore sustained by the muscle for a longer time than during the control afterloaded contractions with the same load (see Figs. 7A and 8A), and at supra-isometric levels at still later times during relaxation.

Muscle characteristics: length at l_{max} 7.00 mm; mean cross-sectional area 1.00 mm²; R/T 9.0%; temperature 29°C; frequency of stimulation 0.2 Hz.

high-calcium, low-sodium media or after addition of ouabain (Le Carpentier *et al.*, 1979), the phenomenon of load-induced three-step yielding could then be demonstrated again.

REFERENCES

Amemiya, Y., Sugi, H. & Hashizume, H. (1979). X-ray diffraction studies on the dynamic properties of cross-bridges in skeletal muscle. This volume, pp. 425–440.

Armstrong, C. M., Huxley, A. F. & Julian, F. J. (1966). Oscillatory responses in frog skeletal muscle fibres. *J. Physiol.* **186**, 26–27 P.

Barden, J. A. & Mason, P. (1978). Muscle crossbridge stroke and activity revealed by optical diffraction. *Science* **299**, 1212–1213.

Blangé, T. & Stienen, G. J. M. (1979). Isometric tension transients in skeletal muscle before and after inhibition of ATP synthesis. This volume, pp. 211–222.

Brady, A. J. (1974). Mechanics of the myocardium. In *The Mammalian Myocardium*, eds. Langer, G. A. & Brady, A. J., pp. 163–192. New York: John Wiley and Sons.

Brutsaert, D. L., Claes, V. A. & De Clerck, N. M. (1978). Relaxation of mammalian single cardiac cells after pretreatment with the detergent Brij-58. *J. Physiol.*, **283**, 481-491.
Brutsaert, D. L., De Clerck, N. M., Goethals, M. A. & Housmans, P. R. (1978). Relaxation of ventricular cardiac muscle. *J. Physiol.*, **283**, 469-480.
Brutsaert, D. L. & Housmans, P. R. (1977). Load clamp analysis of maximal force potential of mammalian cardiac muscle. *J. Physiol.* **271**, 587-603.
Delay, M. J., Vassallo, D. V., Iwazumi, T. & Pollack, G. H. (1979). Fast response of cardiac muscle to quick length changes. This volume, pp. 71-81.
Edman, K. A. P., Elzinga, G. & Noble, M. I. M. (1978). Enhancement of mechanical performance by stretch during tetanic contractions of vertebrate skeletal muscle fibres. *J. Physiol.* **281**, 139-155.
Flitney, F. W. & Hirst, D. G. (1978a). Crossbridge detachment and sarcomere "give" during stretch of active frog's muscle. *J. Physiol.* **276**, 449-465.
Flitney, F. W. & Hirst, D. G. (1978b). Filament sliding and energy absorbed by the crossbridges in active muscle subjected to cyclical length changes. *J. Physiol.* **276**, 467-479.
Ford, L. E., Huxley, A. F. & Simmons, R. M. (1977). Tension responses to sudden length change in stimulated frog muscle fibres near slack length. *J. Physiol.* **269**, 441-515.
Henderson, A. H. & Cattell, M. R. (1976). A study of mechanical oscillations in strontium-mediated contractions of cat and frog heart muscle. *Circulation Res.* **38**, 289-296.
Henderson, A. H., Forman, R., Brutsaert, D. L. & Sonnenblick, E. H. (1971). Tetanic contraction in mammalian cardiac muscle. *Cardiovasc. Res.* **5**, suppl. 1, 96-100.
Housmans, P. R. & D. L. Brutsaert (1976). Three step yielding of load-clamped mammalian cardiac muscle. *Nature, Lond.* **262**, 56-58.
Huxley, A. F. (1974). Muscular contraction. *J. Physiol.* **243**, 1-43.
Huxley, A. F. & Simmons, R. M. (1973). Mechanical transients and the origin of muscular force. *Cold Spring Harb. Symp. quant. Biol.* **37**, 669-680.
Julian, F. J. & Moss, R. L. (1976). The concept of active state in striated muscle. *Circulation Res.* **38**, 53-59.
Julian, F. J. & Sollins, M. R. (1975). Variation of muscle stiffness with force at increased speeds of shortening. *J. gen. Physiol.* **66**, 287-302.
Julian, F. J., Sollins, K. R. & Sollins, M. R. (1973). A model of muscle contraction in which crossbridge attachment and force generation are distinct. *Cold Spring Harb. Symp. quant. Biol.* **37**, 685-688.
Katz, B. (1939). The relation between force and speed in muscular contraction. *J. Physiol.* **96**, 45-64.
Le Carpentier, Y. C., Chuck, L. H. S., Housmans, P. R., De Clerk, N. M. & Brutsaert, D. L. (1979). Nature of load dependence of relaxation in cardiac muscle. *Am. J. Physiol.* in press.
Noble, M. I. M. & Pollack, G. H. (1977). Molecular mechanisms of contraction. *Circulation Res.* **40**, 333-342.
Podolsky, R. J. & Nolan, A. C. (1973). Muscle contraction transients, crossbridge kinetics and the Fenn effect. *Cold Spring Harb. Symp. quant. Biol.* **37**, 661-668.
Pollack, G. H. & Krueger, J. W. (1976). Sarcomere dynamics in intact cardiac muscle. *Eur. J. Cardiol.* **4**, suppl., 53-65.
Rack, P. M. H. & Westbury, D. R. (1974). The short range stiffness of active mammalian muscle and its effect on mechanical properties. *J. Physiol.* **240**, 331-350.
Rüegg, J. C., Güth, K., Kuhn, H. J., Herzig, J. W., Griffiths, P. & Yamamoto, T. (1979).

Muscle stiffness in relation to tension development of skinned striated muscle fibers. This volume, pp. 125–143.

Steiger, G. J. (1979). Kinetic analysis of isometric tension transients in cardiac muscle. This volume, pp. 259–271.

Sugi, H. & Tsuchiya, T.(1975). Effect of isotonic shortening on the load-sustaining ability and the instantaneous elasticity in molluscan smooth muscle. *Proc. Japan Acad.* **51**, 712–715.

Tsuchiya, T., Sugi, H. & Kometani, K.(1979). Isotonic velocity transients and enhancement of mechanical performance in frog skeletal muscle fibers after quick increases in load. This volume, pp. 225–238.

Discussion

Rall: As I interpret your results, they look very much like what Emil Bozler has shown over the past few years (Bozler, 1975, 1977), and I think that he would say that as the load is reduced during shortening, there is an activation, or prolongation, of contraction. I think he calls that "shortening activation." Then, if the load is increased during shortening, there is an abrupt ending of contraction which he calls "lengthening deactivation." Your results look very much like Bozler's results. Would you comment on that?

Brutsaert: Bozler simply described his data without giving any further interpretation. Here, what we have tried to do is to link all these phenomena together, and we then tried to look for a basic mechanism which could explain all of them, because they are all linked together. But Bozler's two publications of the past few years have been simply descriptive studies.

Rall: Could I follow up on that? One thing that is always confused me is that if you decrease the load during the shortening, or if you have a shortening activation, it seems to be inconsistent with Edman's and Brady's shortening deactivation (Edman, 1975; Brady, 1965). Would you clarify that for me?

Brutsaert: Well, to some minor extent you certainly have shortening deactivation. Indeed, when afterloaded contractions are superimposed on a length-tension diagram, the afterloaded contraction during shortening never reaches the length-peak isometric tension point. Hence, this "deactivation" must be due to the shortening *per se* (Brutsaert & Housmans, 1977). But from our data it appears that this "shortening-deactivation" is certainly not enough to explain the marked abbreviation of the isotonic contractions, especially during their relaxation phases. During the isotonic lengthening phase of relaxation, once the muscle starts to elongate, the length passes through a soft shoulder at first; then lengthening proceeds at an increasingly faster speed (Fig. 8). We have postulated that this fast extension of the muscle in relaxing afterloaded contractions occurs when the load exceeds the number of force generating sites or cross-bridges, which then start to rotate backwards and to detach at a faster and faster rate, allowing for the filaments to slide backwards to their original position (Brutsaert, De Clerck, Goethals & Housmans, 1978). Now if you unload the muscle and, by doing so, you would correct, in an appropriate way, for the number of still attached active sites or cross-bridges, then force potential and load would be balanced again, so that the muscle can sustain this new smaller load for a prolonged period of time during relaxation. This interpretation would also easily explain the multiple unloading experiments (Fig. 9).

Pollack: Regarding the compliance measurements you made in intact cardiac muscle toward the end of your presentation, we showed in the previous session that most of the compliance, at least in the preparations we use, resided at the damaged ends rather than in the viable, central region. Our releases were done at high speed. That might accentuate the difference between the compliance at the ends and at the sarcomeres. Your compliance measurements were made at somewhat lower speed, I think. But still, do you believe the properties of the ends are in any way influencing the results that you obtain?

Brutsaert: I am not excluding it. At present we cannot entirely exclude the problem of damaged ends. However, several arguments should make us quite confident that this problem is less important in experiments in which load, and not length, is altered, perhaps less with respect to compliance measurements, but certainly with respect to all other described phenomena, including the three-step yielding, supra-isometric force potential and load-dependence during relaxation. First of all this is done in tetanized muscle, and we are dealing with load clamps, not length changes, so that most of the compliance should be manifested in phase one; moreover, we are working with heavy afterloads, where any length change in the series compliance has already taken place. So we are working on the stiff portion of the series elastic extension curve.

Still, you could argue that there is force redistribution during phase two, but, I do not know, whether this is still the case in a phase two lasting about 100–150 msec, which is quite long.

We have recently shown (Le Carpentier, De Clerck & Brutsaert, submitted) that in normal conditions relaxation of frog ventricular muscle is load-independent, similarly as in single cardiac cells which were pretreated with a detergent (Brutsaert, Claes & De Clerck, 1978). In these studies we have demonstrated that relaxation is load-independent when relaxation is governed solely by activation. Yet, relaxation of frog myocardium can be made load-dependent when the duration of the contraction is shortened, as *e.g.* at higher calcium concentrations or by lowering external sodium or by adding ouabain. Accordingly, coming back to the problem of tissue elasticity due to damaged ends, it appears that the very complex frog ventricular cardiac strip, in which the fibers lay in all different directions, can be made load-pependent in conditions where all other conditions including initial muscle length, are unaltered. With the appearance of load dependence, the three-step yielding can then also be demonstrated, hence providing evidence that these phenomena are unrelated to the structural complexity of the preparation.

Sagawa: You mentioned something about a decreasing compliance in a certain range of amplitude. I would like to know your speculation regarding the mechanism behind that decrease.

Brutsaert: Although it may be somewhat premature to speculate, if for a moment we could think in terms of cross-bridge mechanisms, the argument could be as follows: You are diminishing the number of attached cross-bridges and nevertheless you see an increase in stiffness; so perhaps this could be an example in which you could distinguish between stiffness relating to the number of cross-bridges on one hand and the properties of cross-bridges on the other, because the cross-bridges are already detaching all along, during phases 1, 2, and 3. So, despite the fact that some cross-bridges are already detaching, you see an increased stiffness. This might even further emphasize the possibility of back-rotation with an increased strain on the remaining cross-bridges.

REFERENCES

Bozler, E. (1975). Mechanical control of the time course of contraction of the frog heart. *J. gen. Physiol.* **65**, 329–344.

Bozler, E. (1977). Mechanical control of the rising phase of contraction of frog skeletal and cardiac muscle. *J. gen. Physiol.* **70**, 697–705.

Brady, A. J. (1965). Time and displacement dependence of cardiac contractility: Problems in defining the active state and force-velocity relations. *Fedn. Proc.* **24**, 1410–1420

Brutsaert, D. L. & Housmans: Load clamp analysis of maximal force potential of mammalian cardiac muscle. *J. Physiol.* **271**, 587–603, 1977.

Brutsaert, D. L., N. M. De Clerck, M. A. Goethals & P. R. Housmans (1978). Relaxation of ventricular cardiac muscle. *J. Physiol.* **283**,469–480

Brutsaert, D. L., V. A. Claes & N. M. De Clerck. (1978). Relaxation of mammalian single cardiac cells after pretreatment with the detergent Brij-58. *J. Physiol.* **283**,481–491

Edman, K. A. P. (1975). Mechanical deactivation induced by active shortening in isolated muscle fibres. *J. Physiol.* **246**, 255–275.

Le Carpentier, Y. C., N. M. De Clerck & D. L. Brutsaert (1979). Nature of load dependence of relaxation in cardiac muscle. Submitted.

Kinetic Analysis of Isometric Tension Transients in Cardiac Muscle*

Gerhard J. STEIGER

Department of Physiology and the American Heart Association, Greater Los Angeles Affiliate Cardiovascular Research Laboratories, University of California, Los Angeles, Center for the Health Sciences, Los Angeles, California, U.S.A.

ABSTRACT

The kinetic analysis of the tension transients following small step perturbations of muscle length revealed three significant phases which are common to at least 3 different kinds of muscles, and which can be approximated by single exponential rate constants. The fastest component, κ_1, differs from the other components in its lack of sensitivity to temperature. In cardiac muscle at room temperature and under the given experimental conditions the rate constants of the three phases of the tension transients were approximately one order of magnitude apart. Only κ_2 was clearly related to the amplitude and direction of length change, increasing with increased amplitudes of stretch and decreased amplitudes of release.

The mechanism of force generation proposed by Huxley & Simmons (1971) seems to be unable to account for the results presented. I have also presented evidence that k_3, the delayed tension changes, are very unlikely to be due to a reduction of overlap between adjacent actin filaments, leading to the conclusion that the current sliding filament model has no possibility of explaining the phenomenon of stretch activation. An attempt was made to correlate directly the mechanical and biochemical rate constants. The surprisingly good agreement of the very preliminary data available at the present gave rise to the following speculations: that κ_1, the fastest mechanical rate (following stretch) could be linked to the biochemical dissociation rate of actomyosin following the binding of ATP; that κ_2 might be linked to the recombination and dissociation of myosin-products with actin; and that κ_3 might be strongly correlated to V_{max}, the maximum rate of the actomyosin ATPase and the generation of tension.

---

* The studies were supported by a Grant from the U.S. Public Health Service, HL 11351-11.

INTRODUCTION

Small step perturbations have been applied to various cardiac muscle preparations. The resulting tension transients have been studied at various temperatures and amplitudes of length change. An attempt was made to describe the tension transients following step perturbations by a set of exponential functions. We found that, in most cases, the tension-time course could be fitted by a maximum number of 4 exponentials.

Stretch activation is a relatively slow response and has a considerable delay with respect to the length change. It is preceded by at least two phases of faster responses, the fastest of which is rather temperature insensitive.

The purpose of this symposium is to discuss the validity of current sliding filament/cross-bridge models, which have been essentially formulated by A.F. Huxley & Niedergerke (1954), H.E. Huxley & Hanson (1954), A.F. Huxley (1957), H.E. Huxley (1969) and A.F. Huxley & Simmons (1971).

The objective of this presentation is therefore to check carefully whether the results of the experiments we conducted can satisfactorily be explained in the current framework of sliding filament/cross-bridge models.

METHODS

Stretch activation in cardiac muscle was first demonstrated in 1971 (Steiger, 1971). Since then, the methodological approach basically has not changed, but some components have been significantly improved (Fig. 1). The electromechanical muscle puller (Link Dynamic Systems) which allowed us to apply step perturbations in about 3 msec (Steiger, 1971) and later in ≤ 1 msec (Abbott & Steiger, 1977) has recently been replaced by a piezo-elec-

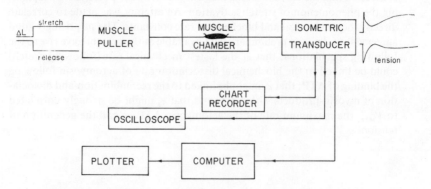

FIG. 1. Scheme of the electromechanical apparatus.

tric-crystal (Ricchiuti *et al.*, in preparation) allowing step perturbations of 75 μm in much less than 300 μsec (the limiting factor is on the tension recording side). On the tension recording side the initial RCA-transducer-valve (RCA 5734), with a resonance frequency of $\sim$ 800 Hz, gave way to a capacitance transducer with a resonance frequency of $\sim$ 5 kHz, a sensitivity of 2V/g and a compliance of $\sim$ 3 μm/g. The length and tension output was displayed on a Tektronix oscilloscope (D 11). A continuous chart record was kept of each experiment (Brush Chart Recorder). The data were also sampled, averaged and stored on a Linc PDP8 computer. For each sweep, up to 200 points could be sampled. The sweep could be divided into a maximum of ten intervals (the usual number used was 2 or 4) each with a different speed of the time base. The shortest sampling rate was 33 μsec/point. In this way it was possible to gather enough information in the first few msec to estimate the rate constant of the fastest transient as well as the slower tension changes. After on-line storage the data have been analyzed on an IBM 360 (University Central Computer Network). The program was provided by Provencher (1976). It is based on an eigenfunction expansion method, runs completely automatically and does not need any initial guesswork.

Sustained, isometric contractures in living rabbit papillary muscles were induced either by a high potassium-low sodium solution, by high rate electrical stimulation or by Ba^{++}, instead of Ca^{++}, in the usual Tyrode's solution (Table 1). The experimental procedures have been described in detail in Steiger, Brady & Tan (1978).

TABLE 1. Solutions components measured in mM

	NaCl	$BaCl_2$	Caffeine	Choline	$CaCl_2$	KCl	$MgCl_2$	NaH_2PO_4	$NaHCO_3$	Dextrose	pH
Standard Tyrode's solution	130	—	—	—	5	4	1	0.435	10	5.56	7.4
Ca-free solution	130	—	—	—	—	4	1	0.435	10	5.56	7.4
Ba contraction solution	130	0–1.0	—	—	—	4	1	0.435	10	5.56	7.4
HKS solution	65	—	5	—	5	70	1	0.435	10	5.56	7.4

RESULTS

During a sustained contracture, the living cardiac muscle develops a steady isometric tension. Superimposed on this steady-state situation are small square-wave length perturbations, which cause the multiphasic tension response (Fig. 2). Figure 2 shows the recorded portion of a stretch cycle on two different time scales at room temperature. It can be seen clearly that the relatively slow, delayed tension change (phase 3, "stretch activation") is preceded by at least 2 phases which are considerably faster. In order of their appearance they are called phase 1, phase 2 and phase 3. The initial elastic tension rise has been omitted in Fig. 2. The isometric tension level is represented by the bottom line of each square. The solid line shows the equivalent exponential curve as found by the eigenfunction expansion method. The rate constants for the different phases of the tension response are given in the legend for Fig. 2. It should be pointed out, however, that an exponential equation to provide a satisfactory fit could not be found for all muscles. However, in all cases the two fast phases of the initial tension recovery were readily visible at low temperatures.

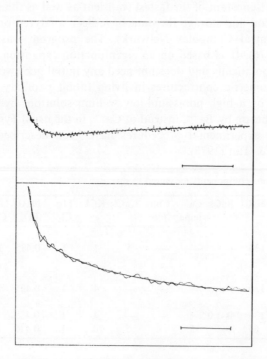

FIG. 2. Tension transients to small step perturbations of muscle length in living cardiac muscle.
The top trace shows the entire sampling period for stretch. Time bar = 1.0 sec.
The bottom traces show, on an expanded time scale, phase 1 and phase 2 only. Time bar = 40 msec; $\kappa_1 = 381$ sec^{-1}; $\kappa_2 = 42.9$ sec^{-1}; $\kappa_3 = 2.1$ sec^{-1}.
Exp. cond.: rabbit papillary, HPS solution, $\Delta L \sim 0.3\%$, room temperature.

KINETIC ANALYSIS OF TENSION TRANSIENTS

For the following it is important that we first thoroughly discuss the assumption that the observed tension transients indeed reflect cross-bridge properties.

A. How can it be excluded that, in particular, the fastest tension transients

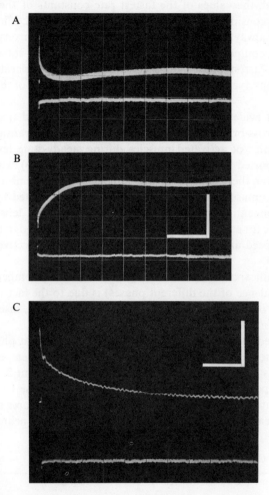

FIG. 3. Tension transients in glycerinated chicken ALD muslces.
A: following stretch
B: following release
C: following stretch—expanded time scale
Exp. cond.: tension bar—55 mg
 temperature ~ 18°C
 time bar—A + B = 1 sec
  C = 100 msec

are an artifact of the mechano-electrical system? First, the tension transducer system was only lightly damped and has a resonant frequency too high to have an exponential time constant of about 300–400 msec. Second, the values of the fastest rate constants of the tension response varied considerably with the type of muscle we used. If it were an artifact it would always have had the same speed. Third, after very large releases the fastest component had a small amplitude and was sometimes not visible. If the fastest phase had been caused by the apparatus it would have been observed to be proportional to the amplitude of the length perturbation.

B. What evidence can we present that the observed transients indeed reflect cross-bridge properties? First, these tension transients are *only* characteristics of activated muscles during steady-state levels of activation. As reported previously (Steiger *et al.* 1978, Fig. 3), without cross-bridge turnover these transients are almost completely absent. Responses of relaxed muscle or of muscle in rigor cannot be fitted by a small number of exponential expressions. Second, the observed tension transients are specific for each kind of muscle, being closely related to the actual contraction speed and metabolic rate *in situ* of the respective muscle (Steiger, 1977a).

The significance of the following results (namely temperature and amplitude dependence of the different phases) is due to the fact that they were observed not only in cardiac muscle, but in rabbit psoas, insect fibrillar muscle and chicken ALD muscle as well.

A detailed temperature analysis (Table 2) shows that phase 1 and phase 2 are characteristically different with respect to their temperature sensitivity. Phase 2 has a high temperature coefficient, Q_{10} of about 3, whereas phase 1 is hardly temperature-sensitive at all. It follows that phase 1 and 2 can easily be separated at low temperatures, but at higher temperatures this separation becomes increasingly difficult. The critical temperature is near body temperature

TABLE 2A. Glycerol extracted rabbit psoas muscle

Temperature (°C)	$\kappa_1(\text{sec}^{-1})$	$\kappa_2(\text{sec}^{-1})$	$\kappa_3(\text{sec}^{-1})$
3	794 ± 354	55 ± 21	—
5	955 ± 425	66 ± 17	12 ± 6
9	832 ± 427	76 ± 29	19 ± 13
11	1202 ± 178	91 ± 24	23 ± 17
15	1259 ± 326	174 ± 35	44 ± 14
18	1148 ± 297	191 ± 98	63 ± 47
22	794 ± 354	288 ± 148	72 ± 42

TABLE 2B. Cardiac muscle

Temperature (°C)	$\kappa_1(\text{sec}^{-1})$	$\kappa_2(\text{sec}^{-1})$	$\kappa_3(\text{sec}^{-1})$
9	156.2 ± 24.9	9.4 ± 3.8	0.4 ± 0.25
16	259.5 ± 95.6	15.7 ± 2.7	0.8 ± 0.08
24	276.0 ± 31.2	37.4 ± 5.8	4.0 ± 0.8
36	350.0 ± 21.2	170.0 ± 16.3	27.9 ± 2.4

Temperature dependence of the mechanical rate constant for rabbit psoas and rabbit cardiac muscle.
(For experimental conditions, see Abbott & Steiger (1977) and Steiger (1977a)).

in the case of the cardiac preparation but is less than 20°C for rabbit psoas and insect flight muscle and might be still lower in frog muscles.

In cardiac muscle only phase 2 shows a distinctive amplitude dependence. The rate constant of phase 2, κ_2, has its lowest value for large amplitudes of release, increases with decreasing amplitudes of release and increases with increasing amplitudes of stretch. κ_1 and κ_3 are not clearly affected by different amounts of stretch and release (Steiger, 1977a, Fig. 8).

DISCUSSION

A. F. Huxley & Simmons (1971) suggested that the early tension recovery (phase 2 in their terminology) might be due to a rotation of the myosin head and they proposed a model of force generation in striated muscle. Their key assumptions were:
 a) That the movement by which a cross-bridge performs work takes place in a small number of steps of progressively lower potential energy, and
 b) That there is a virtually instantaneous elasticity within each cross-bridge, allowing it to shift from one stable position to the next without simultaneous displacement of the main shaft of the thick and thin filaments with respect to one another.

Accordingly, the quick tension recovery is due to the tendency of the myosin head to rotate to positions of lower potential energy, while the fact that the recovery occurs at a finite speed is a manifestation of the rate constant for movement of the system from one of the stable positions to the next.

Their model predicts:
 a) That the rate should become slower as the amplitude of stretch is increased and faster as increasing amplitudes of release are applied, and
 b) That the minimal activation energy needed assuming 3 stable positions of progressively lower potential energy would transform to an expected temperature coefficient, Q_{10}, of approximately 2.2.

Comparing the results I have presented with the predictions of the force generation mechanism of Huxley & Simmons (1971), it seems obvious that neither κ_1 nor κ_2 fulfill the requirements of the Huxley-Simmons model. κ_1 fulfills neither of the two predictions. It does not increase with increasing amplitudes of release nor does it decrease with increasing amplitudes of stretch. κ_1 also differs from all the components of the tension transients due to its lack of sensitivity to temperature. The measured Q_{10} is between 1.1 and 1.3. This applies to cardiac muscles as well as rabbit psoas and insect flight muscles (Abbott & Steiger, 1977). We also can exclude κ_2. κ_2 has a sufficiently high Q_{10}, but its amplitude dependence is just opposite to the predictions made in the Huxley and Simmons model. We must therefore conclude that the mechanism proposed by Huxley and Simmons is unable to account for any of the features of our results.

Having discussed phase 1 and 2 of the observed tension transients, I would like to proceed to discuss the mechanism of delayed tension changes (phase 3). The rate constant, κ_3, has the highest temperature coefficient but does not seem to change significantly with the amplitude of the length perturbation. It has been argued that, in particular, the delayed tension rise following a stretch might be due to the reduction of double overlap of the actin filaments, and hence reflects a property of the ascending limb of the length tension curve. As demonstrated in a previous paper (Steiger *et al.*, 1978, Fig. 4), however, the delayed tension changes are *not* limited to the ascending limb of the length tension curve. On the contrary, the response is maximal at a length where the twitch tension is also maximal. If the response were due to a reduction of the overlap of the actin filaments, we would expect the delayed effects to be larger at short muscle lengtn and to diminish and finally to disappear at lengths in the optimal overlap region. The observation that delayed tension changes are most pronounced in the optimal overlap region where further muscle lengthening no longer augments active twitch is inconsistent with what might have been expected from the Frank-Starling relationship.

In skeletal fibers, it was shown that stretch activation exists even at sarcomere lengths where actin-myosin overlap decreases (Heinl, 1972), thus the current sliding filament model is unable to account for the phenomenon of stretch activation.

In summary, I have tried to give an explanation of our results in terms of the current framework of the sliding filament/cross-bridge model. Obviously, this is not possible. However, the invitation to contribute to a symposium such as this gives the opportunity to discuss some ideas which are somewhat speculative, but serve as a working hypothesis for further experiments.

The most recent formulation of the actomyosin ATPase cycle has been made by White & Taylor (1976) and Sleep & Taylor (1976), based on the original cycle proposed by Lymn & Taylor (1971).

$$\begin{array}{cccc}
AM \rightleftarrows AMATP & AMADPPi \rightleftarrows AMADP \rightleftarrows AM \\
\updownarrow \quad\quad \updownarrow & \updownarrow \quad\quad\quad \updownarrow \quad\quad\quad \updownarrow \\
M \rightarrow MATP \rightleftarrows & MADPPi \rightleftarrows MADP \rightarrow M
\end{array}$$

At ATP saturating concentrations the cycle may be simplified to a minimum of four steps (Marston & Taylor, 1977).

$$\begin{array}{ccc}
A + M.ATP & \xrightleftharpoons{k_{+2}} & A + M.ADP.P. \\
K_1 \updownarrow & & \updownarrow K_3 \\
AM & \xrightleftharpoons[k_{+4}]{} & AMADP.Pi
\end{array}$$

It is probable that steps 1 and 3 are in rapid equilibrium, while step 4 is the slowest in the cycle (White & Taylor, 1976; Marston & Taylor, 1977). This scheme immediately suggests a correlation with mechanical experiments. For instance, step 4 could correspond to the tension-producing step; indeed White & Taylor (1976) have shown that at least half the total free energy change of ATP hydrolysis by acto-S1 occurs in this step.

Step 1 could correspond to the fastest mechanical rate constant, κ_1, and step 3 could correspond to κ_2; whereas step 2, the ATP cleavage step, essentially does not directly have a mechanical equivalent.

With few exceptions, detailed kinetic studies have been confined to fast skeletal actomyosin of rabbit. Only recently some data became available on cardiac and slow skeletal actomyosin (Taylor & Weeds, 1976; Marston & Taylor, 1978). As shown by Marston and Taylor, the cardiac muscle and the chicken ALD muscle seem to satisfy the same kinetic scheme. The systems differ quantitatively in the rates of dissociation and recombination of actin and myosin and in the V_{max} of the actomyosin ATP-ase.

With the few and preliminary data now available, a tentative attempt was made to compare both the biochemical as well as the mechanical rate constants side by side.

Table 4 compares the fastest mechanical rate constants with the fastest biochemical rate constant, the dissociation rate of acto-S1 with ATP. The rate of the actomyosin dissociation is a fast process and a function of the ATP concentration. The maximum rate for rabbit psoas exceeds 800 sec^{-1} and could not therefore be measured very accurately. The estimate given by Mar-

TABLE 3. Amplitude dependence of the mechanical rate constant for rabbit cardiac muscle

Amplitude % Stretch	$\kappa_1(\text{sec}^{-1})$	$\kappa_2(\text{sec}^{-1})$	$\kappa_3(\text{sec}^{-1})$
0.9	201	20.5	0.8
0.7	253	18.2	0.9
0.5	221	17.0	0.8
0.3	334	15.6	0.8
0.1	279	11.8	0.9

Amplitude % Release	$\kappa_1(\text{sec}^{-1})$	$\kappa_2(\text{sec}^{-1})$	$\kappa_3(\text{sec}^{-1})$
0.1	142	10.8	0.5
0.3	236	8.3	0.6
0.5	158	7.4	0.8
0.7	156	6.7	0.42
0.9	164	6.8	0.75

Exp. cond. —: rabbit papillary, HPS solution, 16°C.

TABLE 4. κ_1, the fastest mechanical rate constant and the dissociation rate of acto-S1 with ATP

	Rabbit fast skeletal muscle	Cardiac muscle	Chicken ALD slow skeletal muscle
Maximum disociation rate[a] [sec^{-1}]	~1100	330 ± 60[d]	125 ± 20
Mechanical rate κ_1 [sec^{-1}]	1140 ± 300[b]	276 ± 30[c]	< 200

a) Marston & Taylor (1978)
Exp. conditions: 25°C, 1 mM ATP, pH 7
b) Abbott & Steiger (1977)
c) Steiger (1977)
d) Taylor & Weeds (1976) found a maximum rate for bovine heart of 1500–2000 sec^{-1} using however, a pH of 8.

ston and Taylor is 1100 sec^{-1} ± 30%. For slow muscles however, the maximum dissociation rate could be measured and was determined for cardiac muscle to be ~ 300 ± 60 sec^{-1} and for chicken ALD muscles to be ~ 125 ± 20 sec^{-1}. The mechanical rate constant for the fastest mechanical phase as obtained by fitting the tension transients to a set of exponential functions as described (Abbott & Steiger, 1977; Steiger, 1977) coincide surprisingly well for both cardiac and rabbit psoas muscles.

There is also a second feature they have in common which strengthens our suspicion that the rates might indeed be correlated, even more than does the double coincidence of the right numbers; namely, that both the biochemical as well as the mechanical rate are virtually independent of temperature (Marston & Taylor, 1978; Steiger, 1977a).

Table 5 compares the mechanical κ_2 with the dissociation rate of the myosin product with actin. It has to be pointed out that the biochemical values in Table 5 are calculated values, which are based on the following assumptions. White and Taylor (1976) found that the product K_3k_{+4} at 20°C is $\sim 1 \times 10^5$ at 40 mM KCl. k_{+4} is ~ 40 sec^{-1} for rabbit skeletal (see Table 6), correspondingly, K_3 is approximately $2 \times 10^3 M^{-1}$. The apparent second order rate constants for the association of the myosin products given by Marston and Taylor for fast skeletal and cardiac muscle are $5.2 \times 10^5 M^{-1}$ sec^{-1} and $8 \times 10^4 M^{-1}$ sec^{-1}, respectively. In order to get the values given in Table 5 the assumption was made that the effective actin concentration in muscle might be 5×10^{-4} M. The effective actin concentration in muscle might be as high as 1×10^{-3} M (Sleep, personal communication). Evidently much more information is needed to make a definite statement on whether κ_2 and K_3 correspond to each other. At the moment it remains merely speculation.

TABLE 5. The mechanical rate κ_2 and the dissociation rate of myosin produced from actin

	Rabbit fast skeletal muscle	Cardiac muscle	Chicken ALD slow skeletal muscle
k_{-3} [sec^{-1}][a]	250	40	55
Mechanical rate κ_2 [sec^{-1}]	288 ± 148[b]	37.5 ± 6[c]	>10

a) Marston & Taylor (1978)
   Exp. conditions: double mixing technique, 25°C, pH 7, 10 mM KCl
b) Abbott & Steiger (1977); Solution contained 20 mM phosphate buffer
c) Steiger (1977)

Table 6 finally compares κ_3, the rate constant of the delayed tension changes, with the rate of the actomyosin ATPase of S1 as measured by V_{max}. The values for V_{max} are identical with k_{+4} in the 4 state scheme. Again we find that the mechanical and the biochemical rates show similar numbers for fast skeletal, as well as for cardiac muscles, and share a high temperature sensitivity (White, personal communication; Steiger, 1977). Since at least half the total free energy change of the ATP hydrolysis by acto-S1 occurs in this step and since κ_3, the delayed tension rise, has always been associated with the attachment of new cross-bridges and the development of tension, k_{+4}, could well be the tension-generating step.

Experiments with glycerol-extracted chicken ALD muscle, a slow skeletal muscle (Fig. 3), which are now in progress (Steiger & Abbott, in preparation)

TABLE 6. κ_3, the rate constant of the delayed tension compared with the actomyosin ATPase of S1

	Rabbit fast skeletal muscle	Cardiac muscle	Chicken ALD slow skeletal muscle
V_{max} [sec^{-1}]	40$^{d)}$	4.2$^{a)}$	4.5$^{a)}$
Mechanical rate κ_3 [sec^{-1}]	72 ± 42$^{b)}$	4.0 ± 0.65$^{c)}$	1.0

a) Marston & Taylor (1978)
  Exp. conditions: 25°C, pH 7, 10 mM KCl
b) Abbott & Steiger (1977)
  Solution contained 20 mM phosphate buffer
c) Steiger (1977)
d) Taylor & Weeds (1976)

seem also to fit in the scheme outlined above and therefore strengthen the proposed working hypothesis. Preliminary estimates for the mechanical rates in chicken ALD muscles are given in Tables 4 through 6.

REFERENCES

Abbott, R. H. & Steiger, G. J. (1977). Temperature and amplitude dependence of tension transients in glycerinated skeletal and insect fibrillar muscle. *J. Physiol.* **266**, 13–42.

Heinl, P. (1972). Mechanische Aktivierung und Deaktivierung der isolierten contractilen Struktur des Froschsartorius durch rechteckförmige und sinusförmige Längenänderungen. *Pflügers Arch. ges. Physiol.* **333**, 213–226.

Huxley, A. F. & Niedergerke, R. (1954). Interference microscopy of living muscle fibres. *Nature, Lond.* **173**, 971–973.

Huxley, A. F. (1957). Muscle structure and theories of contraction. *Prog. Biophys. biophys. Chem.* **7**, 255–318.

Huxley, A. F. & Simmons, R. M. (1971). Proposed mechanism of force generation in striated muscle. *Nature, Lond.* **233**, 533–538.

Huxley, H. E. & Hanson, J. (1954). Changes in the cross-striations of muscle during contraction and stretch and their structural interpretation. *Nature, Lond.* **173**, 973–976.

Huxley, H. E. (1969). The mechanism of muscular contraction: Recent structural studies suggest a revealing model for cross-bridge action at variable filament spacing. *Science, N. Y.* **164**, 1356–1366.

Marston, S. B. & Taylor, E. W. (1979). The mechanism of actomyosin ATPases of white, red, cardiac and smooth muscles. submitted.

Provencher, S. W. (1976). An eigenfunction expansion method for the analysis of exponential decay curves. *J. chem. Phys.* **64**, 2772.

Ricchiutti, N. V., Steiger, G. J. & Brady, A. J. (In preparation)

Sleep, J. A. & Taylor, E. W. (1976). Intermediate states of actomyosin triphosphatase. *Biochemistry, N. Y.* **15**, 5813–5817.

Steiger, G. J. (1971). Stretch activation and myogenic oscillation in isolated contractile structures of heart muscle. *Pflügers Arch. ges. physiol.* **330**, 347–361.

Steiger, G. J. (1977a). Stretch activation and tension transients in cardiac, skeletal and insect flight myscle. *In: Insect Flight Muscle.* (Tregear, R. T., ed.), pp. 221–268, North Holland, Amsterdam.

Steiger, G. J. (1977b). Tension transients in extracted rabbit heart muscle preparation. *J. molec. Cell. Cardiol.* **9**, 671–685.

Steiger, G. J., Brady, A. J. & Tan, S. T. (1978). Intrinsic regulatory properties of contractility in the myocardium. *Circulation Res.* **42**, 339–350.

Steiger, G. J. & Abbott, R. H. (1978). Model calculations of muscle tension and ATPase activity based on biochemical data. *Biophys. J.* **21**, 87a.

Taylor, R. S. & Weeds, A. G. (1976). The magnesium-ion-dependent adenosine triphosphatase of bovine cardiac myosin and its subfragment-1. *Biochem. J.* **159**, 301–315.

White, H. D. & Taylor, E. W. (1976). Energetics and mechanism of actomyosin triphosphatase. *Biochemistry, N. Y.* **15**, 5818–5826.

Discussion

Winegrad: Your rejection of the sliding filament hypothesis on the basis of length activation assumes that the only thing that controls tension happens to be the filament overlap or number of cross-bridges, which is clearly not so in cardiac muscle. Secondly, you are assuming that the rising phase of the length-tension curve in cardiac muscle is due to the double-overlap of the filaments. However, I think the evidence is against that. As far as I know, nobody has ever been able to produce double-overlap in a resting striated muscle, either in cardiac or in skeletal muscle, and since cardiac muscle normally operates on the rising phase of the length-tension curve, in order to explain that by double-overlap, you must have double-overlap of thin filaments in the unloaded, resting state. And if there is anything consistent in the sliding filament story, I think it is not possible to produce double-overlap in an inactive muscle fiber.

Steiger: Yes, I see that. The point I wanted to make is just that we do not have any way to explain the stretch activation by the current sliding filament theory.

Brutsaert: I am more familiar with tetanized cardiac muscle to reach a steady state. In those conditions the muscle gets fatigued extremely fast. You can maintain a tetanus for a couple of seconds only. With barium contracture, on the other hand, the steady state lasts much longer. But you assume that you have cyclic, periodic activity with barium. How sure are you about this?

Steiger: As soon as we get the fatigue phenomenon, the delayed tension disappears and the muscle gets very stiff.

Brutsaert: My question is, how sure are you that you have cyclic, periodic activity of cross-bridges during barium contracture?

Steiger: The delayed tension, and the fact that the muscle is able to produce oscillatory work, are indications that the cross-bridges are cycling. If you do not have cycling cross-bridges, you either get a rigor-type phenomenon or you get transients which you usually see in a high tension state, which we described in insect flight muscle and also in rabbit psoas muscle. So the transient response should change dramatically. You should not see a delayed tension. The initial recovery phase should change in character very rapidly. And we know from insect flight muscle, where we measured the ATPase activity, that in every case where we do not have those transients, we do not have good cycling in the preparation.

Sugi: If you apply an isotonic release, or in other words allow the muscle to shorten against an isotonic load during the delayed tension rise, what happens? Does the velocity increase? Has someone done this type of an experiment?

Steiger: I have not done that experiment yet.

Saeki (added after the discussion): In Dr. Sagawa's lab., I once performed isotonic release experiments on cat papillary muscle during Ba contracture, and observed isotonic velocity transience analogous to that reported by Civan & Podolsky (1966) in frog skeletal muscle fibers.

Tregear: It seems from the last part of your talk, Dr. Steiger, that you are making a very good correlation between certain biochemical rate constants and mechanical rate constants. I would ask you to comment on whether you would therefore deduce that the molecules within the tissue are behaving largely in an independent fashion. In other words, you use this same rate constant, and then the cooperative behavior, if there be any, is not greatly influencing the rate of this behavior.

Steiger: The way we measure the mechanical rate constant, at the moment, I would consider to be very preliminary; but the separation of the rate constants of each phase by about one order of magnitude allows us to distinguish them. But we still might have some variation of a factor of two within each rate constant, for example, if there is any variation with the applied step changes or with the amplitude of the length change. We are not precise enough to resolve amplitude dependence within the rate constant.

Tregear: If you can accept the correlation, you obviously are at least thinking of going a long way further, and saying that these processes are due essentially to dissociation. I mean, two of your rate constants are being interpreted in terms of dissociation of the proteins, actin from myosin, and these are the two fastest processes. This is totally different from the conventional way of thinking about them, I would say, isn't it? I mean, it is a hypothesis which is totally at variance with the Huxley-Simmons model.

Steiger: That is right.

Tregear: So, again I wonder how far that needs to be debated around the room. I mean, it is a very important conclusion if it is right, and it therefore deserves a thorough hearing. While I am still asking the question, a rather involved question, in the third of your correlations between k_3 and V_{max}, you give no biochemical meaning to V_{max}, do you? What does it mean in your scheme?

Steiger: In the scheme I showed that V_{max} is equivalent to k_{+4}.

Tregear: The final ending of this long question is, does all this series of logic lead you to the conclusion that the effective, or energetic, use of the hydrolysis of ATP can be maintained even though the proteins part and come together again in between the hydrolysis of ATP? In other words, can you go back through k_{-3} on your scheme, and come back and still get the energy out? That, to me, is the most important conclusion of the whole thing. It is a very interesting, specific hypothesis which I should think has one rather obvious geometrical constraint. It will have to come back onto the same actin; otherwise, if it went on to another, I would suppose that it might well not have the same energy.

Steiger: I am not sure about that.

Tregear: No. But it is an interesting way the speculation might go.

Kushmerick: I would like to have Dr. Kawai's comment. It would seem to me that his important conclusion should be contrasted with the important conclusion that you make. And it seems to me that you are saying that the biochemical rate constants obtained in solutions, and presumably the affinities, the K_m's, and all the rest, do apply to a three-dimensionally intact fiber generating force, whereas it seems to me, an important point that came out, or at least was raised in Kawai's results was that, in fact, some of the values——at least the apparent affinities——are apparently changed presumably because of the three-dimensional structure. Is this just an apparent contradiction, or is it a flat-out contradiction?

Steiger: I think Kawai's data and mine are relatively consistent with one another. The problem is the fastest rate in Kawai's experiment. He is working at room temperature, 20°C or higher. In the psoas muscle step experiments, it is difficult to separate those two processes. On the other hand, at low temperature you can nicely see the kink in the tension recovery phase, and you can separate the two phases because they have a different temperature coefficient. Therefore, I would assume that the rate which Kawai is measuring is a mixture of the two phases, and also, the value he gets is a mixture out of that. It is about 600 or something at 20°C, whereas in the two phases I had, probably 300 and 1,100.

Kawai: As you have said, there is not much disagreement as far as the mechanical rate constants are concerned. The measurement of K_m is a different measurement; we have disagreement and I do not know why, but the disagreement between the biochemical literature and what we measure has to be resolved in the future.

REFERENCE

Civan, M. M. & Podolsky, R. J. (1966) Contraction kinetics of striated muscle fibres following quick changes in load. *J. Physiol.* **184**, 511–534.

IV. LENGTH-TENSION RELATION

IV. LENGTH-TENSION RELATION

The Length-Tension Relation in Skeletal Muscle: Revisited

H. E. D. J. TER KEURS,* T. IWAZUMI and G. H. POLLACK

Department of Anesthesiology and Division of Bioengineering, University of Washington, Seattle, Washington, U. S. A.

ABSTRACT

Tension development during isometric tetani in single fibers of frog semitendinosus muscle occurs in three phases: (i) an initial fast-rise phase; (ii) a slow-rise phase; and (iii) a plateau, which lasts more than ten seconds. The slow-rise phase has previously been assumed to arise out of a progressive increase of sarcomere length dispersion along the fiber (Gordon, Huxley & Julian, 1966a, b). Consequently, the "true" tetanic tension has been considered to be the one existing prior to the onset of the slow-rise phase; this is obtained by extrapolating the slowly rising tension back to the start of the tetanus. In the study by Gordon et al. (1966b), as well as in the present study, the relation between this extrapolated tension and sarcomere length gave the familiar linear descending limb of the length-tension relation.

We tested the assumption that the slow rise of tension was due to a progressive increase in sarcomere length dispersion. During the fast rise, the slow rise and the plateau of tension, the sarcomere length dispersion at any area along the muscle was less than 4% of the average sarcomere length. Therefore, a progressive increase of sarcomere length dispersion during contraction appears unable to account for the slow rise of tetanic tension.

A sarcomere length-tension relation was constructed from the levels of tension and sarcomere length measured during the plateau. Tension was independent of sarcomere length between 1.9 and 2.6 μm, and declined to 50% maximal at 3.4 μm. While this result can be reconciled with the cross-bridge model of force generation if additional assumptions are invoked, it appears that the length-tension relation no longer offers unequivocal support for the theory.

* Present Address: Department of Experimental Cardiology, University of Leiden, Leiden, The Netherlands.

INTRODUCTION

In skeletal muscle, particularly at low temperature, the development of isometric tetanic tension does not occur smoothly. It occurs in three distinct phases: initially, it rises rapidly, then climbs slowly, and finally reaches a plateau which lasts many seconds. The phase of slow tension rise lasts progressively longer with increasing sarcomere length. These features have been observed consistently in many laboratories and with a variety of preparations (Ramsey & Street, 1940; Abbott & Aubert, 1952; Deléze, 1961; Edman, 1966; Gordon et al., 1966b; Sugi, 1972).

The presence of a slow rise of tension interspersed between the fast rise and plateau raises some question as to what value of tension to plot when constructing a length-tension relation. On the basis of certain experimental observations, Gordon, Huxley & Julian (1966a, b) found convincing rationale for assuming that the slow rise of tetanic tension, or in their terms, the "creep," was solely due to the development of progressive dispersion of sarcomere lengths. On the basis of this assumption they felt justified in considering only the value of tension reached prior to the onset of the slow rise of tension. This could be obtained by extrapolating both the fast rise phase and the slow rise phase, and considering the tension at the intersection of these tangents.

The plot of "extrapolated" tension against sarcomere length gave the familiar length-tension relation. The plateau and "descending limb" of this curve indicated that tension was proportional to the number of cross-bridges able to generate tension, and consequently this result has constituted one of the most dramatic pieces of evidence in support of the cross-bridge theory.

The objective of our study was to test whether or not the slow rise of tension was indeed associated with a progressive dispersion of sarcomere length. Evidently, if this were not so, then the rationale for having plotted the extrapolated tensions would be questionable, and one would need to reconsider whether the steady state, plateau tension, would be more appropriate to plot in a length-tension relation. The differences between the two length-tension relations is considerable, as we show below. The methodology, results and conclusions of these experiments are summarized here, but are presented in more detail in a recent paper (ter Keurs et al., 1978).

METHODS

We used single fibers of semitendinosus muscles of *R. pipiens*, carefully dissected from the dorsal head of the muscles. Particular care was exercised to pare the tendinous debris from the ends of the fiber so that striation spacing could be measured right up to the tendons (Fig. 1). Criteria for acceptability of specimens were based on the uniformity of sarcomere length along the fiber. If the peak to peak variation of sarcomere length along the fiber exceeded 0.2 μm in runs at three degrees of stretch (nominally 2.2 μm, 2.8 μm and 3.4 μm) the fiber was discarded. Many of the fibers showed the kind of nonuniformity observed by others (*e.g.*, Huxley & Peachey, 1961), *i.e.*, relatively short sarcomeres near the ends of stretched fibers. These were discarded.

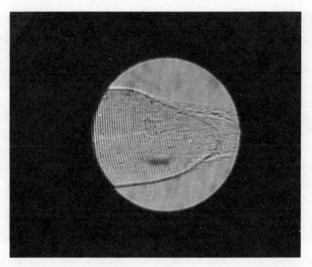

FIG. 1. Tendinous end of a single fiber. The figure illustrates the visibility of the striation pattern right up to the tendinous insertion.

The advantage of this selection procedure was that it obviated the need to length clamp a uniform fiber segment; the entire fiber could be held isometrically, and was effectively length-clamped.

We used light diffraction to measure sarcomere length and sarcomere length distribution. The fiber was illuminated by a laser beam, which, in most of these experiments, was compressed to a diameter of about 180 μm. The cross-striated muscle acts as a grating and therefore diffracts the incident laser light into a zeroth order band and multiple higher order band pairs. The spacing between bands is uniquely related to the striation spacing in the fiber.

The Fraunhofer diffraction pattern was collected with a 40× water immersion objective lens, and was projected onto three sensors. First, a video system was used for monitoring the gross features of the diffraction pattern. Second, the diffraction pattern was projected onto a translucent screen and the image on the screen was filmed at 2 frames per sec. Third, the pattern was compressed along the length of the bands with a cylindrical lens and projected on a 128 element photodiode array.

The array was scanned 5 times per msec. Each scan generated a profile of intensity *vs.* distance along the photodiode array. From this information the sarcomere length was computed. The spread of light intensity was used as an

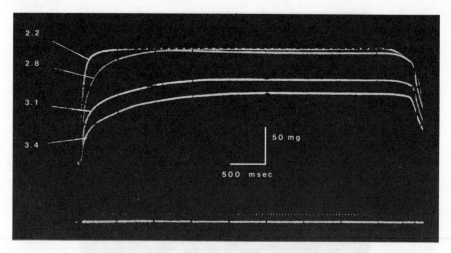

FIG. 2. Time course of isometric tension development at various sarcomere lengths. Fiber diameter 90 μm, slack length 16 mm. Record of stimuli shown at bottom. Sarcomere lengths shown at left were those measured at rest. Sarcomere lengths during the plateau of tetanus were 2.09 μm, 2.77 μm, 3.15 μm, 3.55 μm (top to bottom). Tension development shows a fast initial rise, a slow rise and a plateau. A slight overshoot of tension was often noted in contractions at sarcomere lengths beyond 2.7 μm. The slow rise is most prominent at the longer sarcomere lengths.

index of sarcomere dispersion in the optical field. However, the spread of light results not only from sarcomere length nonuniformity, but from low angle scatter and limited band-width of the optical system as well. Thus the "half-width" of the first order gives an overestimate of sarcomere length dispersion. Nevertheless, this technique was convenient for determining the time course of *changes* of local sarcomere length dispersion.

In addition to the single fibers, we also studied intact muscles. A number of experiments were repeated on the long extensor muscle of the 4th toe of frogs.

RESULTS

Figure 2 shows the time course of tension development in single fibers at various degrees of stretch. Tension rose rapidly at first at all sarcomere lengths. A plateau of tension was always reached by 3 sec following the start of contraction (at 8°C.). However, a slight overshoot of tension following the slow rise was often noticed before the tension settled at the plateau level. Using longer lasting tetani than those illustrated in Fig. 2, we found that the tension remained at the plateau level for about ten seconds, and then gradually decayed with a time constant on the order of twenty seconds. The time course of tension in the intact fibers was similar.

Figure 3 shows the time course of sarcomere length changes that occurred during isometric tetani in a representative fiber. At all initial lengths the sarcomeres shortened during the first 50 msec of tetanus by $0.15 \mu m$ or less. In some of the fibers there was a slight amount of lengthening instead of shortening. (Either could be found in a given fiber, depending upon which region was selected for scrutiny.) Following the initial length change the sarcomere length varied less than $0.1 \mu m$ during the subsequent 4 sec of contraction. In one fiber no sarcomere length changes were detectable during seven successive tetani which started as sarcomere lengths ranging from 2.2 to 3.4 μm; in this muscle, therefore, tension development in the region under observation was truly isometric.

We used several methods of measuring both local and global dispersion. First, local dispersion was measured from the intensity distribution incident on the photodiode array. Figures 3 C' and D' show the distributions photographed during rest and during the plateau phase of two tetani. As the figure illustrates, there is little change between rest and the plateau phase of isometric tetanus.

A second measure of local dispersion was obtained by examining the time course of the half-width of the first order. Generally there was some fluctuation during the rapid rise phase of tetanus (during which time there was usually a small amount of sarcomere shortening or lengthening); beyond this time ther was usually little change of local dispersion. A third, independent measurement of dispersion was obtained by analysis of photographic records of the direct diffraction pattern; Fig. 4 illustrates that only minor changes of the diffraction pattern took place during the course of tetani lasting several seconds. Densitometric analysis of data such as those shown in Fig. 4 showed that local dispersion most often increased modestly from rest to plateau, but sometimes showed no change, or decreased.

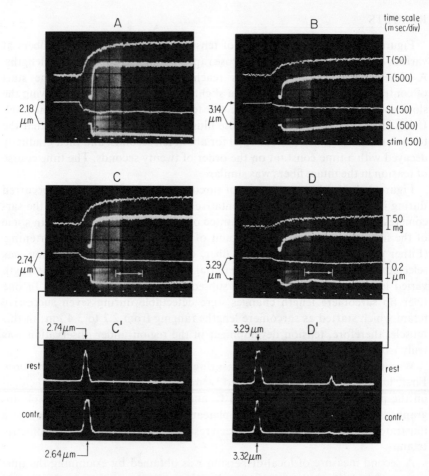

FIG. 3. Sarcomere length changes, tension development and local dispersion. A to D show records of isometric tetani starting from various sarcomere lengths. The upper pair of traces in each record shows tension development on a time scale of 50 and 500 msec/div., respectively. The lower pair of traces shows the sarcomere length changes during these tetani on the same time scales. Stimulus pulses are shown at bottom. Panels C' and D' show records of the intensity profile of the first order of the diffraction pattern at rest and during tetani for the contractions shown in C and D. Each record consists of an exposure of the film to 500 consecutive scans of the intensity profile. The exposure period during the tetanus is indicated by the white arrowed bar above the lower sarcomere length trace. C' shows slight shortening of the sarcomeres in the illuminated area with a small increase of the local dispersion. D' shows slight lengthening of the sarcomeres and a decrease of the local dispersion. The second order is visible in panel D'.

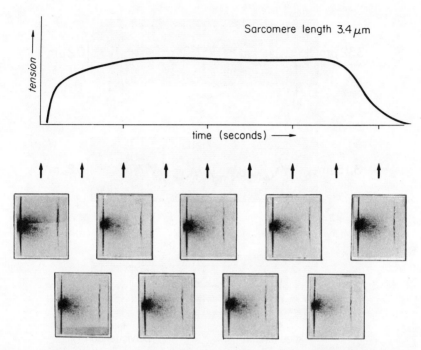

FIG. 4. Tension development and the changes of the direct diffraction pattern during contraction. The upper graph depicts the tension development (reconstructed from the original tension record) during a four-second isometric tetanus. The series of photographs shows the direct diffraction pattern taken at 0.5-second intervals during the tetanus at the moments indicated by the arrows. Only the zeroth and one of the two first orders is shown in each frame. The diffraction patterns show no change in the value of the local sarcomere length dispersion, 0.08 μm, during contraction, and no measurable change of the median sarcomere length. Despite this, the tension record shows a phase of slow tension development.

Global dispersion was measured by rapidly translating the stage during the plateau phase of eight-second tetani, and measuring the local sarcomere length along the fiber axis. A representative result is shown in Fig. 5. The root mean square variation of sarcomere length always increased from rest to plateau, but in no instance did the latter value ever exceed 4% of the mean.

In summary, the results show that during tetani at sarcomere lengths between 2.0 and 3.6 μm, only small changes in sarcomere length dispersion occurred in the fiber both locally and globally. Despite the absence of large increases in dispersion, the tension development at sarcomere lengths above 2.3 μm always showed clearly the presence of a slow-rise phase. These results thus fail to corroborate the assumption made by Gordon *et al.* (1966b).

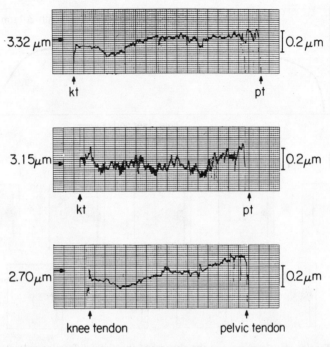

FIG. 5. Variation of the median sarcomere length along the fiber during the plateau of tension. Each record was made by translating the microscope stage in the axial direction during the plateau of a tetanus of 8 seconds' duration, and recording the computed median sarcomere length on a chart recorder. Since the rate of translation may not have been steady, there is not necessarily a direct correspondence between points on the abscissa and position along the muscle. The sarcomere lengths noted on the records were the mean values along the fiber at rest. The arrows correspond to these values. The sharp downward deflections on the right side of the records are artifacts caused by bubbles on the fiber surface.

Figures 6 and 7 show the relation between tension and sarcomere length. In Fig. 6 we plot the relation between tension and the sarcomere length as obtained by extrapolation of the slowly rising tension back to the start of the tetanus. The sarcomere length values were those existing at the time the extrapolated tension was reached. Previous studies using servo control of the length of segment of fiber (Gordon *et al.*, 1966b) show a similar correspondence between sarcomere length and extrapolated tension. The relation, therefore, does not seem to depend critically upon the experimental procedure.

In Fig. 7 we plotted the relation between sarcomere length and plateau tension. Tension was independent of sarcomere length between 1.9 and 2.6 μm,

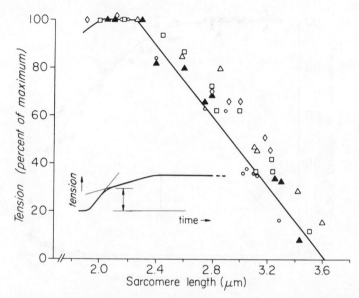

FIG. 6. The relation between tension and sarcomere length. Results obtained from "extrapolated" tension measurements and the sarcomere length at the moment of tension measurement in 7 single fibers. Each symbol indicates the results from one fiber. The curve through the data points is the one obtained by Gordon et al. (1966b). Most points were obtained in contractions in which there was a small amount of initial shortening or lengthening (usually less than 0.1 μm); however, in one fiber (□) seven of the thirteen data points were obtained in contractions in which there were no detectable changes of sarcomere length.

declined to 50% maximal at 3.4 μm, and declined further at longer sarcomere lengths.

The length-tension relations for intact toe muscles were similar to those of Figs. 6 and 7, except that there was considerably more passive tension in the intact muscles.

DISCUSSION

We have obtained two main results. First, the slow rise of tension is not necessarily accompanied by a progressive dispersion of sarcomere length; and second, when steady-state levels of developed tension are plotted against sarcomere length, a length-tension relation is obtained which is flat over a considerable range of sarcomere length.

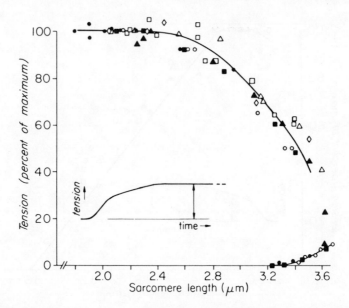

FIG. 7. Similar to Fig. 6, except that tensions obtained during the tension plateau are plotted. The curve in the lower right hand corner gives the passive tension of the single fibers.

The question arises whether our fibers behaved in some anomalous way. As best we could tell the results of our experiments were similar to those obtained by Gordon et al. (1966b). The tensions reached by the end of the fast-rise phase were comparable, as the length-tension relations constructed using extrapolated tensions were similar (Fig. 6). This similarity extended to our experiments with whole toe muscles as well. Furthermore, assuming the few illustrative examples provided by Gordon et al. (1966a, b) were representative, the rates of rise of tension during the slow rise phase of tension development were also similar in the two sets of experiments (see ter Keurs et al., 1978).

Our results are also in excellent agreement with those of Carlsen et al. (1961) who used plateau tensions obtained from tetani lasting three to four seconds to construct the length-tension relation. The curve remained relatively flat well beyond rest length. The sarcomere length at which tension fell to 50% maximal was approximately 3.4 μm, as found here (Fig. 7). Long-lasting contractions have also been obtained with skinned fibers. While the results of Schoenberg and Podolsky (1972) do not resemble "steady-state" length-tension relation obtained here, the work by Endo (1972) and more recently by Fabiato and Fabiato (submitted for publication) indicate that for fully activated skeletal or cardiac muscle fibers, the length-tension relation is flat over a range of sarcomere length similar to that found here.

Thus, it appears that our results are similar to those obtained by others.

Given two well-documented length-tension relations—steady state and extrapolated—the question arises as to which one best describes the primary contractile mechanism. To approach this question one must first consider the rationale for plotting extrapolated tensions. Gordon et al. (1966b) argued that the slow rise of tension was likely to arise out of "progressive development of irregularities of striation spacing," and consequently, that the tensions extrapolated back to the time prior to the onset of such irregularities were the appropriate values to consider. A mechanism by which this could occur is shown in Fig. 8. Sarcomeres with slightly larger initial overlap of thick and thin filaments could conceivably contract more strongly and stretch sarcomeres in series with less initial overlap. The small initial nonhomogeneity would therefore increase during contraction. This could result in the situation illustrated in Fig. 9: the tension in the shortening population would increase as the sarcomeres "creep up" the descending limb. The lengthening population could sustain the tension by stretching along its passive length tension relation, as illustrated in the figure, or by some other mechanism based on active tension.

If the slow rise of tension were mediated by a population of sarcomeres "creeping up" the linear descending limb of the length-tension curve and stretching a weaker population, one should be able to document the existence of two progressively diverging populations. While our records generally showed some increase of nonhomogeneity of sarcomere length during the slow rise, they gave no indication of any separation into widely divergent populations (Figs. 3, 4 and 5).

The expected separation of the populations should have been large enough to be detected. Consider a contraction in a fiber whose initial sarcomere length

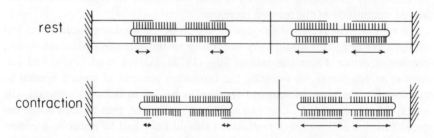

FIG. 8. Hypothetical distribution of sarcomere lengths at rest and during contraction. A small initial disparity of thick and thin filament overlap could give rise to a difference of contractile strength, which could increase the length disparity.

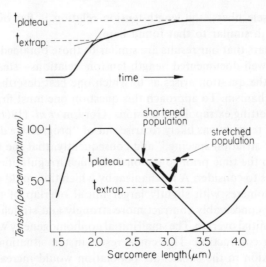

FIG. 9. Interpretation of the slow rise of tension based upon the mechanism shown in Fig. 8. Tension increases as the shortening population creeps up the descending limb. The lengthening population supports tension either by stretching along the passive length-tension relation or by some active mechanism.

is about 3.0 μm (Fig. 6). The tension developed during the rapid-rise phase is about 45% of P_0, i.e., 45% of the plateau tension at 2.2 μm. The tension then climbs and settles at about 80% of P_0 (Fig. 7). To account for this extra tension would require shortening up the descending limb by about 0.5 μm. Shortening of this magnitude was never observed. In regions of our fibers in which shortening was found, it was typically of the order of 0.1 μm and often less (Figs. 3,4 and 5); in no instance did shortening in any region of any fiber exceed 0.25 μm. This includes regions right up to the tendons (Fig. 5). Thus, the observed magnitude of sarcomere shortening would have been inadequate to account for the observed magnitude of the tension increase.

To estimate the required amount of stretch of the weaker population is less straightforward. This could vary according to the mechanism assumed, i.e., passive or active. From the data of Sugi (1972), Edman et al. (1976) and Edman et al. (in press), we estimate the minimum amount of stretch needed to account for the slow increase of tension to be 0.1 to 0.2 μm, although the exact value is uncertain. One can safely say, however, that at least some degree of stretch of the weaker population should have had to occur as a consequence of the shortening of the stronger population.

Therefore, in the example presented above it follows that the sarcomeres should gradually have diverged from the 3.0 μm initial length into two distinct clusters whose lengths were centered about the values 2.5 µm and somewhat beyond 3.0 µm. A cluster which had shortened to 2.8 µm, for example, would not be allowed by the hypothesis since it could not sustain a high enough tension. Thus a "forbidden zone" is created between the two predicted clusters. Such a zone is expected for any contraction in which the initial length is between 2.2 and 3.6 µm. Figure 5 shows that by the time the plateau of tension had been attained the majority of sarcomeres had lengths in the forbidden zones. This rules out the possibility that the slow rise of tension could be generated by a population of sarcomeres which, for one reason or another, was transparent to our technique (Julian et al., 1978); the sarcomeres which were detected should not have had lengths in the forbidden zone.

Another kind of irregularity of striation spacing which could conceivably mediate the slow rise of tension is unbalanced overlap of thick and thin filaments within sarcomeres. Displacement of thick filaments from their central position within the sarcomere would increase the overlap in one half-sarcomere, possibly increasing the tension. However, at sarcomere lengths of 3 µm and above, to achieve the required increase of overlap in one half-sarcomere would necessitate a decrease of overlap in the other half-sarcomere to the extent that the thick and thin filaments would no longer overlap. One might then postulate that the tension could still be supported by some connecting link between the end of the thick filament and either the Z-disc or the thin filament, but such an element would need to be stiffer than the parallel elastic element of the fiber (Iwazumi, 1970).

Theoretical arguments aside, unbalanced overlap of thick and thin filaments could not have occurred consistently during contraction, for a nonhomogeneity of this type would have substantially diminished the intensity of the diffracted orders. Intensity diminution was observed often but not consistently (Fig. 3), whereas the slow rise of tension proved to be a highly consistent phenomenon.

On the basis of these considerations we conclude that it is unlikely that progressive development of irregularities of striation spacing among or within sarcomeres causes the slow rise of tension. Consequently, we question the justification used by Gordon et al. (1966b) in plotting extrapolated tension.

In support of this conclusion it is relevant to point out here that a distinct slow-rise phase of tension development appears not only on the descending limb of the length-tension relation, but also on the ascending limb (Gordon et al., 1966b, Figs. 4 and 6). If one asserts that the descending limb is unstable because of its negative slope, it follows that the ascending limb ought to be stable, since it has a positive slope. Thus progressive dispersion of sarcomere lengths is not expected to occur on the ascending limb; yet the slow rise of tension is clearly in evidence there.

In the preceding paragraphs we have argued that it is improbable that the slow rise of tension is due to progressive dispersion of striation spacing. On the other hand, it remains possible that the slow rise is still due to some secondary process, distinct from the primary contractile process. If so, there is a continuing rationale for considering the length-tension relation obtained by extrapolation to reflect the primary contractile process. The alternative approach is to assume there is a single contractile process, but its time course shows a fast and a slow phase before the steady state is achieved. In this approach the length-tension relation obtained using the plateau tensions best describes the primary contractile process, though the extrapolated length-tension relation could still have significance.

It is worthwhile examining the implications of the two approaches to determine what constraints each might impose on present views of molecular contractile mechanisms, and perhaps also to determine which of the two is most likely to be correct. In the first approach it is assumed that full activation of the myofilaments is achieved by the end of the fast rise of tension. The length-tension relation derived from tensions obtained at this time characterizes cross-bridge tension mechanisms *per se*. The additional tension results from a potentiating mechanism, which increases cross-bridge force over and above the full activation levels.

This approach preserves the excellent correlation between overlap and tension which is demanded by the cross-bridge theory in its simplest form. On the other hand, since the theory must also account for the higher plateau tension, additional constraints need be imposed upon the theory. First it must be assumed that the state of "full activation" is not the one in which maximum force is generated; a mechanism must be identified by which the cross-bridges are capable of generating supramaximal force. Second, the potentiation must vary with sarcomere length. The ratio of plateau tension to extrapolated tension is 1 at sarcomere lengths near 2.0 μm, but increases at longer sarcomere lengths. At 3.4 μm the potentiated tension is about five times the tension at "full activation."

The alternative approach assumes that the activation process consists of a fast phase followed by a slow one, and that full activation is achieved when the tension plateau is reached. In this approach the extrapolated length-tension relation characterizes the length dependence of the fast phase of tension development, while the one obtained with steady-state tensions characterizes the fully activated cross-bridge mechanism. This approach leads immediately to the difficulty that tension is not directly proportional to the degree of overlap of thick and thin filaments. Again, additional constraints must be placed on the basic mechanism. Agreement between data and theory can be preserved if, for example, it is assumed that the cross-bridges within 0.3 μm of the center of the thick filament do not contribute to tension generation (this gives the flat region of the length-tension relation), and if the probability of cross-bridge attachment or the strength of each bridge increases with distance from the center of the thick filament (this accounts for the large tensions with little overlap).

Neither approach provides a straightforward interpretation of the length-tension data. While it is evident from the above discussion that the cross-bridge theory can be accommodated in either of the two approaches, it is clear that the length-tension data no longer offer unqualified support for the cross-bridge theory.

REFERENCES

Abbott, B. C. & Aubert, X. M. (1952). The force exerted by active striated muscle during and after change of length. *J. Physiol.* **117**, 77–86.

Carlsen, F., Knappeis, G. G. & Buchthal, F. (1961). Ultrastructure of the resting and contracted striated muscle fiber at different degrees of stretch. *J. Biophys. biochem. Cytol.* **11**, 95–117.

Deléze, J. B. (1961). The mechanical properties of the semitendinosus muscle at lengths greater than its length in the body. *J. Physiol.* **158**, 154–164.

Edman, K. A. P. (1966). The relation between sarcomere length and active tension in isolated semitendinosus fibres of the frog. *J. Physiol.* **183**, 407–417.

Edman, K. A. P., Elzinga, G. & Noble, M. I. M. (1976). Force enhancement induced by stretch of contracting single isolated muscle fibres of the frog. *J. Physiol.* **258**, 58P–59P.

Edman, K. A. P., Elzinga, G. & Noble, M. I. M. Enhancement of mechanical performance by stretch during tetanic contractions of vertebrate skeletal muscle fibers. *J. Physiol.*, **281**, 139–155.

Endo, M. (1972). Length dependence of activation of skinned muscle fibers by calcium. *Cold Spring Harb. Symp. quant. Biol.* **37**, 505–510.

Gordon, A. M., Huxley, A. F. & Julian, F. J. (1966a). Tension development in highly stretched vertebrate muscle fibres. *J. Physiol.* **184**, 143–169.

Gordon, A. M., Huxley, A. F. & Julian, F. J. (1966b). The variation in isometric tension with sarcomere length in vertebrate muscle fibres. *J. Physiol.* **184**, 170–192.

Huxley, A. F. & Peachey, L. D. (1961). The maximum length for contraction in vertebrate striated muscle. *J. Physiol.* **156**, 150–165.

Iwazumi, T. (1970). A new field theory of muscle contraction. Ph.D. Thesis, Univ. of Pennsylvania.

Julian, F. J., Sollins, M. R. & Moss, R. L. (1978). Sarcomere length nonuniformity in relation to tetanic responses of stretched skeletal muscle fibres. *Proc. R. Soc.* (B) **200**, 109–116.

Ramsey, R. W. & Street, S. F. (1940). The isometric length-tension diagram of isolated skeletal muscle fibers of the frog. *J. cell. comp. Physiol.* **15**, 11–34.

Schoenberg, M. & Podolsky, R. J. (1972). Length-force relation of calcium activated muscle fibers. *Science, N. Y.* **176**, 52–54.

Sugi, H. (1972). Tension changes during and after stretch in frog muscle fibres. *J. Physiol.* **225**, 237–253.

ter Keurs, H. E. D. J., Iwazumi, T. & Pollack, G. H. (1978). The sarcomere length tension relation in skeletal muscle. *J. gen. Physiol.* **72**, 565–592.

Discussion

Noble: I tried doing this sort of analysis which you showed, where you extrapolate back. On the tension records that we obtained, at any rate, the slowly rising portion before the plateau was so curved that I could not draw a line for back extrapolation at all. You have not really presented any criteria; is that extrapolation really valid anyway?

ter Keurs: For us, this was a practical point. In order to compare our data with the data available in the Gordon, Huxley, Julian's paper (Gordon, Huxley & Julian, 1966), we chose an extrapolation procedure which was essentially very comparable to theirs. We took a tension slope at 400 msec after the start of the tetanus. Then we could get an intersection with the extrapolated rising phase of tension. Now, as you have seen in Fig. 2, it is indeed true that the slow rise of tension development is more an exponential process, certainly more than a linear process. Therefore, it heavily depends on where you put your ruler. If you put your ruler just before the moment the plateau is reached, then you arrive at a much higher tension at long sarcomere lengths than when you put your ruler along the tension development trace early in the contraction, so is is indeed an arbitrary choice where to take the tension.

Wilkie: I am very familiar with this argument about instabilities and inequalities, because it was used frequently 20 years ago by A. V. Hill when Brian Jewell and I were doing our mechanical experiments. We satisfied ourselves of the truth of what you just said: that, in fact, on the falling part of the tension-length curve, you do not get instabilities in practice. We did a very simple experiment of taking paired frog sartorii, and attaching them to either end of a lever pivoted in the middle. By this means you can produce graded inequalities by moving the point of the attachment of one of the muscles (Fig. IV–1).

We stretched both of them until they were both on the descending limb of the tension-length curve, and we found that when we stimulated the muscles nothing happened until we altered the lever arm to a ratio of about two to one, and then we got a very slow movement in the direction of stretching the disadvantaged muscle. So I think the kinetic, or time-dependent, argument

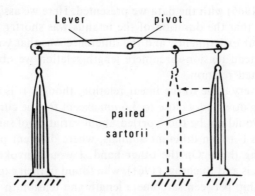

FIG. IV-1. An experiment demonstrating the stability of the descending limb of the length-tension relation.

that you produced is a very good one, as anyone can satisfy for themselves by repeating that experiment. The mechanism involved may be that there is a virtual discontinuity in the force-velocity curve between shortening and stretching (*e.g.* Katz, 1939).

ter Keurs: Thank you for that comment, which implies that you will not expect nonuniformities to develop.

Wilkie: That is right.

Edman: I fully agree with your argument about the stability of the descending limb. One thing which puzzles me, however, is that when I do the same type of measurements in *Rana temporaria* fibers, we do not get the same plateau as you seem to get. The curve starts to descend at about very close to 2.2 μm even in very long tetani. I wonder if you have done experiments on other types of muscle?

ter Keurs: I know that you use *Rana temporaria* and we used *pipiens*. I am testing it at the moment in our own setup in Leiden. The other point is that I assume that, in your measurements, you find a slow phase of tension development as well.

Edman: Yes.

ter Keurs: Well then, I wonder what the length-tension diagram at the inflection point looks like?

Edman: Yes, well, the curve is sloping somewhat less, you know, than if you take the measurement at say 1 sec, as I did in my 1966 paper (Edman, 1966), but it is not very much different; so the data on these fibers certainly differ a great deal from what you present.

ter Keurs: Well, that is surprising, because we compared the data in your paper (Edman, 1966) with the data we presented. Here we assigned the difference to the fact that the duration of the tetanus was shorter in your experiments than in our experiments, and the difference was that your curve lay in between the plateau tension-sarcomere length relation we obtained and the linear extrapolated relation.

Edman: It is very close to the linear relation, though. It is slightly curved, but it is pointing down very close to 3.6 μm except for the curved region.

ter Keurs: I would not be too worried about variation of sarcomere length-tension relations between different animals, where different procedures were used in obtaining them. On the other hand, I would invoke the results of Carlsen, Knappeis and Buchthal (1961), who found in their experiments a very similar relationship between sarcomere length and tension to the one we obtained plotting plateau tensions. They measured plateau tensions as well. And it is very surprising, indeed, that the data of Endo (1973), obtained in maximally activated skinned fibers, produced very similar sarcomere length-tension relations as our fibers.

Winegrad: I think it is appropriate to mention the work of John Blinks and his co-workers. Implicit in this kind of study is the assumption that activation is maximal at all sarcomere lengths, because you are tetanizing the muscle. Blinks and his co-workers (Blinks, Rüdel & Taylor, 1978) have used aequorin to measure the intensity of the light signal, and they found that the intensity of the light signal, which is proportional to the free calcium concentration, is not uniform at all sarcomere lengths. It decreases at long sarcomere lengths, and actually the maximum does not occur at 2.1 μm or 2.2 μm but closer to something like 2.6 or 2.7 μm. I think that if you take this sort of curve and impose it on the Gordon, Huxley, Julian length-tension relation, you would get the final curve that you have come up with, which shows a longer plateau and then a very sharp falloff at long sarcomere lengths.

ter Keurs: That case questions another statement about skeletal muscle, that activation at all lengths is full from very early after the start of the contraction. So, if you interpret the data—either one of the two length-tension relations—you need additional assumptions in order to make cross-bridges work in such a way that they can produce both length-tension relations. In the one case you would assume full activation by the end of the fast rise of tension, and a subsequent process of potentiation to account for the slow rise. This process would have to be length-dependent, time-dependent and temperature-dependent. Or, you could assume that full activation is not attained early in contraction, but is attained after the fast phase and the slow phase are over. Then the plateau tension-length curve could be explained only by assuming that the properties of the cross-bridges are a function of their position along the myosin, which also is quite an appreciable, additional hypothesis.

REFERENCES

Blinks, J. R., Rüdel, R. & Taylor, S. R. (1978) Calcium transients in isolated amphibian skeletal muscle fibres: detection with aequorin. *J. Physiol.* **277**, 291–323.

Carlsen, F., Knappeis, G. G. & Buchthal, F. (1961) Ultrastructure of the resting and contracted striated muscle fiber at different degrees of stretch. *J. Biophys. biochem. Cytol.* **11**, 95–117.

Edman, K. A. P. (1966) The relation between length and active tension in isolated semitendinosus fibers of the frog. *J. Physiol.* **183**, 407–417.

Endo, M. (1973) Length dependence of activation of skinned muscle fibers by calcium. *Cold Spring Harb. quant. Biol.* **37**, 505–510.

Gordon, A. M., Huxley, A. F. & Julian, F. J. (1966) The variation in isometric tension with sarcomere length in vertebrate muscle fibers. *J. Physiol.* **184**, 170–192.

Katz, B. (1939). The relation between force and speed in muscular contraction. *J. Physiol.* **96**, 45–64.

The Effect of Stretch of Contracting Skeletal Muscle Fibers

K. A. P. EDMAN,* G. ELZINGA** and M. I. M. NOBLE†

* Department of Pharmacology, University of Lund, Lund, Sweden
** Laboratory for Physiology, Free University of Amsterdam, Amsterdam, The Netherlands
† Midhurst Medical Research Institute, Midhurst, England

ABSTRACT

Single fibers from the semitendinosus and tibialis anterior muscles of *Rana temporaria* were stretched during tetanic contractions. Force rose to a plateau value during stretch and failed to return within 2 sec after stretch to the value expected from isometric contractions at the stretched length. The possibility that this was due to nonuniformity of sarcomere length was excluded because (1) no evidence of such nonuniformity was obtained from laser diffraction studies, (2) the development of nonuniformity in some fibers was not associated with increased force enhancement by stretch, (3) force enhancement after stretch was independent of velocity of stretch, (4) there was a shift to the right of the force-velocity curve by stretch, *i.e.*, velocity enhancement as well as force enhancement.

The effects of stretch were divided into three components: (1) a velocity-dependent force present only during stretch, (2) a component which was independent of velocity but was dependent on amplitude of stretch up to 18 nm per half-sarcomere during stretch and 11.5 nm per half-sarcomere after stretch; at greater amplitudes of stretch, no further rise in force was found. This component decayed away completely about 2 sec after the end of stretch. (3) A component only present at sarcomere lengths above 2.3 μm. This component was dependent on the amplitude of stretch over the range studied and its force-extension curve depended on the starting sarcomere length. This component decayed to a value which remained higher than control isometric tension at the stretched length in tetani lasting up to 8 sec.

Force enhancement after stretch has features suggesting that it is a property of the contractile mechanism. However, component 2 had an optimum sarcomere length of about 2.7 μm and component 3 an optimum of about 2.8–2.9 μm, *i.e.*, the optima did not correlate with filament overlap.

INTRODUCTION

The detailed characteristics of the length-tension curve of muscle in single muscle fibers has been established for some years (Ramsey & Street, 1940; Gordon, Huxley & Julian, 1966; Edman, 1966). Between sarcomere lengths of 2.2 and 3.65 μm, tension decreases if the sarcomere length at which the tetanus starts is increased; this correlation between tension and the overlap of thick and thin filaments forms the principle basis for cross-bridge theories. However it is also known in whole muscle (Fenn, 1924; Abbott & Aubert, 1951; Hill & Howarth, 1959) and in single fibers (Hill, 1977; Edman, Elzinga & Noble, 1976, 1978) that increase in sarcomere length during contraction results in an increase in tension.

The purpose of the present study was to elucidate in more detail this apparent paradox of tension increase with overlap decrease (Deleze, 1961) during tetanic contractions of frog single muscle fibers.

METHODS

Single fibers from the semitendinosus and tibialis anterior muscles of *Rana temporaria* were studied at temperatures between 1 and 3° C. A detailed description of the apparatus and method of measuring sarcomere length has recently been published (Edman, Elzinga & Noble, 1978). In the present paper, experiments were carried out in short tetani of 1 sec duration using trains of pulses at a frequency of 16–22 Hz. In addition, long tetani of 5–8 sec duration were studied with 5 min intervals between contractions. In order to enable the fiber to sustain this level of activity, frequencies of stimulation in the range 8–12.5 Hz were used.

RESULTS AND DISCUSSION

Force enhancement during stretch

The application of a steady stretching ramp at constant velocity caused a sharp initial increase in force (Fig. 1). After a certain amount of stretch, there was a break in the force record so that the initial sharp rising phase was followed by a plateau or slow rise of tension (Fig. 1). The changes in sarcomere length are shown in Fig. 2, which consists of densitometric scans across the laser diffraction pattern as recorded by streak photography on Gevapan film (Edman, Elzinga & Noble, 1978). There was a smooth increase in sarcomere length during the whole of the stretching period.

The force recorded at the end of the sharp rising phase was dependent upon

EFFECT OF STRETCH ON MUSCLE

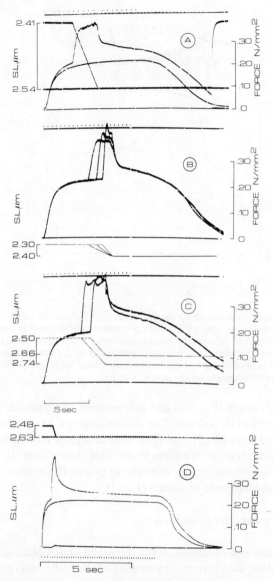

FIG. 1. Force and displacement records from single fibers stretched during tetanus. A. Comparison of stretch during activity (from 2.41 μm to 2.54 μm sarcomere length) with isometric tetanus at 2.54 μm. B. Comparison of three different velocities of stretch ending at the same time during the tetanus. C. Comparison of two different amplitudes of stretch at the same velocity ending at the same time during the tetanus. D. Long tetani on slower time base to show persistence of increased force after stretch compared to isometric control.

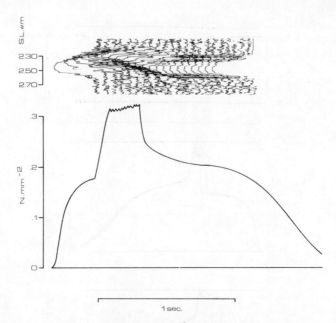

FIG. 2. Densitometric scans across the first order laser diffraction pattern. The zero order is off scale below these scans. The initial outward movement of the peak is due to internal shortening of sarcomeres. The subsequent inward movement is caused by the stretch. Note the constancy of the intensity distribution during the remainder of the contraction following stretch. The force record is drawn in below on the same time base but is not synchronized because of the x modulation of the scans by the intensity signal.

the velocity of stretch (Fig. 1B) and independent of amplitude of stretch (Fig. 1C). The velocity of stretch could be chosen to give a flat-topped second phase at a given sarcomere length, e.g. (Fig. IB); an increase in sarcomere length to above 2.3 µm then caused the force to increase slowly (Fig. 1C). When such a rising force was present during this second phase, the amount of rise was dependent on the amplitude of stretch (Fig. 1C).

Force enhancement after stretch

Following the end of stretch, force remained higher than in isometric contractions at either the short or long length (final length). In Figs. 1A and D the isometric contraction of the long (final) length is superimposed as the control. This force enhancement after stretch decayed with time (Fig. 1). It died away completely at approximately 1 sec after the end of the stretch. At sarcomere lengths above 2.3 µm, there was a similar phase of rapid decay but to a value that was above the control; this value (of residual force enhancement) was reached in about 2 sec (Fig. 1D).

Force enhancement after stretch was completely independent of the velocity with which the stretch had been applied. Thus, in the case of various stretches of the same amplitude but different velocities (Fig. 1B), the force followed a common path after the end of stretch. The dependence on stretch amplitude and sarcomere length will be described later.

Is force enhancement by stretch a property of the sarcomeres?

In our previous paper we state, "We cannot rule out the development of some degree of nonuniformity of a random or highly localised nature which cannot be detected with the method used" (Edman, Elzinga & Noble, 1978). Obviously, if the phenomenon of force enhancement by stretch is due to nonuniformity of sarcomere length it is of little interest in shedding light on more basic properties of muscle. Our studies of nonuniformity have proceeded further (see below) but the statement made above can never be denied; uniformity can never be perfect. Nevertheless, any important nonuniformity can be excluded as the cause of the force enhancement for the following reasons:

1. Nonuniformity has been studied by streak photography of the first order laser diffraction pattern (Cleworth & Edman, 1972) using 1 sec and 6 sec tetani and recording at many locations along the fiber. An example of the densitograms of such photographs is shown in Fig. 2. It can be stated (a) that no portion of the first order beam ever shows an outward movement during stretch, *i.e.*, there is no detectable sarcomere shortening during stretch. (b) The pattern after stretch remains extremely stable throughout the remainder of the contraction even in long tetani, *i.e.*, there is no detectable sarcomere length change of any sort after the end of the stretch. (c) These results, (a) and (b) (Fig. 2), were obtained from streak photographs recorded at 0.5 mm intervals from end to end of the fiber, *i.e.*, from the moment of applying the stretch onwards; there is no detectable shortening of one part of the fiber at the expense of another.

2. Some fibers which are uniform at the beginning of an experiment can become nonuniform, *e.g.*, by stretching them to extremely long sarcomere lengths. Nonuniformity was characterized by increased internal shortening, variable internal shortening along the fiber, spread and instability of the first order laser diffraction pattern, increased variance of sarcomere length along the fiber. When stretch was applied during contraction in such fibers, it was found that the development of nonuniformity never caused greater force enhancement than when the fiber was uniform (Fig. 3).

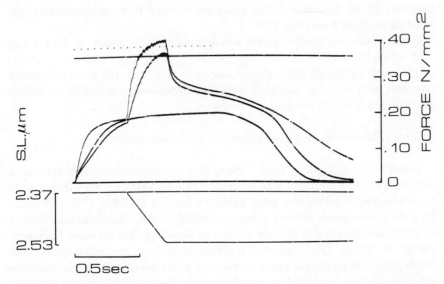

FIG. 3. Comparison of stretch during contraction in the same fiber before and after the development of nonuniformity. Nonuniformity causes decreased rate of rise of tension at the beginning of the tetanus but no increase in tension after stretch.

3. Nonuniformity is supposed to cause force enhancement because some sarcomeres shorten to lengths near the plateau of the length-tension curve and occupy the cross-section of a small part of the length of the fiber. It is then assumed that the sarcomeres in all other parts of the fiber can "hold" the tension generated by these short sarcomeres which determine the total force recorded. Since force enhancement is found immediately after the end of stretch (Fig. 1), this undetectable shortening occurs during stretch and must occur at a finite velocity. By increasing the velocity of stretch (Fig. 1B), we (a) reduced the time for shortening and (b) reduced the velocity of shortening by an increase in force during stretch. Thus much less sarcomere shortening occurs during the faster stretch. Nevertheless, the force enhancement after stretch is identical for all stretch velocities studied (Fig. 1B).

4. Force velocity curves were recorded by releasing the fibers to isotonic loads after the end of the stretch. They were compared with similar curves derived from releases at the same time during control isometric contractions. At sarcomere lengths above 2.3 μm, where residual force enhancement after stretch was found, there was a distinct separation of the force-velocity curves derived after stretch and without preceding stretch (Fig. 4). This difference increased in magnitude with increasing load. This result confirms the findings of Cavagna

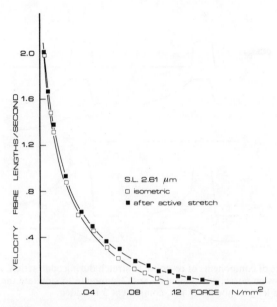

FIG. 4. Force-velocity relationships during ordinary isometric tetanus (open symbols) and after stretch during tetanus (filled symbols) at sarcomere length 2.61 μm.

& Citterio (1974) in whole muscle. Thus at any finite load, the fiber shortens faster after stretch. It is mathematically impossible to obtain such a result from nonuniformity of sarcomere length. The force-velocity data (Fig. 4) thus shows (a) that above 2.3 μm sarcomere length, stretch of the actively contracting fiber causes enhancement of total mechanical performance at finite loads and (b) such enhancement of mechanical performance is not caused by nonuniformity.

Is force enhancement by stretch a property of the contractile machinery?

For convenience, in trying to distinguish the various components of the effects of stretch, a simplified scheme has been adopted (Fig. 5). This divides the effects into three components. At sarcomere lengths on the plateau of the length-tension curve (1.95–2.25 μm) only the first two components are present (Fig. 5A).

Component 1 is a velocity-dependent force which is only present during stretch. This is a purely viscous phenomenon and can be attributed to some sort of friction between the thick and thin filaments when they are actively

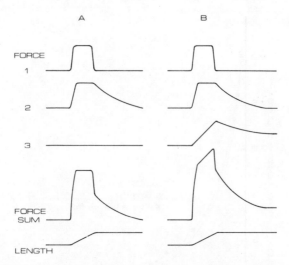

FIG. 5. Scheme of components of response to stretch during contraction. A. At the plateau of the length-tension curve, components 1 and 2 summate to simulate the response found. B. At sarcomere lengths above 2.3 μm, component 3 is added.

interacting. Such sliding of filaments by stretch is likely because (1) Hill (1977) has shown that the A-band remains of constant width under these circumstances and (2) we have shown that the grid spacing increased (Fig. 2). Therefore, there is either a stretch of the non-overlapping part of the thin filaments and Z-disc of the order of 10% or the thin filaments slide outwards. Component 1 will not be considered in any further detail.

Component 2 appears to increase in amount up to a certain fixed amplitude of stretch. Amplitudes of stretch beyond this critical value result in no further increase in force. The enhanced force is present after stretch but decays in about 1 sec. The "break" in the force records (Fig. 1) between the steeply rising part and the plateau during stretch is a feature of this component. The amplitude of stretch at which this break occurs was measured and found to be fairly constant with a mean value of 17.95 ($\pm$2.81) nm per half-sarcomere. In any given experiment, this value was independent of sarcomere length (Fig. 6A), suggesting that there is a certain distance along the thick filament beyond which no further increase in force can be exerted on the thin filament as it is pulled past.

The features of component 2 after stretch are seen by plotting force against sarcomere length at the end of stretch. In Fig. 7, the length-tension curve is drawn in for comparison. At the plateau of the length-tension curve, force

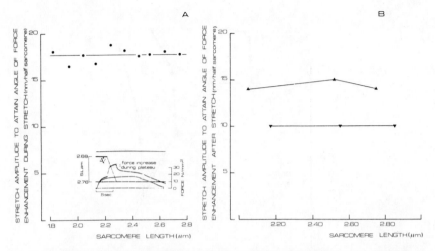

FIG. 6. A. Relationship between the stretch required to reach the "break" in the force record during stretch and sarcomere length. B. Relationship between the stretch required to reach the angle on the force extension curve after stretch and sarcomere length. Inset: Methods of obtaining values on Y axis of A (Δl) and the force increase during the plateau discussed in the text.

increased with amplitude of stretch up to a certain critical value and was then independent of amplitude. At sarcomere lengths above 2.3 µm, components 2 and 3 are combined but the angle at the critical value for component 2 is still discernible; this was measured and found to be fairly constant with a mean value of 11.5 ± 1.78 nm per half-sarcomere. In any given experiment, this value was independent of sarcomere length (Fig. 6B), suggesting that there is a certain relative movement between the thick and thin filaments associated with this component. The similarity between the critical amplitude of component 2 during and after stretch is obvious, but why is the value during stretch 6 nm greater than after stretch?

Component 3 is only present at sarcomere lengths above 2.3 µm. This component appears to be dependent on amplitude of stretch over the range studied. It causes the rising force of the "plateau phase" during stretch and does not decay completely after stretch, being responsible for the residual force enhancement after stretch. There was a tight correlation between the rise of force during the plateau during stretch and the residual force enhancement after stretch (r = 0.83). Component 3 is associated with the enhancement of the force-velocity curve (Fig. 4) which is not present below sarcomere lengths of 2.3 µm.

Component 3 can be likened to some sort of spring which is stretched. This

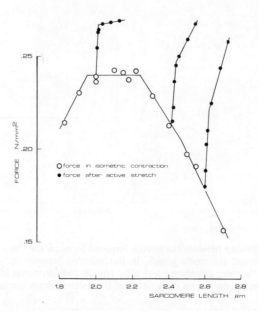

Fig. 7. Effect on total force after stretch of increasing amplitude of stretch at three different starting points on the isometric length-tension curve.

idea was explored by plotting the residual force enhancement (force after stretch above isometric control at the same final length) against sarcomere length (Fig. 8). If component 3 were a spring, all the points should fall on the same force-extension curve. This is not the case. There appear to be different springs which get progressively stiffer with increasing sarcomere length. The characteristics of the apparent elastic element depend upon the starting sarcomere length for the stretch. This property points to the possibility that some elastic structure is formed, reorganized or realigned during activation and argues against the idea that component 3 is a passive phenomenon.

Force enhancement by stretch and the overlap of thick and thin filaments

The features outlined above suggest that force enhancement by stretch is indeed a property of the contractile machinery. We would then expect the phenomenon to be correlated with the overlap of the thick and thin filaments in the same manner as isometric tension. This is not the case.

In Fig. 9, forces during and after stretch are plotted against sarcomere length

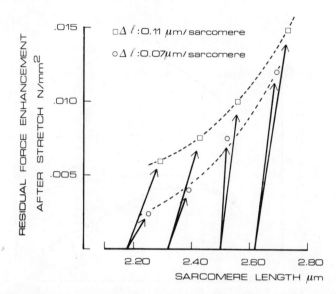

FIG. 8. Relationship between residual force enhancement after stretch and sarcomere length for three different starting points on the isometric length-tension curve. Dashed lines indicate behavior expected for elastic elements.

and compared with the ordinary length-tension curve. Whereas ordinary isometric tension reaches an optimum at sarcomere lengths of 1.95–2.2 μm, force after stretch has an optimum at about 2.4 μm. There is marked overall deviation from the shape of the usual length-tension curve. In fact, these values depend on both component 2 and component 3 because the measurements were made early after the end of the stretch. When force enhancement is calculated by subtracting the isometric from the after-stretch values in Fig. 9, no optimum is obtained due to the influence of component 3 (see below). An attempt was made to separate force enhancement due to component 2 by subtracting the values 4.5 sec after stretch (after component 2 decay) from the 200 msec after-stretch values. This yields an apparent optimum for component 2 at 2.5–2.7μm.

The sarcomere length dependency of component 3 is shown in Fig. 8 over this same range and shows no optimum. More recent studies in progress show that the magnitude of this component declines at sarcomere lengths above 2.9 μm. This decline as the sarcomere length approaches the no-overlap range is consistent with the idea that component 3 is a property of the interaction between actin and myosin filaments. However, the marked deviation from the behavior predicted proportionally from overlap requires explanation.

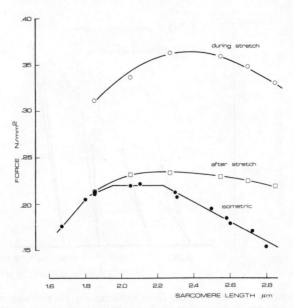

FIG. 9. Relationship between total force produced by stretches of constant velocity and amplitude compared with the isometric values.

REFERENCES

Abbott, B. C. & Aubert, X. M. (1951). Changes of energy in a muscle during very slow stretches. *Proc. R. Soc. B.* **139**, 104–117.

Cavagna, G. A. & Citterio, G. (1974). Effect of stretching on the elastic characteristics and the contractile component of frog striated muscle. *J. Physiol.* **239**, 1–14.

Cleworth, D. R. & Edman, K. A. P. (1972). Changes in sarcomere length during isometric tension development in frog skeletal muscle. *J. Physiol.* **227**, 11–17.

Deleze, J. B. (1961). The mechanical properties of the semitendinosus muscle at lengths greater than its length in the body. *J. Physiol.* **158**, 154–164.

Edman, K. A. P. (1966). The relation between length and active tension in isolated semitendinosus fibres of the frog. *J. Physiol.* **183**, 407–417.

Edman, K. A. P., Elzinga, G. & Noble, M. I. M. (1976). Force enhancement induced by stretch of contracting single isolated muscle fibres of the frog. *J. Physiol.* **258**, 95–96P.

Edman, K. A. P., Elzinga, G. & Noble, M. I. M. (1978). Enhancement of mechanical performance by stretch during tetanic contractions of vertebrate skeletal muscle fibres. *J. Physiol.* **281**, 139–155.

Fenn, W. O. (1924). The relationship between the work performed and the energy liberated in muscular contraction. *J. Physiol.* **58**, 373–395.

Gordon, A. M., Huxley, A. F. & Julian, F. J. (1966). The variation in isometric tension with sarcomere length in vertebrate muscle fibres. *J. Physiol.* **184**, 170–192.

Hill, A. V. & Howarth, J. V. (1959). The reversal of chemical reactions in contracting muscle during an applied stretch. *Proc. R. Soc. B.* **151**, 169–193.
Hill, L. (1977). A-band length, striation spacing and tension change on stretch of active muscle. *J. Physiol.* **266**, 667–685.
Ramsey, R. W. & Street, S. F. (1940). The isometric length-tension diagram of isolated skeletal muscle fibers of the frog. *J. cell. comp. Physiol.* **15**, 11–34.

Discussion

Mohan: Can you explain the stress relaxation phenomenon on the basis of any molecular mechanism?

Noble: I would hesitate to do that at this stage. I hope you will.

Rüegg: Your stretches were comparatively slow. Have you ever done fast stretches, and if you have, does it look like stretch activation? In other words, do you get a very quick transient increase in tension followed by a fall and then by a delayed activation which would then give a peak tension similar to the peak tension you got?

Noble: No. We have not done that. Prof. Sugi has done very rapid stretches (Sugi, 1972), and he should describe them; but I think that, with rapid stretches, the force after stretch is sometimes higher and sometimes lower, much lower than the control. Is that correct?

Sugi: At some moderately rapid speed, there was a delayed rise in tension whose magnitude increased with increasing amount of stretch.

Noble: Yes, that is essentially what we find here; but in your paper (Sugi, 1972) you did show that sometimes, with very rapid stretches of particular amplitude, you got a tension lower than isometric after you stretched. We did not find that. We did not stretch fast enough.

Pollack: On some of your records during the stretch, the tension waveform shows what appears to be noise, but as I think we discussed previously some time ago, it may not be noise, because it does not appear during all phases of stretch. Do you have any idea what causes this, or what mechanism is involved?

Noble: No, we do not always see that. It may be due to the presence of some series viscosity, or something like that, but I would very much hesitate to attribute this to anything other than wobble of the lever, because the whole fiber is so extremely stiff at these higher forces that very, very tiny, almost nondetectable movements could produce that amount of ripple. I am not saying that is the cause, because we could not detect that kind of length oscillation, but we cannot exclude it.

Kawai: I think you excluded, as the cause of the extra elastic component (Component 3), the effect of heterogeneity or the contribution from the passive elements which could be in parallel with the cross-bridges. Now, do you consider this extra component is somehow involved in the cross-bridge?

Noble: No, I do not exclude it. All we are saying is that if the arragnement of this elastic element is in some way from Z line to Z line, you have to explain why the mechanical characteristics of that elastic element are different if we activate the fiber at different sarcomere lengths, so that if we activate the fiber at 2.2 μm its characteristics are of a fairly weak, almost non-existent elastic element, whereas if we go to 2.8 to 2.9 μm it is maximally stiff and if we go to 3.4 μm it is weaker again.

REFERENCE

Sugi, H. (1972) Tension changes during and after stretch in frog muscle fibres. *J. Physiol.* **225**, 237–253.

Length-Tension Relation in Frog Skeletal Muscle Activated by Caffeine and Rapid Cooling

Toshio SAKAI

Department of Physiology, Jikei University School of Medicine, Tokyo, Japan

ABSTRACT

The isometric tension of the caffeine-rapid cooling contracture (caffeine-RCC) in a frog toe muscle is proportional to caffeine concentration. The maximum tension of caffeine-RCC is generated in the presence of 1.0 mM caffeine, and the dose-response relation shows an S-shaped curve in the range from 0.1 to 1.0 mM.

The length-tension relation of caffeine-RCC was investigated in the presence of various concentrations of caffeine, and our results were compared with the schematic summary proposed by Gordon, Huxley & Julian (1966). The effect of muscle stretching on potentiation of caffeine-RCC was observed. In addition, the rate of rise of tension in caffeine-RCC was examined at various sarcomere lengths.

INTRODUCTION

The skeletal muscle which is treated with a lower concentration of caffeine than the threshold for contracture produces a sustained contraction when the temperature of the solution is rapidly lowered, and it immediately returns to the resting state when the temperature is raised (caffeine-Rapid Cooling Contracture) (Sakai, 1965). The maximum tension of caffeine-RCC of the frog toe muscle is generated in the presence of 1.0 mM caffeine, and the dose-response relation shows an S-shaped curve in the presence of 0.1–1.0 mM. According to X-ray diffraction studies on caffeine-RCC, the ratio of the intensities of the 1,1 and 1,0 equatorial reflections, $I_{1,0}/I_{1,1}$, appeared to increase with tension development. Hart, Yu & Podolsky (1977) have concluded that the intensity ratio is a measurement of cross-bridge number during isometric contraction. In caffeine-RCC, it has been postulated that activation occurs mainly as the result of two kinetic processes of the Ca-pump in the sarcoplasmic reticulum: release of Ca from SR, and suppression of Ca-uptake by SR at a low temperature. It might be assumed that the caffeine-RCC technique as shown in the dose-response curve is able to control the intracellular concentration of calcium by using various concentrations of caffeine.

The responses in twitch and potassium contracture were potentiated with increase in sarcomere length (Close, 1972; Gonzalez-Serratus, Valle & Cillero, 1973; Gordon & Ridgway, 1976). According to these results, it has been suggested that the amount of Ca released from the intracellular Ca store during activation increases in proportion to the degree of stretching of muscle within 2.9 μm of striation spacing. In skinned muscle fibers, also, the active tension generated by low concentrations of calcium increased as the muscle was stretched from the optimum length (Endo, 1972).

In the present experiments, the length-tension relation and effect of muscle stretching on caffeine-RCC were studied in both normal Ringer and 95 mM K_2SO_4 solutions. A part of this work has been described elsewhere (Sakai, Matsubara & Hashizume, 1977; Hashizume, 1977).

METHODS

The experiments were done in isolated toe muscles (M. ext. digiti longus IV) of frog (*Rana nigromaculata*). The experimental procedure on activation of the caffeine-RCC has been described in previous papers (Sakai, 1965). The sarcomere lengths before caffeine-RCC were adjusted by a manipulator and calculated from the first order of laser diffraction line given by the muscle fiber, using the equation: $d = n \lambda \sin \theta n$. The laser beam (He-Ne, = 632.8 nm) was adjusted on the middle portion of the muscle and oriented perpendicular to its long axis. The sarcomere lengths were measured before and after the caffeine-RCCs. In some experiments, the sarcomere lengths were measured on many points along the muscle fibers.

RESULTS AND DISCUSSION

The length-tension relation of caffeine-RCC

Caffeine-RCCs in Ringer solution were produced at different sarcomere lengths in the presence of 1.0 mM caffeine. A peak tension of sustained contracture was continued but tension declined gradually with time though the preparation was kept at a low temperature. In the present experiment, the temperature of the solution, as soon as the peak tension was obtained by rapid cooling, was raised in order to aid relaxation. The resting tension of muscle developed with increase in sarcomere length, so that the active tension of caffeine-RCC was determined as the difference in the peak and resting values at each sarcomere length. Using results obtained from 13 muscles, the length-tension diagram of caffeine-RCCs in the presence of 1.0 mM caffeine was derived (Fig. 1). The active tension of caffeine-RCC was obtained as the relative value relative to the maximum tension in each fiber. The tension-length relation of caffeine-RCCs at the sarcomere length from 2.2 μm to 2.5 μm was shown empirically to be constant, but when the muscle was stretched above 2.5 μm, the active tension decreased gradually at the sarcomere length of 2.5–3.0 μm and declined steeply at above 3.0 μm in the manner which was observed in tetanus by Gordon et al. (1966). However, the tension-length relation on caffeine-RCCs in comparison with the schematic summary by Gordon, Huxley and Julian's experiment was shifted to the right and upward; that is, at the sarcomere length of 2.9 μm the active tension developed to about 80% of the maximum, which was observed at the sarcomere length of 2.3 μm. Furthermore, variation of the active tension at the sarcomere length of above 3.0 μm was

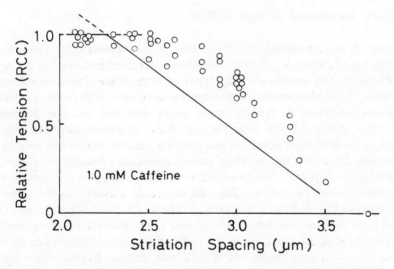

FIG. 1. The length-tension relation of caffeine-RCC. Caffeine-RCC was carried out in the presence of 1.0 mM caffeine. The active tension of caffeine-RCC was normalized relative to that which was produced at 2.2 μm. The schematic summary by Gordon et al. (1966) was superimposed with a line. This diagram was constructed by use of 13 frog toe muscles (Hashizume, 1977).

large, so that the descending limb of the length-tension relation did not cross the zero level when the sarcomere length was increased above 3.65 μm.

There are several uncertainties about this explanation: (1) since the frog toe muscle was used in a bundle, no accurate determination of the sarcomere length was made, and (2) the time course of the tension development in caffeine-RCC was different from that in tetanus.

Measurement of sarcomere length in stretched frog toe muscle

The sarcomere length was mainly measured in the middle point of the muscle, as a parameter of muscle stretching. In order to measure the nonuniformity of the sarcomere length, the variation of the sarcomere length along the muscle fibers at several degrees of stretch was examined. As pointed out by Huxley & Peachey (1961), the uniformities of the sarcomere length in stretched muscle were recognized (Fig. 2).

Measuring the shortening of the sarcomere length on the caffeine-RCC was difficult because transportation of the laser beam was disturbed by water droplets which were produced on the outside wall of the glass chamber at the low temperature.

Isometrtc contractions of caffeine-RCC

In Fig. 3, tension records of caffeine-RCC in the presence of 0.5 mM at varying degrees of stretch are shown. The tension increased and reached a steady-state level in a few seconds after rapid cooling. In the range of sarcomere lengths less than 2.8 μm, the time required for tension development to reach the steady-state level decreased as the sarcomere length increased and was prolonged by stretching over 2.8 μm. Also, the rate of rise of tension increased with increase in the sarcomere length of less than 3.0 μm. At sarcomere lengths of more than 3.0 μm, the rate of rise of tension decreased obviously in proportion to tension development. The rate of rise of tension increased in proportion to the concentration of caffeine. The rate of rise of tension was measured as illustrated in Fig. 3. From the results of caffeine-RCCs in the presence of 0.5 and 0.8 mM caffeine, the length-tension and the length-rate of rise of tension relations were investigated. As shown in Fig. 4, the active tension and the rate of rise of tension were normalized at each relative value. In these experiments, both relations were observed in the normal and depolarized muscles. In the case of depolarized muscles, the preparations were immersed in 95 mM K_2SO_4 solution for 60 min and then the caffeine-RCCs were observed. As shown in Fig. 1, the length-tension relation of caffeine-RCC was remarkably different

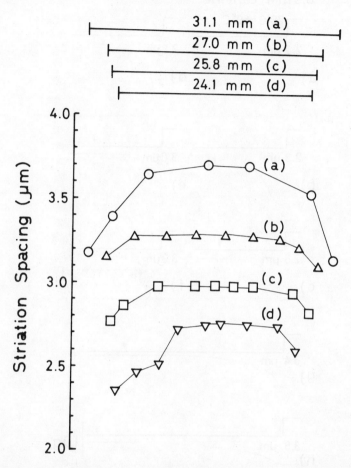

FIG. 2. Variation of sarcomere length along an isolated frog toe muscle at rest. Each sarcomere length was calculated from the first order of laser diffraction line. The horizontal lines at the upper part show the total length of stretched muscle.

from Gordon, Huxley & Julian's schematic summary. In another experiment, even when the sarcomere length on the middle portion of the muscle was more than 3.7 μm, a remarkable tension development was observed as well as the result illustrated in Fig. 4. These phenomena might be accounted for by uniformities of the stretched muscle as mentioned above. Such a possibility could be prospected in the present experiments, using the whole toe muscle.

On the other hand, the descending limb in the length-rate of rise of tension

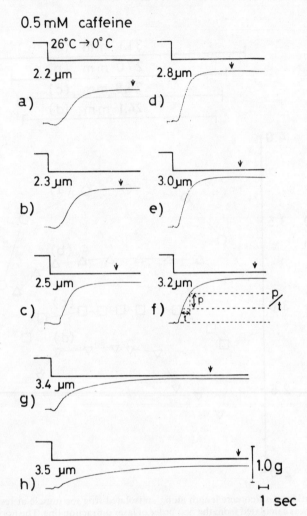

FIG. 3. Time course of caffeine-RCC in the presence of 0.5 mM caffeine at various sarcomere lengths. Rapid cooling was done by lowering from 26° to 0°C. Tension was recorded on a pen-writing oscillograph (San-Ei) through the strain gauge (VL type, Shinko) in normal Ringer solution. Arrow marks show the steady-state tension of caffeine-RCCs at each sarcomere length. In (f), the way of measuring the rate of rise of tension is shown.

relation on stretching crossed the zero level at the sarcomere length near 3.65 μm. This result seems to suggest that the initial reaction between the thick and thin filaments could be detected with minimum changes of each sarcomere length in activation of caffeine-RCC. However, the maximum value of the

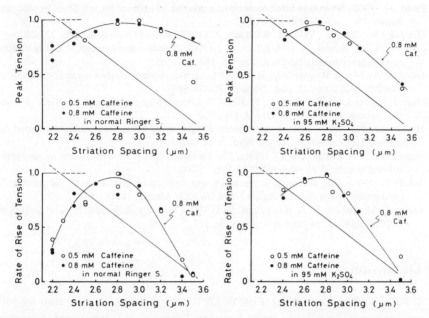

FIG. 4 Length-tension and length-rate of rise of tension relations on caffeine-RCCs in normal Ringer and 95 mM K_2SO_4 solutions. Caffeine-RCCs were observed in less than maximum concentration of caffeine, 0.5 and 0.8 mM. All points were normalized in relative value. Gordon, Huxley & Julian's schematic summary was superposed by a straight line in each diagram.

rate of rise of tension against the sarcomere length was located near 2.8 μm.

In the range of sarcomere lengths of 2.4 to 2.8 μm, both the tension and the rate of rise of tension increased as the sarcomere length was stretched. In the presence of less than maximum concentration of caffeine, the caffeine-RCC was below maximal tension. As reported in previous papers (Sakai et al., 1977; Hashizume, 1977) when the concentration of caffeine is 1.0 mM, tension output of caffeine-RCC reaches its maximum value at sarcomere lengths of 2.4 to 2.8 μm. According to these results, it might therefore be assumed that the amount of calcium released from SR is dependent upon the degree of muscle stretching.

REFERENCES

Close, R. I. (1972). The relations between sarcomere length and characteristics of isometric twitch contractions of frog sartorius muscle. *J. Physiol.* **220**, 745–762.

Endo, M. (1972). Stretch-induced increase in activation of skinned muscle fibres by calcium. *Nature (New Biol.)* **237**, 211–213.
Gonzalez-Serratus, H., Valle, R. & Cillero, A. (1973). *Nature (New Biol.)* **246**, 221–222.
Gordon, A. M., Huxley, A. F. & Julian, F. J. (1966). Tension development in highly stretched vertebrate muscle fibres. *J. Physiol.* **184**, 143–169.
Gordon, A. M., & Ridgway, E. B. (1976) Length-dependent electrochemical coupling in single muscle fibres. *J. gen. Physiol.* **68**, 653–669.
Hart, J. E., Yu, L. C. & Podolsky, R. J. (1977). Cross-bridge number and the X-ray diffraction pattern of muscle. *Biophys. J.* **17**, 171a.
Hashizume, K. (1977). Length dependence of tension in a frog toe muscle activated by caffeine rapid cooling. *Jikeikai Med. J.* **24**, 145–154.
Huxley, A. F. & Peachey, L. D. (1961). The maximum length for contraction in vertebrate striated muscle. *J. Physiol.* **156**, 150–165.
Sakai, T. (1965). The effects of temperature and caffeine on activation of the contractile mechanism in the striated muscle fibres. *Jikeikai Med. J.* **12**, 88–102.
Sakai, T., Matsubara, S. & Hashizume, K. (1977). Effect of muscle stretching on tension of caffeine-RCC. *J. Physiol. Soc. Japan* **39**, 287.

Discussion

Pollack: Just a comment. First of all, it is very nice to see that your length-tension curve is similar to ours; however, I noticed, at submaximal activation with lower concentrations of caffeine, that the length-tension curve shows a peak, and I think that the sarcomere length at which the peak occurs is rather similar to what Dr. Endo found in skinned fibers (Endo, 1973).

Sakai: That is right. It should, however, be kept in mind that Dr. Endo used skinned fibers, while I used intact muscles.

REFERENCE

Endo, M. (1973) Length dependence of activation of skinned muscle fibers by calcium. *Cold Spring Harb. Symp. quant. Biol.* **37**, 505–510.

Contribution of Connectin to the Parallel Elastic Component in Muscle

Koscak MARUYAMA and Kayoko YAMAMOTO
Department of Biology, Faculty of Science, Chiba University, Chiba, Japan

ABSTRACT

1. A simple rapid method to estimate the connectin content in myofibrils of rabbit psoas muscle is described. Myofibrils were directly extracted with either 0.1 M NaOH or 1 N acetic acid at room temperature. Care had to be taken to remove collagen debris. The amino acid composition of the alkali- or acid-extracted residues was very similar to that of connectin.

2. Rabbit psoas myofibrils mildly treated with trypsin fell to pieces on addition of Mg ATP, although the myofibrils retained their continuity regardless of the disappearance of the Z lines. The connectin content was markedly decreased by the trypsin treatment.

3. It is concluded that the connectin nets play a role in the transmission of active tension; as well as in passive elasticity, as the parallel elastic component in muscle.

INTRODUCTION

In skeletal muscle, active tension developed in sarcomeres of myofibrils is transmitted to bone through tendon, although the details of the tension transmission system at the myotendinous junction are still obscure (cf. Nakao, 1976; Maruyama & Shimada, 1978). What makes the tension transmission possible in myofibrils? This is easily understood, if some elastic structure connects adjacent Z-lines along the entire length of myofibrils to the myotendinous junction. The elastic structure must also be responsible for passive elasticity, when myofibrils are stretched (Natori, 1954).

Physiologists had long assumed a "parallel elastic component" in muscle to explain the phenomena described above. Since 1976 we have claimed that an elastic protein called connectin plays an essential role in the passive tension generation of myofibrils (Maruyama, 1976; Maruyama et al., 1976; 1977; Matsubara & Maruyama, 1977). In the present communication we show evidence that the connectin structure is also involved in the contraction process in myofibrils as the parallel elastic component.

METHODS

Myofibrils Rabbit psoas muscle was homogenized in 50 mM KCl, 1 mM EDTA, 1 mM $MgCl_2$, and 1 mM $NaHCO_3$. After filtration through a sheet of gauze, the suspension was centrifuged for 5 min at 5,000 g. This was repeated five times, and the solution was finally suspended in 0.1 M KCl and 1 mM $NaHCO_3$ after a wash with the same solution.

Trypsin treatment The myofibrillar suspension was treated with trypsin (Sigma, Type 1; weight ratio, 1/500) in 0.1 M KCl and 0.01 M Tris-HCl buffer, pH 7.2 at 25°. The reaction was terminated by adding a sufficient amount of soybean trypsin inhibitor (Sigma, Type I-S).

Amino acid analysis The proteins were hydrolyzed in 6 N HCl for 24 hr at 110° and subjected to an automatic amino acid analyzer (Hitachi KLA-3B).

Microscopic observations Contraction of myofibrils attached to a cover glass was observed under a phase contrast microscope (Nikon Biophot).

RESULTS

A simple method of connectin preparation

Connectin has been prepared from muscle residues from which all the salt-soluble proteins had been extracted (Maruyama et al., 1977). Because the procedures were rather tedious, a simpler method of preparing connectin was developed using isolated psoas myofibrils.

Myofibrils were extracted with 50 volumes of 0.6 M KI, 1 N acetic acid and 0.1 M NaOH, respectively, for 2 hr at room temperature. The suspension was filtered through a sheet of gauze to remove collagen debris, and this procedure was repeated once more. After centrifugation for 2 hr at 100,000 g, the sediment was washed with each extraction medium followed by a wash with water. The yield was approximately 200–300 mg per 50 g of fresh muscle, but only 10 mg or less with SDS.

In an SDS gel electrophoresis using 10% acrylamide, the acid- or alkali-treated residue had an immovable band, 90,000 and 43,000 dalton bands as in purified connectin (Maruyama et al., 1977). The KI-treated residue showed several bands of chain weights lower than 40,000 in addition to the above. The SDS-insoluble material did not have any movable bands in 10% acrylamide gels.

The amino acid compositions of various preparations are listed in Table 1 in comparison with that of connectin prepared by an SDS method of Maruyama et al. (1977). The SDS-insoluble material consisted almost entirely of collagen. The KI residue had an amino acid composition similar to connectin except for valine (cf. Kimura et al., 1979). This was rather surprising, because Z-lines remained almost intact after KI extraction. The materials insoluble in dilute acid or alkali were also very similar to connectin in their amino acid composition, although the lysine content was appreciably lower.

TABLE 1. Amino acid composition of myofibrillar residues extracted with 0.6M KI, 1 N acetic acid, 0.1 M NaOH and 1% SDS*

	Connectin	KI	Acid	Alkali	SDS
Hyp	1	1	3	4	90
Asp	94	92	90	95	64
Thr	61	63	61	64	25
Ser	57	64	61	60	44
Glu	124	116	120	118	93
Pro	66	64	59	53	115
Gly	80	75	88	94	341
Ala	83	80	86	80	34
Cys/2	3	2	2	2	3
Val	67	87	68	64	27
Met	25	24	29	30	12
Ile	63	57	58	60	20
Leu	75	73	77	72	30
Tyr	31	33	31	38	9
Phe	31	33	39	37	5
Lys	68	71	56	57	27
His	18	17	17	19	7
Arg	53	54	55	53	54

* Given as number of residues per 1000 residues

Direct extraction of myofibrils with dilute alkali or acid may be used for a rapid preparation of connectin, especially to estimate its content under various conditions. However, great care must be taken to avoid contamination with collagen fibers; otherwise the contents of hydroxyproline, glycine, and proline in the preparation become higher because of the coexistence of insoluble collagen (cf. Kimura et al., 1978).

As described above, 1% SDS completely solubilizes connectin in rabbit psoas myofibrils, when directly applied. However, after the extraction with KI, the residue becomes less soluble in 1 % SDS, even if thoroughly washed with water and finely homogenized in a Waring blender. Cardiac connectin is not easily soluble in 1 % SDS (Maruyama et al., 1977). Furthermore, it is of much interest that connectin in sarcolemma is not soluble in 0.5–1 % SDS, whereas that in myofibrils is very easily solubilized (Maruyama & Natori, 1978), as clearly seen in Fig. 1. The solubility in SDS may be dependent on the extent of aggregation and also of cross-linking of the connection peptides.

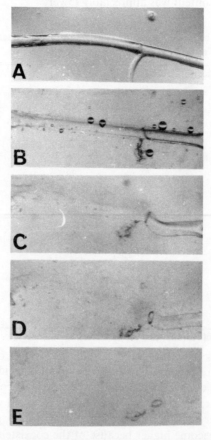

FIG. 1. Effect of an SDS solution on a skinned muscle fiber of the bullfrog. A, skinned fiber in liquid paraffin. B, immediately after the addition of 0.5 % SDS. C, 1.5 min. D, 2.5 min. E, 5 min (after Maruyama & Natori, 1978).

The role of connectin nets in tension transmission

Connectin forms nets linking adjacent Z-lines. The diameter of the connectin filament is as small as 2 nm, and the distance between the knots of the net does not exceed 100 nm (Toyoda & Maruyama, 1978). The connectin nets are shown in Fig. 2. Recent studies using high-resolution scanning electron microscopy have clearly revealed the net structure of connectin filaments (Sawada *et al.*,

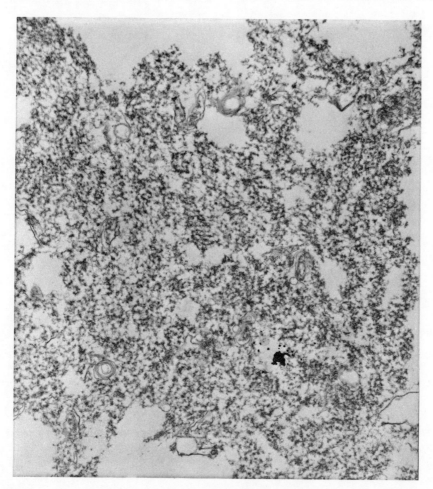

FIG. 2. Electron micrograph of a thin section of frog atrial muscle treated with 1% SDS for 2 hr at room temperature. × 21,000 (after Toyoda & Maruyama, 1978).

1979). It appears that connectin nets cover the myofibrils like a stocking; they also are present within the myofibrils, between thick and thin filaments.

These connectin nets have been already shown to be responsible for the passive elasticity of myofibrils and also of glycerinated muscle fibers; using Natori's skinned fiber and glycerinated muscle fiber treated with dilute alkali and SDS, respectively, reversible tension generation was measured upon stretch (Maruyama et al., 1976; 1977; Matsubara & Maruyama, 1977). However, it

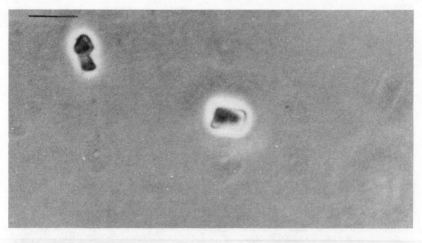

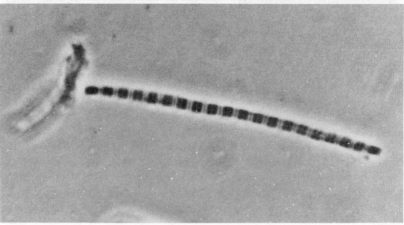

FIG. 3. Contraction of an intact myofibril from rabbit psoas muscle. Before and after the addition of MgATP.

was not easy to demonstrate that the same structure is involved in transmitting tension produced by the ATP-actin-myosin interaction. In the present study, we have utilized the fact that the connectin structure is very sensitive to proteolytic action in intact myofibrils (Maruyama et al., 1977).

Rabbit psoas myofibrils contract into a small mass, when allowed to do so, in the presence of MgATP. This implies the presence of an elastic connection along the whole length of a myofibril. Figure 3 shows this drastic shortening of a myofibril. The situation was different in trypsin-treated myofibrils (Fig. 4). As is well known, the Z-lines quickly disappear after a mild treatment with trypsin (e.g., 1/500 by weight ratio) for several min at room temperature. The trypsin-treated myofibrils still retained their continuity after the disappearance of the Z-lines. However, on addition of MgATP, the myofibrils fell into short pieces, as seen in Fig. 4. When both ends of a myofibril were by chance attached to a cover glass, the intact myofibril did not shorten with MgATP (isometric contraction), whereas the trypsin-treated one was torn off into small segments.

The connectin content of trypsin-treated myofibrils was measured by the simple method described in the previous section. The myofibrils were extracted with 0.1 M NaOH three times; each time the suspension was filtered through a gauze before centrifugation. Tendons attached to a stirring rod were also re-

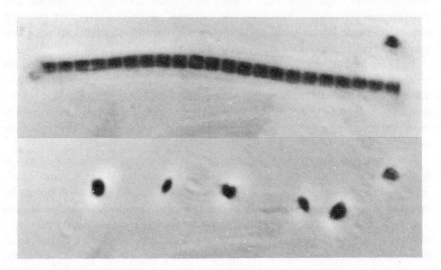

Fig. 4. Contraction of trypsin-treated myofibril from rabbit psoas muscle. Before and after the addition of MgATP.

TABLE 2. Effect of trypsin treatment on the connectin content of rabbit psoas myofibrils

Incubation time (min)	0	3	6	15
Connectin content (mg/g)	46	24	18	8

Myofibrillar suspension (protein, 20 mg/ml, total volume 25 ml) in 0.1 M KCl and 0.01 M Tris-HCl buffer, pH 7.2, was treated with trypsin (1/500 weight ratio) at 25°. Two times amount of soybean trypsin inhibitor was added to stop the reaction, followed by the addition of solid NaOH to give a final concentration of 0.1 M.

moved each time. The final suspension was homogenized in 1 M NaOH, and aliquots were taken for measuring the protein concentration. Duplicate runs were performed starting from 500 mg of myofibrillar proteins (approximately 5 g of fresh muscle). The results are summarized in Table 2. As seen in the table, only 3 min of treatment with trypsin (1/500 weight ratio) resulted in 50% decrease in the connectin content, where fragmentation of myofibrils occurred in the presence of MgATP (see Fig. 4). It should be noted that the fragmental contraction took place after 15 min of trypsin treatment, when most of the connectin was freed from the insoluble nets.

DISCUSSION

The nature of the parallel elastic component of muscle remained obscure until recently, while numerous studies concentrated on the contraction and excitation mechanisms. The insoluble "stroma" were not attractive research topics to muscle biochemists, and electron microscopic observations were also centered on the contractile and excitation systems.

In 1976, Maruyama, Natori and Nonomura first reported an elastic protein in muscle which is involved in generation of passive tension. Since then, efforts have been made to elucidate the nature and function of the elastic protein, connectin. The net-like structure of connectin has been revealed by transmission and scanning electron microscopy (Toyoda & Maruyama, 1978; Sawada et al., 1979).

The elastic nature of the connectin nets has been ascribed to cross-links of lysine derivatives such as lysinonorleucine (Fujii et al., 1978; Kimura et al., 1979). If the polypeptide chains of connectin are cut off at several points by proteolytic enzymes, the elasticity of the connectin nets will be easily lost, and this is indeed the case. A very mild treatment of skinned muscle fibers with trypsin results in disappearance of passive tension development. If stretched, the treated fibers are torn off. At the same time, the content of insoluble connectin is remarkably decreased (Maruyama et al., 1977).

The present study showed that a brief treatment of isolated myofibrils with trypsin caused fragmentation of contracted sarcomeres on addition of MgATP. The trypsin-digested myofibrils retained continuity before the addition of MgATP, although the Z-lines were no longer visible under a phase contrast microscope. Again, the connectin content was decreased by the mild trypsin treatment. It should be noted that the connectin nets disappeared after trypsin treatment (Maruyama et al., 1977). Hence it is likely that the tension developes by the sliding mechanism in each sarcomere is transmitted by the linking nets of connection in an extremely contracted case; the elastic nets cut by trypsin attack are not strong enough to maintain the structural continuity of contracting myofibrils so that "local" breakages of the myofibrils take place.

Thus it is suggested that the connectin nets can transmit the tension generated by the ATP-actin-myosin system to the myotendinous junction (cf. Maruyama & Shimada, 1978). The mechanism of the tension transmission from the cell membrane to collagen fibers via "connecting filaments" and external lamina is a subject for future study (cf. Nakao, 1976).

Acknowledgements

We wish to thank President Reiji Natori of the Jikei University for his helpful advice and warm encouragement. This work was supported by grants from the Ministry of Education, Science and Culture, the Ministry of Health and Welfare, the Muscular Dystrophy Association and the Yamada Science Foundation.

REFERENCES

dos Remedios, C. G. & Gilmour, D. (1978). Is there a third type of filament in striated muscles? *J. Biochem. Tokyo* **82**, 235–238.
Fujii, K., Kimura, S. & Maruyama, K. (1978). Cross-linking of connectin, an elastic protein in muscle. *Biochem. biophys. Res. Commun.* **81**, 1248–1253.
Kimura, S., Akashi, Y. & Kubota, M. (1978). Carp connectin: amino acid composition. *J. Biochem. Tokyo* **83**, 321–323.
Kimura, S., Fujii, K., Kubota, M. & Maruyama, K. (1979). Connectin crosslinking: reducible crosslinks in native fibrils. *Bull. Jap. Fish. Soc.* (in press).
Maruyama, K. (1976). Connectin, an elastic protein from myofibrils. *J. Biochem. Tokyo* **80**, 405–407.
Maruyama, K., Natori, R. & Nonomura, Y. (1976). New elastic protein from muscle. *Nature, Lond.* **262**, 58–59.
Maruyama, K., Matsubara, S., Natori, R., Nonomura, Y., Kimura, S., Ohashi, K., Murakami, F., Handa, S. & Eguchi, G. (1977). Connectin an elastic protein from muscle. Characterization and function. *J. Biochem. Tokyo* **82**, 317–337.
Maruyama, K. & Natori, R. (1978). Connectin an elastic protein from myofibrils and sarcolemma. *Zool. Mag.* **87**, 162–164.

Maruyama, K. & Shimada, Y. (1978). Fine structure of the myotendinous junction of lathyritic rat muscle with special reference to connectin a muscle elastic protein. *Tissue & Cell* **10,**741–748.

Matsubara, S. & Maruyama, K. (1977). Role of connectin in the lengthtension relation of skeletal and cardiac muscles. *Jap. J. Physiol.* **27,** 589–600.

Nakao, T. (1976). Some observations on the fine structure of the myotendinous junction in myotomal muscles of the tadpole tail. *Cell & Tissue Res.* **166,** 241–254.

Natori, R. (1954). The property and contraction process of isolated myofibrils. *Jikeikai Med. J.* **1,** 119–126.

Sawada, H., Ishikawa, H. & Maruyama, K. (1979). The net structure of connectin in myofibrils as revealed by high resolution scanning electron microscopy. *Tissue & Cell* (in press).

Toyoda, N. & Maruyama, K. (1978). Fine structure of the connectin nets in cardiac myofibrils. *J. Biochem. Tokyo* **84,** 239–241.

Discussion

Pollack: I was wondering about the size of these connectin filaments. As you know, it is often difficult to detect the presence of a regular structure of crossbridges either in cardiac muscle or in some types of skeletal muscle, and I wonder to what extent it is possible that the stubs one sees in some instances are actually pieces of connectin filaments rather than cross-bridges? Is the size of the connectin filaments such that it might be possible to mistake a segment of connectin for a cross-bridge?

Maruyama: I agree with your very interesting remarks, especially in cardiac myofibrils, where connectin filaments are abundant. The width of the connectin filaments is approximately 2 nm; therefore, it would be difficult to distinguish them from cross-bridges in the A-band region. However, in the I-band region, you can recognize them in the presence of thin (actin) filaments.

ter Keurs: In many people's results, and in my results as well, it was apparent that cardiac muscle cannot be stretched beyond 2.4 μm, at least rat cardiac muscle. Frog cardiac muscle can be stretched beyond 2.4 μm. Are there differences in elastic content of the species?

Murayama: No, I have not worked with rat cardiac muscle. I will try to determine the connectin content of rat heart in the near future.

Kawai: Is there any indication of the calcium sensitivity of connectin?

Maruyama: No, not at all.

Kawai: So, we can consider it totally stable on addition of calcium. Right?

Maruyama: Right.

Regulation of the Contractile Proteins in Cardiac Muscle*

Saul WINEGRAD and George B. McCLELLAN

Department of Physiology, School of Medicine, University of Pennsylvania, Philadelphia, Pennsylvania, U. S. A.

ABSTRACT

The regulation of the properties of cardiac contractile proteins has been studied in rat ventricular fibers that have been made hyperpermeable by a soak in cold 10 mM EGTA. These fibers are very permeable to small ions and molecules but not to the soluble proteins. The range of concentration of Ca ions that activate contraction can be varied up to six-fold by cyclic nucleotide-sensitive reactions. A cyclic AMP-regulated phosphorylation probably of the inhibitory subunit of troponin decreases Ca sensitivity, and this reaction requires the function of the sarcolemma. A cGMP-dependent reaction inhibits the decreased Ca sensitivity produced by the cAMP-dependent reaction. Although the amount of Ca-activated force is not changed by exposure to the cyclic nucleotides in relaxing or contracting solution, exposure to either cAMP, cGMP or epinephrine in combination with a non-ionic detergent can increase contractility. The extent to which Ca-activated force is increased depends on the existing amount of epinephrine stimulation of the preparation: the greater the epinephrine effect the greater the increase in contractility from cGMP. The effect of added cAMP does not depend on epinephrine. These observations can be explained by a model in which the β receptor is coupled with adenylate cyclase by a guanine nucleotide-dependent reaction, and the contractile force can be regulated by a cAMP-dependent reaction.

Although the force of contraction changes with the length of the muscle fiber as a result of a difference in the number of myosin cross-bridges that can interact with the actin-containing thin filaments, this mechanism is unlikely to be physiologically important in the regulation of the muscle's force-generating capabilities. The capacity for controlling the length of the fiber in the intact

* This work was supported by U. S. P. H. S. Grants HL 16010 and HL 18900.

animal is too limited. In order to use the rising phase of the sarcomere length-tension curve for regulating force, the resting cell must have short sarcomere lengths with double overlap of the thin filaments (Gordon et al., 1966), but double overlap in an unloaded resting vertebrate striated muscle has never been unequivocally demonstrated (Brown et al., 1970; Winegrad, 1974). The intensity of activation by Ca^{++} has been extensively studied as a mechanism for modulating contractility, particularly in cardiac muscle (Wood et al., 1969; Reuter & Scholz, 1977), and the results indicate that a difference in the transmembrane Ca^{++} current during the action potential is important in determining the concentration of activator Ca^{++} during the associated and subsequent contractions. All modulation of contractility in cardiac muscle cannot, however, be explained by changes in Ca flux across the sarcolemma.

The demonstration of enzymatically controlled phosphorylation sites on the contractile proteins, particularly on the inhibitory subunit of troponin (TNI) (Solaro et al., 1976; Ray & England, 1976; Stull & Buss, 1977) and one of the light chains of myosin (Frearson et al., 1976; Cole et al., 1978), has raised the possibility that changes in the contractile proteins themselves may be important in regulating the contractile system. The range of Ca concentration that is required for activation of the ATPase of isolated cardiac contractile proteins can be altered by a factor of about three by phosphorylation of the inhibitory subunit of troponin (Cole et al., 1978; Ray & England, 1976), but it is not clear how this potential regulation might operate in the intact cell. We have used a hyperpermeable preparation of rat ventricle to study regulation of contractility by modification of the contractile proteins (McClellan & Winegrad, 1977, 1978). This preparation, which is produced by soaking the tissues overnight in 10 mM EGTA at 0°C, has the useful property of high permeability to small ions and molecules, but not to large ones. A calcium buffer system can be used to activate the contractile proteins directly without washing out soluble cytoplasmic proteins. In addition, at least some of the enzymatic functions of the membrane are retained. In this preparation, both Ca sensitivity (that is, the concentration of Ca required for activation) and the maximum Ca force of the contractile proteins can be modified by what appear to be physiologically important regulatory mechanisms.

Regulation of Ca sensitivity

The hyperpermeable rat ventricular bundle generally needs a pCa (-log {Ca}) of about 4.5 for the production of maximum force, but this pCa requirement can be increased by four different substances. Non-ionic detergents in concentrations of 0.5–1.0% in a relaxing solution increase the pCa requirement for maximum Ca activation to about 5.2, which is equivalent to a decline of about 80 % in the concentration of Ca (Fig. 1). This change in Ca requirement involves merely a shift of the relation between force and Ca concentration along the Ca concentration axis, and it is not accompanied by a change in shape of the relation. After the treatment with detergent, the relation between tension and pCa is the same as that observed with mechanically skinned rat ventricular fibers (Fabiato & Fabiato, 1975), in which the surface membrane has been completely removed. Since the use of detergent with mechanically skinned rat heart cells, which have a functioning sarcoplasmic reticulum, does not alter Ca sensitivity (Fabiato & Fabiato, 1975), the detergent effect in the hyperpermeable cells must be on the sarcolemma. It is the sarcolemma and not the sarcoplasmic reticulum that is involved in the regulation of Ca sensitivity.

The nucleotide cytosine triphosphate is an excellent substitute for ATP in the reaction between actin and myosin, but it is a very poor phosphate donor. These properties make it a useful agent in determining the role of phosphorylation reactions in the regulation of the contractile proteins. Replacements of

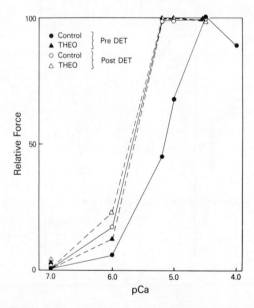

FIG. 1. The relation between developed tension and pCa is shifted to higher pCa (lower Ca concentrations) by addition of 5 mM theophylline or treatment with a non-ionic detergent (DET). There is no significant difference between the tension-pCa relation in theophylline or after 1 % detergent in relaxation solution. Curves are normalized to the value for maximum force. Each point is the mean of results from at least five different experiments.

all of the ATP in the bathing solutions with CTP results in the same increase in Ca sensitivity as treatment with detergent, but the change produced by CTP substitution is completely reversible (Fig. 2). Phosphorylation is therefore involved in maintaining a lower Ca sensitivity, and inhibition of phosphorylation raises Ca sensitivity.

Cyclic nucleotides are involved in the regulation of the phosphorylation since the addition of a phosphodiesterase inhibitor, which should raise the concentration of cyclic nucleotides by blocking their hydrolysis, reversibly increases Ca sensitivity about as much as CTP or detergent treatment (Fig. 1). The effect can be repeated several times until the preparation has been treated with detergent and the sarcolemma inactivated. The phosphodiesterase inhibitor is then no longer effective in changing Ca sensitivity, presumably because the adenylate cyclase in the membrane is no longing functioning.

The influence of cyclic nucleotides themselves on Ca sensitivity can be observed in hyperpermeable cardiac cells, although it is smaller than the response

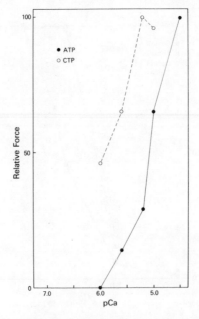

FIG. 2. Replacement of ATP with 5 mM CTP produces a shift in the tension-pCa relation to higher pCa. The CTP can support the generation of force but it cannot act as a phosphate donor.

to detergent, CTP, or phosphodiesterase inhibitors. Preparations with certain characteristics show the effects best. Although the responses of all preparations are qualitatively the same, the initial Ca sensitivities are not identical. The Ca sensitivity becomes similar in all preparations after treatment with detergents or theophylline; the more sensitive to Ca the cells were initially, the smaller the shift in Ca sensitivity with CTP, detergents or theophylline. Cyclic AMP produces a definite decrease in Ca sensitivity in the more sensitive preparations, and 10^{-6} M cGMP causes an increase in Ca sensitivity in preparations with a very low Ca sensitivity. Both changes are reversible. The cyclic nucleotides alter the response of the Ca sensitivity of the cells to a phosphodiesterase inhibitor in a way that is consistent with the effects of the cyclic nucleotides by themselves. The increase in Ca sensitivity that is produced by theophylline is, respectively, increased or decreased by cGMP or cAMP.

Cyclic nucleotides apparently regulate Ca sensitivity in an antagonistic way, but the key reaction is a cAMP-dependent phosphorylation that decreases sensitivity and requires the function of the sarcolemma. The regulatory mechanism is sensitive to catecholamines. When the isolated heart has been perfused with catecholamines before EGTA treatment has made the sarcolemma hyperpermeable, the cells have a lower Ca sensitivity. After treatment with detergent, however, these cells have the same calcium sensitivity as cells from unperfused hearts, indicating that the catecholamine perfusion modified the states but not the fundamental nature of the contractile proteins. The membrane reaction that regulated Ca sensitivity through phosphorylation of the contractile proteins is itself controlled by epinephrine (Fig. 3).

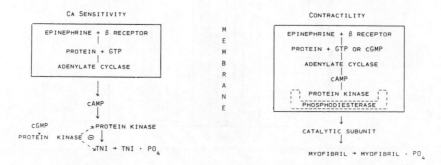

FIG. 3. Proposed models for the regulation of Ca sensitivity and contractility by catecholamines. Interrupted lines are used for the antagonistic, cGMP-dependent reactions to indicate the lack of definite evidence for the nature of the reaction. Note the difference in location of the protein kinases involved in the two types of regulation.

Regulation of the force of contraction

Cyclic nucleotides can also modify the ability of the contractile proteins to develop force, although this is more difficult to demonstrate than their effects on Ca sensitivity. No change in contractility is produced by the addition of either cAMP or cGMP to the bathing solution over the concentration range of $10^{-9} - 10^{-5}$ M. The presence of a phosphodiesterase inhibitor does not change the results. These nucleotides become effective regulators, however, if 0.5–1 % non-ionic detergent is present with the nucleotide in a relaxing solution (Fig. 4). Although the detergent itself reversibly inhibits development of force, the enhancement of contractility by exposure to the combination of nucleotide and detergent remains after their removal, and it can be demonstrated by adding Ca to the bathing solution. Five minutes in 10^{-6} M cGMP and 1 % Triton X-100 in relaxing solution followed by a 30-minute washout of the detergent and nucleotide with relaxing solution produces an average increase in maximum Ca activated force of about 80 %. A second exposure to the combination of nucleotide and detergent has no effect on subsequent development of force, and an exposure to detergent without nucleotide prevents the positive response to the combination from occurring. The detergent seems to facilitate certain reactions involving nucleotides and at the same time to slowly remove the active reactants from the preparation.

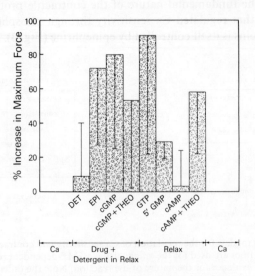

FIG. 4. Bar graph indicating the change in maximal Ca activated force produced by an exposure to the stated transmitter or cyclic nucleotide in the presence of detergent. ± 1 S.D.

There is considerable variability in the size of the response to the nucleotides and detergent (Fig. 4). This variability can be reduced substantially if the existing influence of catecholamines on the cells is taken into account. The trauma of death in an animal causes a release of a considerable amount of catecholamines (Depocas & Behrens, 1977; Frearson et al., 1976), which should enhance the contractility of the heart. The intensity of catecholamine stimulation of the heart at the time it is removed from the animal is likely to vary among the different preparations; consequently the contractile state of the hyperpermeable cells will vary as well since this property depends on the inotropic state of the isolated heart when it is first chilled before incubation in cold EGTA. The extent of the epinephrine stimulation of the hypermeable cells can be quantitated. Since the Ca sensitivity is regulated by a cAMP-dependent phosphorylation that itself is sensitive to epinephrine, the extent to which Ca sensitivity increases when the cAMP-regulated reaction is inactivated with detergent should provide a reasonable estimate of the relative degree of initial epinephrine stimulation. When the effect of the difference in epinephrine stimulation is reduced or eliminated in this way, the variability in the contractile response to cyclic nucleotides is greatly diminished, and some interesting correlations emerge.

Cyclic GMP increases Ca-activated force in relation to the amount of epinephrine stimulation (Fig. 5). In preparations where the detergent produces little shift in Ca sensitivity and presumably there is little epinephrine stimulation, cGMP has no effect on contractility, but in tissues where the degree of epinephrine stimulation is considerable, the Ca-activated force is increased 2 to 3-fold by the cyclic GMP.

The response to cAMP and detergent differs in three important ways from cGMP, although an increase in contractility occurs in each case (Fig. 6): 1) the relation of the change of contractility to the degree of epinephrine stimulation is not strongly positive, it is moderately negative; 2) the extent of the increase in contractility is less with cAMP than with cGMP; and 3) whereas the addition of theophylline to the cGMP and detergent has essentially no effect on the size of the response, theophylline enhances the cAMP response and is generally required for any increase in contractility to occur. Epinephrine with detergent generally increases contractility without any dependency on the existing level of epinephrine stimulation of the tissue. As much as a three-fold increase can occur, considerably more than is produced by cAMP.

The responses to cAMP, cGMP and epinephrine, and their relation to the existing level of epinephrine stimulation, can be incorporated into a model based primarily on already demonstrated properties of cyclic nucleotide-regulated systems (Fig. 3). Since the β receptor is linked to adenylate cyclase in

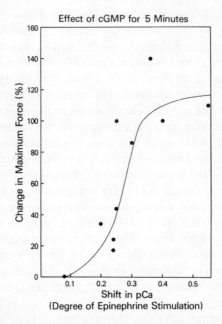

FIG. 5. The effect of 10^{-6} M cGMP in detergent on maximal Ca-activated force as a function of the original degree of epinephrine stimulation. The latter is inferred from the shift in Ca sensitivity from treatment with detergent.

the surface membrane by a step that requires a guanine nucleotide (Rodbell et al., 1975), a single mechanism involving a cAMP-regulated reaction can explain the enhancement of contractility produced by all three agents. Epinephrine binds to the β receptor, which is coupled by a cGMP-binding protein to adenylate cyclase (Pfeuffer, 1977). The combination of epinephrine and cGMP induces the synthesis of cAMP which in turn activates a protein kinase that regulates the force capability of the contractile proteins. Cyclic AMP added to the bath therefore should not require epinephrine for its effect on hyperpermeable fibers, but cGMP should. The greater effect of cGMP than cAMP suggests that the protein kinase can be activated more easily by cAMP synthesized in the membrane than by cAMP from the bath. Apparently an appropriately located phosphodiesterase guards the protein kinase from cAMP from the bath or, in the case of the intact cell, from the cytoplasm, while allowing cAMP produced in the membrane to activate the kinase. Under these circumstances, cyclic AMP could regulate both membrane and cytoplasmic reactions without mutual interference.

Still lacking, however, is precise knowledge of the reaction modifying the

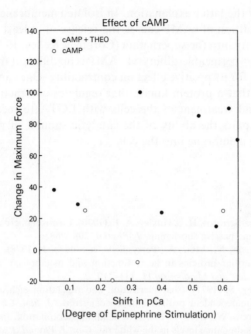

FIG. 6. The effect of 10^{-6} M cAMP in detergent on maximal Ca-activated force. (○) without theophylline and (●) with theophylline.

contractile proteins themselves. The only cAMP-regulated phosphorylation of cardiac contractile proteins that has been well studied so far is the phosphorylation of TNI, but phosphorylated TNI does not increase ATPase activity in isolated protein systems as might be expected if it were increasing contractility. ATPase activity in the isolated protein, however, is not a rigorous analog of force generation in an intact cell. Another phosphorylation reaction, in which one of the light chains of myosin is modified, is regulated by Ca^{++} through the modulator protein (Cole et al., 1978), but it does not appear to be cAMP-sensitive. This reaction is important in the contraction of smooth muscle, but its role in cardiac muscle has not yet been clarified.

Although the role of the detergent in producing the change in the contractile properties is not clear, its requirement does indicate that different protein kinases regulate force and Ca sensitivity. In view of the detergent's lipid solubility it is reasonable to assume that the protein kinase that regulates contractility is either in a lipid phase or isolated by a lipid phase, and the importance of the detergent is in facilitating either the accessibility of the protein kinase to cAMP or the release of the catalytic subunit from the holoenzyme. Two observations favor the latter explanation. In isolated membranes from mammalian heart the addition of cAMP causes the release of at least 50% of the protein kinase activity into the supernatant (Corbin & Kelley, 1977). Secondly, the effect of the lipid-permeable dibutyryl cAMP is no different from cAMP; it requires detergent for its positive effect on contractility. One can therefore tentatively conclude that a protein kinase that regulates contractility is located in a membrane, and treatment of the cells with EGTA to increase membrane permeability impairs the ability of the catalytic subunit of protein kinase to diffuse from the membrane into the cell.

REFERENCES

Brown L., Gonzalez-Serratos, H. & Huxley, A. F. (1970). Electron microscopy of frog muscle fibers in extreme passive shortening. *J. Physiol.* **208**, 868.

Cole, H. A., Frearson, N., Moir, A., Perry, S. V. & Solaro, J. (1978). Phosphorylation of cardiac myofibrillar proteins in heart function and metabolism. *Recent Advances in Studies on Cardiac and Metabolism* **11**, 111–210.

Corbin, J. & Kelley, S. (1977). Characterization and regulation of heart. Adenosine 3':5'-monophosphate-dependent protein kinase isoenyzmes. *J. biol. Chem.* **252**, 910–918.

Depocas, F. & Behrens, W. A. (1977). Effects of handling, decapitation, anesthesia and surgery on plasma neoadrenaline levels in the white rat. *Can. J. Physiol. Pharmacol.* **55**, 212.

Fabiato, A. & Fabiato, F. (1975). Contractions induced by a calcium triggered release of calcium from the sarcoplasmic reticulum of single skinned cardiac cells. *J. Physiol.* **249**, 469–498.

Frearson, N., Solaro, J. & Perry, S. V. (1976). Changes in the phosphorylation of the P light chain of myosin in perfused rabbit heart. *Nature, Lond.* **264**, 801–802.

McClellan, G. B. & Winegrad, S. (1977). Membrane control of cardiac contractility. *Nature, Lond.* **268**, 261–262.

McClellan. G. B. & Winegrad, S. (1978). The regulation of calcium sensitivity of the contractile system in mammalian cardiac muscle. *J. gen. Physiol.* **72**, 737–764.

Pfeuffer, T. (1977). GTP-binding proteins in membranes and the control of adenylate cyclase activity. *J. biol. Chem.* **252**, 7724–7734.

Ray, K. & England, P. (1976). Phosphorylation of the inhibitory subunit of troponin and its effects on the calcium dependence of cardiac myofibril adenosine triphosphatase. *FEBS Lett.* **70**, 11–17.

Reuter, H. & Scholz, J. (1977). A study of the ion selectivity and the kinetic properties of the calcium dependent slow inward current in mammalian cardiac muscle. *J. Physiol.* **264**, 17–47.

Rodbell, M., Lin, M., Salomon, Y., Londos, C., Harwood, J., Martin, B., Rendell, M. & Berman, M. (1975). Role of adenine and guanine nucleotides in the activity and response of adenylate cyclase systems to hormones: Evidence of multisite transition states. *Advances in Cyclic Nucleotide Res.* **5**, 3–29.

Solaro, J., Moir, A. and Perry, S. V. (1976). Phosphorylation of troponin I and the inotropic effect of adrenaline in the perfused rabbit heart. *Nature, Lond.* **262**, 615–617.

Stull, J. & Buss, J. (1977). Phosphorylation of cardiac troponin by cyclic adenosine 3′:5′-monophosphate-dependent protein kinase. *J. biol. Chem.* **252**, 851–857.
Winegrad, S. (1974). Resting sarcomere length tension relation in living frog heart. *J. Physiol.* **208**, 868.
Wood, E., Heppner, R. L. & Weidmann, S. (1969). Inotropic effects of electric currents. *Circulation Res.* **24**, 409–445.

Discussion

Dawson: The cyclic nucleotides of course affect metabolism as well. Can I take it that you can certainly rule out an effect of nucleotides on contractility by changes in glycolysis or something such as that in this preparation?

Winegrad: Yes, this is a functionally hyperpermeable preparation in which the ATP is supplied. The preparation is bathed in a simulated intracellular solution containing 5 mM ATP, 15 mM creatine phosphate, plus creatine phosphokinase. We also know that the intracellular creatine phosphokinase is still retained inside the cell. So I do not think that is a problem.

Rüegg: Can you say something about the possible site of phosphorylation? Do you have different sites for, say, the shift of calcium sensitivity and for the change in contractility? Is it the light chains in one case and Troponin I in the other?

Winegrad: We are in the process of doing the gel studies at the present time, so I cannot give you any rigorous information from our own work. However, from the very close parallelism with the observations of England and Ray (Ray & England, 1976), and of Perry's group (Solaro, Moir & Perry, 1976), I think we can say that the calcium sensitivity regulation is a phosphorylation of Troponin I, but as yet, I cannot say anything about where the contractility site is.

Mohan: Is the receptor fixed, or is it able to move around on the membrane?

Winegrad: I don't know. It would be interesting to try to find out. I should say that we have done studies to try to determine what the nature of the detergent requirement is, and it is clearly not an impairment of the entry of cyclic nucleotide, because you can use a lipid-soluble cyclic nucleotide, like dibutryl cyclic nucleotide, and that behave exactly the same way as a non-soluble cyclic nucleotide, but as far as the mobility is concerned, I cannot say.

This question concerning the organization of the membrane is really very important and very interesting. The increase in force produced by cyclic AMP was something like 80% or 100% of the maximum; yet you could increase the force by twice as much with cyclic GMP. Cyclic GMP is operating presumably, only by enhancing the synthesis of cyclic AMP. Since the end reaction is an activation of protein kinase by cyclic AMP, you can ask why cyclic AMP added to the solution is much less effective than cyclic AMP generated through the sequence of reactions in the membrane (Fig. 3). I think the implications of this difference are important, because the data suggest that cyclic AMP produced in the membrane has much better access to protein kinase than cyclic AMP in the bath. Inasmuch as the bath is analogous to the cytoplasm, there must be a phosphodiesterase that guards this protein kinase and prevents access to it, except from a highly organized sequence of reactions producing cyclic AMP in the membrane.

Consequently, there can be a cyclic AMP-regulated reaction in the cytoplasm inside the cell, and another cyclic AMP-regulated reaction in the membrane, without interference between these two. In other words, the same messenger can control two different reactions without necessarily coupling those two reactions, because of the presence of the phosphodiesterase. This is analogous, in a way, to the organization of the phosphorylation chain in mitochondria, and is one of the reasons why we think that disorganization produced by the institution of the hyperpermeable state impairs the reactions, and the presence of a membrane solvent is required for them to occur.

REFERENCES

Ray, K. & England, P. (1976) Phosphorylation of the inhibitory subunit of troponin and its effects on the calcium dependence of cardiac myofibril adenosine triphosphate. *FEBS Lett.* **70**, 11–17.

Solaro J., Moir, A. & Perry, S. V. (1976) Phosphorylation of troponin I and the inotropic effect of adrenaline in the perfused rabbit heart. *Nature, Lond.* **262**, 615–617.

Discussion on the Length-Tension Relation

Noble: Perhaps I could just start the ball rolling by inviting the proponents of classical muscle physiology to comment on the suggested deviations from overlap proportionality in the length-tension curve which several speakers have mentioned.

Kramer: We have seen two presentations, by Dr. ter Keurs and Dr. Sakai, of sarcomere length measurements by laser diffraction in the same muscle with the same technique. Dr. Sakai presented, in the resting muscle, a 20% deviation towards the tendons; *i.e.*, the sarcomeres were about 20% shorter near the tendons, whereas Dr. ter Keurs states that there is nearly no spread in the sarcomere lengths toward the tendon end, and I should like to hear a comment from both authors about the resolution of their measurements, especially in the noisy tendon ends.

ter Keurs: I showed you the sarcomere length data obtained from single fibers, as opposed to the results of Dr. Sakai, who showed data from toe muscles, which consist, on average, of about 40 fibers. Also, the toe muscle has a lot of fibrous tissue, so there is a lot of scatter.

The second point I should make is that we selected the single fibers specifically for their uniformity. The uniformity, thereby, was restricted by our choice to 0.1 μm over the whole length of the fiber.

The third comment is that the resolution of the technique is in the order of 2 nm.

Sakai: As Dr. ter Keurs suggested, my preparation is from muscle bundles, and this muscle consists of both fast and slow fibers. The toe muscle consists of approximately 20 fibers, of which 30% are twitch fibers, and the others are slow fibers. My measure of sarcomere length is an average over all fibers.

Huxley: Just a comment. It seems to me that in a number of the papers there was an assumption, perhaps not being made, though it sounded as though it is being made, that the sarcomere length alone was sufficient to define the pattern of overlaps of the thick and thin filaments. But, for example, in the paper that Dr. Dewey gave, he reminded us that in *Limulus* muscle, when it was stretched to a sarcomere length at which there would be no overlap between the thick and thin filaments, there was irregularity in the alignment of the thick filaments—not necessarily *because* there was, but there *was* irregularity in the alignment of the thick filament—so that the muscles at these sarcomere lengths still seem to be able to develop tension pretty well, and some people might say that was because one end of a thick filament was interacting with one-half sarcomere worth of I filaments, and the other end of another thick filament was interacting with the thin filaments in the other half of the sarcomere, and there was some ·connection of some sort between the thick filaments that could sustain force.

And, it seems to me that it would be very interesting to have some way of measuring the degree of alignment that is maintained in the thick filaments during some of the experiments we heard. There may be ways of doing this by rapidly freezing the muscle and then looking under the electron microscope, and I think it would be very interesting to do some of those experiments and see what they show. I think it is a little dangerous to assume that the thick filaments necessarily maintain perfect alignment in all situations.

Winegrad: I would like to sort of second that. We have made some observations that are very consistent with it. If you stretch frog cardiac trabeculae in which the fibers are aligned to the point where the sarcomere length is 2.6 μm or more, the A-band begins to get broader. From electron microscopy, we have seen that this is due to misalignment of the thick filaments. This broadening is not uniform throughout the whole length of the bundle. If you then stimulate that bundle, what happens is that the broad A-bands get even broader, and then, as the muscle relaxes, they go back to their resting width which was still broader than normal, but not as broad as at the peak of contraction, showing that the degree of misalignment not only increases with resting length, but with change in a given contraction under certain circumstances.

Noble: I agree entirely with what you say as far as the experiments I presented are concerned. The only direct evidence is that by Lydia Hill (1977), who showed that where stretch is applied during contraction, using Andrew Huxley's interference microscope, the A-band width remains constant. Now, with that sort of technique, you cannot see the H-band width, which would also be necessary, and that is why I think that your proposal for doing quick freezing during this kind of experiment and electron microscopic examination is certainly one which should be done, and which we will try to do.

Kushmerick: It seems a little late in the game to make a suggestion of what really does constitute the sliding filament idea, and I do not know whether Dr. Huxley would agree with this definition, but I would like to put forward the review that, instead of one very large protein filament that coils up into a different length, you have a sliding of several filaments as one important element of the sliding filaments idea. The second important element is that there are sites distributed along these filaments which interact. It seems to me that I have heard nothing here that contradicts that idea, but rather, there is some discussion as to whether or not these sites are indeed independent, whether they all have the same properties, or whether there is some distributed kind of property. That is the kind of thing that is being debated; I have not yet heard that there is any contradiction to the fact that the filaments slide, and that there are sites distributed along the filament that have the properties of interacting with each other. It seems to me that we might agree on what constitutes the basic elements of the hypothesis that we are here to discuss.

Pollack: Well, we are here to discuss the sliding filament and the cross-bridge theory, and it seems that the two theories are separable. The first, obviously, has to do with the sliding of thick and thin filaments. But the second has to do with cross-bridges working in parallel, and independently. I think we can consider the two theories separately from one another. There has been nothing, perhaps with the exception of Dr. Dewey's data, to contradict the idea of sliding, and I know that Dr. Dewey does not necessarily suggest that his data do contradict the sliding filament mechanism. But it is the independent action of cross-bridges which does seem to be contradicted, particularly by the length-tension relations.

Dewey: We certainly do not interpret our data as opposed to the sliding filament theory. I thought I made that explicitly clear. There is, in this muscle, and in slow muscles in general, an additional component which we believe is simply the rearrangement of constituent proteins in the filament; nothing like a spring.

Winegrad: I think, in response to Dr. Pollack's comment, one really has to bring biochemists into this. There are studies now in several laboratories in which cross-bridges or heavy meromyosin have been labeled with spin labels, and the changes in the spectra followed during interaction with the thin filaments. There have also been fluorescent relaxation studies to look at these reactions, and I think that the answers, or perhaps not answers but important data in making these evaluations, will come from studies like that, and we should not really focus on biophysical approaches of the mechanical variety.

White: Well, Professor Andrew Huxley always claims that the bit of evidence which suggests that the cross-bridges are independent force generators is the fact that V_{max} is independent of sarcomere length, but yet we have heard nothing about that. I do not think that he would say that the tension-length diagram is the main evidence for that feature. Dr. Pollack, I think it would not be fair to say—as a comment also—that your evidence contradicts that. I think there are many explanations for your data in terms, for example, of calcium activation effects, and features along those lines, so I do not think you can say that your data contradicts it.

REFERENCE

Hill, L. (1977) A-band length, striation spacing and tension change on stretch of active muscle. *J. Physiol.* **266**, 677–685.

V. FORCE-VELOCITY RELATION

The Velocity of Shortening at Zero Load: Its Relation to Sarcomere Length and Degree of Activation of Vertebrate Muscle Fibers

K. A. P. EDMAN

Department of Pharmacology, University of Lund, Sweden

ABSTRACT

A method was designed for direct measurement of the speed of shortening at zero load (V_0) in frog single muscle fibers. Any contribution from the recoil of series elastic components could be excluded by this approach. V_0 was higher (approx. 10%) than predicted by hyperbolic extrapolation from velocities recorded (by load-clamp technique) at finite loads.

V_0 remained constant when the sarcomere length was varied between 1.65 and 2.7 μm. There was an abrupt decline in velocity below 1.65 μm, attributable to increased passive resistance to shortening. V_0 increased steeply at sarcomere lengths ($>$ 2.7 μm) where resting tension introduced a passive compressive force on the sarcomeres. V_0 data obtained at long sarcomere lengths plotted against the passive compressive force at each length provided information about the force-velocity relation in the region of negative loads.

V_0 was found to be constant, to within a few percent, when the fiber's ability to produce force was reduced from maximum (tetanic tension) to less than 10% of the maximum (isometric twitch depressed by dantrolene).

The results support the view that the velocity of shortening at zero load is independent of the number of myosin bridges that are able to interact with the actin filament. A change in activation does merely affect the number of cross-bridges formed between the A and I filaments, not the cycling rate of the cross-bridges.

INTRODUCTION

A great deal of interest has been focused on the force-velocity relation in recent years, both in skeletal and cardiac muscle, since it has become clear that this relation contains information concerning the cross-bridge mechanism of muscle contraction (Huxley, 1957). Most of the earlier work, however, has been performed on whole muscle, and it is only recently that the force-velocity relation has been analyzed in intact muscle fibers (Edman, Mulieri & Scubon-Mulieri, 1976). So far there has been no convenient method for determining the *maximum* speed of shortening, *i.e.*, the velocity at zero load. With the standard techniques used this value has been approximated by measuring the shortening velocity at a small load (Gordon, Huxley & Julian, 1966), or, alternatively, by extrapolating to zero tension from measurements at finite loads (*e.g.*, Hill, 1938; Aubert, 1956; Jewell & Wilkie, 1960; Mashima, Akazawa, Kushima & Fujii, 1972; Edman, Mulieri & Scubon-Mulieri, 1976). This difficulty in previous studies has recently been overcome by using an experimental approach (Edman, 1978) which makes it possible to directly measure the speed of unloaded shortening in single fibers. The following account describes experiments, based on the new technique, that have been aimed at elucidating the force-velocity relation at varied sarcomere length and varied degree of activation of the contractile system.

METHODS

The essentials of the experimental arrangement are illustrated in Fig. 1. For a more detailed description, see Edman (1975) and Edman & Hwang (1977). Single fibers dissected from the semitendinosus and tibialis anterior muscles of *R. temporaria* were used. The fiber was mounted horizontally in a thermostatically controlled bath at 1–2°C between a force transducer and an arm. The force transducer (RCA 5734 or an AME 801) was provided with a stainless steel hook to which one tendon was attached close to the insertion of the fiber. The other tendon was attached to another steel hook which was secured in a holder on the free end of the arm. The arm was operated by an electromagnet and could be moved in the longitudinal direction of the trough. In these experiments the puller was used: 1) to produce a quick release of varied amplitude, or 2) to produce load clamp, *i.e.*, to constrain the fiber to keep a preselected constant tension during activity (Edman & Hwang, 1977). The movements of the arm were monitored by means of a displacement transducer. The signals from the tension and displacement transducers were displayed on a Tektronix 5103N oscilloscope and photographed on 35 mm film.

The fiber was stimulated by passing current between two platinum plate electrodes placed on either side of the fiber. Single twitches or 1-sec fused tetani were studied.

The sarcomere length was determined from the diffraction pattern that was obtained by passing a laser beam (He-Ne, 1 mm diameter) through the fiber.

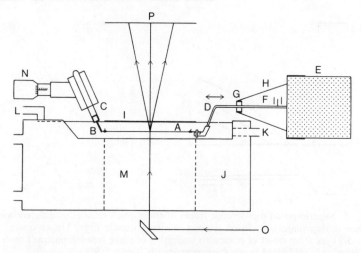

FIG. 1. A, muscle fiber. B, trough containing Ringer solution. C, tension transducer. D, arm connected to puller. E, electromagnetic vibrator. F, displacement transducer. G, teflon bearing. H, brackets. I, glass slide. J, jacket for circulation of water-glycol mixture. K, inlet for bath solution. L, suction drain. M, air channel for passage of laser beam. N, micrometer screw. O, laser beam. P, screen for laser diffraction pattern.

The details of this technique have been described previously (Cleworth & Edman, 1972).

Measurements from the oscilloscope records were carried out in a Nikon model 6C profile projector using 3.4 × magnification from oscilloscope to projector screen.

RESULTS AND DISCUSSION

Approach for measuring the speed of shortening at zero load

The approach used for measuring the velocity of unloaded shortening is illustrated in Fig. 2. The puller was quickly released at a given moment during a tetanus so as to slacken the fiber and allow it to shorten and redevelop tension at a new length. There was a drop in tension to zero, and the tension remained at zero level until the fiber had taken up the slack. Three or more amplitudes of release were used, and in each case the time, Δt, from the onset of release to the onset of force redevelopment was measured and plotted against the respective release movement, ΔL. Figure 2B shows a typical plotting of ΔL against Δt for a series of measurements where each point is the mean of 5 repeated release

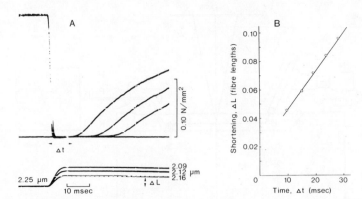

FIG. 2. A. Superimposed oscilloscope traces of three quick releases of different amplitude performed during plateau of fused tetanus in single muscle fiber. Upper traces: tension records. Δt, time from onset of release to beginning of force redevelopment. Lower traces: release steps, ΔL, calibrated in μm sarcomere length. Temp. 0.9°C.
B. Relationship between ΔL and Δt. Releases performed from 2.25 μm sarcomere length during plateau of fused tetanus. Each data point is the mean of five release recordings. Straight line is the least squares regression of Δt upon ΔL. The slope of the line is V_0 (see text). Temp. 2.7°C.

recordings. The slope of the straight line (regression analysis) relating ΔL to Δt provides a measure of the *speed of shortening at zero load*, V_0. It is essential to point out that the slope of the regression line in Fig. 2B is determined from the *differences* between the various Δt's; i.e., V_0 refers to the shortening that occurs above the control release. Using this approach it is thus possible to exclude any contribution from the recoil of series elastic elements which occur during the initial fall in tension.

The force-velocity relation in the single fiber

The complete force-velocity relation at single fiber level (sarcomere length 2.1 μm) is depicted in Fig. 3. Here the filled circles are velocity data at various loads obtained by load-clamp technique. The open circle is the V_0 value in the same fiber. Typically, the force-velocity relation can be fitted well with Hill's (1938) hyperbolic equation in the range 5–80% of the isometric force (P_0). There is a reversal of curvature between 0.8 P_0 and P_0, and it can be seen that the hyperbola (fitted between 0.05 and 0.8 P_0) intersects the abscissa at a force that is about 25% higher than P_0. This non-hyperbolic nature of the force-velocity relation has been investigated in detail recently (Edman, Mulieri & Scubon-Mulieri, 1976) and there is reason to believe that it represents the true behavior of the contractile system. The phenomenon could not be explained by

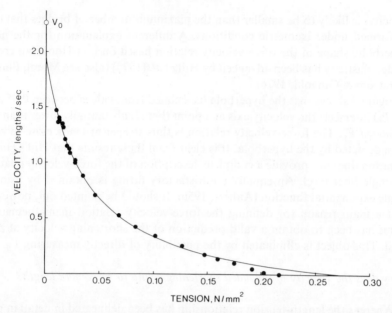

FIG. 3. Force-velocity relation in single muscle fiber. Filled circles: load clamp recordings. Open circle: V_0 measurement. All data refer to plateau of fused tetanus at 2.1 μm sarcomere length. Solid line: least squares fitting of Hill's (1938) hyperbolic equation to data points within the range 5–80% of the measured isometric force. Temp., 1.7°C.

redistribution of sarcomere lengths in the fiber as studied by laser diffraction technique (Edman et al., 1976).

The departure from a hyperbola in the high-force region may indicate that not all of the myosin cross-bridges in the overlap zone are able to interact with the actin filament simultaneously. Such an upper limit of cross-bridge formation may be set by geometric factors, as both the actin and myosin subunit repeats and the helical repeats of the actin and myosin filaments are different (Huxley & Brown, 1967). Since the measured P_0 is 24 ± 3% (mean ± S.E. of mean, $n = 16$) lower than the predicted P_0 it is possible that during an isometric contraction only 76% of the myosin bridges in the overlap zone are able to make proper contact and interact with the actin filament at any given time (Edman et al., 1976). This limitation of cross-bridge formation would not, however, influence the force-velocity relation at moderate and low loads for the following reason. When the velocity of sliding is increased, the number of cross-bridges formed will be progressively reduced due to a *temporal* factor that determines the rates of making and breaking of cross-bridges (Huxley, 1957). Thus, the actual number of cross-links formed at a low tension (high velocity) is likely to be smaller than the maximum number of bridges that can be formed under isometric conditions. A different explanation for the non-hyperbolic shape of the force-velocity relation based on non-Hookeian cross-bridge elasticity has been advanced by Hill et al. (1975) (also see Morel, Pinset-Härström & Gingold, 1976).

Figure 3 shows that the hyperbola (calculated from values between 0.05 and 0.8 P_0) intersects the velocity axis at a point that is substantially lower than the *measured V_0*. The force-velocity relation is thus steeper at loads close to zero than depicted by the hyperbola. It is clear from these results that Hill's (1938) equation does not provide a complete description of the force-velocity relation at single fiber level. An equally unsatisfactory fitting is obtained by using a single exponential function (Aubert, 1956). It should be pointed out, however, that a main reason for defining the force-velocity relation in mathematical terms has been to obtain a valid prediction of the shortening velocity at zero load. This object is eliminated by the possibility of directly measuring V_0.

How does the velocity of unloaded shortening relate to sarcomere length?

Whereas the length-tension relationship has been delineated in detail in previous studies, there is relatively little information concerning the length dependence of the speed of unloaded shortening. Gordon et al. (1966) reported that the speed of shortening against a small load decreased by approximately 20% as the fiber shortened from 3.0 to 2.0 µm sarcomere length. The velocity declined more steeply below 2.0 µm together with a decrease of the tetanic force. The length dependence differs a great deal, however, if the velocity of shortening of the completely unloaded fiber is considered instead, as is evident from the present experiments.

Figure 4 shows representative V_0 data from two single fibers. For comparison the length dependence of tetanic tension and of resting tension measured in one of the fibers is also illustrated. V_0 was determined as described above. In these experiments, however, the releases started at different sarcomere lengths to allow the fiber to redevelop tension at the same length after a small and a large release. It can be seen that V_0 is quite constant between 1.65 and approximately 2.7 µm sarcomere length. Below 1.65 µm there is a rather abrupt decline in velocity. Above 2.7 µm sarcomere length, on the other hand, V_0 increases quite steeply. It is evident that V_0 is constant over a range where the tetanic tension varies considerably. This constancy of V_0 would seem to make clear that the velocity of unloaded shortening is not dependent on the number of myosin bridges that project onto the thin filament.

It is of interest to note that the velocity of shortening at zero load does not

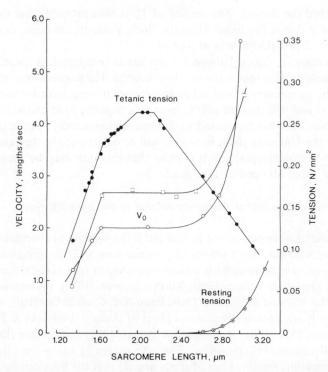

FIG. 4. Relationship between maximum velocity of shortening (V_0) and sarcomere length measured in two different fibers (○, □). The length dependence of tetanic (●) and resting (⊙) tensions is also shown. Symbols ○, ● and ⊙ refer to the same fiber, temperature: 1.4°C. Temperature for □, 2.1°C.

decrease between 2.0 and 1.65 μm sarcomere length where there is a substantial decline in tension. This gives a relevant clue to the nature of the decrease in tension in this range. The results almost certainly mean that there is no increase of the passive resistance to shortening as the sarcomere length is reduced from 2.0 to 1.65 μm as previously assumed (Gordon et al., 1966; Edman, 1966; Julian, Moss & Sollins, 1978). An increased resistive force, if it existed, would cause V_0 to decline. The decrease in tension that is recorded in this range is therefore to be attributed to a reduced *capacity* of the fiber to produce tension rather than to a passive force that neutralizes some of the active force.

The decrease in velocity below 1.65 μm sarcomere length suggests that there is an increased passive resistance to shortening in this range. The angle of the length-velocity curve coincides with a distinct bend of the length-velocity curve and it occurs at a point where the tips of the A filaments can be assumed to have reached the Z-discs. The decline of V_0 is thus probably due to the opposing force that is produced when the thick filaments are being compressed against the Z-disc (Gordon et al., 1966).

The increase in velocity above 2.7 μm sarcomere length is closely related to the passive tension that occurs at these lengths. The passive force will tend to shorten the sarcomeres. It will act as an external driving force for the shortening process and will thus, in effect, serve as a negative load on the contractile system when the fiber is released to shorten during activity. The actual speed by which the filaments slide, however, will be determined by the rate of turnover of the cross-bridges, and it is clear that this rate may be increased far above the value attained at zero load.

Is the velocity of unloaded shortening related to the isometric force?

Experiments were performed to find out if the velocity of shortening at zero load varies with the fiber's ability to produce tension. This question has previously been investigated, with different results, in frog skeletal muscle preparations (Jewell & Wilkie, 1960; Mashima et al., 1972), mammalian papillary muscles (Edman & Nilsson, 1969; Brutsaert, Claes & Goethals, 1973) and mechanically and chemically skinned fibers of skeletal (Podolsky & Teichholz, 1970; Julian, 1971; Wise, Rondinone & Briggs, 1971) and cardiac (De Clerck, Claes & Brutsaert, 1977) muscles. The advantages of using the intact single fiber preparation for this kind of study are: 1) that the fiber can be activated uniformly along its entire length (Cleworth & Edman, 1972), and 2) that the passive resistance to shortening is negligible (Edman & Hwang, 1977).

Measurements of V_0 were performed at the peak of the isometric twitch and during the plateau of the tetanus. In addition, the peak twitch force was depressed by dantrolene, a substance which has been shown to reduce the calcium release from the sarcoplasmic reticulum (Putney & Bianchi, 1974; Van Winkle, 1976; Desmedt & Hainaut, 1977; Morgan & Bryant, 1977) without affecting the excitation process (Ellis & Bryant, 1972). V_0 was found to be constant in spite of the large differences in tension between twitch and tetanus. Thus, whereas the peak twitch force varied between 38 and 85% of the tetanic tension in 8 different fibers, V_0 during the twitch was 99 ± 2 (S.E. of mean) % of the value recorded during the tetanus. Furthermore, depression of the isometric twitch by dantrolene to about 10% of the control twitch tension did not affect V_0 significantly.

These results support the idea that the individual cross-bridge is activated in an all-or-nothing manner; *i.e.*, a change in activator-calcium concentration may be assumed to alter the *number* of cross-bridges formed but not the *rate* at which the cross-bridge goes through a cycle of activity. This would account for the finding that the isometric force may be varied over a very wide range without any detectable change in V_0. The conclusion that the velocity of shortening at zero load is independent of the number of cross-bridges formed is further supported by the finding that V_0 remains constant as the area of overlap between the A and I filaments is changed.

REFERENCES

Aubert, X. (1956). La relation entre la force et la vitesse d'allongement et de raccourcissement du muscle strié. *Arch. int. Physiol.* **64**, 121–122.
Brutsaert, D. L., Claes, V. A. & Goethals, M. A. (1973). Effect of calcium on force velocity-length relations of heart muscle of the cat. *Circulation Res.* **32**, 385–392.
Cleworth, D. R. & Edman, K. A. P. (1972). Changes in sarcomere length during isometric tension development in frog skeletal muscle. *J. Physiol.* **227**, 1–17.
De Clerck, N. M., Claes, V. A. & Brutsaert, D. L. (1977). Force velocity relations of single cardiac muscle cells. *J. gen. Physiol.* **69**, 221–241.
Desmedt, J. E. & Hainaut, K. (1977). Inhibition of the intracellular release of calcium by dantrolene in barnacle giant muscle fibres. *J. Physiol.* **265**, 565–585.
Edman, K. A. P. (1966). The relation between sarcomere length and active tension in isolated semitendinosus fibres of the frog. *J. Physiol.* **183**, 407–417.
Edman, K. A. P. (1975). Mechanical deactivation induced by active shortening in isolated muscle fibres of the frog. *J. Physiol.* **246**, 255–275.
Edman, K. A. P. (1978). Maximum velocity of shortening in relation to sarcomere length and degree of activation of frog muscle fibres. *J. Physiol.* **278**, 9–10P.
Edman, K. A. P. & Hwang, J. C. (1977). The force-velocity relationship in vertebrate muscle fibers at varied tonicity of the extracellular medium. *J. Physiol.* **269**, 255–272.
Edman, K. A. P., Mulieri, L. A. & Scubon-Mulieri, B. (1976). Non-hyperbolic force velocity relationship in single muscle fibres. *Acta physiol. scand.* **98**, 143–156.
Edman, K. A. P. & Nilsson, E. (1969). The dynamics of the inotropic change produced by altered pacing of rabbit papillary muscle. *Acta physiol. scand.* **76**, 236–247.
Ellis, K. O. & Bryant, S. H. (1972). Excitation-contraction uncoupling in skeletal muscle by dantrolene sodium. *Naunyn-Schmiedeberg's Arch. Pharmacol.* **274**, 107–109.
Gordon, A. M., Huxley, A. F. & Julian, F. J. (1966). The variation in isometric tension with sarcomere length in vertebrate muscle fibres. *J. Physiol.* **184**, 170–192.
Hill, A. V. (1938). The heat of shortening and the dynamic constants of muscle. *Proc. R. Soc.* B **126**, 136–195.
Hill, T. L., Eisenberg, E., Chen, Y. & Podolsky, R. J. (1975). Some self-consistent two-state sliding filament models of muscle contraction. *Biophys. J.* **15**, 335–372.
Huxley, A. F. (1957). Muscle structure and theories of contraction. *Prog. Biophys. biophys. Chem.* **7**, 255–318.
Huxley, H. E. & Brown, W. (1967). The low-angle X-ray diagram of vertebrate striated muscle and its behaviour during contraction and rigor. *J. molec. Biol.* **30**, 383–434.
Jewell, B. R. & Wilkie, D. R. (1960). The mechanical properties of relaxing muscle. *J. Physiol.* **152**, 30–47.

Julian, F. J. (1971). The effect of calcium on the force-velocity relation of briefly glycerinated frog muscle fibres. *J. Physiol.* **218**, 117–145.
Julian, F. J., Moss, R. L. & Sollins, M. R. (1978). The mechanism for vertebrate striated muscle contraction *Circulation Res.* **42**, 2–14.
Mashima, H., Akazawa, K., Kushima, H. & Fujii, K. (1972). The force-load-velocity relation and the viscous-like force in the frog skeletal muscle. *Jap. J. Physiol.* **22**, 103–120.
Morel, J. E., Pinset-Härström, I. & Gingold, M. P. (1976). Muscular contraction and cytoplasmic streaming: A new general hypothesis. *J. theor. Biol.* **62**, 17–51.
Morgan, K. G. & Bryant, S. H. (1977). The mechanism of action of dantrolene sodium. *J. Pharmac. exp. Ther.* **201**, 138–147.
Podolsky, R. J. & Teichholz, L. E. (1970). The relation between calcium and contraction kinetics in skinned muscle fibres. *J. Physiol.* **211**, 19–35.
Putney Jr., J. W. & Bianchi, C. P. (1974). Site of action of dantrolene in frog sartorius muscle. *J. Pharmac. exp. Ther.* **189**, 202–212.
Van Winkle, W. B. (1976). Calcium release from skeletal muscle sarcoplasmic reticulum: site of action of dantrolene sodium? *Science, N.Y.* **193**, 1130–1131.
Wise, R. M., Rondinone, J. F. & Briggs, F. N. (1971). Effect of calcium on force velocity characteristics of glycerinated skeletal muscle. *Am. J. Physiol.* **221**, 973–979.

Discussion

Sugi: I have a question about your force-velocity curve in the negative force region. In this case, the muscle shortens very rapidly while it is still in the active state. It seems to me to imply that the thick and thin filaments can be made to slide past each other very rapidly in the direction of shortening, while actin and myosin tend to interact, *i.e.*, tend to inhibit very rapid shortening. It seems somewhat surprising, doesn't it?

Edman: I think it is just a continuation of the situation when you have positive loads. At any load there is a balance between the driving force, *i.e.*, the number of pulling cross-bridges on the one hand, and those which are opposing plus the external load on the other hand. In the case of a negative load, you have effectively an external driving force, and there is then some resistance to this formed by the contractile system.

Sugi: Do you mean, then, that each cycle becomes faster if there is a compressive force?

Edman: I don't know, but I imagine that there must be some interaction, some contact between the bridges when the filaments slide in a shortening direction. I think it is the opposite of what you see when you stretch the fiber; there is contact there, too.

Sugi: I would have thought that the maximum velocity of shortening should limit the relative sliding of filaments. This was my expectation.

Edman: You mean at zero load?

Sugi: Right.

Edman: I do not think so. The system would be expected to work, in principle, the same way on the other side of zero.

Huxley: Continuing the same argument, in the A.F. Huxley '57 paper (A.F. Huxley, 1957), there is a definite model for what determines the velocity at zero load, and as I recall there is a diagram there with areas with pushing and pulling bridges (Fig. V-1B), so presumably, from that, one could make predictions about what kind of a force you needed to, say, accelerate the shortening by a factor of two.

I wondered from your measurements, when you have the fiber shortening very rapidly with the resting tension helping it, what sort of agreement you get with the predictions of the model?

Edman: Well, I would think that the modeling people would be able to do that; I certainly have not pursued that problem.

White: I think that must be right. You only need very small change in the way that g (the rate constant for breaking links) changes on the left side of the diagram shown in Fig. V-1A to give yourself anything you want; it is a very insensitive test of the model. If you are going to take the precise relationship for g on the negative side of x (the distance from the equilibrium position where a cross-bridge exerts zero force), then you may well be able to say something. But if all you have to do is to modify that function by a very small amount to get very large differences of V_{max}, it seems to me not a very worthwhile thing to do.

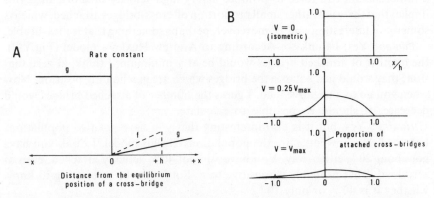

FIG. V-1. A. F. Huxley's contraction model (A. F. Huxley, 1957, by permission of *Prog. Biophys. biophys. Chem.*). A: rate constants for making links (f) and for breaking links (g) as a function of distance (x) from the equilibrium position. B: distribution of attached cross-bridges at three different values of shortening velocity (V).

Hill: Certainly any more complete model than Andrew Huxley's will automatically give the sort of curve you had; in principle, there is no problem at all. There is continuity from the model, almost any model. In fact, I do not know of one that would not give continuity.

Rall: In Fig.2B, when you plotted Δl vs. Δt, you had a non-zero intercept. Is that significant?

Edman: Yes, that is the amount of compliance of the series elastic element. The intercept is, on the average in these experiments, very close to 2% of muscle length.

Sugi: I keep wondering about your (Fig.4) result in the negative force region. Why can the thick and thin filaments be made to slide past each other so rapidly in the direction of shortening. I think it might not be due to sliding, but maybe to simple crumpling.

Edman: No. That can be ruled out because I measure the onset of active shortening, so it is not a matter of a simple recoil of the passive, parallel elements. The parallel elements will produce a compressive force on the contractile system, but the velocity by which the contractile unit is shortening is determined by the properties of that contractile unit. The way I measure it guarantees shortening of the active unit because I measure the onset of force redevelopment.

Huxley: We could continue on the same point. It may be that you can explain any results by altering g, or whatever it is, but that will lead to some experimental consequences in terms of the proportion of cross-bridges attached at maximum velocity of shortening. It seems to me that if you only need a rather small extra force to produce a very high velocity of shortening, this implies that there is a rather small proportion of cross-bridges attached, which is something interesting and is, moreover, perhaps something that is measurable.

Edman: Yes, I think so. According to Andrew Huxley's model (Fig.V-2), the number of attached bridges would be at a minimum, I think, at zero tension; they would just balance the bridges which are in a holding position. Now when you go to a negative load, I guess the number of attached bridges would decrease. It should be possible to calculate.

Huxley: Yes, but it is also interesting that you have got two populations balanced, but how large are the populations? In his model, I think you have got about 20% either way. You have still got 40% actually attached, even at zero load. That probably is sensitive to g. But it will be interesting to know whether it is 40% or only 10%.

Tregear: Can you deduce already from what you have got whether the number actually goes down in the negative load region, or whether it stays constant?

Edman: No. I certainly have not been quantifying that yet.

Tregear: Well, it is a lot more interesting when you can make a prediction and then test it.

Sagawa: I have a very simple-minded interpretation of this negative force. At some stage you are bound to think about the effect of mass. If one end of the muscle is fixed, and the other end is moving, it is absolutely necessary that the sarcomeres near the fixed end have to have some force to pull and cause motion of the rest of the muscle. This massive, *i.e.*, accelerative, load decreases as you go more toward the moving end, so there is an inhomogenous load for the various sarcomeres in series, and although the total force you can measure from outside is zero, there is internal distribution of forces which is in balance with these viscous and accelerative force loads. By adding external "help" to cancel out these forces, you can get a truer, or more evenly distributed, zero-load sarcomere shortening velocity, and that is why you get a large V_0. That is what I thought of as I was listening to your talk.

Edman: Yes, I am sure there is some counter-force, but I think it is very small. There is, of course, some passive resistance to shortening, and I tried to estimate that. We presented some data in a recent paper in collaboration with Dr. Hwang (Edman & Hwang, 1977), where we simply stretched resting fibers at a velocity corresponding to V_0 and measured the force under those conditions. It turns out to be less than 1% of P_0, and of course the weight of the muscle in solution is very small compared to its capacity to produce tension.

Sagawa: It is not enough to be significant. Is that what you are saying?

Edman: Yes. It would be a force, I think, of less than 1% of P_0, but nevertheless the force-velocity curve is very steep, so this 1%, if we allow that, would correspond to about 5% change of V_0.

Sagawa: Whereas your negative force magnitude was something like 15%, wasn't it?

Edman: About 10%.

Pollack: A number of people, including you, have shown that after a muscle is stretched, the muscle has a larger capacity to develop force or to shorten, while the opposite is true if the muscle has shortened first. I wonder if you have done any measurements after a previous stretch or release, or, if you have not, whether you have some predictions as to what might happen to V_0 in such cases?

Edman: You know if you produce shortening during activity, there is a depression of the capability to produce tension in the fibers, which stays over about 800 msec (Edman, 1975). That does not affect V_0 and also, I think, there is a very small change, if any, when we measure the maximum velocity after stretch enhancement (Fig.4 in Edman, Elzinga & Noble's paper). At that time we did not study V_0. We have not any actual measurements of that yet.

Rüegg: It would be interesting to know the effect of calcium ion concentration on the shortening velocity of skinned fibers. There is still this ambiguity of the results of Podolsky vs. Julian. Julian (1971) found, I think, an effect on V_{max} and Podolsky (Podolsky & Techholz, 1970) did not. Could you comment on that?

Edman: Well, there are to my knowledge three groups who have looked at this. Podolsky & Teichholz (1970) found the same independence of activation of their V_{max} which they extrapolated from data which were rather scattered, but nevertheless, they did not find any difference of V_{max}. Then, Julian (1971) found some changes. Then, we have Wise, Rondinone & Briggs (1971), I think. They also found some changes of V_{max} in glycerinated psoas muscle, the old type of zent-Györgyi glycerinated fibers. You can pick out data from whole muscle studies which seem to show that there is no change of V_{max}, but it is very difficult, really, to determine from the standard measurements, because you have to extrapolate, and if you get a slight change in shape of the curve you may very easily misinterpret the results.

Rüegg: Herzing in our laboratory used a very similar method to the one you used for determining V_{max} in briefly glycerinated cardiac muscle and there he found a very large effect of calcium ion concentration on V_{max}.

Edmon: Yes, there may be a difference, although I would be very much surprised if it turns out to be so. In cardiac muscle you do have internal forces, connections between cells, *i.e.*, you may have intracellular forces which produce a counterforce, so the situations are perhaps not quite comparable.

Pollack: You mentioned some time ago that there was a difference between the values of V_0 and V_{max}, something on the order of 20%.

Edman: There is a difference, and it is of the order of about 7%. It depends upon how close to zero load you are when you take up the standard load clamp, because the force-velocity curve is very steep at the very small loads, and if you have data say down to 10% of P_0, and if you fit a hyperbola to extrapolate to V_{max}, you would be about 15% lower than my V_0. But if you go to smaller loads and determine your hyperbola from those data, then you get close to it.

Pollack: So, your prediction is if you could reduce the load sufficiently, V_0 would be equal to V_{max}. Is that right?

Edman: Yes, then you would really visualize that very steep portion of the force-velocity relation.

Sugi: As far as I understand, V_{max} only means the extrapolation of data points towards the Y axis according to Hill's equation. Is that right?

Edman: Well, yes. I have chosen to use V_0 just to distinguish it from that extrapolated term.

Sugi: So, V_0 may not necessarily be the same as V_{max}.

Edman: Well, it depends upon on how well you can determine V_{max}. If you can take up velocities against very small loads, which is a bit difficult, technically, then you will go up to V_0, no doubt. I could show it experimentally.

Sagawa: Could I ask you a question which might deviate somewhat? I would like to ask those people who do skinned muscle fiber experiment whether they see the restoring forces that Dr. Mashima talked about. I remember Carl Honig of Rochester mentioned in one meeting that in skinned fibers you do not see this restoring force, although you do see it in single fibers. I asked him recently if he still believes in this, and he does.

Rüegg: Yes, you do see negative forces. Güth and Kuhn did very quick release, within some 0.25 msec, and measured tension and length during the release with very large quick releases of about 1%. At the end of release, the tension definitely went into the negative region; so negative forces could actually be recorded if they are not an artifact.

Sagawa: That negative force you mentioned is a negative force relative to that at the unstressed length of the muscle. Is that correct?

Rüegg: No. It is relative to zero force.

Huxley: I think you are talking about two different things. You are talking about the restoring force when the muscle shortens below, say, sarcomere spacings of $2\mu m$ and elongates itself, and the question is whether the elongation takes place in the skinned fiber.

Winegrad: In the hyperpermeable cardiac fiber, it does as long as the shortening is not to less than approximately 1.5 μm sarcomere length. At that point there seems to be a disorganization of the filament lattice, and then there is no restoring force. If you measure ATP splitting as a function of sarcomere length, you find an interesting parallelism that at the point at which the restoring force disappears, about less than 1.5 μm—these are measurements made in the light microscope so the precision is not all that great—but at that point, you see also a reduction in the rate of ATP splitting, which is consistent with the notion that the filament lattice has now been disrupted, and the access of crossbridges on the thick to thin filaments has been reduced.

Blangé: I would like to go a little further into the similarity between V_0 and V_{max}. In terms of cross-bridges, I would think that in your method you may shift the cross-bridge distribution in a non-steady way. But if you try to measure V_{max}, what you try to do is to measure velocity during a steady state. That is, when the muscle is continuously shortening there is a certain velocity. So do you have any idea whether, indeed, your V_0 represents some steady state? Can you have enough cross-bridge turn-over in the time that you are going over from isometric force to zero load?

Edman: I can not see why there should be a difference between V_{max} and V_0, other than pure terminology. There are only two different approaches to that particular point. When I measure V_0, the fiber is unloaded and I measure the time it takes to shorten over a certain distance. When you measure V_{max}, it is just an extrapolation from data you have at finite loads.

White: I think V_{max} is measured under conditions in which you have a steady state, whereas your V_0 experiments are done under the kind of conditions that Prof. Sugi showed in his talk when you get in the oscillatory velocity transients.

Edman: Well, there is a steady state in this case, also. Why should there not be?

White: No. You certainly have not got a steady state by that stage. It takes something of the order of 50 msec to get to the steady state.

Edman: No. You see I exclude that first portion. I measure the shortening which occurs in excess of that.

White: I think that is probably right. How long is it before the tension starts to rise again?

Edman: For the smallest release it is about 12 msec.

White: So the question really arises as to how long it takes to get to the steady state, and I would have thought that 12 msec was far too little.

Noble: Well, surely you have got that in Fig.2. With increasing amplitude of release, you have increasing time to redevelopment of tension (Fig.2A). When you plot that you get this unsteady state in the beginning, and then it becomes a straight line (Fig.2B). It is that straight line which goes on for as long as you make your measurments that you are using for V_0. So, you are in a steady state by your own measurements.

Edman: Yes, that is my interpretation of it. The straightness of that line, I think, assures that you are in a steady state.

REFERENCES

Edman, K. A. P. (1975). Mechanical deactivation induced by active shortening in isolated muscle fibres of the frog. *J. Physiol.* **246**, 255–275.

Edman, K. A. P. & Hwang, J. C. (1977). The force-velocity relationship in vertebrate muscle fibres at varied tonicity of the external medium. *J. Physiol.* **269**, 255–272.

Huxley, A. F. (1957). Muscle structure and theories of contraction. *Prog. Biophys. biophys. Chem.* **7**, 255–318.

Julian, F. J. (1971). The effect of calcium on the force-velocity relation of briefly glycerinated frog muscle fibres. *J. Physiol.* **204**, 475–491.

Podolsky, R. J. & Teichholz, L. E. (1970). The relation between calcium and contraction kinetics in skinned muscle fibres. *J. Physiol.* **211**, 19–35.

Wise, R. M., Rondinone, J. F. & Briggs, F. N. (1971). Effect of calcium on force-velocity characteristics of glycerinated skeletal muscle. *Am. J. Physiol.* **221**, 973–979.

Effect of "Viscosity" of the Medium on Mechanical Properties of Skinned Skeletal Muscle Fibers

M. ENDO, T. KITAZAWA, M. IINO and Y. KAKUTA

Department of Pharmacology, Tohoku University School of Medicine, Sendai, Japan.

ABSTRACT

The effect of raising viscosity of the medium on tension development of skinned skeletal muscle fibers was examined. Viscosity was altered by adding polyvinylpyrrolidone (PVP) K15 or various non-electrolytes such as ethylene glycol, glycerol, glucose or sucrose. Tension production was inhibited in high-viscosity media. Except the case of PVP, tension produced by lower concentrations of Ca was more strongly inhibited. The mechanism of inhibition of tension production by the high viscosity media was unclear. The magnitude of inhibition was qualitatively in parallel with viscosity in that a concentration of sugars was less effective at higher temperatures, but quantitatively not in parallel with the magnitude of viscosity of the solutions determined by an Ostwald viscometer, when the effects of different substances were compared.

INTRODUCTION

In a current sliding hypothesis (Huxley, 1957), maximum shortening velocity, V_{max}, is considered to be limited by cross-bridges that remain attached in a region where they exert negative tension. Theoretically, however, viscous force should also be working during shortening in the direction of resisting the sliding movements of filaments. Therefore, if viscosity of the medium is increased to a level high enough, V_{max} is expected to be reduced. From the magnitude of viscosity that just begins to affect V_{max}, then, one might be able to deduce the magnitude of negative force exerted normally by cross-bridges persisting in the negative force region during unloaded shortening.

A preliminary study along this line immediately revealed that not only V_{max} but also isometric tension was strongly depressed in high viscosity media. The present paper reports some results of the effect of raising viscosity of the medium on tension development of skinned skeletal muscle fibers.

METHODS

Single fibers were isolated from iliofibularis muscles of African clawed toads, *Xenopus laevis*, and skinned in a relaxing solution as previously described (Endo & Nakajima, 1973). Fibers were mounted by pressing the ends between small pieces of Scotch double-stick tape and suspended in a covered trough through which solutions could be perfused rapidly. Isometric tension was recorded with a strain-gauge transducer (U-gauge, type UL, Shinkoh Co. Ltd., Tokyo). Sarcomere length in the middle of the fibers was determined under a light microscope and adjusted usually to about 2.8 μm, or to some other values when necessary.

The relaxing solution consisted of 4 mM ATP Na_2, 4 mM Mg methanesulfonate, 2 mM EGTA (ethyleneglycol-bis-(β-amino-ethylether)-N-N'tetraacetic acid), 111 mM K methanesulfonate, 20 mM Tris (hydroxymethyl)aminomethane and 20 mM maleic acid brought to pH 6.8 with KOH. Calcium-containing solutions were different from the relaxing solution only in that they contained 10 mM (sometimes 20 mM) EGTA and an appropriate concentration of Ca methanesulfonate, and further that the concentration of K methanesulfonate was appropriately reduced so as to keep ionic strength of the solutions the same as that of the relaxing solution. Free Ca ion concentrations of solutions were calculated using Ogawa's binding constant of CaEGTA (Ogawa, 1968). Viscosity of the media was altered by adding polyvinylpyrrolidone (PVP) K15 (average molecular weight 10,000; Tokyo Kasei Kogyo Co. Ltd.) or various non-electrolytes such as ethylene glycol, glycerol, glucose or sucrose. These substances were added to concentrated normal solutions, dissolved or mixed completely, and then the final volume of the solutions was adjusted by adding distilled water slowly with stirring. All the substances used were of the reagent grade.

Experiments were carried out usually at about 2°C, and pH of all solutions were adjusted to pH 6.8 at this temperature. When it was necessary to conduct experiments at room temperature, however, pH of solutions was adjusted to 6.8 also at room temperature.

To avoid possible interference with uniform distribution of Ca throughout the fiber space by the sarcoplasmic reticulum, most fibers were treated before experiments with 5 mg/ml Brij-58 for 30 min in relaxing solution to destroy the function of the sarcoplasmic reticulum. However, results were not different with or without the Brij-58 treatment.

Viscosity of solutions was determined by using an Ostwald viscometer. Viscosity-raising substances were added to distilled water in various concentrations, and falling time of each solution was measured at 2°C. Relative viscosity, η/η_0, of each solution was calculated by using following equation

$$\frac{\eta}{\eta_0} = \frac{\rho}{\rho_0} \cdot \frac{t}{t_0},$$

where η, ρ, t, η_0, ρ_0 and t_0 are the viscosity, density, and falling time of the test solution and those of distilled water, respectively.

RESULTS

Figure 1 shows the effect of 2.1 M glycerol on tension development of a skinned fiber. As clearly seen, tension produced by 2×10^{-6} M Ca was almost abolished by 2.1 M glycerol, while tension by 3×10^{-5} M Ca was practically unchanged by the same concentration of glycerol. If glycerol concentration was increased, say to 6.8 M, however, tension produced by 3×10^{-5} M Ca was also strongly inhibited (Fig. 2). The effect of glycerol was completely reversible. The same effect was obtained irrespective of the order of application of Ca and glycerol.

Magnitude of inhibition of tensions produced by a low (2×10^{-6} M) and a high (3×10^{-5} M) concentration of Ca by various concentrations of glycerol

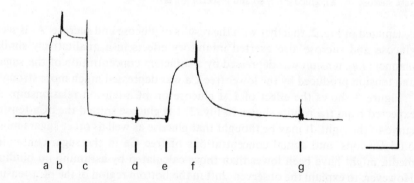

FIG. 1. Tension record of a skinned muscle fiber showing an inhibitory effect of 2.1 M glycerol. The main characteristics of each solution applied were as follows. a, 3×10^{-5} M Ca; b, 3×10^{-5} M Ca and 2.1 M glycerol; c and g, 10 mM EGTA; d, 2 mM EGTA; e, 2×10^{-6} M Ca; f, 2×10^{-6} M Ca and 2.1 M glycerol. Fiber 780824. Sarcomere length 3.3 μm. Scale bars 1 min and 1 mN.

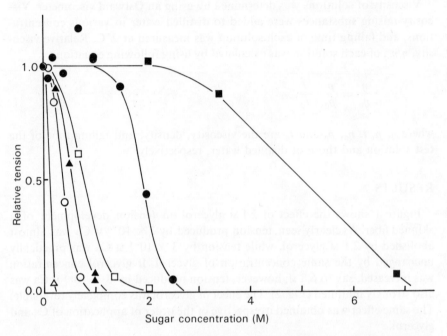

FIG. 2. Effect of various sugars on tension development of skinned muscle fibers. Free Ca ion concentrations were either 2×10^{-6} M (open symbols) or 3×10^{-5} M (filled symbols). Steady tensions at each Ca concentration in the presence of various concentrations of sugars are plotted in values relative to tension at each Ca concentration without sugars. Sugars used were sucrose (△, ▲), glucose (○, ●) and glycerol (□, ■).

is depicted in Fig. 2, together with the results of glucose and sucrose. As is seen, glucose and sucrose also exerted inhibitory effects in a qualitatively similar manner; *i.e.*, tension was depressed by a sufficient concentration of the sugars and tension produced by the lower free Ca was depressed much more strongly.

Figure 3 shows the effect of 1 M glucose on pCa-tension relationship. As expected from the results shown in Fig. 2, 1 M glucose shifted the pCa-tension curve to the right. It may be thought that glucose as well as other sugars might bind Ca ions, and actual concentrations of free Ca in the sugar-containing media might have been lower than those calculated by assuming no binding. However, to explain the observed shift in the bottom region of the pCa-tension curve of Fig. 3 along this line, one must assume an association constant of Ca-glucose complex of about 10^3 M^{-1}, which is an unreasonably high value. Such a binding has not been found. Furthermore, if one assumes this magni-

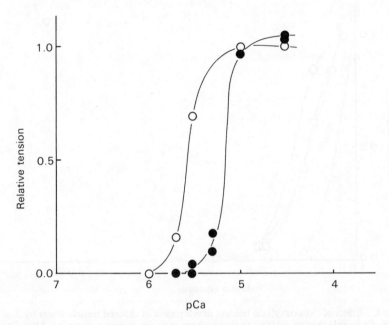

FIG. 3. Effect of 1 M glucose on the relation between pCa and tension. Tension was plotted in values relative to that at 3×10^{-5} M Ca without glucose. Fiber 780710. ○, control; ●, glucose 1 M.

tude of Ca-glucose binding, actual free Ca concentration of the solution of "3×10^{-5} M" Ca is calculated to be about 3×10^{-6} M in the presence of 1 M glucose. This means that tension developed by "3×10^{-5} M" free Ca in the dreence of 1 M glucose should be only about 70% of the tension without glucose. However, actual steady tension produced by "3×10^{-5} M" free Ca in the presence of 1 M glucose was not smaller than that by 3×10^{-5} M Ca without glucose (Fig. 3), indicating failure of the binding hypothesis. Further evidence against the binding hypothesis was as follows. If there exists substantial Ca binding to glucose, the magnitude of reduction of actual free Ca concentrations should be smaller when the free Ca was buffered with a higher total EGTA concentration. However, with 20 mM total EGTA, the magnitude of inhibition by glucose was exactly the same as in Fig. 3 where 10 mM total EGTA was used. It is concluded, therefore, that the lower apparent Ca sensitivity in the presence of 1 M glucose is due not to Ca binding of glucose but to genuine

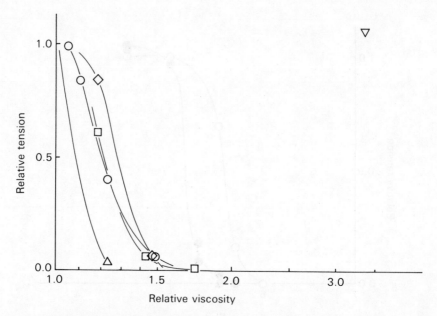

FIG. 4. Effect of "viscosity" on tension development of skinned muscle fibers by 2×10^{-6} M Ca. Ordinate: tension relative to that without viscosity-raising substances. Abscissa: relative viscosity determined by using an Ostwald viscometer. △, sucrose; ○, glucose; □, glycerol; ◇, ethylene glycol; ▽, PVP K15.

decrease in the apparent Ca sensitivity of the contractile system somehow produced by the sugar.*

All the sugars so far tested decreased apparent Ca sensitivity of the contractile system (Figs. 2, 4 and 5). In the case of PVP, however, although tension production was inhibited in an essentially similar way, the apparent Ca sensitivity was not decreased at all but rather increased slightly (Figs. 4 and 5).

Inhibition of tension development at a saturating concentration of Ca by the sugars is not a result of a reduction in free Ca ion concentration either, because an addition of Ca up to 10 mM or more to sugar-containing 10^{-4} M Ca solutions did not increase tension at all.

All the sugars used in this study are small molecules and are expected to

* In the Symposium, Dr. Tregear pointed out the possibility that sugars may alter the apparent binding constant of CaEGTA complex, and thus decrease free Ca ion concentration to bring about results such as Fig. 3. However, after the Symposium we measured free Ca ion concentrations of CaEGTA buffer solutions in the absence and presence of glucose by using Arsenazo III or Antipyrylazo III, and found that the presence of 1 M glucose did not change the free Ca ion concentrations to any detectable degree.

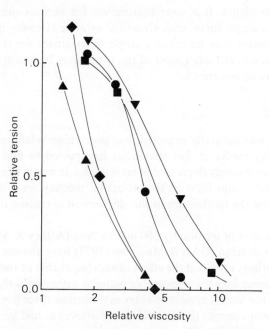

FIG. 5. Effect of "viscosity" on tension produced by 3×10^{-5} M Ca in skinned muscle fibers. Ordinate, abscissa and symbols are as in Fig. 4, except that all symbols are filled.

enter into myofibrillar space freely. Therefore, no osmotic effect would be expected with the sugars. In fact, the width of skinned fibers measured under a microscope was not altered at all by an application of even very high concentrations of sugars, whereas it was decreased on an application of PVP K15, which probably did not easily enter into the myofibrillar space because of its molecular size. The inhibitory effect of sugars is, therefore, very unlikely to be due to their osmotic effect.

If the inhibition of tension production by the sugars is due to an increase in viscosity of the medium, the effect should be weaker at higher temperatures where viscosity is smaller. In fact, at least with 1 M sucrose that almost abolished high Ca tension at 2°C (Fig. 2), a substantial tension was obtained at 20°C. However, of course this does not prove that the viscosity is the responsible factor.

Viscosity of solutions was determined by using an Ostwald viscometer at 2°C and its values relative to that of distilled water were calculated as described in Methods. In Figs. 4 and 5, steady tensions produced by a low and a high free Ca, respectively, in various high-viscosity media were plotted against relative viscosity of the media. It is clear that curves for various substances do not converge into a single curve, and, therefore, relative viscosity measured by an Ostwald viscometer is at least not a single determinant for the inhibition of tension production, although most of the substances behave similarly and the curves are more or less parallel.

DISCUSSION

Although it was naturally expected that shortening velocities were reduced in high-viscosity media, it was somewhat unexpected to find that isometric tension was also strongly depressed in such media. It would be very interesting, and might throw some light on the molecular mechanism of tension generation, to elucidate the mechanism of this depression of tension in high viscosity media.

Small inhibitions of tension by 360 mM sucrose (Ashley & Moisescu, 1977) and by PVP or dextran (Godt & Maughan, 1977) have already been reported. The former authors regarded the effect of sucrose as that of osmolarity, but at least in the present case the responsible factor is rather unlikely to be osmotic pressure since the sugars appear to enter myofibrillar space freely; as indicated by the absence of changes in fiber width (see above) as had been described by Godt & Maughan (1977). However, to make this point clearer, it is desirable to measure filament spacings in high viscosity media with these sugars by using an X-ray diffraction technique.

Godt & Maughan (1977) attributed the decrease in tension in PVP- or dextran-containing solutions to the shrinkage of skinned fibers. However, since high-viscosity solutions that do not change fiber width cause a similar decrease in tension, and since PVP or dextran might also enter into myofibrillar space to some extent, it must also be considered as a possibility that the responsible factor may be raised viscosity, or some other changes of the solutions in parallel with the viscosity, rather than shrinkage of the fibers.

It is unclear at present which of the factors, viscosity or some other parallel change, is actually responsible for depression of tension in high viscosity media. It certainly cannot be concluded that high viscosity is responsible, especially because relation between tension and viscosity measured by an Ostwald viscometer did not converge into a single curve (Figs. 4 and 5). However, this does not seem to exclude the possibility completely either, since possible viscous force exerted on cross-bridges to reduce tension might not be in parallel with viscosity measured by an Ostwald viscometer. On the other hand, it is conceivable that a number of other factors such as changes in dielectric constants, those in activity of water and so on might be responsible for the inhibition in high viscosity media. Further studies are clearly necessary to elucidate the mechanism.

Ando & Asai (1977) reported that raised viscosity increased turnover time of myosin and H-meromyosin ATPase. Such a reduction in ATPase activity may result in a decrease in shortening velocity but not necessarily in a reduction of isometric tension. Furthermore, inhibition of tension by sucrose (Figs. 4 and 5) is much stronger than the inhibition of myosin ATPase. Thus, the inhibition of tension production described above may not be directly related to the ATPase inhibition. They also reported that K_m of myosin and H-meromyosin ATPase was altered in high viscosity media. However, since the concentration of ATP used in the present experiments was much higher than these K_m values, the effect of viscosity on K_m may not be related to the inhibition of tension reported here.

It is interesting to note that, with a lesser degree of raised viscosity, the rate of tension development was decreased, but steady tension reached was slightly increased rather than decreased (Fig. 5). This phenomenon should be examined in more detail.

Smaller concentrations of sugars decreased the apparent Ca sensitivity of skinned fibers without any reduction in maximum tension produced by a saturating concentration of Ca (Figs. 2 and 3). Whether this is a result of the same effect of "high viscosity" as that on actin-myosin interaction to reduce maximum tension in higher sugar concentrations, or is due to some other effect(s) of "high viscosity" on, for example, the troponin-tropomyosin system, is not clear. The fact that the inhibitory effect of ethylene glycol was much stronger on the basis of viscosity at higher Ca (Fig. 5), but rather weaker at lower Ca (Fig. 4) compared with glycerol and glucose, might be favorable for the idea that the sites involved in low and high Ca may not be exactly the same. PVP K15 did not share this effect. Whether or not this is related to the relatively non-penetrating nature of PVP into the filament lattice should be examined.

Acknowledgement

This study was supported in part by grants from the Ministry of Education, Science and Culture, Japan, and the Muscular Dystrophy Association of the United States.

REFERENCES

Ando, T. & Asai, H. (1977). The effects of solvent viscosity on the kinetic parameters of myosin and heavy meromyosin ATPase. *J. Bioenerg. Biomembr.* **9**, 283–288.

Ashley, C. C. & Moisescu, D. G. (1977). Effect of changing the composition of the bathing solution upon the isometric tension-pCa relationship in bundles of crustacean myofibrils. *J. Physiol.* **270**, 627–652.

Endo, M. & Nakajima, Y. (1973). Release of calcium induced by "depolarisation" of the sarcoplasmic reticulum membrane. *Nature (New Biol.)* **246**, 216–218.

Godt, R. E. & Maughan, D. W. (1977). Swelling of skinned muscle fibers of the frog. Experimental observations. *Biophys. J.* **19**, 103–116.

Huxley, A. F. (1957). Muscle structure and theories of contraction. *Progr. Biophys. biophys. Chem.* **7**, 255–318.

Ogawa, Y. (1968). The apparent binding constant of glycoletherdiamine tetraacetic acid for calcium at neutral pH. *J. Biochem. Tokyo* **64**, 255–257.

Discussion

Wilkie: You just touched lightly on the possibility that it might be osmolarity rather than the viscosity of the solutions that was important. I believe that you can get closer packing of the filaments with high osmolarity even in the absence of the surface membrane. Have you actually looked at electron micrographs to see whether you have got closer lattice spacing, and the development of vesicles in the sarcoplasmic reticulum, and all those things that come with the high osmolarity?

Endo: I have not measured the lattice spacing. I am going to measure it with X-ray diffraction instead of electron microscopy, but I have been observing the widths of the fiber under the light microscope. If we add PVP, for example, the width of the fiber shrinks dramatically and almost instantaneously. But if we add sucrose, for example, fiber width does not change at all under the light microscope, but from this measurement I cannot be certain whether, actually, the filament spacing is altered or not. I should check that.

Edman: May I continue on that? I think there is a lot of evidence that the spacing between the filaments would not effect the tension output, nor, as I showed here, the velocities of unloaded shortening.

Wilkie: Well, certainly in intact muscle fibers, raising the osmolarity to about double certainly reduces the force.

Edman: Yes, that can be thought to be a change of the ionic strength, instead, because if you take up the length-tension curves in normal Ringer, and in hypertonic or hypotonic Ringer, and you normalize these curves with respect to the plateau tension, you would see that they superimpose very nicely. We have published this in 1968 (Edman & Andersson, 1968), and I think they are very superimposable. That would seem to indicate that width, as such, does not matter. It is likely to be the ionic strength.

Wilkie: The problem is that you have at least three things altering, don't you? You have filament spacing, you have ionic strength, and you have the development of large, Swiss cheese-looking vesicles.

Edman: Yes, but it has been shown, I think, by April and co-workers (April, Brandt, Reuben & Grundfest, 1968), that injecting certain doses of calcium intracellularly gives the same response, whether you have the fibers swollen or shrunk. So, I think there is a lot of evidence against the idea that variations of width alter the tension.

Pollack: But aren't you talking about the shape of the length-tenison curve, and isn't Dr. Wilkie talking about the absolute values of tension?

Wilkie: Yes, I am.

Edman: Yes, but you would not get superimposable curves like that if width had an effect. There is the question of angles between the filament axis and the cross-bridge. These would shift with fiber width and would give differently shaped length-tension curves if width really affected the tension output.

Wilkie: But equally, if it were the ionic strength as you were saying, then, there is no reason why injecting calcium should produce tension; I mean it is very difficult to know which of the many things that change is actually the cause of the diminished or abolished tension development.

Tregear: It is clear how you excluded binding by glucose as a cause, but I was not quite clear how you excluded a change in the affinity of EGTA for the calcium. I expect there is a reason, but I did not understand it. In other words, is the higher concentration of glucose and sucrose changing the affinity of the buffer for calcium, making it bind the calcium more strongly, in fact?

Endo: In other words, maybe EGTA binds calcium more strongly?

Tregear: Yes, and perhaps one could find a direct test of that.

Endo: Yes. I was going to check the free calcium by means of a dye or something like that. If I could get that kind of evidence that would be useful.

Kushmerick: I think, along the same line, the high sugar concentration and the reduced activity of water could influence the pK_a of EGTA, and by an indirect means affect the binding constant of calcium.

Endo: Yes, certainly. I had not considered that possibility, so that I will have to recalculate. But I have shown that with certain sugar concentrations, the steady tension developed by high calcium is not altered at all, and that would probably be very difficult to explain along that line. If you increase the apparent binding constant of EGTA, since the total calcium does not increase very much from 10^{-5}M to above, maximum tension would probably be decreased.

Edman: Have you any idea about the passive forces in these fibers? Have you tested the resistance to extension for instance?

Endo: No, I have not measured that.

Huxley: Do you have any idea about the effect of the substances you are using on actomyosin ATPase in solution?

Endo: I think somebody from Prof. Ebashi's laboratory is measuring the actomyosin under a rather similar condition. And I gather, in very preliminary results, actomyosin ATPase is decreased rather similarly as tension development.

REFERENCES

April, E., Brandt, P. W., Reuben, J. P. & Grundfest, H. (1968). Muscle contraction: the effect of ionic strength. *Nature, Lond.* **220**, 182–184.

Edman, K. A. P. & Andersson, K. E. (1968). The variation in active tension wtih sarcomere length in vertebrate skeletal muscle and its relation to fibre width. *Experientia* **24**, 134–136.

Force-Velocity Relation in Tetanized Cardiac Muscle

Hidenobu MASHIMA

Department of Physiology, School of Medicine, Juntendo University, Tokyo, Japan

ABSTRACT
A frog ventricular muscle strip was tetanized by transverse AC field at 10Hz and 20 V/cm in Ringer's solution containing 9 mM Ca^{2+} at 20°C without adding caffeine. The tension-length relation for this complete tetanus was similar to that for the skeletal muscle, but the tension declined almost linearly at lengths shorter than 90% L_m, where L_m is the optimal length at which the maximum isometric force, F_m, is generated. The tetanic force was not potentiated by adrenaline. The load-velocity relations determined at various isometric forces can be described by the following single equation:

$$F - P = F\left(1 + \frac{a}{F_m}\right)\frac{v}{v + b} \qquad (1)$$

where P is the load, F the isometric force, v the velocity, a and b are constants. This equation implies that each cross-bridge generates a proper force and moves against proper viscosity. Although the dynamic constants $a/F_m = 0.51$, $b = 0.75$ L/sec and $V_{max} = 1.47$ L/sec (L is the muscle length) did not alter between 90–100% L_m, the internal load increased and V_{max} decreased at lengths shorter than 90% L_m. The internal load defined as the difference between the external load and calculated load at a given velocity was a linear function of the velocity and the rate of increase in the internal load increased linearly with decreasing muscle length between 70–90% L_m.

Previous studies on the force-velocity relation of the cardiac muscle faced the difficulty of the inability to tetanize the muscle. All measurements were made at a certain instant during the twitch contraction, in which the active state was changing more or less. The first attempt to tetanize the cardiac muscle was made by Henderson et al. (1971) in rat papillary muscles. They applied a train of electrical pulses at 25–50 Hz in the presence of 0.6–7.5 mM of external calcium and obtained a small tetanic force preceded by the initial twitch. Forman et al. (1972) effected repetitive electrical pulses in cat papillary muscles and found that smooth tetani could be obtained with repetitive electrical stimulation in the presence of both 10 mM caffeine and 10 mM calcium. In the frog ventricular muscle, however, strong and smooth tetanic contraction was generated in the presence of 9 mM of calcium without caffeine. This paper is intended to summarize the studies done in the tetanized frog cardiac muscle (Mashima, 1977a, b) and to discuss the force-velocity relation from the viewpoint of the cross-bridge theory.

METHODS

The strip was prepared from the ventricle of the frog (*Rana nigromaculata*), about 10 mm in length and less than 0.8 mm in diameter, near the coronary sulcus. Both ends of the preparation were ligated with thin nylon threads. The whole apparatus was set on a table designed to absorb external mechanical disturbance. The muscle preparation was mounted horizontally in a polystyrol bath (3 × 7 × 1.5 cm) which contained 10 ml of Ringer's solution. A pair of platinum foil electrodes (7 × 1.5 cm) were placed on opposite walls of the bath in parallel with the muscle, and the whole length of the muscle was stimulated simultaneously by a transverse electric field between the electrodes. The alternating current (AC) was effected to generate the tetanic contraction for 3–4 sec through a high-current stimulator, inserting 2 min rests between stimuli. It was confirmed that the maximum tetanic tension was developed by AC stimulation at 10 Hz and 20 V/cm in the solution containing 9 mM Ca^{2+}. The temperature of the Ringer's solution (110 mM NaCl, 12 mM $NaHCO_3$, 2 mM KCl, 1.8 mM $CaCl_2$, pH 7.2) was maintained at 20°C by a thermoelectric heat exchanger throughout the experiments. The solution was always bubbled with 95% O_2 and 5% CO_2 gas mixture.

One end of the preparation was penetrated at the ligated portion by a stainless steel needle attached at the tip of the isometric lever, by which the muscle tension was conveyed to the anodal pin of an RCA 5734 tube, and the other end was also penetrated by a needle attached at the tip of the isotonic lever. The muscle was connected tightly to the needle by thread. The length of the wooden vertical arm of the isotonic lever was 10 cm and that of horizontal arm (made of thin aluminum plate) was 1.5 cm, and the equivalent mass was 320 mg. As the external load was hung from 0.5 cm from the pivot, the load on the muscle was 1/20 of the external load. The movement of the lever was detected photo-

FORCE-VELOCITY RELATION

electrically. The tension and displacement were displayed simultaneously on an ink-writing rectigraph or sometimes on a cathode-ray oscilloscope. The compliance of the whole mechanical system was about 5 μm/g. Muscle length was varied with an accuracy of 0.1 mm by moving the isometric lever, which was mounted on a sliding scale with a vernier.

The quick release was made during the isometric contraction and the velocity of the release was adjusted by a piston-type velocity controller in order to minimize the oscillation of the lever after the release.

RESULTS

1. Tetanic contraction generated by repetitive electrical stimulation

When the muscle was stimulated by repetitive pulses or an alternating current at more than 5 Hz in Ringer's solution, tetanic contraction was observed. This tetanic tension increased with increasing external Ca^{2+} concentration up to 9 mM (Fig. 1, series A). Therefore, the experiments in this section were performed in solution containing 9 mM Ca^{2+}.

An alternating current (AC) was more effective for obtaining high tetanic tension, as shown in Fig. 1, series B. At less than 3 Hz, the summation was incomplete, and at more than 20 Hz, the tension declined. Therefore, the optimum frequency was around 10 Hz, similar to that for the repetitive pulses. The threshold intensity at 10 Hz was about 1.5 V/cm in peak-to-peak voltage. The tetanic tension increased only slightly with increasing voltage up to 17 V/cm, but declined considerably at more than 20 V/cm. Then, the optimum conditions for AC stimulation were 10 Hz frequency, 17–20 V/cm intensity, and a smooth plateau tension was obtained in 2 sec. It was possible to obtain identical tetanic tension curves repeatedly, provided the interval between stimuli was 2–3 min. The tetanic tension thus obtained was extremely constant for several hours, although the twitch tension declined rather quickly with time. It is notable that 10^{-6} g/ml adrenaline markedly potentiated the twitch tension but never the maximum tetanic tension, although the rate of rise of tension was increased.

When the external K^+ concentration was raised to more than 8 mM, various tetanic tensions less than the maximum were obtained by reducing the stimulus intensity.

2. Tension-length relation for tetanic contraction

Isometric tetanic tensions generated by maximum stimulus were measured at

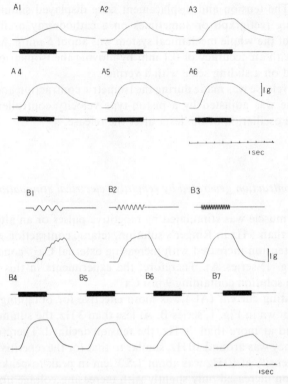

FIG. 1. Effects of external Ca^{2+} concentration and stimulus frequency on the tetanic tension of frog ventricle. Series A, Ca^{2+} concentration; A1: 1.8 mM, A2: 3.6 mM, A3: 7.2 mM, A4: 9 mM, A5: 12.6 mM, A6: 16.2 mM; stimulus intensity: AC 10 Hz, 17 V/cm. Series B, AC 20 V/cm; B1: 1 Hz, B2: 2 Hz, B3: 5 Hz, B4: 10 Hz, B5: 20 Hz, B6: 40 Hz, B7: 80 Hz.

various muscle lengths, and the length at which the maximum tension, F_m, was obtained was defined as the optimum length, L_m. After the contraction the muscle length was always returned to the initial length (usually 90% L_m), at which the resting tension was less than 0.3 g (7% F_m), throughout the interval of stimuli. The developed tension-length relations obtained in 11 muscles are summarized in Fig. 2.

The tension-length diagram for the tetanic contraction of cardiac muscle shown in Fig. 2 is similar to that of the skeletal muscle (broken line in Fig. 2) described by Gordon et al. (1966). However, in the region shorter than 90% L_m, the tension decreased almost linearly. It was difficult to stretch the cardiac muscle beyond 140% L_m because of the danger of rupturing the preparation. Curve W shows the tension-length relation for the twitch response of the

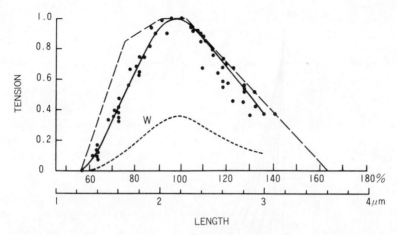

FIG. 2. Developed tension-length curve for the tetanic contraction of frog ventricle. W: twitch tension; broken line: tension-length curve for the skeletal muscle; bottom scale: estimated sarcomere length.

same preparations. After the experiments, some preparations were fixed at L_m and the sarcomere spacing was measured with an electron microscope. Usually, the sarcomere length was 2.0–2.2 μm at L_m. In the other preparations the diameters at L_m were measured directly, or after cutting the frozen preparation into serial sections under a microscope. As a result, the average value of F_m was 4.6 g/mm².

3. Force-load-velocity relation

During the plateau of isometric tetanus, controlled release was made and the shortening or lengthening velocity against the load was measured. The initial length was fixed at $0.9 L_m$, where the resting tension was sufficiently small and the relation between the load, P, and the velocity, v, was determined at various isometric tetanic tensions, F. One of the results is shown in Fig. 3, in which the load-velocity points are determined at six different isometric forces ($F_1 = 2.8$ g, $F_2 = 2.4$ g, $F_3 = 2.0$ g, $F_4 = 1.55$ g, $F_5 = 1.1$ g, $F_6 = 0.6$ g). The hyperbolic curves 1–6 are drawn by calculation so as to satisfy the relation

$$(P + a_i)(v + b_i) = b_i(F_i + a_i) \qquad (i = 1\text{–}6) \tag{1}$$

where $a_i/F_i = 0.32$ and $b_i = 0.52$ cm/sec are the dynamic constants. All measured load-velocity points fit well with the calculated curves. That is, the

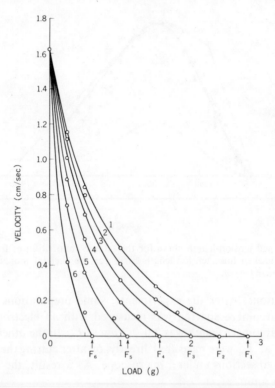

FIG. 3. Load-velocity curves at various isometric forces for a constant initial length (10 mm = 0.9 L_m). Isometric force, F_1: 2.8 g, F_2: 2.4 g, F_3: 2.0 g, F_4: 1.55 g, F_5: 1.1 g, F_6: 0.6 g. All curves (1–6) are drawn, using the same dynamic constants of $a/F_m = 0.32$ and $b = 0.52$ cm/sec.

relative load-velocity curves are identical, irrespective of the force of the contractile component.

When the initial length was L_m, it was necessary to subtract the resting tension at the instant of velocity measurement from the external load in order to estimate the real load to the contractile component. The resting tension at this instant was estimated from the X-Y record of the resting tension and the muscle length. By this correction it was confirmed that the load-velocity curve at L_m was also hyperbolic and the dynamic constants a/F_m and b were equal to those obtained at 0.9 L_m. Therefore, the general force-load-velocity equation can be written as follows:

$$(P + A)(v + b) = b(F + A), \qquad A = (F/F_m)a \tag{2}$$

or $$F - P = F\left(1 + \frac{a}{F_m}\right)\frac{v}{v + b} \tag{3}$$

at least at the initial length between 0.9–1.0 L_m. Average values of the constants are $a/F_m = 0.51$, $b = 0.75$ L/sec, and $V_{max} = 1.47$ L/sec ($n = 34$, at L_m, 20°C), where L is the muscle length.

When the load was larger than the isometric force, the muscle was lengthened by the load. The relation between the load and the lengthening velocity was also hyperbolic and expressed as follows:

$$(2F - P + A')(-v + b') = b'(F + A'), \quad A' = (F/F_m)a' \tag{4}$$

or $$P - F = F\left(1 + \frac{a'}{F_m}\right)\frac{v}{v + b'} \tag{5}$$

where $a'/F_m = 0.39$, $b' = 0.75$ L/sec ($n = 5$).

4. The load-velocity relation at an initial length shorter than L_m

When the muscle length was shortened below 0.9 L_m, the observed force-velocity points did not fit well with the calculated force-velocity curve. One of a group of load-velocity curves at 0.8 L_m (10 mm) is shown in Fig. 4. In this case L_m was 12.5 mm. Curve 1 was obtained at 12 mm after the correction of the resting tension where the isometric force was 3.7 g and the resting tension was 0.6 g. The dynamic constants of curve 1 are $a/F_m = 0.47$ and $b = 1.02$ cm/sec ($= 0.82$ L/sec). Curves 2, 3 and 4 are calculated using the same dynamic constants at the isometric forces of 2.45 g, 2.0 g and 1.1 g, respectively. The load-velocity curves at various isometric forces at 0.8 L_m are shown by dotted curves in Fig. 4. Obviously, the observed curves differ systematically from the calculated curves. And the maximum velocity decreased as the isometric force decreased. It is assumed that this slower than expected velocity is resulted from an increase in the internal load. The internal load is defined as the difference between the calculated and external loads at a given shortening velocity. In Fig. 5B, the internal load measured from Fig. 4 is plotted against the velocity. Clearly, the internal load at 10 mm is directly proportional to the velocity, regardless of the isometric forces. The internal loads at 9 mm and 11 mm are also plotted against the velocity in Figs. 5 A and C. The results are similar to that of Fig. 5B, and the only difference is the slope of linearity, which increases as the muscle length becomes shorter. From these results the following equation can be introduced:

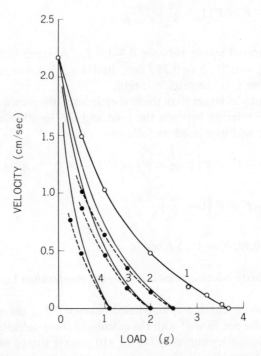

FIG. 4. Load-velocity curves at various isometric forces for a constant initial length of 0.8 L_m. Isometric force, curve 1: 3.7 g at L_m, curve 2: 2.45 g, curve 3: 2.0 g, curve 4: 1.1 g. All curves are calculated by using the same dynamic constants of $a/F_m = 0.47$ and $b = 1.02$ cm/sec. Observed load-velocity relations illustrated by dotted curves are not on the calculated curves.

$$P_i = \gamma v \tag{6}$$

where P_i is the internal load, v is the shortening velocity and γ is the factor of internal load. It was found that γ increases almost linearly with a decrease in the muscle length between 0.7–0.9 L_m and that γ is zero at lengths longer than 0.92 L_m. Namely,

$$\gamma = c(d - L/L_m), \quad (L \leqq 0.92\ L_m), \tag{7}$$

where c and d are constants. Average values are $c = 0.64\ (g/F_m)/(\text{cm/sec})$ and $d = 0.92\ L/L_m$ $(n = 5)$. It was concluded that the internal load is a linear function of the shortening velocity and that the internal load at a given

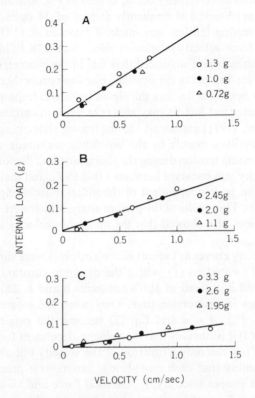

FIG. 5. Relation between the internal load and the shortening velocity at different isometric forces. A, at 9 mm (0.72 L_m); B, at 10 mm (0.8 L_m), obtained from data in Fig. 4; C, at 11 mm (0.88 L_m). Inserted figures indicate the isometric forces.

velocity increases also linearly by 6.4% F_m with a decrease of 10% L_m in the muscle length at least between 0.7–0.92 L_m, although no internal load is observed at lengths longer than 0.92 L_m.

DISCUSSION

During the plateau of tetanic contraction the active state is steady and the initial length of the contractile component can always be kept constant, so that the force-velocity curve for cardiac muscles must be essentially hyperbolic, obeying Hill's equation. However, it was noticed that the velocity at the load close to the isometric force was frequently higher than the value expected from the calculated force-velocity curve, as seen in Fig. 3. Similar high velocity at heavy load was observed as frequently as half of 34 cases, even after correction for the resting tension was made. Edman et al. (1976) observed the non-hyperbolic force-velocity relation at more than 0.78 P_0 in the single fiber of frog skeletal muscle and suggested that due to the geometric factors not al of the myosin cross-bridges in the overlap zone were properly oriented to allow interaction with nearby actin sites during an isometric response. It is possible that similar geometrical factors may also take place in cardiac muscles.

Brutsaert et al. (1971) estimated the maximum shortening velocity in the twitch of cat papillary muscle by the unloading technique which permitted removal of the resting tension during the shortening and demonstrated that the maximum velocity was constant between 1.0–0.88 L_m and that the force-velocity-length relation was independent of the initial muscle length and the time after the stimulus. The results of the present study, which were obtained during tetanic contraction, also showed that the maximum velocity was constant between 1.0–0.9 L_m.

The load-velocity curves at various isometric forces were summarized by the generalized Hill's equation (2), where the dynamic constants are written as $A = (F/F_m) a$ and b, instead of Hill's constants a and b. This means that the value of A varies in proportion to F. Only when the isometric force is the maximum $(F = F_m)$, $A = a$ and Eq. (2) become the original Hill's equation. From Eq. (3) it is obvious that the velocity-dependent force-loss (viscous-like force), $(F-P)$, is not only a function of the velocity but also a linear function of F. Assuming that each cross-bridge generates a proper force, f, and moves against a proper viscosity, f_v, the total force and viscosity should become $F = nf$ and $F_v = nf_v$, respectively, where n is the number of working cross-bridges, because all cross-bridges are parallel.

The internal load is assumed to be a kind of simple viscosity or friction which works within the muscle cell with decreasing muscle length, because the internal load is directly proportional to the shortening velocity. The site or origin of the internal load is not clear, but it is extremely suggestive to consider that the internal load is the result of the overlap of thin filaments, as the fine structure of sarcomere in cardiac muscle is almost the same as that of skeletal muscle. According to Gordon et al. (1966), the sarcomere length is 2.2 μm at L_m. On shortening, the thin filaments which slide toward the center of a sarcomere from both sides begin to overlap with each other at 2.05 μm (0.93 L_m), and the amount of overlap length increases in proportion to the shortening. At length shorter than 1.65 μm (0.75 L_m), both ends of the thick filament collide with Z-bands, so that another internal resistance against sliding will be added.

REFERENCES

Brutsaert, D. L., Claes, V. A. & Sonnenblick, E. H. (1971) Velocity of shortening of unloaded heart muscle and the length-tension relation. *Circulation Res.* **29**, 63–75.
Edman, K. A. P., Mulieri, L. A. & Scubon-Mulieri, B. (1976). Non-hyperbolic force-velocity relationship in single muscle fibres. *Acta physiol. scand.* **98**, 143–156.
Forman, R., Ford, L. E. & Sonnenblick, E. H. (1972). Effect of muscle length on the force-velocity relationship of tetanized cardiac muscle. *Circ. Res.* **31**, 195–206.
Gordon, A. M., Huxley, A. F. & Julian, F.J. (1966). The variation in isometric tension with sarcomere length in vertebrate muscle fibres. *J. Physiol.* **184**, 170–192.
Henderson, A. H., Forman, R., Brutsaert, D. L. & Sonnenblick, E. H. (1971). Tetanic contraction in mammalian cardiac muscle. *Cardiovasc. Res.* **5** (Suppl. 1), 96–100.
Mashima, H. (1977a). Tetanic contraction and tension-length relation of frog ventricular muscle. *Jap. J. Physiol.* **27**, 321–335.
Mashima, H. (1977b). The force-load-velocity relation and the internal load of tetanized frog cardiac muscle. *Jap. J. Physiol.* **27**, 483–499.

Discussion

Pollack: You measure the total tension and subtract the resting tension from it to estimate the active tension. My question is, as the sarcomeres shorten at the central segments, presumably the parallel elastic element is unloaded, so that the contribution of resting tension is actually smaller than you think, and the active tension at long sarcomere lengths may therefore be larger than you think. The length-tension curve could actually be flat, or could even be higher at long sarcomere lengths then at, say, 2.2 μm.

ter Keurs: May I phrase the question in a slightly different way? Do you have an estimate of internal shortening in your preparation? Do you have an estimate of how much the center of your preparation shortens during isometric contractions?

Mashima: I did not measure that.

VI. X-RAY DIFFRACTION STUDIES

Time Resolved X-Ray Diffraction Studies on Muscle

H. E. HUXLEY

MRC Laboratory of Molecular Biology, Cambridge, England

ABSTRACT

The central problem of muscular contraction now is to establish with certainty the exact mechanism by which the myosin cross-bridges develop the sliding force between the actin and myosin filaments. Although a number of plausible models exist, and although detailed kinetic studies have been made on actin-myosin-ATP interactions in solution, it has proved much more difficult to obtain direct evidence about the actual behaviour of the cross-bridges in muscle. Low angle X-ray diffraction studies offer the possibility of obtaining structural information about the contractile mechanism in its normal working state, although this information must be interpreted with care. The use of various types of one- and two-dimensional position sensitive X-ray photon counters and of fast electronic switching and storage devices has made it possible to record time-resolved patterns from contracting muscle, and the availability of higher intensity X-ray sources, including synchrotron radiation has enabled this time resolution to be extended below the 10 msec level. These techniques and results will be reviewed. The results so far are in general agreement with the concepts of the cycling cross-bridge model, where force is generated during attachment to actin, but a number of unexpected features have been found and will be described.

The mechanism of muscular contraction is not yet well understood in detail at the molecular level, and it is to investigate how the protein molecules involved are behaving during contraction that we are engaged in the present work.

In one sense, we already know what happens during contraction—sliding, changes in filament overlap. What we are concerned with now is what produces the sliding force. It is generally accepted that the sliding force is produced in some way by repetitive action of cross-bridges. It is perfectly possible to put forward a plausible model, involving myosin head subunits rotating about their attachment region on actin and attaching and detaching cyclically, which brings together the structure, the physiology and the biochemistry in a very reasonable way. However, one must not forget that it is just a model, a hypothesis; we have not merely to say "It *might* work like this," but we have to actually prove to the hilt that it does work that way: since nature is very subtle and unexpected and can often catch us out when we make one unsupported assumption too many. So we have to look very hard at the actual experimental evidence for the model, and distinguish between evidence that very strongly indicases that some particular interpretation is correct and evidence which is merely consistent with it but might conceivably be consistent with something rather different. And we have to look at the actual numbers that are required by the physiology, and the rates that are measured by biochemical kinetics and by, as we will see, structural kinetics. All this work has still a long way to go before this reasonable working hypothesis can be taken as firmly established fact—or as firmly as most other "scientific beliefs" are. Since it is such an important mechanism, operating in both muscle and non-muscle tissue, it is very important to get it right.

First, what sort of rate constants are we talking about here (in frog muscle at 2° C for example)? Different parts of the process obviously have different rate constants, but the basic on/off rate for attachment/detachment might be expected to have rates of the order of tens of times per second—one can tell that crudely, from the lengths of the structural repeats in the actin filaments and the approximate shortening velocities of the muscle. Then there are other structural transitions, changes in orientation, which in the Huxley-Simmons model might be very much faster—say in the millisecond range. There may be other ones within the molecules even faster still. But to begin with, we are concerned with rather crude, elementary questions of getting cross-bridges on and off, and for this purpose we need to be able to make the measurements with a time resolution of about 10 msec, if possible, and to see if the structural changes do conform to the requirements of the model.

X-ray diffraction of static muscle has already been very informative in supplying one part of the picture which, together with electron-microscope evidence, built up the overlapping, sliding filament model, so it was natural to wonder if that technique could be extended to dynamic studies. In a way, of course, that had already begun when we first took pictures of tetanically contracting muscles on film, but these measurements had two limitations: 1) The examination was of a steady state pattern, and gave no information as to how quickly it developed after stimulation nor whether it persisted the whole time that the muscle was active. 2) The picture was built up over a very long contraction series. To what extent, then, were we seeing long-term changes as the muscle fatigued or altered? Would the same changes occur in the first few contractions of a freshly dissected muscle?

At the previous Biophysics Congress in Copenhagen in 1975 I described how we had speeded up the registration of the low-angle diagrams from muscle, so that the changes in the intensity of the stronger spots could be followed with a time resolution of about 10 msec, and had begun to collect data, so this time I will summarize the techniques used and recent improvements in them only very briefly. Basically, most of the work has been done using a "Big Wheel" type Elliot fine focus rotating anode tube (60 mA, 40kV) and a Huxley-Holmes mirror-monochromator X-ray camera, the pattern being recorded by either a normal proportional counter set on the reflection of interest or (more usually now) a one-dimensional position-sensitive counter looking at the whole of the equator or meridian. The detector output is fed into the memory of a multichannel scalar, and switched as a function of time, from one memory position (or one set of 128 position for P.S.C. pattern) to the next. Thus with a P.S.C. one obtains a series of snapshots of the pattern at successive time intervals. The switching sequence can be repeated in synchronism with stimulus of muscle (electronics due to Dr. Wasi Faruqi), and the patterns summed in the computer and examined at leisure both during and after the experiment.

Most recently, we have begun to use the storage ring at DESY, Hamburg, as an X-ray source 10–100 times more intense than normal laboratory X-ray tubes. Used with a TV 2-d image intensifier (Milch), patterns can be recorded in a tenth of a second, but usable beam from the ring has been available only a very small proportion of the time so far.

It's convenient to think of the X-ray diagram in two main parts, the equatorial pattern which is derived from structures repeating sideways across the width of the muscle, and the axial pattern which derives from repeats along the length of the muscle—*i.e.*, along the long axes of actin and myosin filaments.

The equatorial pattern arises from the double hexagonal array of actin and myosin filaments, and the intensities of the reflections give us information about the distribution of mass between those filaments. In rigor, and during contraction, the 1,0 reflection decreases in intensity and the 1,1 increases, each by a factor of 2 or more, so it is a big effect; and I have interpreted this change as arising from a substantial proportion of the cross-bridges moving laterally to a position close to the actin filaments, as one would expect them to do if they produce sliding force by a process in which they attach to actin.

The first questions which we wanted to answer with the rapid X-ray measurement were the following. First of all, do these changes in pattern really take place in a fresh muscle the first few times it is stimulated after dissection, or are they something which progressively builds up over a long period of time? This was answered in the affirmative. Using a position-sensitive detector, one can see the characteristic changes in the equatorial reflections directly in real time, while the muscle is undergoing the first tetanic contraction immediately after dissection, and so there is no doubt that the change in pattern is characteristic of the contracting state itself, in the absence of fatigue or other secondary factors. To record the time course with 10 msec time resolution necessitates repeating the contraction perhaps 1000 times to accumulate enough counts in each time channel, but during this time the overall pattern during rest and during activity remain quite stable, and the time course of the tension development during the onset of tension also remains remarkably constant.

As I described a few years ago (Huxley, 1975), the changes in the equatorial pattern do indeed take place rapidly enough to be consistent with the very rapid change in the average position of the cross-bridges after stimulation which would be required by the swinging cross-bridge model. The structural changes in the muscle can first be detected about 15–20 msec after stimulation and proceed with a rate constant of about 20 per sec at 2°C. Indeed, they run ahead of tension development (Table 1) by about 10–15 msec. The 1,0 and 1,1 reflections change with slightly different time courses from each other, due to small amounts of internal shortening of the muscle even

TABLE 1. [10] Data at 2°C

Date 1977	Tension half-time (msec)	[1,0] Change half-time (msec)	Δ (msec)
7/2	57	40	17
14/2	62	49	13
17/2	60	44	16
25/2	45	43	2
28/2	51	33	18
1/3	50	39	11
2/3	55	38	17
3/3	59	52	7
4/3	53	42	11
5/3	61	49	12
7/3	45	33	12
8/3	54	44	10
9/3	65	51	14
10/3	46	38	8
		Average	12

under isometric conditions. During relaxation (in twitches), the records are somewhat variable, probably due to different rates of relaxation of different parts of the muscle, but when tension has disappeared, the 1,0 and 1,1 reflections have usually returned to within 10% of their resting values.

The delay between the extent of structural change and the development of isometric tension would be most straightforwardly explained if the change we are detecting represents attachment of cross-bridges to actin filaments, and if some subsequent "chemical" step in the reaction cycle has to take place (with a *not* extremely fast rate constant) before an attached cross-bridge can develop tension.

The question arises, however, as to whether the cross-bridge movement we see, or part of it, could arise from some activation mechanism within the myosin filaments, rather than being entirely due to combination with actin sites made accessible during activation by the tropomyosin-troponin system. (In the latter case, we would suppose that Brownian movement allowed the cross-bridges always to move between the myosin and actin filaments with a time constant of 1 msec or less and thus to spend a certain small proportion of their time near actin, even in a resting muscle, and that the distribution became altered when attachment began, trapping more and more bridges near actin.) I have attempted to gain further evidence on this issue by examining the equatorial diagrams given by frog semitendinosus muscle, stretched to lengths at which the arrays of thick and thin filaments should no longer overlap. Such muscles no longer give a 1,1 reflection, but the inner 1,0 reflection is still perfectly readily visible, if somewhat broader than at rest length. If a radial movement of cross-bridges takes place during the onset of activity— and such would seem to be necessary to account for the results near rest length (Haselgrove, Stewart & Huxley, 1976), this would be detected as a decrease in the 1,0 intensity, even if the bridges do not cluster around the trigonal positions. In the stretched muscles, however, during contractions of 0.5- or 1-sec duration, no significant intensity change was observed in the 1,0 reflection (Table 2), even though a large decrease was seen to develop rapidly in the same preparations at shorter lengths, both before and after the experiments at the stretched lengths, and even though the stretched muscles were developing significant ($P_0/10$) tensions, presumably from shorter sarcomeres near the ends of the fibers. This would seem to rule out an activation mechanism in the thick filaments which caused outward bridge movement and had a rapid response time.

During tetani of longer duration (5–10 sec) a decrease in the 1,0 intensity was seen consistently; it will be recalled that such long duration tetani were used in previous experiments (Huxley, 1972) in which a decrease in the

TABLE 2. Semitendinosus stretched beyond overlap

Experiment no.	Type of stimulus	Resting counts	Stimulated counts	Error (%)	Change (%)
1	100 × 1 sec	5,999	5,920	±3	−1.3
2	100 × 1 sec	12,167	12,005	±2	−1.3
3	100 × 1 sec	8,701	8,907	±3	+2.4
4	100 × 1 sec	14,079	14,132	±2.4	+0.4
5	100 × 1 sec	10,900	10,862	±2.3	−0.3
6	100 × 1 sec	8,082	7,744	±3.3	−4.2
7	100 × 1 sec	12,189	12,216	±2.9	+0.2
8	100 × 1 sec	9,909	9,338	±3.5	−5.8
9	100 × 500 msec	6,661	6,905	±3.1	−3.6
10	100 × 500 msec	6,882	6,903	±3	+0.3
11	150 × 500 msec	8,188	8,172	±3	−0.2
12	100 × 500 msec	7,001	7,044	±2.7	+0.6
				Average	−0.5%
	c.f. Change in [1,0] for 6 muscles at R. L.			Average	−44.6%
	and change in [1,0] for 4 muscles, stretched, after 5 sec stim.			Average	−16.6%

intensities of the axial pattern were observed. Whether both these effects stem from a progressive disorder of the muscle structure during the "creep" phase of tension development can only be settled by further experiments with higher intensity X-ray sources.

If we accept, provisionally, therefore, that the changes in the equatorial X-ray diagram do give us a measure of the extent of cross-bridge attachment, we can attempt to interpret the observations made on muscles during isotonic contraction (*i.e.*, in which filament sliding was taking place while the equatorial X-ray diagram was recorded). This was first done by Podolsky and his colleagues (Podolsky, St. Onge, Yu & Lymn, 1976) in muscles shortening under moderate load, and they found diagrams which differed little from those given under isometric conditions. I have confirmed that result, and I have also made observations on very rapidly shortening muscles (loads of $P_0/10$ and less) and found that, although the results show a considerable amount of scatter, the patterns do appear to change significantly (about 30% of the way) in the direction of the relaxed state (Table 3). Thus, as the actin and myosin filaments slide past each other at moderate speeds, the cyclical process of detachment and reattachment appears to be able to keep up the population of attached bridges. At higher velocities, however, where any bridges can only remain attached to a given site for a shorter time, the attachment rate appears to be insufficient to maintain the full population of attached bridges. Nevertheless, the proportion attached remains quite high, and, as was observed during initial tension development, a

TABLE 3. Fast shortening experiments $P_0/10$.
Intensity change from isometric as % of change between isometric and relaxed

Exp. no.	Counts			Areas		
	[1,1]	[1,0]	Bal.	[1,1]	[1,0]	Bal.
1	−40	+47	43	−30	+46	38
2	−15	+26	20	−47	+20	34
3	−40	+22	31	−6	+56	32
4	−16	+61	39	−59	+37	48
5	−23	+47	35	−29	+17	23
6	−24	+30	27	−43	+15	29
7	−33	+19	26	−26	+21	24
8	−23	+33	28	−25	+30	27
9	−27	+37	32	—	—	—
10	−37	+16	26	−48	+62	55
11	+5	+41	18	−10	+24	17
12	−20	+51	36	−19	+79	49
13	−32	+31	31	—	—	—
14	−14	+29	21	−18	+8	13
15	−3	+26	15	−20	+31	25
16	−19	+2	10	−20	+6	13
17	−8	+34	21	+2	+19	9
18	−18	+50	34	−35	+30	32
19	−21	+54	38	−29	+25	27
20	−3	+38	20	−3	+36	20
21	−31	+36	33	−43	+22	32
22	−16	+37	27	−20	+20	20
23	−29	+40	34	−45	+28	37
24	−28	+10	19	−22	+16	19
25	−3	+24	14	−8	+23	15
Average	−21%	+34%	27%	−26%	+29%	28%
S.E.M.	±2	±3	±2	±3	±4	±3
Av. Error	±9	±11	±7	±12	±13	±9

substantial population of cross-bridges appear to be attached but not developing tension. Some of these may be ones which detach late and develop negative tension (see A. F. Huxley, 1957), but since 70% of the maximum number of attached bridges appear to be developing only 10% of the maximum tension, some additional factors must be involved. If bridges can remain attached over a distance of about 100 Å, then under these conditions the attached time per cycle would only be about 5 msec. Clearly, if there is a delay in tension development after a bridge becomes attached, as suggested previously, the proportion of active bridges could be quite small.

Thus the time resolved results on the equatorial pattern support the basic features of the cross-bridge mechanism rather strongly and suggest additional features which are relevant to the mechanical and biochemical-kinetic results. The axial X-ray diagrams provide very important information about the position and orientation of the cross-bridges with respect to the lengths of the filaments. They show a helical arrangement of the bridges about the backbone of the thick filaments in resting muscle, with groups of bridges repeating at 143 Å intervals, giving rise to a fairly strong meridional reflection, and with a helical repeat of 429 Å, giving rise to weaker off-meridional layer lines with that repeat spacing. There are also some "forbidden" meridional reflections, including a strong second order 215 Å reflections. All these reflections are considerably weaker than the equatorials, and it is therefore less easy to study their behavior during contraction.

Earlier results on film showed that, in a long series of tetani, the layer line pattern became very much weaker than at rest, and that the meridional 143 Å reflection was also reduced in intensity, but to a lesser extent (Huxley & Brown, 1967). These changes were interpreted as being brought about by the attachment of a proportion of the cross-bridges to actin. Those bridges could then no longer conform to the helical pattern about the myosin backbone and might also disturb other bridges not yet attached, thus causing the layer lines to become weaker. The decrease in intensity of the meridional 143 Å reflection would, on this basis, indicate an increased longitudinal disorder of the bridges and/or an increase in their average angle of tilt.

In order to study the behavior of the layer lines during contraction series of short duration, it has been necessary to use the DESY electron-storage ring at Hamburg (facilities provided by the EMBL outstation) as a more intense X-ray source. In collaborative work with Dr. J. Milch, using a two-dimensional detector of the image-intensifier-TV type developed at Princeton and built by him, it was possible to record the 429 Å layer line in the summation of series of 6 exposures of 150-millisecond duration, each exposure series made either at the peaks of 6 successive twitches of a freshly dissected sartorius muscle or at other defined times before and after stimulation. It was found that the layer lines almost completely disappeared when the muscle was contracting, and reappeared again as the muscle relaxed. Thus the helical arrangement of the cross-bridges is indeed disrupted as a specific accompaniment of contraction. Because the reflection is a weak one, we have not yet measured the time course of these changes accurately.

The meridional 143 Å reflection presents a more complicated picture. In all muscles in which the series of contractions is prolonged (though there may be little loss of tension), the 143 Å reflection decreases in intensity during activity, and the decrease follows a very rapid time course after stimulation, similar to that observed on the equatorials. This decrease is also seen in many muscles even in the first 30 twitches or the first 5 tetani, even though, of course, its time course cannot be studied in such short series. In some muscles, however, the decrease in the 143 Å meridional reflection, during relatively short series of twitches or tetani, immediately after dissection, may be very small and in some cases the reflection may even become stronger. (This appears not to be due to a narrowing of the reflection about the meridian, since it shows up just as strongly in tilted specimens.) With more prolonged contraction series, the large decrease in intensity reappears. This shows that the drop in 143 Å intensity is not an essential concomitant of contraction and is most likely brought about by progressive longitudinal disorder in the muscle. While the result may seem paradoxical at first, it simply indicates that during contraction the attachment of bridges to actin does not increase their average angle of tilt and may indeed reduce it. Indeed, it has been shown recently by Haselgrove (in press) that the X-ray diagram of resting frog sartorius muscle indicates that the cross-bridges in resting muscle are likely to be tilted by about 40° to the long axis of the muscle, so that a smaller degree of tilt during attachment is quite plausible, as can also be seen from the following considerations.

It has been argued previously that the myosin S_1 subunits are likely to be able to move laterally and circumferentially about the backbone of the thick filaments, using the S_2 portion of the myosin rod to do so. This would entail the presence of flexible regions in the rod at either end of S_2. Such movement would permit attachment to actin over a range of orientations of the actin monomers near any particular cross-bridge. However, on this model, the proximal ends of the S_1 subunits (*i.e.*, the ends attached to S_2) would still remain confined fairly closely to a series of planes at right angles to the filament axis and 143 Å apart. Thus if cross-bridges attached when they were tilted away from the center of the A-band (say by 40°) and detached when their tilt had changed over to the opposite sense, their average tilt could be less than in the resting state. Given that the S_1 is an elongated structure, this could lead to an increase in intensity of the 143 Å reflection, at the same time as the layer line reflections were lost by the breaking up of the helical order. Thus the changes seen, and the specific character of the cross-bridge disorder which they imply, support rather strongly our model of cross-bridge attachment during contraction.

Another feature of the axial pattern detected in the earlier experiments using film was the approximately 1% increase in the spacing of the 143 Å meridional reflection during contraction. In recent experiments, initiated by Dr. John Murray, we have confirmed that the spacing change does occur, even in muscles which show little intensity change in the 143 Å reflection during contraction. This rules out the possibility that the spacing change arises from some sort of measurement artifact caused by an intensity change, or that the structure contains two different repeats differing by 1 %, the shorter of which is lost during contraction. We have measured the time course of the spacing change (in longer contraction series) and find that it manifests itself very rapidly and reaches its half-maximum value in about 25 msec after stimulation in twitches at 10°C when the time to half maximum tension was about 40 msec. During relaxation the spacing returns to its resting value with a time course similar to the tension decay. The fact that the time course of the change relative to tension is different during the rising and the falling phases of the twitch indicates that the spacing increase is not produced by tension directly, acting on a structure with non-linear stress-strain relationship. Changes in pH or ionic strength were not found to produce any spacing change in resting muscle. Thus the results are consistent with Haselgrove's suggestion (1970) that the changes arise as a direct effect of activation itself on the thick filaments. Conceivably, this could be related to the phosphorylation of myosin observed by Barany and Barany(1977) to take place during activation.

In general, then, the time course results accord rather well with our expectations from the cycling cross-bridge model, and they certainly show very dramatically the rapid cross-bridge movement upon stimulation. However, they have shown up some quite unexpected features, and what is particularly surprising and interesting is the evidence that attachment is too rapid to be the main rate-limiting step in tension development and that a subsequent slower transition may have to occur in the attached state. I hope to explore these questions further by studying the X-ray pattern during various types of mechanical transients, using more intense X-ray sources including synchrotron radiation.

REFERENCES

Barany, K. & Barany, M. (1977). Phosphorylation of the 18,000–Dalton light chain of myosin during a single tetanus of frog muscle. *J. biol. Chem.* **252**, 4752–4754.
Haselgrove, J. C. (1970). X-ray diffraction studies on muscle. Ph. D. thesis, University of Cambridge.
Haselgrove, J. C., Stewart, M. & Huxley, H. E. (1976). Cross-bridge movement during muscle contraction. *Nature, Lond.* **261**, 606–608.
Huxley, A. F. (1957). Muscle structure and theories of contraction. *Prog. Biophys. biophys. Chem.* **7**, 255–318.
Huxley, H. E. (1972). Structural changes in the actin- and myosin-containing filaments during contraction. *Cold Spring Harb. Symp. quant. Biol.* **37**, 361–376.

Huxley, H. E. (1975). The structural basis of contraction and regulation in skeletal muscle. *Acta Anatomica Nipponica* **50**, 310–325.
Huxley, H. E. & Brown, W. (1967). The low-angle X-ray diagram of vertebrate striated muscle and its behaviour during contraction and rigor. *J. molec. Biol.* **30**, 383–434.
Podolsky, R. J., St. Onge, R., Yu, L. & Lymn, R. W. (1976). X-ray diffraction of actively shortening muscle. *Proc. natn. Acad. Sci. U. S. A.* **73**, 813–817.

Discussion

Sugi: I am very much interested in your finding that the 143Å periodicity increases rapidly on stimulation. Can this be taken to imply that the filament is longer, *i.e.*, lengthened by 1%?

Huxley: Well, that is one possibility. There is another one which, I think, Elizabeth Rome (1973) first suggested that there may be an interference in the optical sense between the patterns given by the cross-bridges in either half of the sarcomere, and that this might produce interference fringes on the actual 143Å reflection itself. For some reason during contraction, possibly because the mean position of the cross-bridge is changing, those interference lines are shifted. It is an ingenious and interesting idea, but I have not been able to actually make it agree with the detailed observations of the character of the change. But I cannot exclude it.

Sugi: I would like to mention that the increase in the 143Å periodicity may be taken as evidence that the instantaneous elasticity may largely originate from the filaments, not from the cross-bridge elasticity.

Huxley: It is possible that some of it is in the thick filaments. But the fact that the spacing change occurs much faster than tension during tension rise, and at a similar rate during tension decay indicates that only a small part of the change can be an elastic effect.

Pollack: You showed that during isometric contraction the time course of the $I_{1,0}$ and $I_{1,1}$ reflections did not follow one another; the $I_{1,1}$ led the $I_{1,0}$. If the changes of intensity represented movements of cross-bridges from one filament to another, in the simplest case, one would expect the two intensities to be mirror images of one another. You attributed the fact that the $I_{1,1}$ preceded the $I_{1,0}$ to some internal shortening. How do you know that perhaps the entire change of $I_{1,1}$ is not caused by internal shortening?

Huxley: Well, because we measured the internal shortening by laser diffraction. It is normally of the order of about 3% or so, and we can see what a 3% length change will have on the X-ray diagram both in the contracting and in the relaxed states. The difference in the intensities is significant; I mean, it is on the order of 5% or so, but it is very much smaller than the factor of two or so change in each of the two reflections that one sees during contraction.

Pollack: Is the increase of $I_{1,1}$ due to the fact that the amount of thick and thin filament overlap increases by shortening?

Huxley: Yes.

Pollack: What happens then if, for example, the amount of overlap of the filaments diminishes during contraction? Dr. Fujime has shown that there is axial misregistration of thick filaments during contraction; this would tend to decrease the thick filaments effectively in the overlap zone. Should not that diminish the $I_{1,0}$ by similar reasoning?

Huxley: No. It is a question of what the projected density of the thick and thin filament lattice is. It seems that the thin filaments are ordered where they lie between thick filaments and that, as you withdraw the thin filaments, you decrease the ordered region, but the thick filaments are ordered throughout the A-band, and if you longitudinally displace them, their projected densities will still remain the same. That is, unless the longitudinal displacement becomes very great, when you might begin to get to some disorder. But in normal contractions around rest length, which is where I am working, the amount of misregistration of the thick filaments I think is still relatively very small. So there is no reason to believe that it is causing a big decrease in the $I_{1,0}$. And then you have got an increase by a factor of two or three times in the $I_{1,1}$. I think you really have got to make a big change in the amount of extra material at the trigonal points to account for it.

Rüegg: Would you predict that the immediate stiffness changes go parallel to the changes in the $I_{1,0}$?

Huxley: No—not necessarily. What I am suggesting is that there may be an attached state which is different from the normal tension-producing attached state, and since I am of course aware that the stiffness measurements do not agree with the X-ray patterns, then obviously what I say is that my early attached state is one which has a lower stiffness. Possibly, it is one where maybe there is some flexibility in the head of the cross-bridge; *i.e.*, when it first attaches it is still flexible, so it is still able to rotate relatively freely, and then perhaps the next step in the biochemical sequence is a change that rigidifies that link. And then the next one is the movement. So I have to make some such postulate, but I do not see anything terribly difficult about making such a postulate.

Kawai: How does your X-ray time course compare with so-called active state measurements?

Huxley: Well, I think you could define the active state in many different ways. If it is the capacity of the cross-bridges to develop tension, they still have to go through the attachment phase, and you might say you were measuring it by the degree to which the thin filaments were ready to receive the cross-bridges. Or, you could say that the active state was fully developed when the thin filaments were fully active, and you have to wait for shortening to take place for tension to be generated. But if there is shortening taking place, and the tension has not yet reached its plateau level, on the cross-bridge model this must be because there is relative movement of filaments taking place, and you never build up the full population of attached bridges. So even during the onset of tension, there should still be approximate proportionality between the number of attached bridges and the tension development, although, of course, you have got to make allowance for any bridges carried into the negative tension area. But there should be some relationship between number attached and tension developed.

Sagawa: What is the error range or reproducibility range of this measurement? I ask this question since you discuss very small percentage changes.

Huxley: In the case where I am measuring the intensity, one has to draw the background under the curve as best as one can, either by eye or by fitting some formula to it, and you do get, as I mentioned, quite a lot of scatter in the results. I probably would not put individual measurements to better than 10 or 20% or so. If one averages over 20 or 30 muscles, one can reduce that error, but of course the total change of intensity one see is very much larger than the uncertainty in the individual tension measurements.

Sagawa: How about in one muscle? Is the scatter much smaller if you repeat the measurements?

Huxley: I think the difficulty of the measurements is usually in knowing where to draw in the background under the curve. I think it gives you about a 10 or 20% uncertainty.

Edman: Have you any quantitative comparisons of the $I_{1,0}/I_{1,1}$ ratio during isotonic shortening at various loads? It seems to me that this might be an approach to use to get some idea about how many bridges are actually attached.

Huxley: Yes. Well, I have a lot of data at moderate loads, that is at about half P_0, which are similar to those that Podolsky (Podolsky, St. Onge, Yu & Lymn, 1976) published. I did not talk about these because of lack of time; but at moderate loads it is difficult to see any substantial deviation from the isometric case. I have already described what happens at small loads like 0.1 P_0. I hope to get more data about this, but they are not easy experiments to do; you get a lot of scatter and one would like to have a higher intensity source. Hopefully one can get better data. At the moment we have got rather poor data, but we can see the general direction it is going, I think.

Pollack: Regarding the interpretation of the $I_{1,0}/I_{1,1}$ ratio, there have been a number of presentations here during the past week or so, particularly among the Japanese, indicating that the thin filaments are not rigid, but have some flexibility. If one were to suppose that in the resting state the thin filaments were not simple rods, but actually had some small wiggles in them, and as they were tugged upon during contraction they simply straightened out, do you think that would affect the 1,1 reflection? Would that increase it, and if so, would you be able to distinguish that increase from an increase due to attachment of cross-bridges or to cross-bridges approaching the thin filament?

Huxley: No. That is a very good point. The sort of temperature factor that you associate with thin filaments does have a large effect on the intensity of the 1,1 reflection, and is something, again, that we have worried a lot about. One piece of evidence that bears on this is the time-course of the $I_{1,1}$ change in relation to tension. If tension of the thin filaments was producing a better orientation or better localization of them, then you might expect that to show up in a change in the $I_{1,1}$ which was simply related to tension in some way, but what I find is that during the rising phase of tension the 1,1 intensity increases very rapidly so that one would have to say that a relatively small amount of tension applied to the thin filaments holds them in position. But during relaxation the 1,1 intensity falls to within about 10% of its resting value with the same sort of time course as tension, so in this case it looks as though on that basis the tension is not able to hold the thin filaments in the same way. So you would have to put in some sort of extra assumptions to explain it that way. It seems to me at present, anyway, more straightforward simply to explain it on the basis of attached cross-bridges which, although of course you cannot see them in the actively contracting muscle, you can actually see in the rigor muscle whose diagrams are so similar to the contracting muscle diagrams.

Pollack: Regarding the early phase of contraction, though, isn't it reasonable that perhaps only a small amount of tension would be required to straighten out the thin filaments, and once they straightened they remain straightened? Could that account for the fact the intensity change of the 1,1 occurs very early in contraction?

Huxley: In that case I would expect it, as I explained, to persist for a long time during relaxation, that is if you use only a small amount of tension to do it. And the fact of the matter is that the $I_{1,1}$ returns quite rapidly during relaxation.

REFERENCES

Podolsky, R. J., St. Onge, R., Ya, L. & Lymn, R. W. (1976). X-ray diffraction of actively shortening muscle. *Proc. natn. Acad. Sci. U.S.A.* **73**, 813–817.

Rome, E. (1973). Structural studies by X-ray diffraction of striated muscle permeated with certain ions and proteins. *Cold Spring Harb. quant. Biol.* **37**, 331–339.

The Use of Some Novel X-Ray Diffraction Techniques to Study the Effect of Nucleotides on Cross-Bridges in Insect Flight Muscle

R. T. TREGEAR,* J. R. MILCH,** R. S. GOODY,† K. C. HOLMES† and C. D. RODGER‡

* A. R. C. Unit, Zoology Department, South Parks Road, Oxford, England
** EMBL, Hamburg, Germany[a]
† Max-Plank Institute, Heidelberg, Germany
‡ Laboratory of Molecular Biophysics, Oxford, England

ABSTRACT

For several years we have studied the diffraction pattern from insect flight muscle when the chemical conditions in the muscle are altered. We have looked in particular at the effect of the unhydrolyzable ATP analog, β,γ-imido-ATP. This compound binds to the enzymatic site of myosin and changes the diffraction pattern without reducing the muscle's mechanical stiffness. Recently we have employed a novel two-dimensional detector on the intense X-ray source at Hamburg to compare the imido-ATP pattern to that obtained from Ca-activated muscle. Preliminary results from this investigation are described and interpreted in terms of attachment of the cross-bridges.

INTRODUCTION

A decade ago Yount and his colleagues synthesized β,γ-imido-ATP (Yount, Babcock, Ballantyne & Ojala, 1971). At the same time Holmes's group began to build an X-ray camera on the Hamburg synchrotron (Rosenbaum, Holmes & Witz, 1971). These two innovations generated a pattern of work which has begun to illuminate the structural basis of contractility. We describe here the current position of that work. We start with a critique of the X-ray methodology involved, since techniques have been critical in this field and are still rapidly developing. We then consider the knowledge of the insect flight muscle's diffraction pattern that has been generated. This knowledge is incomplete, but it is sufficient for a limited structural analysis to be made and hence a hypothesis for the mechanism of force generation to be proposed.

a) Present Address: Department of Physics, Joseph Henry Laboratories, Princeton University, Princeton, New Jersey, U.S.A.

TECHNICAL DEVELOPMENT

(a) Tissue handling

All our work has been on glycerol-extracted flight muscle of *Lethocerus*, whose properties have been exhaustively investigated (Tregear, 1975; Pringle, 1978). For the present study we have used well-aligned bundles of between 20 and 80 fibers, attached to the now-conventional mechanical apparatus and immersed in a low ionic-strength, neutral ($\mu = 0.06$ to 0.1, pH 7.0) solution at between 4 and 16°C. Large fiber bundles were used for most studies, but in active muscle, where both diffusion and mechanical control are critical, small fiber bundles were employed.

The conditions for rigor, relaxation and activation of insect flight muscle are well known (Tregear, 1975). The use of β,γ-imido-ATP (hereafter termed imido-ATP) is more recent; early errors due to the presence of ADP and consequent formation of ATP in the analog solutions have been avoided in the present study both by the use of highly purified imido-ATP and also by the inclusion of both a myokinase inhibitor (diadenosyl pentophosphate) and an ATP-dephosphorylating system (hexokinase plus glucose) in the analog solution; both nucleotides were synthesized by R.S.G.

(b) The use of high-energy X-ray sources

Those with experience of rotating-anode X-ray generators will be keenly aware of the limitations that their complications and unreliability place on biological experiments. Similar limitations apply at the high-energy sources. Four factors may be distinguished which cause problems:

(i) Biologists are at present parasitic users of a highly expensive machine. They cannot determine the machine's operating mode and so can work only when the main user of the machine happens to require conditions which generate X-rays at high intensity. In practice this means that the source is available only infrequently and for short periods (usually only a few days), which imposes tight planning on the experimenter; he must know in advance exactly what he is going to attempt.

(ii) The machine is large, complex, and often pushed to its limit; it therefore tends to break down. Such breakdowns, lasting minutes or hours and occurring within an already tight time schedule, frustrate the exact plans of the experimenter and put a high premium on his own adaptability under stress.

(iii) The mirror-monochromator camera is complicated to adjust. As a result, setting-up can consume a considerable part of the time available for the experiment.

(iv) The X-ray beam is intense and therefore potentially dangerous, so that it has to be approached with caution. This can make a simple manipulation of the biological apparatus also very slow; the biggest single improvement in the current operation of the Hamburg storage ring has been the permission to work directly on the apparatus while the beam shutter is open.

All these factors conspire to reduce the biological yield; it requires an intense and continuous effort to maintain the system as reliably as possible. The effort can be worthwhile, however, as the results from muscle show.

(c) Available X-ray source-camera combinations

The technical appraisal given below is *as found* for the present experiments. The theoretical performance of the instruments concerned has been discussed elsewhere (Stuhrmann, 1978; Rosenbaum et al., 1971; Goody, Barrington-Leigh, Mannherz, Tregear & Rosenbaum, 1976). The present specimens are only 0.2 to 0.5 mm wide; thicker specimens favor in particular the storage ring camera's use (due to beam geometry) and to a lesser extent the synchrotron camera over the rotating-anode generator.

(i) Rotating-anode generator and camera: An Elliott GX6 generator with a Franks mirror-bent quartz monochromator camera set to give an 0.8 m specimen-detector distance allows detection of the inner equatorials from our specimens in some 20 min on film (or 20 sec on a counter; see below for comparison of detectors). These peaks can therefore be detected fast enough for all our experiments. The weaker parts of the pattern take up to 2000 min to collect, and therefore can only be obtained for maintained conditions. The meridional focus of the camera is excellent so that layer line spacings can be accurately specified; on the other hand, the equatorial focus is poor and obscures details of the lattice sampling.

(ii) The synchrotron camera: This camera, now dismantled, was used for the film work described below. It increased detection speed approximately 50 times over that from the rotating-anode generator and thus allowed collection of the entire diffraction pattern on film in 120 min; comparative exposures of different conditions were therefore possible. The equatorial focus was very good so that adjacent lattice peaks could all be differentiated. The instrument was also used with a line counter for rapid collection of the inner equatorial peaks.

(iii) The storage ring camera: This camera is still under development. In

its present form, and with the current single-bunch mode on the storage ring, it already gives higher X-ray intensity than the synchrotron camera. The equatorial focus is at present inferior to that on the synchrotron but adequate for separation of nearly all the lattice-sampled peaks. This instrument has been used in conjunction with the TV detector described below. The combination allowed rapid collection of the sampled part of the diffraction pattern and hence a comparative study under different chemical conditions (exposures approximately 30 sec).

(d) Detectors

We have used three detectors, and present here an appraisal of their relative performance, again *as found*.

(i) The line counter: Line counters have adequate position sensitivity (typical FWHM 0.3 mm) to differentiate the sampled peaks of the muscle diagram at 0.8 m from the specimen. When filled with Xe, a line counter is highly efficient in detecting 1.5 Å X-rays and has a negligible noise level. It is the simplest detector for a sampled layer line since it can be aligned so as to accept the entire width of the line. No further processing of the data is necessary (Fig. 1a); the sampled peaks can be separated from the unsampled layer line by eye. It is also readily switched between time-slices by alteration of the voltage level entering the pulse height analyzer (Fig. 1a). The line counter is unsuitable for the study of unsampled layer lines because of the extreme difficulty of background subtraction.

We have used line counters chiefly for the study of the inner equatorials (Goody *et al.*, 1976).

(ii) Film: This, the classical area detector, allows an excellent analog display and simple data storage. Its major disadvantage is its intrinsic slowness. Film has a pre-exposure "fog" of silver grains which represents an unavoidable noise level (Milch, 1976). Enough X-rays must be collected on the film to rise above this noise before detection becomes accurate. In practice an optical density greater than 0.1 is desirable, which increases exposure times relative to either of the other detectors by approximately a factor of fifty (dependent on the focal spot size). A major difficulty in film used to be the collection of layer line data. Now that films can be automatically densitometered and the data analyzed into layer lines (Fig. 1b; software by K.C.H.), this difficulty no longer exists.

Despite its slowness, film is still the only satisfactory way of collecting unsampled layer lines. All our outer layer line data were collected on it. Because they are weak and the film slow they have only been recorded a few times, in cases where the biological conditions can be held constant for a long time.

(iii) Image-intensifier area detector: This instrument consists essentially of a photon-electron transducer followed by an electron multiplier and charge detector. The present device was developed at Princeton University and brought by J.R.M. to Hamburg for trial on muscle specimens in the high-flux X-ray beam. It uses a silicon vidicon TV tube with a built-in intensifier (Milch, 1976; Milch, 1979). The diffraction pattern is integrated in analog form on the silicon target and read out with an electron beam in slow-scan mode.

The TV detector is highly efficient and has no pre-exposure noise, unlike film. It is therefore similar in detection speed to a line counter; the sampled inner layer lines of our muscle pattern were collected in 30 sec. Its position sensitivity is just adequate to resolve the sampled peaks (Fig. 2). An analog display of the data can be obtained either in real time or after storage (Figs. 1c, 2). The stored data can be analyzed in the same manner as a digitally-densitometered film (Fig. 1c). The principal disadvantage of the current device is the rapid development of noise during the exposure, which limits its usable duration to

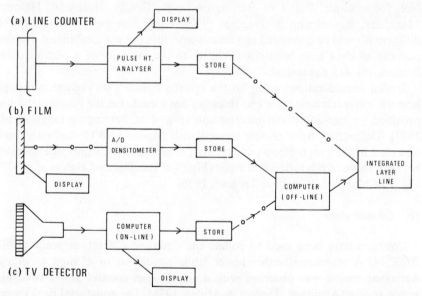

FIG. 1. Methods of recording diffraction patterns: Line counter (a), film (b) and TV area detector (c). The stages of producing integrated layer line records are shown for each in block form. The point at which each method goes off-line is indicated by the circled lines (—O—); each method contains an analog display stage in parallel with the flow of data through digital analysis. The same software may be used for analysis of film or TV detector data.

some 50 sec. The TV detector was extremely successful at data collection from the sempled layer lines. The weaker unsampled layer lines could also be detected, but only just; higher diffraction intensity is needed for their efficient collection by this means.

The TV detector is still being developed and shows great promise. It may be that it will eventually be replaced by a direct area counter (the TV detector contains an analog buffer), but for the moment it represents the best form of X-ray detection for many purposes. We hope that it will become more generally available.

EXPERIMENTAL RESULTS

(a) Film data

The diffraction pattern from insect flight muscle in rigor was first obtained 13 years ago (Reedy, Holmes & Tregear, 1965). The films from the synchrotron camera separated the adjacent lattice-sampled peaks of the inner layer lines (see, for example, Fig. 1 of Barrington-Leigh, Goody, Hofmann, Holmes, Mannherz, Rosenbaum & Tregear, 1977). Layer line analysis has allowed different films to be compared and an average rigor pattern established; for this purpose all data have been cited relative to the intensity of a particular reflection, the 41$\bar{5}$ equatorial.

Similar considerations apply to the relaxed muscle's diffraction, although here the lower intrinsic layer line intensity has meant that the synchrotron has provided greater additional information (Fig. 4 of Barrington-Leigh et al., 1977). Diffraction from muscle treated with Mg-imido-ATP has only been observed in the synchrotron camera. Its pattern shows features intermediate between those of relaxation and rigor (Fig. 3 of Barrington-Leigh et al., 1977; Fig. 7 of Marston, Rodger & Tregear, 1976).

(b) Counter data

Counters have been used to collect the strongest diffraction peaks (10$\bar{1}$0, 20$\bar{2}$0, 145 Å meridional) either under labile conditions or at high accuracy. Activated muscle was observed with a simple spot counter and a rotating-anode source (Armitage, Tregear & Miller, 1975). The equatorial peaks were seen to change slightly in the direction of rigor, while the 145 Å meridional fell as the active tension rose. Later a line counter was used on the synchrotron camera to relate the binding of imido-ATP to the equatorial changes; they were found to be proportional to the amount bound (Goody et al., 1976).

The counter data on the equatorials were used to cross-calibrate the films obtained under different chemical conditions.

(c) TV detector data

These data come from a series of 10 experiments conducted in 8 experimental days. Unfortunately the detector is no longer available so we report the necessarily sparse results here. The TV detector faithfully recorded the diffraction pattern from the sampled layer lines; the pattern was similar to that from film although some detail was lost (Fig. 2). The relative intensities of the various peaks were all within the statistical bounds of those recorded on film and, apart from the notoriously variable 145 Å meridional reflection, varied little between exposures (in terms of our standard, the intensity of the 41$\bar{5}$0 peak in rigor).

In each experiment the fiber-bundle was aligned in the X-ray beam by

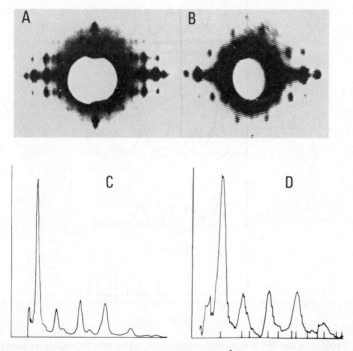

FIG. 2. Comparison of analog display (A, B) and 385 Å layer line output (C,D) from film (A,C) and TV detector (B,D) of insect flight muscle in rigor. The distortion and limited tone resolutions of C are properties of the particular display and are not inherent in the data.

maximizing the real-time display of its principal diffraction peaks. Timed exposures of between 1 and 30 sec were then collected in a variety of chemical conditions until the preparation showed signs of degradation; up to 30 exposures were obtained on each preparation. Apart from the "standard" conditions of rigor, relaxation and Mg-imido-ATP, the effect of H-imido-ATP was investigated by adding imido-ATP in the presence of EDTA to a muscle in rigor, and that of activation by adding Ca^{2+} to a relaxed muscle either in a concentration sufficient to allow high-amplitude work to be performed

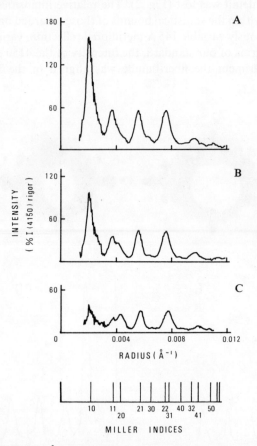

FIG. 3. Profiles of the 385 Å layer line obtained from TV detector exposures under various conditions: rigor (A), H-imido-ATP (B), Mg-imido-ATP (C). Abscissa marked with the positions of the lattice-sampled peaks whose h, k indices are shown below; ordinate in terms of the intensity of the 41$\bar{5}$0 reflection in rigor.

(Pringle & Tregear, 1969) or in higher concentration so that static tension built up in the "hypertension" response (Abbott & Mannherz, 1970). The latter only occurred in large fiber bundles.

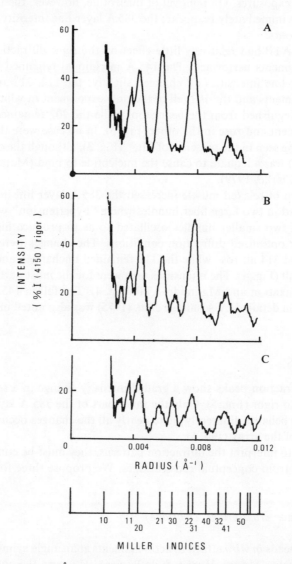

FIG. 4. Profiles of the 385 Å layer line in Mg-imido-ATP (A), isometric Ca-activation (B) and relaxation (C). TV detector exposures; note the change in vertical scale from Fig. 3.

The diffraction pattern changes consequent on the addition of nucleotide or Ca^{2+} were completed within the time necessary to add the solution and start the exposure (approx. 2 min); there was no indication of any further change in subsequent exposures. On removal of nucleotide, however, the full rigor pattern did not immediately reappear; the 385 Å layer line intensity rose over the ensuing 2 hours.

H-imido-ATP had relatively little effect on the rigor diffraction pattern in the 2 experiments performed. The 145 Å meridional remained low while the 385 Å layer line intensities all changed slightly; the $11\bar{2}$, $21\bar{3}$ and $31\bar{4}$ fell in both experiments and the $10\bar{1}$ fell in the only experiment in which it could be clearly distinguished from the beamstop, while the $20\bar{2}$ remained constant in one experiment and rose in the other (Fig. 3). In no case were the changes as large as those seen in the presence of Mg^{2+} (Fig. 3), although the concentration used (1 mM) was adequate to cause the nucleotide to bind (Marston, Tregear, Rodger & Clarke, 1979).

Activation of relaxed muscle increased the 385 Å layer line intensities; this was observed in two large fiber bundles where "hypertension" was produced, and also in two smaller bundles oscillated so as to produce high-amplitude work under optimized diffraction conditions. The changes were specific: the $20\bar{2}$, $21\bar{3}$ and $31\bar{4}$ all rose while the $11\bar{2}$ remained unchanged and the $30\bar{3}$ appeared to fall (Fig. 4). The intensities seen were for the most part intermediate between relaxation and Mg-imido-ATP (Fig. 4). The fall in 145Å meridional on activation detailed by Armitage et al. (1975) was also noted in these experiments.

ANALYSIS

Most diffraction peaks show a graded intensity change in a sequence from relaxation to rigor (Fig. 5). Only the inner part of the 385 Å layer line ($10\bar{1}2$, $11\bar{2}2$, $20\bar{2}2$) behaves differently; here nearly all the changes occur in the right-hand part of the array (Fig. 5).

In order to interpret the diffraction patterns, they must be compared to the predictions from conceptual cross-bridges. We propose three forms of cross-bridge:

(a) The rigor cross-bridge (r-bridge)

Myosin heads *in vitro* attach to actin filaments at an angle azimuthally across the 385 Å helix (Moore, Huxley & de Rosier, 1970), and the outer layer lines of our rigor films show the existence of similarly-angled cross-bridges (Miller

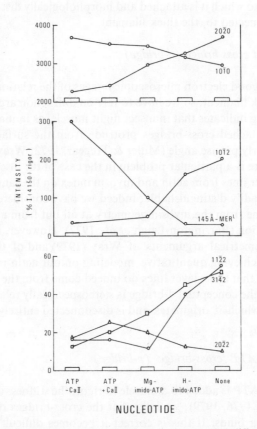

FIG. 5. Comparison of the intensities of various lattice-sampled diffraction peaks under different chemical conditions; all data are cited relative to the 41$\bar{5}$0 (equatorial) intensity in rigor. Please note that this is a histogram and not a graph. The data for the inner equatorials (10$\bar{1}$0 and 20$\bar{2}$0) were obtained from spot or line counters (Armitage et al., 1975; Goody et al., 1976; Marston et al., 1979). The other data were obtained either from film (first, third and fifth columns) or from the TV detector (second and fourth columns); the two methods were cross-checked.

& Tregear, 1972). According to quantitative modeling, between two- and three-seventh of the actin monomers bind cros-sbridges grouped at each turn of the 385 Å helix (Holmes, Tregear & Barrington-Leigh, 1979). Such groups have been seen in the electron microscope (Reedy, 1968), and a similar grouping has been inferred in crab muscle from X-ray diffraction analysis (Namba, Wakabayashi & Mitsui, 1979).

The conceptual r-bridge is a myosin head, stereospecifically related to the actin monomer to which it is attached and morphologically disconnected from, though still connected to, the thick filament.

(b) The relaxed cross-bridge (d-bridge)

There are no good electron microscopic images of the relation of the myosin head to the thick filament, so we have to rely on diffraction arguments. Quantitative modeling indicates that in insect flight muscle, as in many other invertebrates, the detached cross-bridges protrude from the surface of the thick filament at a fairly precise angle (Miller & Tregear, 1972; Wray, Vibert & Cohen, 1975). There is a particular problem in the case of the insect flight muscle because the layer lines from actin and myosin index on the same pitch and are therefore not readily distinguishable; indeed we have considered the idea that they do not come from the myosin geometry at all but from a modulation of the actin attraction (Barrington-Leigh et al., 1977). However, in view both of the general geometrical arguments of Wray (1979) and of the difficulty we have found in achieving quantitative modeling on our actin-modulation argument, we infer that these layer lines do indeed come from the thick filament.

On this basis the conceptual d-bridge is stereospecifically related to the thick filament from which it originates and is disconnected entirely from the actin filament.

(c) The imido-ATP cross-bridge (p-bridge)

When imido-ATP is added to a muscle in rigor, the stiffness does not change (Marston et al., 1976, 1979); we infer that the cross-bridges remain attached when the analog binds. If this is correct it becomes difficult to explain the imido-ATP diffraction pattern on a mix of r- and d-bridges. It is possible; for example one of each pair of myosin heads might detach, but then it is difficult to see how it could orient on to the thick filament while the other remained oriented on the actin. Alternatively, the cross-bridges which apparently do not attach even in rigor (Squire, 1972; Offer & Elliott, 1977; Holmes et al., 1979) could become myosin-oriented when they bind imido-ATP, but then it is necessary to explain why the coupled tension and nucleotide affinity phenomena occur (Kuhn, 1977; Marston et al., 1979). If these suggestions are rejected, it is necessary to postulate a third form of cross-bridge.

The conceptual p-bridge is attached to actin but still stereospecifically related to the myosin filament. Since it is attached we must suppose the contact region takes up the actin symmetry while the S1-S2 junction retains the myosin symmetry.

(d) Cross-bridge interpretation

The observed diffraction patterns can now be qualitatively interpreted in terms of these three elements. A relaxed muscle contains d-bridges. Mass is clustered around the thick filaments so the $10\bar{1}0$ is strong and the $20\bar{2}0$ weak. The d-bridges sit totally on the myosin symmetry so that the 145 Å meridional is strong and the 385 Å peaks are characteristic of the myosin helix; in particular the $20\bar{2}2$, which is excluded on the actin lattice space group (Holmes et al., 1979), is strong (Fig. 5).

On activation some p-bridges are formed so the equatorials shift, while the 145 Å meridional falls and the general actin-based 385 Å layer line ($21\bar{3}2$, $31\bar{4}2$) rises. In Mg-imido-ATP the p-bridges dominate, producing the characteristic "mixed" pattern. When MgII is withdrawn the H-imido-ATP has less effect at the enzymatic site and an appreciable proportion of r-bridges is formed, preferentially at a particular point along the 385 Å helix; this causes a sharp rise in the $10\bar{1}2$ reflection. Finally the empty enzymatic site of rigor allows r-bridges to dominate. The 145 Å meridional and its associated layer lines vanish while the actin-based layer lines reach their maximum strength.

DISCUSSION

Our deductions lead us to an unusual postulate: that the conformational state of a protein's enzymatic site affects its relative adherence to two incompatible geometric systems. This could generate contraction; if the cross-bridge follows the d- to p- to r-sequence as the ATP hydrolysis cycle proceeds, the mechanical strain energy generated from the change of conformation will provide the chemo-mechanical transduction demanded.

The evidence is obviously weak at three points. First, we depend entirely on the mechanical stiffness measurements to postulate the existence of the p-bridge at all. If stiffness does not represent attachment, our argument fails. There are other ways of detecting attachment, notably by the mobility of the myosin head. EPR can now measure myosin head mobility in rabbit fibers (Thomas, Seidel & Gergely, 1978). We hope that it will be applied to the insect flight muscle.

Second, we depend entirely on the diffraction measurements for the postulated structure of the p-bridge; indeed our choice of a quantal state rather than a continuous spectrum of states is based primarily on a prejudice in favor of molecular switches. Again, this may be investigated in other ways and particularly by electron microscopy. The images identified by the Reedys and their colleagues (personal communication) as most faithful to the imido-ATP X-ray diffraction pattern contain in the same sarcomere both perpendicular bridges at 145 Å intervals and angled bridges at 385 Å intervals. Examination of filtered, averaged images indicated a regular hybrid of the two structures. These results are consistent with, though they do not prove, the quantal hypothesis.

Third, our own evidence has been entirely based on one muscle type. Again, this can be overcome. Rabbit psoas muscle behaves mechanically like the insect flight preparation when imido-ATP is added (Marston *et al.*, 1976, 1979) and its X-ray diffraction pattern also alters in a similar manner, the equatorial shift and 145 Å meridional rise being particularly apparent (Lymn, 1975). The preparation merits investigation on the high-energy source.

In summary, we postulate a third cross-bridge state, attached to actin but attracted and partially oriented to the thick filament, and propose that during ATP's hydrolytic cycle the progressive reduction of strain at the enzymatic site is coupled with an increased actin orientation of the cross-bridge—and hence energy transduction.

REFERENCES

Abbott, R. H. & Mannherz, H. G. (1970). Activation by ADP and the correlation between tension and ATPase activity in insect fibrillar muscle. *Pflügers Arch.* **321**, 223–232.

Armitage, P. M., Tregear, R. T. & Miller, A. (1975). Effect of activation by calcium on the X-ray diffraction pattern from insect flight muscle. *J. molec. Biol.* **92**, 39–53.

Barrington-Leigh, J., Goody, R. S., Hofmann, W., Holmes, K. C. Mannherz, H. G., Rosenbaum, G. & Tregear, R. T. (1977). The interpretation of X-ray diffraction from glycerinated flight muscle fibre bundles: new theoretical and experimental approaches. In *Insect Flight Muscle*, ed. Tregear, R. T., pp. 137–146, Amsterdam: North-Holland.

Goody, R. S., Barrington-Leigh, J., Mannherz, H. G., Tregear, R. T. & Rosenbaum, G. (1976). X-ray titration of binding of β-γ-imido-ATP to myosin in insect flight muscle. *Nature, Lond.* **262**, 613–615.

Holmes, K. C., Tregear, R. T. & Barrington-Leigh, J. (1979). Interpretation of the low angle X-ray scattering from insect flight muscle in rigor. *Proc. R. Soc.* (in press).

Kuhn, H. J. (1977). Reversible transformation of mechanical work into chemical free energy by stretch dependent binding of AMP-PNP in glycerinated fibrillar muscle fibres. In *Insect Flight Muscle,* ed. Tregear, R. T., pp. 307–316. Amsterdam: North-Holland.

Lymn, R. (1975). X-ray diagrams from skeletal muscle. *J. molec. Biol.* **99**, 567–582.

Marston, S. B., Rodger, C.D. & Tregear, R. T. (1976). Changes in muscle cross-bridges when β-γ-imido-ATP binds to myosin. *J. molec. Biol.* **104**, 263–276.

Marston, S. B., Tregar, R. T., Rodger, C. D. & Clarke, M. L. (1979). Coupling between the enzymatic site of myosin and the mechanical output of muscle. *J. molec. Biol.* (in press).

Milch, J. R. (1976). X-ray detectors for diffraction studies and their use with synchroton radiation. Princeton University. Technical report no. 20, Contract AT(11–1)-3120.

Milch, J. R. (1979). S. I. T. TV detector for X-rays. *I. E. E. Trans. Nuclear Sci.* (in preparation).

Miller, A. & Tregear, R. T. (1972). Structure of insect fibrillar flight muscle in the presence and absence of ATP. *J. molec. Biol.* **70**, 85–104.
Moore, P. B., Huxley, H. E. & DeRosier, D. J. (1970). Three-dimensional reconstruction of F-actin, thin filaments and decorated thin filaments. *J. molec. Biol.* **50**, 279–295.
Namba, K., Wakabayashi, K. & Mitsui, T., (1979). The structure of crab striated muscle in the rigor state. This volume, pp. 445–456.
Offer, G. & Elliott, A. (1978). Can a myosin bind to two actin filaments? *Nature, Lond.* **271**, 325–329.
Pringle, J. W. S. (1978). Stretch activation of muscle: function and mechanism. *Proc. R. Soc.* B.**201**, 107–130.
Pringle, J. W. S. & Tregear, R. T. (1969). Mechanical properties of insect fibrillar muscle at large amplitudes of oscillation. *Proc. R. Soc.* **174**, B.33–50.
Reedy, M. K. (1968). Ultrastructure of insect flight muscle. I. Screw sense and structural grouping in the rigor cross-bridge lattice. *J. molec. Biol.* **31**, 155–176.
Reedy, M. K., Holmes, K. C. & Tregear, R. T. (1965). Induced changes in orientation of the cross-bridges of glycerinated insect flight muscle. *Nature, Lond.* **207**, 1276–1280.
Rosenbaum, G., Holmes, K. C. & Witz, J. (1971). Synchroton radiation as a source for X-ray diffraction. *Nature, Lond.* **230**, 434–437.
Squire, J. M. (1972). Myosin filaments and cross-bridge interactions in vertebrate striated and insect flight muscles. *J. molec. Biol.* **72**, 125–138.
Stuhrmann, H. B. (1978). The use of X-ray synchroton radiation for structural research in biology. *Quart. Rev. Biophys.* **11**, 71–98.
Thomas, D. D., Seidel, J. C. & Gergely, J. (1978). Rotational motion of muscle cross-bridges during contraction in spin-labelled myofibrils. *Biophys. J.* **21**, 43a.
Tregear, R. T. (1975). The biophysics of fibrillar flight muscle. In *Insect Muscle*, ed. Usherwood, P.N.R., pp. 357–404. London: Academic Press.
Wray, J. S. (1979). Structure of the backbone in myosin filaments of muscle. *Nature, Lond.* **277**, 37–40.
Wray, J. S., Vibert, P. J. & Cohen, C. (1975). Diversity of cross-bridge configurations in invertebrate muscle. *Nature, Lond.* **257**, 561–564.
Yount, R. G., Babcock, D., Ballantyne, W. & Ojala, D. (1971). Adenylyl imidodiphosphate, an adenosine triphosphate analog containing a P-N-P linkage. *Biochemistry* **10**, 2484–2489.

Discussion

Gott: Does the Hamburg instrument and conventional X-ray diffraction instrumentation look at approximately the same area of the muscle between the tendon ends?

Tregear: Well, there are no tendon ends in this muscle, but it looks at about the same area.

Steiger: As I remember, you lose some tension if you apply AMP-PNP to the muscle, and you get this tension back if you stretch the muscle back about 0.3%. Is that correct?

Tregear: Yes.

Steiger: If you do that, do you get a change in your X-ray pattern?

Tregear: No, you do not. We have tried to find a change in the X-ray diffraction pattern by stressing the muscle, and have found no change. In fact, theoretically, I am not sure you would expect to, or I am not quite sure what you would expect to see, but practically you do not see a change.

Rüegg: I think you would expect a slight change on stretching, because, as Kuhn (1977) has shown, and you and Marston have shown (Marston, Tregear, Rodger & Clarke, 1979), on stretching, more AMP-PNP is bound to crossbriges, and these should then rotate. So if you stretch under conditions where AMP-PNP is not saturating, but half saturatigg, at about 0.1 mM concentration, then I think you should expect a slight change of the X-ray diffraction pattern.

Tregear: The point had not escaped us. We tried that, but we have not seen anything. But you see, we do not get very much observational time. A negative observation in these circumstances is not really evidence.

O'Brien: You said in the intermediate state you proposed a hypothetical "bent" or modified kind of cross-bridge. Would you also propose that there might be some rearrangement of the bridges along the actin filament, because in the rigor state we have a well-defined, at least on the electron microscope, chevron pattern. What corresponding evidence is there for the electron microscopy of the intermediate stage?

Tregear: There is evidence now from Reedy (1977), which shows in the same region—by optical diffraction; therefore, I am not certain if it is from the same cross-bridges. But it shows chevron and bar patterns, 145 Å bar patterns from the same region apparently interfering optically with one another, which does not prove what I said, but is consistent with it. Electron microscopically, a distinct image is obtained from either relaxation or rigor. That is new evidence.

O'Brien: But could it be a mixed population?

Tregear: Of course it could be on the electron microscope. Yes, I am saying there is interference between the two, which would mean that if it was a mixed population, the occurrence of one influenced the occurrence of the other. O.K.? That is to say, there would have to be a definite spatial relationship. They could be one and the same thing, or they could be two groups of thing which were spatially related, but they are not unrelated.

Steiger: Could you eventually use IDP instead of AMP-PNP? I guess IDP is not attacked by the myokinase.

Tregear: But ADP does not do the job either. You can get ADP there. You can kill the myokinase with AP5A. I mean there is a whole range of things you could use. There is a whole bestiary of this work that sort of goes on forever, you know, with nucleotides. But one does know that they are at the enzymatic site. They are competing with one another; they are doing something at the enzymatic site which produces a variable population of these things.

REFERENCES

Kuhn, H. J. (1977). Reversible transformation of mechanical work into chemical free energy by stretch dependent binding of AMP-PNP in glycerinated fibrillar flight muscle fibers. In *Insect Flight Muscle*, ed. Tregear, R. T., pp. 307–316. Amsterdam: North Holland.

Marston, S. B., Tregear, R. T., Rodger, C. D. & Clarke, M. L. (1979). Coupling between the enzymatic site of myosin and the mechanical output of muscle. *J. molec. Biol.* (in press).

Reedy, M. K. & Garrett, J. R. (1977). Electron microscope studies of *Lethocerus* flight muscle in rigor. In *Insect Flight Muscle*, ed. Tregear, R. T., pp. 115–136, Amsterdam: North Holland.

X-Ray Diffraction Studies on the Dynamic Properties of Cross-Bridges in Skeletal Muscle

Yoshiyuki AMEMIYA,* Haruo SUGI ** and Hiroo HASHIZUME*

* Engineering Research Institute, Faculty of Engineering, University of Tokyo, Tokyo, Japan
** Department of Physiology, School of Medicine, Teikyo University, Tokyo, Japan

ABSTRACT

Using a position-sensitive counter incorporated in a data collection system, time-dependent low-angle X-ray diffraction diagrams were recorded from frog sartorius muscle contracting in various conditions: a) isometric tetanus, b) slow stretch during isometric tetanus, c) isometric twitch, and d) isotonic twitch under a small load. The main results obtained are:

i) A slow stretch does not significantly affect the intensity ratio $I_{1,0}/I_{1,1}$ from isometrically tetanized muscle, although slight increases in both $I_{1,0}$ and $I_{1,1}$ were observed after stretch.

ii) This indicates that a higher level of isometric tension after a slow stretch is not due to the increase in cross-bridge number, but to a greater force exerted by each cross-bridge.

iii) During isotonic twitch under a small load, the intensity ratio decreased to a minimum value of 0.8–1.0, which was significantly greater than that (0.5–0.6) attained during isometric twitch.

iv) This suggests that the number of involved cross-bridges is significantly smaller in actively shortening muscle than in isometrically twitching muscle.

INTRODUCTION

As reported previously (Hashizume, Amemiya, Kohra, Izumi & Mase, 1976; Hashizume, Mase, Amemiya & Kohra, 1978), we developed a data collection system which permits time-resolved studies of X-ray diffraction diagrams produced by active muscle. The system is based on a linear position-sensitive proportional counter which detects low-angle reflections from live muscle in a focusing camera. For data collection in high-time resolution experiments, the processed output of the detector is directed to the computer through a specially designed controller. This controller contains two latch memories of 8 bits which store temporarily the digitized linear positions of two X-ray photons detected. The computer fetches the contents of the latch memories every 40 μsec through a general-purpose interface. The data can be stored in computer memory in two modes. The first mode was described previously (Hashizume et al., 1978). In the second mode, the data area of the computer memory is divided into 40 segments of 256 channels (16 bits/channel), which correspond to separate phases of the muscle contraction to be studied. Within the time duration of each phase, detected events are added to give the current position-count histogram as in the conventional multi-channel pulse height analyzer. This duration can be fixed at any multiple of 20 μsec by the program. When one cycle of data collection is finished, the contents of the core memory are dumped onto magnetic tape to start a new cycle after some interval. Stored data are not integrated over repeated data collection cycles before the whole measurement is completed. This leaves us the greatest flexibility in post-experiment data handling.

Since the last report was published, the system has been progressively improved. The time needed to store one photon event in computer memory has now been reduced to about 20 μsec, which corresponds to a maximum countrate of 10,000 cps with 10% count loss due to coincident events. The speed of our system is not remarkably high, but it is just sufficient to meet the countrate requirement in our muscle work (see Methods). We have also developed computer programs for various data manipulations, including correction for the non-linear response of the detector. The present paper describes studies made using this system to provide information about kinetic properties of cross-bridges in active skeletal muscle.

EFFECTS OF SLOW STRETCH ON CROSS-BRIDGES IN ACTIVE MUSCLE

In the sliding filament model (Huxley, 1957), the isometric tension of striated muscle is determined by the number of cross-bridges formed between the thick and thin myofilaments. This number depends primarily on the amount of overlap of the two filaments. It is known, on the other hand, that when a stimulated muscle is slowly stretched, the tension rises to a peak at the end of stretch and then decays to a new isometric level which is higher than the normal isometric tension at the same length (Abbott & Aubert, 1952). We have undertaken the following experiment to give information about cross-bridges involved in this phenomenon (Sugi, Amemiya & Hashizume, 1977).

METHODS

The sartorius muscle isolated from a bullfrog (*Rana catesbiana*) was mounted in a specimen chamber with thin Mylar windows and a multi-electrode assembly. The pelvic end of the muscle was clamped by a pin near the bottom of the chamber, and the tibial end was connected to a strain guage (Shinkoh, U) carried by a vibrator (Ling, 200). The chamber was filled with oxygenated Ringer's solution (NaCl, 115 mM; KCl, 2.5 mM; $CaCl_2$, 1.8 mM; pH adjusted to 7.2 by $NaHCO_3$ buffer) which was continuously renewed during the experiment. The solution was precooled to 2–4°C. A low-angle camera of the Huxley-Brown type was used with a 1.2 kW output X-ray generator (Rigaku Denki, RU200; 6000 rpm) operated with a line focus (1 × 0.1 mm, take-off angle 6°) on a copper target. A 6-cm-long, gold-coated glass mirror was set 16 cm from the focus. The curved monochromator was of Johansson type and employed a 3°-cut germanium crystal (111 plane) bent to a radius of 144 cm. The monochromator was positioned 26 cm from the X-ray source to give a focus 41 cm from its center. The specimen-to-detector distance was 36 cm. Both the line and the point parts of the focused beam were included in the experiment. Distributions of X-rays scattered from muscle were detected by a multiwire position-sensitive counter (Hashizume & Masaki, 1978) using an external delay line for position sensing. This detector, operated with 90% argon plus 10% methane gas pressurized to 1300 Torr, provided a linear spatial resolution of 0.23 mm and a detection efficiency of 15% for $CuK\alpha$ X-rays.

The sarcomere length of muscle was adjusted to 2.2 μm by means of optical diffraction of a He-Ne laser. The muscle was then tetanized with a 4-sec train of 3 msec pulses at 20 Hz and was stretched by about 2 mm with a velocity of 1.4 mm/sec about 1 sec after the onset of stimulation (see Fig. 1). The extent of stretch corresponded to 4–6% of the initial length (35–52 mm) of muscle. It was observed that the isometric tension developed after stretch was higher than that before stretch by 20–40%. The activation and stretch were repeated regularly at intervals of more than 1 min. X-ray diagrams were sampled for 1 sec at two different phases of the mechanical activity of muscle: (1) at the isometric tension before stretch and (2) at the steady isometric tension after stretch. The data of (1) and (2) were stored in separate memory segments of the multi-channel analyzer. The runs were terminated when the isometric tensions at the two phases fell below half the initial values. We could achieve reasonable photon statistics by accumulating the data over 8–12 repeated runs. Figure 2A shows typical tension traces recorded at the first and the last runs. The diffraction diagrams from the resting state of muscle were recorded at the initial and the stretched lengths at the end of each experiment.

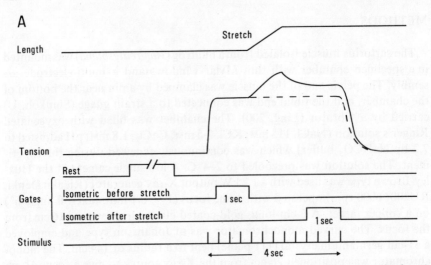

FIG. 1. Diagram showing the experimental procedure. For explanation see text (Sugi et al., 1977).

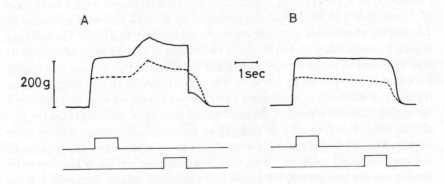

FIG. 2. Typical tension traces (top) and X-ray gates (bottom). Tension at the first run in solid line and that at the 10th run in broken line. A: tetanus and slow stretch. Initial muscle length 37.5 mm, muscle weight 540 mg, amount of stretch 2 mm. B: simple tetanus. Muscle length 44.5 mm, muscle weight 750 mg.

RESULTS

In Fig. 3 are reproduced diffraction diagrams obtained from a typical experiment in which the tension traces of Fig. 2A were recorded. The total exposure times for muscle at rest (Figs. 3A, B) and in isometric contraction (Figs. 3C, D)

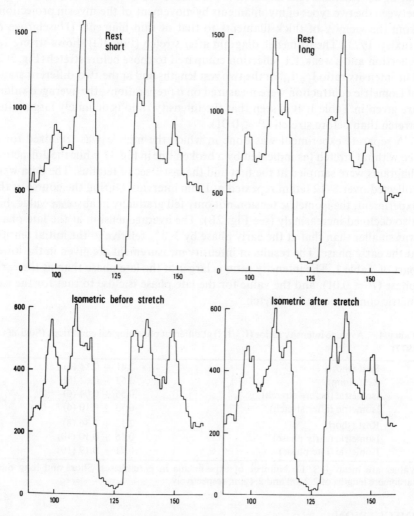

FIG. 3. Equatorial diffraction diagrams recorded at resting states (A and B) and during contraction (C and D). The abscissa shows channel number of the analyzer; the ordinate shows counts per channel; 0.12 mm on the detector/channel. Muscle length was 37.5 mm for A and C, and 39.5 mm for B and D (Sugi et al., 1977).

were 20 sec and 10 sec, respectively. In agreement with previous observations (Elliott, Lowy & Millman, 1967; Haselgrove & Huxley, 1973), the resting diagrams show strong 1,0 and weak 1,1 reflections, while the intensity of the 1,0 reflection decreases and that of the 1,1 reflection increases in the diagrams recorded during contraction. This has been attributed to the formation of cross-bridges between the two types of myofilaments by movement of the myosin projections from the vicinity of thick filaments to that of thin filaments (Haselgrove & Huxley, 1973). The isometric diagram after stretch (Fig. 3D) shows strong 1,0 reflection and a weak 1,1 reflection compared to those before stretch (Fig. 3C). The intensity ratio $I_{1,0}/I_{1,1}$ at the two rest lengths and at the two different states of isometric contraction were measured on 6 preparations; the averaged values are given in Table 1. It is seen that the intensity ratio is definitely larger after stretch than before stretch ($P < 0.01$).

A separate experiment was done in which the muscle was tetanized for 4 sec without stretch (as indicated by a broken line in Fig. 1), while the diffraction diagrams were sampled at the first and the last 1 sec of tetanus. The data were collected over 8–12 tetani repeated at 1 min intervals. During the course of the experiment, the isometric tension not only fell gradually in absolute value, but also declined increasingly (see Fig. 2B). The average tension at the late phase was smaller than that at the early phase by 3–8% relative to the initial tension at the early phase. The results of intensity measurement are given in the lower part of Table 1. The intensity ratio is larger at the late phase than at the early phase ($P < 0.01$), and the value for the late phase is close to that for the isometric contraction after stretch.

TABLE 1. Average intensity ratios ($I_{1,0}/I_{1,1}$) at different physiological conditions (Sugi et al., 1977)

Rest (short)	2.41 ± 0.14 (6)
Rest (long)	2.63 ± 0.62 (6)
Isometric (before stretch)	0.55 ± 0.04 (6)
Isometric (after stretch)	0.73 ± 0.10 (6)
Rest (short)	2.12 ± 0.36 (8)
Isometric (early phase)	0.55 ± 0.10 (10)
Isometric (late phase)	0.71 ± 0.19 (10)

Values are mean ± S. D. Number of experiments in parentheses. Short and long mean sarcomere lengths of 2.2 μm and 2.4 μm, respectively.

DISCUSSION

The quantitative relation between the intensity ratio $I_{1,0}/I_{1,1}$ and the cross-bridge number in stimulated muscle is not yet established. However, if we follow the line of reasoning of Haselgrove & Huxley (1973) and that of Matsubara & Yagi (1978), the present finding implies that 10–20% of the myosin projections which were initially active ceased to transfer to the vicinity of thin filaments at the late phase of 4 sec tetanus without stretch. This seems to be consistent with the decline of isometric tension observed at the late phase. It is worth noting that a small decline (3–8%) in tension is associated with a large decrease (10–20%) in the number of cross-bridges. The decrease in cross-bridge number appears also to be consistent with the time-dependent decrease in the rate of energy expenditure (Jöbsis, 1967) or in the rate of heat production (Fales, 1972) during a prolonged tetanus, since the decrease in the number of cross-bridges undergoing cyclic reaction may result in a decrease in the rate of ATP breakdown.

The observation that the intensity ratio for the isometric contraction after stretch is close to that for the late phase of tetanus without stretch implies that the number of cross-bridges involved is not significantly affected by slow stretch. However, this discussion is valid only if the same relationship between the intensity ratio and the cross-bridge number exists in the stretched and the unstretched states of stimulated muscle. It has been assumed that, after a slow stretch, myosin projections remain attached to thin filaments, in a displaced configuration, to generate a greater force (Huxley, 1960). If this assumption is correct, the intensity of the 1,1 reflection should be greater in the stretched state than in the unstretched state. Table 2 compares the reflection intensities measured in the two separate experiments. We have normalized the observed intensities relative to the 1,0 intensity before stretch in one experiment and to that at the early phase in another experiment. Although the uncertainties are large, one can notice that both $I_{1,0}$ and $I_{1,1}$ for the stretched state are greater than those for the unstretched state (the second and the forth lines of Table 2). It is reasonable to suppose that these intensity changes result from some change in muscle structure introduced by slow stretch, because $I_{1,1}$ before stretch in one experiment agrees fairly well with that at the early phase in another experi-

TABLE 2. Normalized reflection intensities at different physiological conditions

Condition	$I_{1,0}$	$I_{1,1}$
Tetanus & slow stretch		
before stretch (6)	100	185 ± 14
after stretch (6)	116 ± 18	162 ± 29
Simple tetanus		
early phase (8)	100	189 ± 24
late phase (8)	102 ± 13	150 ± 35

Values are mean ± S. D. Number of experiments in parentheses.

ment. In this connection it should be recalled that the intensity changes that accompany transition from the rest state to the isometric state follow a reciprocal relationship (Haselgrove & Huxley, 1973); i.e., the change in the 1,0 intensity is balanced by a nearly equal but opposite change in the 1,1 intensity. Such a reciprocal relationship does not appear to exist in the changes caused by slow stretch. A possible explanation for the increase of $I_{1,0}$ in the stretched state may be that the thick filaments in non-overlap regions become more regularly arrayed in the 1,0 lattice as a result of slow stretch. The intensity gain due to this effect might overcome the loss resulting from locked-on myosin projections to give a net increase. We plan to conduct another experiment in the near future to obtain more quantitative information.

TIME-RESOLVED X-RAY DIFFRACTION FROM MUSCLE DURING ISOTONIC TWITCH

Recently Podolsky, Onge, Yu & Lymn (1976) reported an experiment in which they measured the intensity ratio $I_{1,0}/I_{1,1}$ from frog sartorius muscle allowed to shorten from the tetanized state. According to them, the intensity ratio measured during isotonic shortening under a load of 0.14–0.40 P_0 (P_0 being the maximum isometric tension) does not significantly differ from that observed in isometrically tetanized muscle. This finding has led them to the conclusion that the number of cross-bridges formed between the two types of myofilaments is almost the same in isotonic and isometric contractions. This is in apparent contradiction with the sliding filament model of A. F. Huxley (1957), which predicts that the number of myosin projections attached to thin filaments at any one moment is less the greater the shortening velocity. We have carried out an experiment to study in further detail the kinetic properties of cross-bridges in actively shortening skeletal muscle (Sugi, Amemiya & Hashizume, 1978).

METHODS

The sartorius muscle of *Rana catesbiana* was mounted in the specimen chamber described previously; the pelvic end was clamped while the tibial end was connected to an isotonic lever, the movement of which was sensed with a differential transformer (see Fig. 4). The initial sarcomere length was adjusted to 2.2–2.4 μm. The muscle, kept at 2–4° C by perfusing precooled oxygenated Ringer's solution, was stimulated with a single supramaximal current pulse of 3 msec duration to produce an after-loaded twitch. The low-angle diffraction diagrams during the isotonic twitch were detected with a linear position-

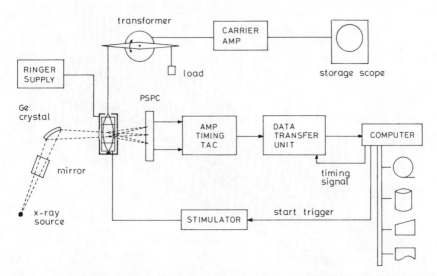

FIG. 4. Experimental arrangement used for recording kinetic diffraction diagrams from isotonically shortening muscle. For further explanation see text.

sensitive counter (Hashizume et al., 1978) specially designed for a focusing camera. This counter, again based on the delay principle and operated with a 90% xenon plus 10% carbon dioxide mixture gas under a pressure of 2280 Torr, provided a spatial resolution better than 0.15 mm and a detection efficiency of 81% for $CuK\alpha$ X-rays. The data were registered in computer memory to give 39 kinetic diffraction diagrams of 25 msec resolution for a 975 msec period of isotonic twitch. The stimulation and the data collection were repeated 120–150 times. The interval between two successive stimuli was 10–20 sec in the early experiments; later it was increased to 40 sec. The resting diagram was sampled for 1 or 2 sec before each stimulus.

To test the performance of the data collection system, diffraction diagrams were first recorded from muscle (sarcomere length 2.4 μm) undergoing isometric twitch. In this test the isotonic lever was replaced by a strain guage (Shinkoh, U) to which the tibial end of the specimen was connected. Figure 5 displays time-resolved equatorial diffraction diagrams. The foremost profile in Fig. 5 represents the resting diagram. The total exposure for this diagram was 240 sec. The second profile was obtained at 0–25 msec after stimulus, the third one at 25–50 msec after stimulus, and so forth. The total exposure for these diagrams was 3 sec. The ordinate of the first profile is normalized to that of the following profiles so that a direct comparison is possible. It is seen that the 1,0

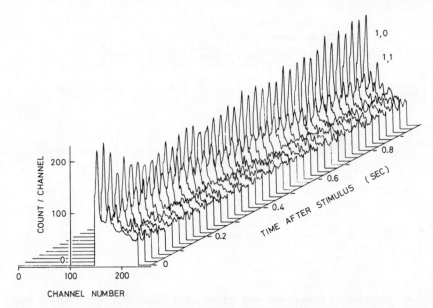

FIG. 5. Time-resolved equatorial diffraction diagrams recorded from isometrically twitching muscle. Only the right halves are shown. The numbers on the abscissa give channel number of the analyzer, while the ordinate represents count in each channel for the second to the 39th diagrams. The total exposure for these diagrams is 3 sec. The foremost diagram was sampled when muscle was at rest between repeated stimuli. The total exposure for this diagram is 240 sec. The data were processed to correct for the nonlinearity of the detector and to give the 1, 2, 1 running average for each diagram. For further explanation see text.

peak decreases on stimulation to a minimum in the sixth profile (*i.e.*, at the 100–125 msec phase). This peak height seems to be maintained for a while before the recovery occurs. The 1,1 peak behaves almost as a reduced mirror image of the 1,0 peak. We measured the intensity ratios ($I_{1,0}/I_{1,1}$) from these diagrams and plotted them against time after stimulus in Fig. 6 together with the isometric tension observed. The intensity ratio starts to decrease on stimulation and reaches the lowest value of 0.5–0.6 at the 100–125 msec phase. The isometric tension, which is still rising at this phase, attains a peak at about 220 msec after stimulation. The intensity ratio appears to remain at the lowest level and then returns gradually toward the resting value of 2.0. At the final (120th) twitch, the peak tension decreased by 10% relative to the initial value, but the time course of tension development remained unchanged, as can be seen in Fig. 6. Similar results were obtained on three other preparations with a sarcomere length of 2.4 μm. On all the preparations examined, the lowest value of the intensity ratio was 0.5–0.6. The general time course of the change

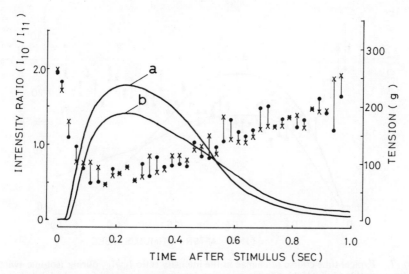

FIG. 6. Typical time course of change in the intensity ratio $I_{1,0}/I_{1,1}$ during isometric twitch. Circles represent data points measured from the left halves of the recorded diagrams, while crosses are data points measured from the right halves. Twitches were repeated 120 times. The initial sarcomere length was 2.4 μm. Curves a and b show the isometric tension at the first and the 120th twitches, respectively.

in the intensity ratio as well as the lowest value reached are in good agreement with other recently reported observations (Huxley, 1975; Matsubara & Yagi, 1978).

Before proceeding to the isotonic experiment, we examined internal shortening of muscle fibers during isometric twitch against the tendons and the tension recording system. For this purpose the compliance of the tendinous parts and of the recording system was measured and it was found that the internal shortening was less than 1.5% of muscle length. This was also the case with tetanized muscle from which a greater tension was developed.

RESULTS

Figure 7 is a typical result obtained from muscle undergoing after-loaded twitch with a small load of 0.02–0.03 P_0. As in the isometric case, stimulation causes a prompt decrease in the intensity ratio from the resting value of 2.1. However, in contrast to the isometric case, the intensity ratio levels off at a value of about 1.0. This occurs at the 100–125 msec phase, which corresponds to the first 20–30% of the shortening phase. The lowest value of about 1.0

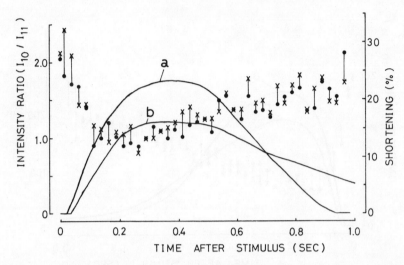

FIG. 7. Typical time course of change in the intensity ratio $I_{1,0}/I_{1,1}$ during isotonic twitch under a small load ($0.025P_0$). Circles represent data points measured from the left halves of the recorded diagrams, while crosses are data points measured from the right halves. Twitches were repeated 120 times. The initial sarcomere length was 2.4 μm. Curves a and b show the degrees of shortening at the first and the 120th twitches, respectively.

seems to be maintained for another 250 msec before the recovery starts. Similar results were obtained on five other preparations with an initial sarcomere length of 2.4 μm and also on two other preparations with initial sarcomere lengths of 2.3 μm and 2.2 μm. In all cases, the intensity ratio decreased to 0.8–1.0, but not to as low as the 0.5–0.6 in the isometric case. On some preparations, oscillatory changes in the intensity ratio were observed in the recovery phases, which need further investigation. Although the extent of shortening was reduced by about 30% at the end of each experiment, the duration of the shortening phase remained unchanged on all preparations examined. As in the isometric case, the change in the 1,0 intensity was reciprocally coupled with that in the 1,1 intensity.

DISCUSSION

The observation that the lowest steady value (0.8–1.0) of the intensity ratio reached during isotonic twitch under a small load is larger than that (0.5–0.6) reached during isometric twitch provides information, in principle, about cross-bridges in the two physiologically different states of muscle. However, during isotonic twitch, the muscle shortened by 10–20% relative to the initial length. This alone possibly affects the intensity ratio, since it is known that the intensity ratio of stimulated muscle depends on the sarcomere length (Elliott et al., 1967; Haselgrove & Huxley, 1973; Podolsky et al., 1976). To estimate this effect, we measured the intensity ratio from muscles tetanized isometrically at various sarcomere lengths. Each muscle was regularly activated for 1 sec at intervals of 1 min, and the data were collected over 8–12 activations. The results are shown in Fig. 8, where the intensity ratio decreases with decreasing sarcomere length in agreement with the previous reports (Haselgrove & Huxley, 1973; Podolsky et al., 1976). However, the dependence shown in Fig. 8 is much less marked than that reported by Haselgrove & Huxley (1973) and is similar to that reported by Podolsky et al. (1976). In the study by Haselgrove & Huxley (1973), a substantial number of myosin projections probably ceased to transfer to the vicinity of thin filaments during the course of the experiment, in which muscles were activated a great number of times to collect data on photographic film. It is worth noting that, in stimulated muscle, the intensity ratio depends much less strongly on the sarcomere length than it does in resting muscle (Haselgrove & Huxley, 1973).

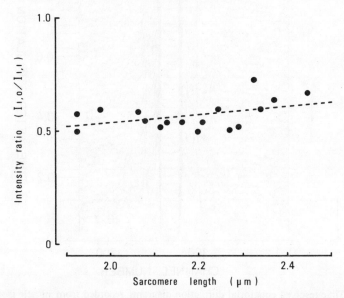

FIG. 8. The intensity ratio $I_{1,0}/I_{1,1}$ for isometrically tetanized muscles plotted against sarcomere length. The data points were obtained from different preparations. The line represents a least square fit (Sugi et al., 1978).

One can expect that the effect of decrease in sarcomere length during shortening is to decrease, rather than to increase, the intensity ratio from its intrinsic value, although it is difficult to determine the extent of decrease on the basis of the present data. Then, if it is assumed that possible changes in the configuration of cross-bridges do not significantly affect the intensity ratio during shortening (Podolsky *et al.*, 1976), the present results imply that the number of attached cross-bridges is about 50% smaller in isotonic twitch under a small load than in isometric twitch.

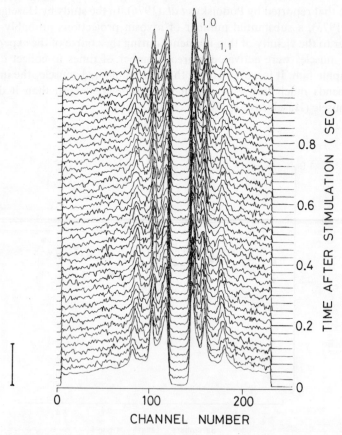

FIG. 9. Time-resolved equatorial diffraction diagrams recorded from muscle isotonically twitching under a small load ($0.03P_0$). The initial sarcomere length was 2.4 μm. The vertical scale mark represents 200 counts/channel for the second to the 39th diagrams. Note the variation of the peak positions as a function of time after stimulus.

Figure 9 shows consecutive diffraction diagrams during an isotonic twitch. The change in the spacing between myofilaments during the isotonic shortening can be estimated by measuring the positions of the reflection peaks. In Fig. 10 the inverse square of the 1,0 spacing is plotted against the corresponding sarcomere length, where the sarcomere length is assumed to change in parallel with the muscle length. The regression line starts from the origin (within experimental errors), indicating that the filament-lattice volume remains constant during the course of isotonic twitch. Since the filament-lattice volume is known to be constant when the sarcomere length of intact muscle is varied (Huxley, 1953; Elliott, Lowy & Worthington, 1963), the above result indicates

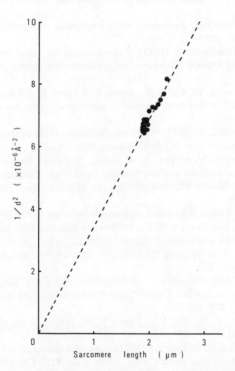

FIG. 10. Inverse square of the 1,0 lattice spacing (d) as a function of sarcomere length during the shortening phase of isotonic twitch. The regression line represents a least square fit (Sugi *et al.*, 1978).

that the isovolumic nature of the filament-lattice exists in both resting and actively shortening muscles.

Acknowledgement

The authors express their sincere gratitude to Professors K. Kohra and S. Ebashi for their interest and encouragement.

REFERENCES

Abott, B. C. & Aubert, X. M. (1952). The force exerted by active striated muscle during and after change of length. *J. Physiol.* **117**, 77–86.
Elliott, G. F., Lowy, J. & Millman, B. M. (1967). Low angle X-ray diffraction studies of living striated muscle during contraction. *J. molec. Biol.* **25**, 31–42.
Elliott, G. F., Lowy, J. & Worthington, C. R. (1963). An X-ray and light diffraction study of the filament lattice of striated muscle in the living state and in rigor. *J. molec. Biol.* **6**, 295–305.
Fales, J. T. (1972). Total heat production of frog sartorius: isometric contractions. *Am. J. Physiol.* **222**, 1085–1090.
Haselgrove, J. C. & Huxley, H. E. (1973). X-ray evidence for radial cross-bridge movement and for the sliding filament model in actively contracting skeletal muscle. *J. molec. Biol.* **77**, 549–568.
Hashizume, H., Amemiya, Y., Kohra, K., Izumi, T. & Mase, K. (1976). Techniques for time-resolved X-ray diffraction using a position sensitive counter. *Japan. J. Appl. Phys.* **15**, 2211–2219.
Hashizume, H. & Masaki, N. (1978). An X-ray diffraction study of frog sciatic nerve myelin in dimethyl sulfoxide. *Arch. biochem. Biophys.* **186**, 275–282.
Hashizume, H., Mase, K., Amemiya, Y. & Kohra, K. (1978). A system for kinetic X-ray diffraction using a position sensitive counter. *Nucl. Instr. Methods*, **152**, 199–203.
Huxley, A. F. (1957). Muscle structure and theories of contraction. *Prog. Biophys. biophys. Chem.* **7**, 255–318.
Huxley, H. E. (1953). X-ray analysis and the problems of muscle. *Proc. R. Soc. B* **141**, 59–62.
Huxley, H. E. (1960). Muscle cells. In *The Cell*, vol. 4. eds. Brachet, J. & Mirsky, A. E., pp. 365–481. New York and London: Academic Press.
Huxley, H. E. (1975). Time-resolved X-ray studies on muscle contraction. *5th Int. Biophys. Cong.*, Copenhagen, S53.
Jöbsis, F. F. (1967). Mechanical activity of striated muscle. *Symp. biol. Hung.* **8**, 151–205.
Matsubara, I. & Yagi, N. (1978). A time-resolved X-ray diffraction study of muscle during twitch. *J. Physiol.* **278**, 297–307.
Podolsky, R. J., Onge, R. St., Yu, L. & Lymn, R. W. (1976). X-ray diffraction of actively shortening muscle. *Proc. natn. Acad. Sci. U.S.A.* **73**, 813–817.
Sugi, H., Amemiya, Y. & Hashizume, H. (1977). X-ray diffraction of active frog skeletal muscle before and after a slow stretch. *Proc. Japan Acad.* **53B**, 178–182.
Sugi, H., Amemiya, Y. & Hashizume, H. (1978). Time-resolved X-ray diffraction from frog skeletal muscle during an isotonic twitch under a small load. *Proc. Japan Acad.* **54B**, 559–564.

Discussion

Huxley: Concerning Table 1, if you were to measure the ratio at the stretched length, but hold isometric, *i.e.*, not stretch during contraction, what would the value of $I_{1,0}/I_{1,1}$ be? Would it still be somewhere around 0.7?

Sugi: As we showed in Fig.8, the dependence of the intensity ratio on sarcomere length is not large. So we omitted consideration of this.

Noble: Still, on Table 1, I did not quite understand your explanation for the difference between early phase and late phase isometric. I think you said it was due to fatigue. Do you mean that during the course of the long experiment there was fatigue going on progressively? In which case, was the tension in the late phase also falling?

Amemiya: When we needed ten tetanic contractions repeated in succession, the late phase of tension grew smaller than the early tension. The average difference was only 5%, so at first I thought this difference was negligible, but we had three time-repetition experiments recently with an improved technique, where there was no difference between the tension in the early phase and the late phase. In those cases, there was no difference of the intensity ratio between the early and the late phases.

Huxley: Concerning the isotonic shortening experiments, I think there are two things one could say about these. First of all, it seems to me it is a striking observation that even though in your case you have the muscle shortening against an extremely small load (2–3% of P_0), the X-ray pattern shows that, according to your interpretation, at least about 50% of bridges are still attached. That may or may not be exactly the same as is predicted by particular models, but I think it is a very interesting thing that when a muscle is shortening at high velocity you find that 50% of the bridges are still attached, and Podolsky (Podolsky, St. Onge, Yu & Lymn, 1976) at somewhat lower velocities finds them nearly all attached, and at an intermediate velocity I find that perhaps 70% are attached. I think there is a considerable measure of agreement between everyone that in a very, very rapidly shortening muscle exerting very little tension, there are still an awful lot of bridges attached.

But then the second thing is exactly how many are attached: is it 30%, or 50% or 70%? What I wanted to know was how did you arrive at the proportion of 50% still attached? Was it by constructing two-dimensional Fourier maps and measuring areas, or was it by just looking at the actual intensities of the reflections and how far they had gone toward the resting state?

Amemiya: There are many assumptions, but the calculation was the same as that used by Matsubara and Yagi (1978).

Hashizume: Up to the present we have not constructed any electron-density maps from our observations. The statement that the number of attached cross-bridges is almost 50% smaller in the isotonic condition than in the isometric condition simply means that the minimum value of the intensity ratio is 50% smaller in the former case than in the latter case.

Sugi: Many people have measured stiffness during isotonic shortening at zero load as an extrapolated value (*e.g.* Julian & Sollins, 1975; Tsuchiya, Sugi & Kometani, 1978). It is about 25% of the maximum, but the number of cross-bridges appears to be larger if the intensity ratio is taken as an index of attached cross-bridges. One possible explanation may be that during very rapid shortening, considerable shortening heat appears, so that there might be a considerable number of "wasted" cross-bridges only producing heat. This is one possibility.

Tregear: You have isotonic and isometric data, in which you have an intensity change in the 1,1 and an intensity change in the 1,0. You have a proportion for the ratio. Do you also have the same proportions for the two individual intensities?

Hashizume: Do you mean the ratio of the 1,0 intensity in the isotonic contraction to that in the isometric condition?

Huxley: Can I put the question another way? You said that the $I_{1,0}$ and the $I_{1,1}$ during the isometric contraction changed symmetrically.

Sugi: Yes.

Huxley: Now, is the same true during isotonic contraction?

Sugi: In this connection, we should examine the data more carefully.

Hashizume: But anyway, as far as I know, the symmetry of the intensity change of 1,0 and 1,1 reflections is almost the same in isometric and isotonic conditions.

Sagawa: I would like to catch up on the point that Dr. Huxley discussed a moment ago. The intensity ratio $I_{1,0}/I_{1,1}$, in my understanding, shows how close the cross-bridge head is located near actin. If so, it does not necessarily show attachment or detachment if the muscle is moving very fast and if it takes time for the cross-bridge to extend and approach actin. So when movement is very fast it is entirely possible that although the $I_{1,0}/I_{1,1}$ ratio is still very high, many of the cross-bridges are in a detached state, but have not had enough time to retract back toward myosin. In that way the ratio may not be in exact proportion to the actual number.

Huxley: That is a very good point and it is a very difficult one, and it is certainly one which we spent a lot of time worrying about. Possibly one way of thinking of it is in terms of diffusion. If a cross-bridge is simply free to diffuse away from the actin filaments when it is detached by ATP, and if it is behaving as a protein with a molecular weight of say, 100,000, then it should diffuse away, I think, quite quickly, in a time on the order of a msec or less. But, of course, we do not know that it is free to diffuse as rapidly as that. Perhaps the only evidence on this is that there are stop-flow kinetic experiments where one adds ATP to actomyosin or actin-S-1 and looks at the change in the light scattering. One can then measure dissociation rates with rate constants on the order of a thousand per second. So it is clear that under those conditions the S-1 is free to diffuse away very rapidly. But whether this is the situation in muscle we do not know at the moment.

O'Brien: If we restrict ourselves to just considering the ratio of intensities rather than the number of cross-bridges, can I take it from Dr. Huxley's summary that there is agreement between the results of Podolsky (Podolsky *et al*, 1976) and the results just presented; in other words, are the experimentally observed differences of intensity ratio consistent with the velocity of shortening, *i.e.,* with the load applied?

Huxley: I think that we have really got to look at the actual numbers to see, becaute I think they are being expressed in different ways. We have got to sit down and look at the tables together. Clearly they are in the same general direction, but whether there is any inconsistency in detail I would not like to say right at this moment.

O'Brien: But Podolsky's was at a lower velocity, wasn't it?

Huxley: There were actually only about four or five experiments in that paper. And it was not clear what loads they were all done at.

REFERENCES

Julian, F. J. & Sollins, M. R. (1975). Variation of muscle stiffness with force at increasing speeds of shortening. *J. gen. Physiol.* **66**, 287–302.

Matsubara, I. & Yagi, N. (1978). A time-resolved X-ray diffraction study of muscle during twitch. *J. Physiol.* **278**, 297–307.

Podolsky, R. J., St. Onge, R., Yu, L. & Lymn, R. W. (1976). X-ray diffraction of actively shortening muscle. *Proc. natn. Acad. Sci. U.S.A.* **73**, 813–817.

Tsuchiya, T., Sugi, H. & Kometani, K. (1978). Isotonic velocity transients and enhancement of mechanical performance in frog skeletal muscle fibers after quick increases in load. In *Cross-bridge Mechanism in Muscle Contraction*, ed. Sugi, H., & Pollack, G. H., Tokyo: University of Tokyo Press.

The Structure of Thin Filament of Crab Striated Muscle in the Rigor State

Keiichi NAMBA, Katsuzo WAKABAYASHI and Toshio MITSUI

Department of Biophysical Engineering, Faculty of Engineering Science, Osaka University, Toyonaka, Osaka, Japan

ABSTRACT

The difference cylindrically symmetrical Patterson function is discussed. Several theorems are mentioned which make the function useful in structure analysis of fiber specimens.

Using the function, the structure of leg striated muscle of the marine crab *Portunus trituberculatus* in the rigor state was determined. X-ray photographs of the muscle gave many distinct layer lines, with a basic period of 765 Å due to the thin filaments, well separated from several relatively weak layer lines with a basic period of 145 Å caused by the thick filaments. No sampling effect was recognized on any of these layer lines. Analysis of the structure of the thin filament used the intensity distributions on the layer lines with the basic period of 765 Å, of which the reflection orders were 2, 4, 6, 7, 9, 11, 13, 15, 17, 22, 24, 26, 28. Trial models were constructed and refined referring to the difference cylindrically symmetrical Patterson function. Results showed that molecules which seemed to be the myosin subfragment-1 were bound to 3, 4, 3', 4' and 10, 11, 10', 11' actin monomers when the actin monomers were numbered as $1 \sim 14$ along one strand of the long-pitch helix and $1' \sim 14'$ along the other, 1' being between 1 and 2. Some molecules having molecular weight of about 100,000 were bound to the 1, 1' and 8, 8' actin monomers.

INTRODUCTION

The first part of this paper concerns methodology in the X-ray structure analysis of fiber materials. The succeeding part describes the determination of the structure of crab striated muscle in the rigor state using the proposed method.

McGillavry & Bruins (1948) pointed out that the cylindrically symmetrical

Patterson function would be useful in X-ray structure analysis of fiber materials. Franklin & Gosling (1953) used this function in studies of DNA, determined the unit cell parameters, and discussed the possible shape of the fundamental structural unit. Since then several authors have used this function in analysis of biological structures, as cited in Vainshtein's text book (1966). In the present paper we introduce a difference cylindrically symmetrical Patterson function and demonstrate its usefulness in studies of muscle structures.

X-ray studies of *Limulus* and crab muscles have been made by several authors (Wray, Vibert & Cohen, 1974, 1975; Yagi & Matsubara, 1977; Maéda, 1978; Maéda, Matsubara & Yagi, 1978; Namba, Wakabayashi & Mitsui, 1978; Wakabayashi & Namba, 1978). We found that leg striated muscle of the marine crab *Portunus trituberculatus* gave very distinct layer-line reflections in the rigor state, and analyzed the structure using the difference cylindrically symmetrical Patterson function.

BASIC THEOREMS USEFUL IN STRUCTURE ANALYSIS

Suppose a fiber specimen has periodic structure along the fiber axis with the period c. When an X-ray beam is incident perpendicularly to the fiber axis, layer-line reflections are caused. Our problem is how to extract information on the fiber structure from these layer-line reflections. Below we shall use the following notations: r, position vector in real space; x, y, z, the orthogonal Cartesian coordinates, z being parallel to the fiber axis; r, ϕ, z, the cylindrical coordinates; R, position vector in reciprocal space; X, Y, Z, the orthogonal Cartesian coordinates; R, Φ, Z, the cylindrical coordinates; $\rho(r)$, the electron density distribution; $Q(r)$, the autocorrelation function of $\rho(r)$; $I(R)$, intensity distribution of scattered X-rays. $Q(r)$ is defined as

$$Q(r) = \int \rho(r') \, \rho(r' + r) \, dv_{r'}, \qquad (1)$$

where $dv_{r'}$ is the volume element. In our case $I(R)$ is nonzero only near the planes of $Z = l/c$ where l are integers. We define $I_l(R, \Phi)$ by

$$I_l(R, \Phi) = \int_{l/c-\varepsilon}^{l/c+\varepsilon} I(R) \, dZ, \qquad (2)$$

where $\varepsilon < 1/(2c)$. We shall use the following quantities:

$$\bar{\rho}(x, y) = \frac{1}{c} \int_0^c \rho(r) \, dz, \qquad (3)$$

$$\Delta\rho(\mathbf{r}) = \rho(\mathbf{r}) - \bar{\rho}(x, y), \tag{4}$$

$$\bar{Q}(x, y) = \frac{1}{c} \int_0^c Q(\mathbf{r}) \, dz, \tag{5}$$

$$\Delta Q(\mathbf{r}) = Q(\mathbf{r}) - \bar{Q}(x, y). \tag{6}$$

Concerning these quantities we have the following theorems:
Theorem 1

$$\Delta Q(\mathbf{r}) = \int \Delta\rho(\mathbf{r}') \, \Delta\rho(\mathbf{r}' + \mathbf{r}) \, dv_{\mathbf{r}'}. \tag{7}$$

Theorem 2

$\Delta Q(\mathbf{r})$ can be calculated without information on $I_0(R, \Phi)$:

$$\Delta Q(\mathbf{r}) = \sum_{\substack{l=-\infty \\ l \neq 0}}^{\infty} \left[\int_{-\infty}^{\infty} I_l(X, Y) \exp\{i 2\pi(xX + yY)\} \, dX dY \right] \cdot \exp(i 2\pi l z/c). \tag{8}$$

Theorem 3

The ϕ-average of $Q(\mathbf{r})$ can be calculated by using the Φ-average of $I_l(R, \Phi)$:

$$Q(r, z) = \sum_{l=-\infty}^{\infty} \left\{ \int_0^{\infty} I_l(R) J_0(2\pi R r) \, 2\pi R dR \right\} \cdot \exp(i 2\pi l z/c), \tag{9}$$

where

$$Q(r, z) = \int_0^{2\pi} Q(r, \phi, z) \, d\phi, \tag{10}$$

$$I_l(R) = \int_0^{2\pi} I_l(R, \Phi) \, d\Phi \tag{11}$$

and $J_0(2\pi Rr)$ is the zeroth order Bessel function of the argument $2\pi Rr$. Equation (9) can be rewritten as

$$Q(r, z) = \int_0^{\infty} I_0(R) J_0(2\pi R r) 2\pi R dR + 2 \sum_{l=1}^{\infty} \left\{ \int_0^{\infty} I_l(R) J_0(2\pi R r) 2\pi R dR \right\} \cdot \cos(2\pi l z/c). \tag{12}$$

Theorem 4

$$\Delta Q(r,z) = 2\sum_{l=1}^{\infty}\left\{\int_0^{\infty} I_l(R)J_0(2\pi Rr)2\pi R\,\mathrm{d}R\right\}\cdot\cos(2\pi lz/c), \quad (13)$$

where

$$\Delta Q(r,z) = \int_0^{2\pi}\Delta Q(r,\phi,z)\,\mathrm{d}\phi. \quad (14)$$

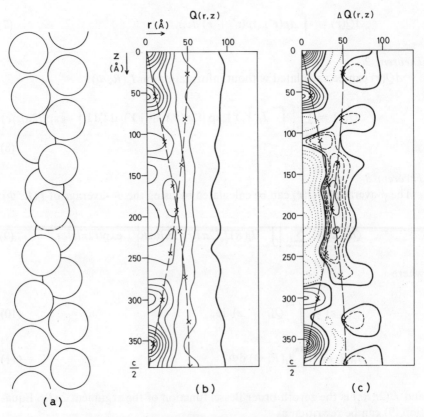

FIG. 1. A simplified model of F-actin filament, and $Q(r, z)$ and $\Delta Q(r, z)$ calculated with it. (a) The model. The circles stand for actin monomers. (b) Contour map of $Q(r, z)$. (c) Contour map of $\Delta Q(r, z)$. In b and c, the peaks at the origins were normalized as 1,000. Full and dashed lines show positive contours, thick full lines zero and dotted lines negative. Contour interval is 100 for the full and dotted lines, and 25 for the dashed lines. Crosses stand for vectors between the actin monomers.

Theorems 1 and 2 can be proved easily. Theorem 3 was first derived by McGillavry & Bruins (1948). Theorem 4 can be derived directly from Theorems 2 and 3.

By Theorem 1, $\Delta Q(r)$ has definite physical meaning, *i.e.*, the autocorrelation function of $\Delta \rho(r)$. $I_l(R)$ defined by Eqs. (2) and (11) can be obtained from the intensity distribution measured on the *l*th layer line. Therefore, these theorems mean that the ϕ-average of the autocorrelation function of $\Delta \rho(r)$ can be obtained from X-ray data without use of the data on the equator ($l = 0$). The usefulness of these conclusions can be demonstrated by the model calculation shown in Fig. 1. Figure 1a shows a simple model of F-actin filament and Figs. 1b and c respectively give contour maps of $Q(r,z)$ and $\Delta Q(r,z)$ calculated for the model. In Figs. 1b and c, the crosses indicate the heads of inter-molecular vectors expected from the model shown in Fig. 1a. It can be seen that more distinct peaks appear on the $\Delta Q(r,z)$ map than on the $Q(r,z)$ map related to the inter-molecular vectors. Therefore, the $\Delta Q(r,z)$ map seems to be a useful tool in structure determination of muscles. Below we apply this method to the crab striated muscle.

STRUCTURE ANALYSIS OF CRAB STRIATED MUSCLE

Glycerol-extracted muscles were prepared from the striated muscle of the marine crab *Portunus trituberculatus* by the procedures described by Yanagida, Taniguchi & Oosawa (1974). A single fiber having the thickness of about 0.3 mm was mounted in a sample holder and immersed in the rigor solution which contained 100mM KCl, 10mM $MgCl_2$, 1mM glycoletherdiamine-*N*, *N*, *N'*, *N'*-tetraacetic acid (GEDTA) and 10mM histidine-HCl (pH 7.0). The temperature of the specimen was maintained at 4°C during X-ray exposure. X-rays were produced by a microfocus rotating-anode X-ray generator made by Rigaku Denki (RU-100 unit) operated at 40kV and 30mA. A mirror-monochromator camera of the Huxley-Brown type (Huxley & Brown, 1967) was used, which consisted of a curved glass mirror and a bent quartz crystal [3° cut to (10$\bar{1}$1)] and gave a focused spot of $CuK\alpha_1$ radiation (1.5405Å). The specimen-to-film distance was 32 cm. Diffraction photographs were taken with Ilford Industrial G films. Figure 2 shows an example.

The fiber exhibits many layer lines. They consist of two sets of layer lines, which will be called M and A layer lines, respectively, below. The M layer lines can be indexed as 1st ~ 4th order reflections with the basic period of 145 Å, as indicated in Fig. 2, and thus seem to be caused by the thick filaments. Many A layer lines are seen in Fig. 2, which can be indexed with the basic period of 765 Å. The orders *l* of the distinctly observed reflections were 2, 4, 6, 7, 9, 11,

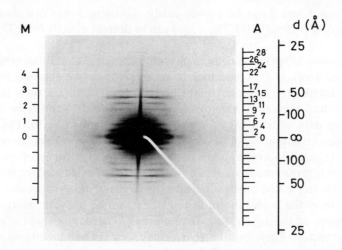

FIG. 2. X-ray diffraction pattern of crab striated muscle in the rigor state; fiber axis vertical. The strong, nearly vertical line was caused by parasitic scattering. Numbers on the scales M and A indicate the order of reflections caused by the thick filaments (M, the basic period 145 Å) and by the thin filaments (A, the basic period 765 Å). d, the spacing.

13, 15, 17, 22, 24, 26, 28. They seem to be due to the thin filaments. Lattice sampling effect was not observed on any of these layer lines, unlike those in insect muscles (Miller & Tregear, 1972). Therefore, we assumed that the thin filaments did not form an ordered three-dimensional lattice (although they were parallel to each other) and treated the A layer lines as if they were caused by a single thin filament. (The M layer lines have not been treated in the present study.) Distribution of X-ray intensities was measured by the densitometer trace of the photographs. Using the data obtained, $I_l(R)$ was calculated by Eq. (2). Figure 3a shows results. The difference cylindrically symmetrical Patterson function $\Delta Q(r, z)$ was calculated with these $I_l(R)$'s of the above-mentioned 13 layer lines by Eq. (13). Figure 4a shows the contour map of $\Delta Q(r, z)$.

Analysis of the structure of the thin filament was made by looking for the model which gives $I_l(R)$ and $\Delta Q(r, z)$ resembling those in Figs. 3a and 4a, respectively. Peaks of various heights, plateaus and shoulders on the map in Fig. 4a were of a great help in constructing the model, as well as characteristic features of $I_l(R)$ in Fig. 3a. (Detailed discussion will be published elsewhere.) Several trial models were constructed, and $I_l(R)$ and $\Delta Q(r, z)$ were calculated for them. Figures 3b and 4b give the curves of $I_l(R)$ and the map of $\Delta Q(r,z)$ which at present bear the strongest resemblance to those in Figs. 3a and 4a, respectively. Figure 5 shows the model used for the calculation. In the model

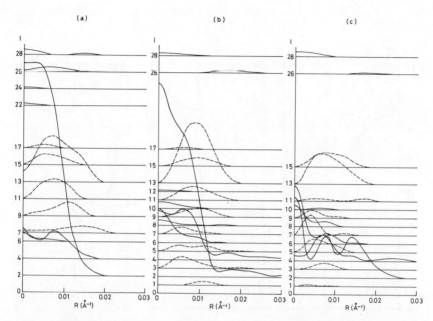

FIG. 3. Intensities $I_l(R)$ defined by Eq. (2) on various layer lines. The layer lines on which $I_l(R)$ was smaller than 1/500 of the peak value of observed $I_2(R)$ are not shown. (For $I_3(R)$ experimental error was large. It was weak but could have been stronger than 1/500 of the peak of $I_2(R)$.) (a) Observed values. (b) Calculated with the model shown in Fig. 5. (c) Calculated with the model shown in Fig. 6.

the actin monomer was assumed to be a sphere of radius of 24 Å, having the molecular weight of 42,000, corresponding to the density of 1.20 g/cm³. The diffraction pattern in Fig. 2 indicated that the period c of 765Å corresponds to 14 actin monomers along one strand of the long-pitch helix. These are numbered as indicated in Fig. 5a. Naturally calculated $I_l(R)$ and $\Delta Q(r,z)$ depend upon assumed molecular weights of the molecules denoted as X and Y in Fig. 5, although only their values relative to that of the actin monomer (42,000) are important since our measurements concerned only relative intensities of X-rays. Values between 50,000 and 350,000 were tested for the molecular weight of X, and the best fit was obtained for 100,000 ∼ 150,000. For the Y molecule the best value was around 100,000. The shapes of the molecules also were adjusted so as to get better agreement between observed and calculated values of $I_l(R)$ and $\Delta Q(r,z)$, keeping the densities of the molecules at about 1.25 g/cm³. Results are depicted in Fig. 5. The model shown in Fig. 5 will be further refined, but calculations made so far on several models have suggested that the fol-

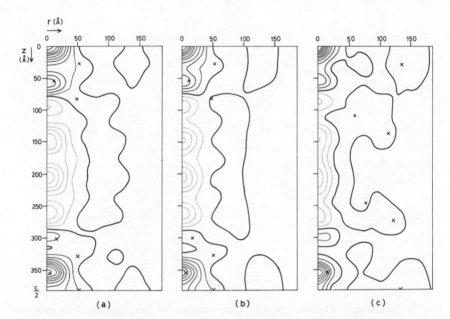

FIG. 4. Contour maps of $\Delta Q(r, z)$. The peak at the origin was normalized as 1,000. Contour interval is 100. Full lines show positive contour, thick full lines zero and dotted lines negative. (a) Calculated with the X-ray intensity distributions on the A layer lines. Crosses stand for vectors between the X molecules in Fig. 5. (b) Calculated with the model shown in Fig. 5. Crosses stand for vectors between the X molecules in Fig. 5. (c) Calculated with the model shown in Fig. 6. Crosses stand for vectors between the S-1 molecules in Fig. 6.

lowing structural features are close to reality.

(1) The X molecules are bound to 3, 4, 3′, 4′ and 10, 11, 10′. 11′ actin monomers as shown in Fig. 5.

(2) The center of gravity of the X molecule sits at the r coordinate of about 30 Å or less.

(3) The Y molecules sit with their centers of gravity at roughly the same z coordinates as the 1, 1′ and 8, 8′ actin monomers and at relatively large r ($\approx$ 65 Å), although there is more uncertainty about their positions and shapes than about the X molecules at present.

DISCUSSION

Preliminary X-ray studies were made on the structure of the crab striated

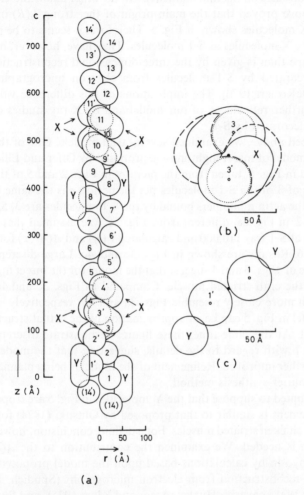

FIG. 5. A model of the thin filament in the rigor state. Actin monomers are numbered. See text for X and Y molecules. (a) Side view. (b) Top view of the part around the actin monomers 3 and 3'. Small filled circle, z axis; small open circle, the center of gravity of the actin monomer. The dashed lines show another possible model. They are added to give an idea on the approximate nature of the model at the present stage. (c) Top view of the part around the actin monomers 1 and 1'.

muscle in the relaxed state. Obtained photographs gave very weak $I_l(R)$ except for $I_{13}(R)$ and $I_{15}(R)$. The remarkable variations of $I_l(R)$ from the relaxed state to the rigor state seem to be caused by the binding of the subfragments-1 (S-1) of myosin molecules on the thin filaments. On the other hand, the calculations described above proved that the main origin of the strong $I_l(R)$ in the rigor state is the X molecules shown in Fig. 5. Therefore, it seems to be reasonable to identify the X molecules as S-1 molecules. They have, however, much more compact shape than is given by the three-dimensional reconstruction of thin filaments decorated by S-1 molecules from electron micrographs (Moore, Huxley & DeRosier, 1970). The implications of this difference will be made clearer by further refinement of our model and by X-ray studies of the filaments fully decorated by S-1.

As described above, we examined several trial models. One of them was a copy of the model of insect flight muscle proposed by Offer and Elliott (1978). It is depicted in Fig. 6, taken from the pictures in Figs. 2 and 5 of their paper. The number of 8 of the S-1 molecules per the period c is the same in Figs. 5a and 6a, but the actin monomers bound by the S-1 molecules are 3, 5, 3′, 5′ and 10, 12, 10′, 12′ in Fig. 6a, different from Fig. 5a. We assumed that the molecular weight of S-1 was 110,000 and calculated $I_l(R)$ and $\Delta Q(r, z)$ for the structure in Fig. 6. Results are shown in Figs. 3c and 4c. Large discrepancies between a and c of Figs. 3 and 4 suggest that the model for the insect flight muscle does not fit the crab striated muscle. Compared to Figs. 3c and 4c, Figs. 3b and 4b much more closely resemble Figs. 3a and 4a, respectively, suggesting that the model in Fig. 5 can be an approximation of the actual structure of the thin filament. At the same time, these figures demonstrate discrepancies between a and b with regard to the details, suggesting that the model in Fig. 5 should be further improved. Refinement of the model is being planned with the difference Fourier synthesis method.

We are tempted to suppose that the Y molecules in Fig. 5 are troponin, since their arrangement is similar to that proposed by Ohtsuki (1974) for troponin molecules in chicken striated muscles. For a definite conclusion, however, more investigation is needed. We examined the contribution to the $\Delta Q(r,z)$ map from tropomyosin by calculations based upon the model proposed by three-dimensional reconstruction from electron micrographs (Spudich, Huxley, & Finch, 1972; Wakabayashi, Huxley, Amos and Klug, 1975), and found it difficult to get any information on its structure at the present stage.

The studies described above have some similarity to structure analysis by the heavy atom method. The X molecules played the role of the heavy atom and made it easy to obtain the approximate structure of the thin filament. The $\Delta Q(r,z)$ functions offer many clues and suggestions for constructing the model, although detailed discussion had to be omitted in this paper due to space limitations. This function seems to be a useful mathematical tool in structure analysis of other muscle fibers or more generally of other fiber specimens.

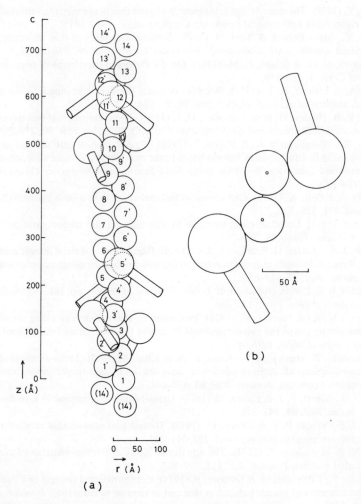

FIG. 6. A model after the structure of insect flight muscle proposed by Offer & Elliott (1978). Large circle with bar represents the S-1 molecule.

REFERENCES

Franklin, R. E. & Gosling, R. G. (1953). The structure of sodium thymonucleate fibres. II. The cylindrically symmetrical Patterson funciton. *Acta Cryst.* **6**, 678–685.

Huxley, H. E. & Brown, W. (1967). The low-angle X-ray diagram of vertebrate striated muscle and its behaviour during contraction and rigor. *J. molec. Biol.* **30**, 383–434.

Maéda, Y. (1978). The cross-bridge arrangement around thin filament in crab striated muscle in rigor. *Sixth International Biophysics Congress. abstr. VII-9-(J)*.

Maéda, Y., Matsubara, I. & Yagi, N. (1978). Structural changes in thin filaments of crab striated muscle. *Sixth International Biophysics Congress. abstr. VII-10-(J)*.

McGillavry, C. H. & Bruins, E. M. (1948). On the Patterson transforms of fibre diagrams. *Acta Cryst.* **1**, 156–158.

Miller, A. & Tregear, R. T. (1972). Structure of insect fibrillar flight muscle in the presence and absence of ATP. *J. molec. Biol.* **70**, 85–104.

Moore, P. B. Huxley, H. E. & DeRosier, D. J. (1970). Three-dimensional reconstruction of F-actin thin filaments and decorated thin filaments. *J. molec. Biol.* **50**, 279–295.

Namba, K., Wakabayashi, K. & Mitsui, T. (1978). X-ray studies of crab skeletal muscle II: Regularity in attachment of myosin heads to the thin filament in rigor state, observed on a modified cylindrical Patterson map. *Sixth International Biophysics Congress. abstr. VIII-4-(J)*.

Offer, G. & Elliott, A. (1978). Can a myosin molecule bind to two actin filaments? *Nature, Lond.* **271**, 325–329.

Ohtsuki, I. (1974). Localization of troponin in thin filament and tropomyosin paracrystal. *J. Biochem. Tokyo* **75**, 753–765.

Spudich, J. A., Huxley, H. E. & Finch, J. T. (1972). Regulation of skeletal muscle contraction II. Structural studies of the interaction of the tropomyosin-troponin complex with actin. *J. molec. Biol.* **72**, 619–632.

Vainshtein, B. K. (1966). *Diffraction of X-Rays by Chain Molecules.* pp. 181–190. Amsterdam, London and New York: Elsevier.

Wakabayashi, K. & Namba, K. (1978). Projection of electron density along the fibre axis obtained by use of the two-dimensional Patterson function. *Sixth International Biophysics Congress. abstr. VIII-3-(J)*.

Wakabayashi, T., Huxley, H. E., Amos, L. A. & Klug, A. (1975). Three-dimensional image reconstruction of actin-tropomyosin complex and actin-tropomyosin-troponin T-troponin I complex. *J. molec. Biol.* **93**, 477–497.

Wray, J. S., Vibert, P. J. & Cohen, C. (1974). Cross-bridge arrangement in *Limulus* muscle. *J. molec. Biol.* **88**, 343–348.

Wray, J. S., Vibert, P. J. & Cohen, C. (1975). Diversity of cross-bridge configurations in vertebrate muscles. *Nature, Lond.* **257**, 561–564.

Yagi, N. & Matsubara, I. (1977). The equatorial X-ray diffraction patterns of crustacean striated muscles. *J. molec. Biol.* **117**, 797–803.

Yanagida, T., Taniguchi, M. & Oosawa, F. (1974). Conformational changes of F-actin in the thin filaments of muscle induced *in vivo* and *in vitro* by calcium ions. *J. molec. Biol.* **90**, 509–522.

Arrangement of Troponin and Cross-Bridges around the Thin Filaments in Crab Leg Striated Muscle

Yuichiro MAÉDA

Department of Pharmacology, Tohoku University School of Medicine, Sendai, and Institute of Molecular Biology, Faculty of Science, Nagoya University, Nagoya, Japan

ABSTRACT

The detailed arrangement of troponin and cross-bridges around the thin filaments in crab *Plagusia* leg striated muscle both in living resting and in rigor state was investigated.

1. Electron micrographs of longitudinal thin sections of the muscle in rigor showed that the cross-bridges were attached in pairs to the thin filaments with a repeating distance of 38 nm. This supported our assignment of the X-ray reflections (a series of meridional and near-meridional ones in the rigor pattern) to the cross-bridges in the preceeding work.

2. Some features of the X-ray reflections both in rigor pattern and relaxed pattern could be interpreted in terms of a set of identical helices which are staggered with respect to each other. With this qualitative analysis, the precise arrangement of troponin molecules could be determined from the relaxed pattern. From the rigor reflections, it was suggested that one or two pairs (the second pair comes in the position next to the first one) of cross-bridges might be included in every 38 nm.

3. Two pairs of cross-bridges in the 38 nm were deduced from the pattern of the cylindrically symmetrical Patterson function, in which a pair of additional materials was also suggested between two groups of cross-bridges (each group consists of two pairs).

4. Ultra-thin sections of the muscle in rigor were examined under the electron microscope. Although the number of pairs in the 38 nm could not be determined, chevron-like patterns of attached cross-bribges were observed, and additional electron-dense materials were frequently observed at around the mid-point of the 38 nm.

INTRODUCTION

In the past five years, several structural features of the contractile apparatus of crab *Plagusia* leg striated muscle have been elucidated (Maéda, 1978a, b).

(1) The length of the thin filament and that of the thick filament are about twice as long as those in vertebrate skeletal muscle: 2.2 μm for the thin filament and 4.2 μm for the thick filament, according to polarizing microscopy.

(2) Electron micrographs showed that this muscle has a well-ordered myofilament lattice in which each thick filament is surrounded by 12 thin filaments; the number ratio of thin filaments to thick filaments in the A-band is 6:1.

(3) The number of the thin filaments per unit cross-sectional area is about two times as great as that in vertebrate skeletal muscle.

Recently, some structural features of the thin filaments have been elucidated by X-ray diffraction studies on this muscle in living resting state and in rigor state (Maéda, Matsubara & Yagi, 1979).

(1) The actin helix changes when the living resting muscle goes to the rigor state.

(2) The arrangement of troponin molecules proposed by Ohtsuki (Ebashi, 1972; Ohtsuki, 1974) has been proved to be the case in this muscle.

(3) In the muscle in rigor, projections of myosin molecules are attached to the thin filaments with a repeating distance of 38 nm.

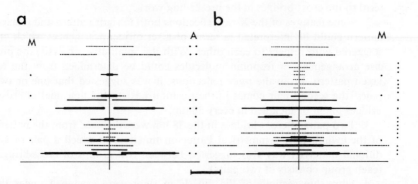

FIG. 1. Schematic representation of X-ray diffraction patterns from crab *Plagusia* muscle (a) in living resting state, and (b) in rigor state (Maéda *et al.*, 1979). In addition to actin layer lines (marked by A) and meridional reflections (marked by M) attributed to myosin projections on the surface of the thick filaments, a series of meridional and near-meridional reflections (including a part of intensities observed inner than 0.07 nm^{-1} on the actin layer lines) were observed and indexed as orders of 2 × 38 nm. In (a), reflections were seen up to the 9th order; in (b), up to the 24th. Orders of reflections are indicated in the right of (a). Thick solid lines, solid lines, and dotted lines represent intensities observed even on the third film, up to the second one, and only on the first film of each set, respectively. The bar, 0.1 nm^{-1}.

Conclusion (2) has been deduced from a series of meridional and near-meridional reflections observed in the X-ray diffraction pattern (Fig. 1a) from the muscle in living resting state. In the pattern from the muscle in rigor (Fig. 1b), a series of reflections was observed with stronger intensities and was seen up to higher orders. The reflections were interpreted as arising from both troponin and cross-bridges, and conclusion (3) has been deduced from these reflections.

In this report, I intend to describe the preliminary results of studies on the detailed arrangement of cross-bridges around the thin filaments in the muscle in rigor. In addition, the detailed interpretation of troponin reflections in the relaxed pattern will also be given here.

ASSIGNMENT OF A SERIES OF MERIDIONAL AND NEAR-MERIDIONAL REFLECTIONS IN THE RIGOR X-RAY PATTERN

These reflections were indexed as orders of 76.6 nm and seen up to the 24th order (Fig. 1b). In a preceding paper (Maéda *et al.*, 1979) these reflections were attributed to both troponin (TN) and cross-bridges. The contribution of cross-bridges to these reflections was strongly supported by electron micrographs as shown in Fig. 2a and Fig. 6. Pairs of bridges were attached to the thin filaments symmetrically with a repeating distance of 38 nm. The mi-

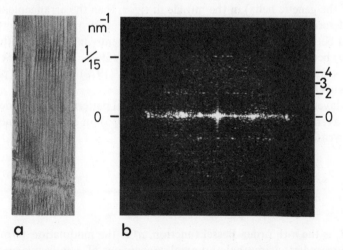

FIG. 2. (a) Electron micrograph of a longitudinal thin section of the muscle in rigor, and (b) an optical diffraction pattern obtained from (a).

crograph in Fig. 2a produced an optical diffraction pattern (Fig. 2b) in which layer lines were observed at 2nd, 3rd and 4th orders of 2 × 38 nm. A weak meridional 6th order was also observed (it is not seen in the reproduced materials). A meridional reflection at around 15 nm corresponds to the 14.4 nm meridional reflection in the X-ray pattern which was attributed to the thick filaments.

It should be noted that there must be some contribution of TN to these reflections, since some low-order reflections survived the treatment with the myosin-extracting solution (Maéda et al., 1979).

QUALITATIVE ANALYSES OF THE X-RAY DIFFRACTION PATTERNS

In this section, I will describe a method which enables us to analyze diffraction by molecules attached in pair to the thin filaments with the repeating distance of 38 nm. With this method of qualitative analysis of X-ray patterns, the TN arrangement was precisely determined and possible models were made for the cross-bridge arrangement in the muscle in rigor.

Figure 3a shows a surface lattice (radial projection) of the thin filament in Ohtsuki's model. It is worth noting that two TN molecules at the same level are not axially aligned but are staggered with respect to each other. Since the thin filaments consist of actin, tropomyosin and TN in a ratio of 7:1:1, and the actin helix has a symmetry of 28/13 (28 subunits are included in every 13 turns of the genetic helix) in the muscle in rigor, then the arrangement of TN can be described with a pair of identical 2/1 helices. If specific interactions are assumed between the molecules, axial displacement (Δz) and azimuthal rotation ($\Delta \psi$) between the two helices are specified by $\Delta \psi = (1 - 1/14)\pi k$ and $\Delta z = c/28 \times k$, where k is an odd integer. Ohtsuki (1974) proposed a plausible model with $k = 1$, which remains to be proved.

Since layer lines were not sampled in the actual patterns, we consider cylindrically averaged intensity (Franklin & Klug, 1955) on each layer line, which can be written as

$$\langle F(R)F^*(R)\rangle l = \sum_n J_n^2(2\pi Rr)|a_l^n|^2 \tag{1a}$$

$$a_l^n = \sum_j \exp i \{-n\psi_j + 2\pi l z_j/c\} \tag{1b}$$

where J_n is the n-th order Bessel function, a_l^n is the modulation term; r, ψ_j, and z_j are radial, azimuthal and axial coordinates of a molecule on the j-th helix. The summation in Eq. (1a) is taken over n, satisfying the selection rule

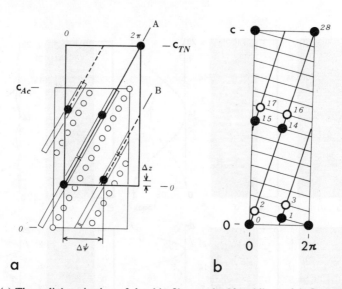

FIG. 3. (a) The radial projection of the thin filament in Ohtsuki's model. Open circles are actin subunits arranged in a helix with a symmetry of 28/13 and the repeating distance of c_{Ac}; rectangules are tropomyosin molecules; closed circles, TN molecules the arrangement of which can be described with two identical 2/1 helices, A and B, with the repeating distance of $c_{TN} = c_{Ac}$. There are axial displacement (Δz) and azimuthal rotation ($\Delta \psi$) between the two helices. (b) The radial projection of a model for cross-bridge arrangement around the thin filament in crab leg muscle in rigor. When a pair of cross-bridges ($L = 1$) occurs symmetrically around the thin filament at intervals of 38 nm, 4 lattice points (closed circles) out of 28 are occupied by cross-bridges. This arrangement is just the same as the TN arrangement shown in (a). When two pairs ($L = 2$) of cross-bridges occur at positions adjacent to each other, four more sites (open circles) are filled. In the latter case, the arrangement can be described with four identical 2/1 helices.

(Cochran, Crick & Vand, 1952) of $l = n + 2m$, where m is an integer.

Single 2/1 helix would give rise to uniformly spaced layer lines at orders of 76 nm. Every even (odd)-order Bessel function contributes to every even (odd)-order layer line, as shown in Fig. 4a. Addition of another 2/1 helix would result in systematic intensity modulation without changing the positions of the reflections. The modulation terms are given as $|a_l^n|^2 = 4\cos^2\{\pi k/28 \times (l + n)\}$ for $l =$ even, and $4\sin^2\{\pi k/28 \times (l + n)\}$ for $l =$ odd. If $k = 1$ is assumed, meridional J_0 reflections would be dominant around the equator and $l = 28$, and near-meridional $J_1 + J_{-1}$ reflections would be dominant around $l = 13$ and 15 as shown in Fig. 4c, in which modulation terms for $J_0(J_1 + J_{-1})$ gradually vary with even (odd) number of l. If $k = 3$ is assumed, modulation terms vary three times faster with l as shown in Fig. 4d.

In the actual resting pattern from the crab leg muscle (Fig. 1a), (1) reflections

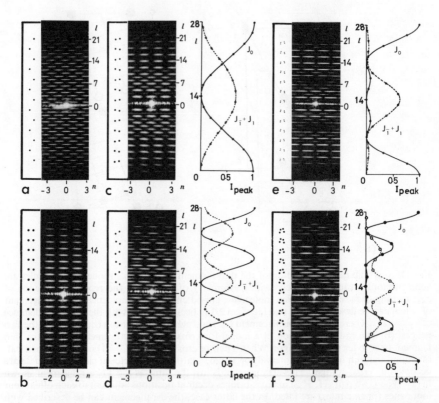

FIG. 4. Models of molecular arrangement around the thin filament and intensity modulations of diffraction patterns predicted from them. Models are: (a) single 2/1 helix with the repeating distance of 76 nm; (b) a pair of subunits in every 38 nm, which can be described with two identical 2/1 helices with $\Delta z = 0$ and $\Delta \psi = 0$; (c) the same as in Fig. 3a and 3b (closed circles only): two identical 2/1 helices with $k = 1$; (d) $k = 3$; (e) the same as in Fig. 3b (closed circles plus open circles); (f) Offer & Elliott model (1978) for the arrangement of cross-bridges in insect flight muscle in rigor. In each set, the left is a model. Closed circles represent molecules such as TN and cross-bridges. The right sides in (a) and (b), and the middle illustrations in (c)–(f), are optical diffraction patterns obtained from these models. In order to make higher order layer lines visible, the size of circles in every model is actually chosen as that in (e). On the right in (c)–(f), the relative intensity maxima of J_0 and $J_1 + J_{-1}$ contributions are plotted against l. These are calculated by Eq. (1), and the contribution of the form factor is assumed to be constant. Note that if two molecules in a pair are axially aligned as in (b), no odd-order reflections can be observed. All the models are based on the actin helix of 28/13.

were observed at every order from 2nd to 9th, except the 5th, of 2×38 nm; and (2) even-order reflections had intensity maxima on the meridian whereas odd orders did not. These features can be explained only by $k = 1$. The presence of a meridional reflection at $l = 4$ and a near-meridional one at $l = 9$ can not be explained by $k = 3$. $k = 5$ and 7 are also excluded. The near-meridional reflection at $l = 5$ may be buried in the strong reflection at 14.4 nm. In the muscle in resting state, the actin helix deviated slightly from 28/13, and accordingly, the TN helix is deformed slightly from 2/1. Nevertheless, the difference was so small that the diffraction patterns would not be affected (Maéda, 1978b). It is concluded that the TN arrangement in Ohtsuki's model has been proved to be the case in this muscle, which shows the usefulness of the method in which intensity modulation is interpreted in terms of the helical arrangement of molecules about thin filament axis.

Figure 3b shows a model of the arrangement of cross-bridges around the thin filaments. If a pair of myosin projections ($L = 1$) are attached specifically to the thin filaments, the arrangement is just the same as that of TN with $k = 1$. When another pair comes to the next position and both pairs are included in every 38 nm, the surface lattice (Fig. 3b) can be described with four identical 2/1 helices. When the number of pairs is increased to L, modulation terms for meridional J_0 reflections are calculated by Eq. (1) as

$$|a_l^{\eta=0}|^2 = \sin^2(2L\pi l/28)/\sin^2(\pi l/28), \tag{2}$$

which is an example of the Laue function. It should be noted that, the more cross-bridges occur, the smaller number of reflections intensities becomes concentrated on.

The actual pattern from the crab leg muscle in rigor (Fig. 1b) had the following characteristics: (1) meridional and near-meridional reflections were observed at orders, from 2nd to 24th, of 76 nm; (2) meridional reflections at $l = 12$, 14 and 16 were lacking, which seemed to be due to a selection rule (if the form factor is diminished there, we cannot explain the fact that near-meridional components at $l = 13$ and 15 were observed in the actual pattern); (3) even-order reflections had intensity maxima on the meridian whereas odd orders did not. These features can be explained both by the model with $L = 1$ and by that with $L = 2$. In the diffraction pattern predicted from the latter model, a few reflections (especially $l = 2$) would be almost four times stronger and the rest would be weaker than in the former case. Nevertheless, we cannot decide between the two with the qualitative analysis. Models with $L \geq 3$ might be excluded, since these models would predict the extinction of a reflection at $l = 4$ which was not the case. No other model could explain the actual pattern. As described previously, TN molecules also contribute to these reflections. However, the TN contribution may be unimportant when we extract the tendency of intensity modulation caused by cross-bridge arrangement from these reflections as a whole, since the contributions of TN might be much smaller than those of cross-bridges.

Recently, Offer & Elliott (1978) proposed a model for the cross-bridge arrangement in insect flight muscle in rigor. In the model, two pairs of cross-bridges are included in every 38 nm; the second pair comes to the position next but one to the first pair (Fig. 4f). Intensities predicted from the model, again based on intensity modulation calculated by Eq. (1), appear to be consistent with the recorded X-ray diffraction pattern (Miller & Tregear, 1972). In this muscle, the thin filaments are themselves arranged in a helical manner about the thick filament (Reedy, 1968; Reedy & Garett, 1977). The absence of meridional reflections at $l = 2$ and 4 may be explained by the large helix. Lack of reflections at $l = 3$ might support the model, though other possibilities cannot be ruled out. Disappearance of several near-meridional components in diffraction pattern (Barrington-Leigh *et al.*, 1977) from the muscle which were immersed in S-1 solution can well be explained by additional attachment of S-1 molecules to the surface lattice with a symmetry of 28/13.

It was reported that, from scallop striated muscle in rigor, a series of reflections were recorded, which are thought to be similar to those from crab leg muscle in rigor (Vibert, Szent-Györgyi, Craig, Wray & Cohen, 1978). They observed an increase in intensities of near-meridional components at 38 nm and at 19 nm, after removal of regulatory light chains. We might explain these observations by the change in arrangement of cross-bridges—from the arrangement with $L = 1$, for example, to that with $L = 2$. If this is the case, the intensities of some reflections would be decreased simultaneously.

CYLINDRICAL PATTERSON FUNCTION

In order to decide between the two possible models for cross-bridge arrangement ($L = 1$ or 2) and to understand the spatial relationship between TN and the cross-bridges on the individual thin filament in the muscle in rigor, a more quantitative analysis was carried out; the cylindrically symmetrical Patterson function (MacGillavry & Bruins, 1948) was obtained from intensity distribution along layer lines at orders of 76 nm in the rigor pattern. Intensities on the equator and on the reflections indexed as orders of 14.4 nm were not incorporated in the calculation. Since the equatorial reflections represent the Fourier transform of the axial projection of the structure on the plane $z = 0$, the Patterson function from the layer lines except the equator should represent the autocorrelation function of deviation of density distribution from the mean value.

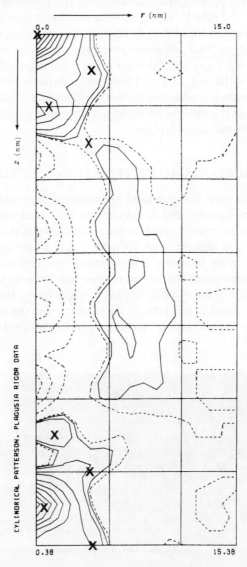

FIG. 5. Cylindrical Patterson function around the axis of the thin filament calculated from the intensity distribution in the rigor pattern. The intensities on the equator were not incorporated in the calculation. When the molecular arrangement of $L = 2$ (Fig. 4e) was assumed around the thin filament, inter-molecular vectors (X) would be expected.

Figure 5 shows 8 peaks and a shoulder near the positions marked with X in the region of $r < 5$ nm. X's represent inter-molecular vectors expected from the molecular arrangement of $L = 2$ (Fig. 3e). These molecules might be two pairs of cross-bridges. Two additional peaks are observed; one is a small one at $r = 10$ nm, $z = 2.7$ nm, and the other is a low but large hill in between $r = 5$ and 10 nm, $z = 10$ and 25 nm. These peaks indicate that there must be another pair of molecules attached to the thin filament at around the midpoint of the 38 nm. From this analysis we could not determine whether the molecules were TN or additional cross-bridges.

ELECTRON MICROGRAPHS OF VERY THIN SECTIONS

In order to get a view of the detailed arrangement of cross-bridges and TN, very thin sections (around 200 Å thick) were examined under the electron microscope. Since this muscle contains more thin filaments than thick filaments (Maéda, 1978a), it is difficult to identify the region where only one thin filament is sampled in its thickness. To overcome this difficulty, two kinds of specimens were examined. The first were fiber bundles which were treated in 50% glycerol solution for one day or two. In these fibers, the filament lattice was well preserved and it was rather easy to determine the thickness and the orientation of the section, taking careful note of the lattice. In sections with a

FIG. 6. Electron micrograph of very thin section of the crab leg muscle in rigor. The Z-line lies beyond the right margin. Bar, 200 nm.

thickness around 200 Å and an orientation slightly slanting to 1,1 plane, some regions were expected to sample only one thin filament in the neighborhood of a thick filament, most of which was included in the section (Maéda, 1978b). The second kind of specimen was used after a long-term glycerol treatment, longer than a month. In these fibers (Fig. 6), the filament lattice was partially disordered and disintegrated. Then ordered cross-bridge arrangements were preserved in a small number of regions. However, single thin filaments were frequently observed in these regions. In several sarcomeres from both kinds of specimens, individual thin filaments might be identified in 15 regions. Although there is need of further investigation, at the present stage observations can be summarized as follows: (1) The repeating distance of 38 nm along individual thin filaments was prominent; one or two pairs of cross-bridges might be attached to thin filaments at the intervals of 38 nm. The number of pairs, one or two, included in one group could not be determined in the micrographs. It should be noted that the cross-bridge arrangement was different from that in insect flight muscle. (2) Frequently, additional electron-dense materials were observed between two groups of cross-bridges. Some might be cross-bridges, some might be not. Quite often, these additional materials were observed at the midpoint. Accordingly, thin filament structures appear to have a periodicity of 19 nm. (3) The bridges were slightly slanted to the filament and the chevrons always pointed away from the Z-lines. (4) Cross-bridges protruded from the surface of the thick filaments at intervals of 28 nm, not 38 nm.

From these investigations, a model may be deduced for the molecular arrangement around the thin filament in the muscle in rigor: (1) Two pairs of cross-bridges are included in every 38 nm; the second pair comes next in position to the first one ($L = 2$). (2) A pair of TN molecules may lie at near the midpoint between two groups of cross-bridges (each group consists of two pairs of cross-bridges). It should be noted that this model is a tentative one, and it is difficult to rule out other possibilities at the present stage. Moreover, the structure might not be a regular one, like a crystal, but there must be some statistical diversities.

Nevertheless, it should be emphasized that the 38 nm repeating distance of cross-bridge occurrence has been proved in this muscle as the second example of such a case; the first one was the insect flight muscle. This poses two problems. The first one is what causes the repeating distance of attachment in this muscle in which the filament lattice is different from that in insect flight muscle. The second problem is whether the interaction of myosin heads with the thin filament occurs at intervals of 38 nm even in the muscle in contracting state.

Acknowledgements

I am grateful to Dr. I. Ohtsuki (University of Tokyo) for electron micrographs and optical diffraction patterns, and to Professors F. Oosawa and M. Endo for their advice and encouragement.

REFERENCES

Barrington-Leigh, J., Goody, R. S., Hofmann, W., Mannherz, H. G., Rosenbaum, G. & Tregear, R. T. (1977). The interpretation of X-ray diffraction from glycerinated flight muscle fibre bundles: new theoretical and experimental approaches. In *Insect Flight Muscle*, ed. Tregear, R. T., pp. 137–146. Amsterdam: North-Holland.

Cochran, W., Crick, F. H. C. & Vand, V. (1952). The structure of synthetic polypeptides. I. The transform of atoms on a helix. *Acta Cryst.* **5**, 581–586.

Ebashi, S. (1972). Calcium ions and muscle contraction. *Nature, Lond.* **240**, 217–218.

Franklin, R. F. & Klug. A. (1955). The splitting of layer lines in X-ray fibre diagrams of helical structures: Application to tabacco mosaic virus. *Acta Cryst.* **8**, 777–780.

MacGillavry, C. H. & Bruins, E. M. (1948). On the Patterson transforms of fibre diagrams. *Acta Cryst.* **1**, 156–158.

Maéda, Y. (1978a). Birefringence of oriented thin filaments in the I-bands of crab striated muscle and comparison with the flow birefringence of reconstituted thin filaments. *Eur. J. Biochem. Tokyo* **90**, 113–121.

Maéda, Y. (1978b). Optical and X-ray diffraction studies on crab leg striated muscle. Ph. D. Thesis, Nagoya University.

Maéda, Y., Matsubara, I. & Yagi, N. (1979). Structural changes in the thin filaments of crab striated muscle. *J. molec. Biol.* **127**, 191–201.

Miller, A. & Tregear, R. T. (1972). Structure of insect flight muscle in the presence and absence of ATP. *J. Mol. Biol.* **70**, 85–104.

Offer, G. & Elliott, A. (1978). Can a myosin molecule bind to two actin filaments? *Nature, Lond.* **271**, 325–329.

Ohtsuki, I. (1974). Localization of troponin in thin filament and tropomyosin paracrystal. *J. Biochem. Tokyo* **75**, 753–765.

Reedy, M. K. (1968). Ultrastructure of insect flight muscle. 1. Screw sense and structural grouping in the rigor cross-bridge lattice. *J. molec. Biol.* **31**, 155–176.

Reedy, M. K. & Garett, W. E. (1977). Electron microscope studies on *Lethocerus* flight muscle in rigor. In *Insect Flight Muscle*, ed. Tregear, R. T., pp. 115–136, Amsterdam: North-Holland.

Vibert, P., Szent-Györgi, A. G. Craig, R., Wray, J. & Cohen, C. (1978). Changes in crossbridge attachement in a myosin-regulated muscle. *Nature, Lond.* **273**, 64–66.

Discussion
(This discussion is for the previous two papers.)

O'Brien: In your analysis, Dr. Mitsui, you spent some time describing the positions of the troponin molecules; I wonder whether you consider that troponin is responsible for at least a portion of this X molecules? You have something at a rather low radius, and since troponin is clearly playing an important part, according to the analysis of Dr. Maéda, I wonder whether it may be the X molecule, or part of it?

Mitsui: I am not sure about the identification of X and Y molecules at present, and I guess we can do the identification later, after our refinement of the structure, but at present the contribution to the X-ray scattering from the Y molecules is rather small, and my guess is that the X molecules correspond to S-1, and the Y molecules corresponds to something including troponin.

O'Brien: I believe the troponin in these muscles is very large. Isn't it, Dr. Maéda? What is the molecular weight?

Maéda: I checked the molecular weight of the troponin from this muscle. It is about 96,000 daltons. Among the three subunits, I can say the corresponding weight of troponin T is very large, about 50,000 daltons; troponin I is 30,000 daltons, and troponin C is about 16,000 daltons. But I cannot say whether or not the troponin molecule is long. But let me make one comment.

As Dr. O'Brien said, it is important whether or not the troponin molecules are located at the same constant distance from the cross-bridges. If this is the case along the length of the thin filament, actin monomers are not homogeneous. There must be some heterogeneity along the thin filament, in which case actins are not equally acceptable to a cross-bridge. It is a very important point, but I cannot say about the spatial relationship between cross-bridges and troponin molecules at the present stage.

O'Brien: Dr. Mitsui showed that the model of Offer and Elliott (1978) for insect flight muscle did not agree with the crab rigor pattern; the bridges are sticking further out from the thin filaments in the insect model than is required to fit the crab pattern. According to King's College group, including myself, their model can explain the insect pattern quite well, so the difference may lie in the structure of crab and insect muscles. I note in Dr. Maéda's electron micrographs you can actually see cross-bridges, so it would be interesting to see whether you could see a substantial difference between that micrograph and those published by Reedy on insect flight muscle (Reedy, 1968), about the angle of attachment. I mean the distribution was evidently the same, *i.e.*, the 385Å intervals. But what about the angle?

Maéda: In our electron micrographs, the chevron-like pattern was not so clear—less clear than Reedy's electron micrographs—but the direction is always the same as in other muscle. Frequently the cross-bridges are slanted, but the angle of slant is somehow less diverse, around 60°.

O'Brien: Is that 60° to the vertical or the horizontal?

Maéda: To the axis of the filament.

O'Brien: So that, if anything, sticking further out than in.

Maéda: No near perpendicular.

Huxley: Could I ask an experimental question? When you are measuring the intensities of the X-ray reflection, one of the most important ones is this meridional 385Å. Do you measure the intensity of that by rocking the specimen through the vertical direction, or are these intensities taken from stationary photographs, where the angle of the muscle to the X-ray beam is kept constant? I think you get a more reliable value for the meridional reflection if you do rock the muscle, and as that is such a strong reflection and it is so important, it may be a good idea to get a value that way.

Mitsui: We didn't do that, but thank you very much for your suggestion.

REFERENCES

Offer, G. & Elliott, A. (1978). Can a myosin molecule bind to two actin filaments? *Nature, Lond.* **271**, 325–329.

Reedy, M. K. (1968). Ultrastructures of insect flight muscle I. Screw sense and structural grouping in the rigor cross-bridge lattice. *J. molec. Biol.* **31**, 155–176.

VII. ENERGETICS

Problems of Muscle Energetics and Some Observations on the Period after a Stretch of Active Muscle

R. C. WOLEDGE and N. A. CURTIN

Department of Physiology, University College London, London, England

ABSTRACT

We have investigated the ATP splitting and heat production by active muscle after it has been stretched. As has been reported earlier, the active tension is considerably greater after stretching than in a purely isometric tetanus at the same muscle length. Despite this enhanced tension, the rate of ATP splitting after stretch is not significantly different than that in the isometric controls. Only a small and transitory increase in the rate of heat production occurs as a result of stretching.

INTRODUCTION

What reactions produce the energy which is liberated during and after muscle contraction? How do the rates of these reactions depend on the pattern of stimulation, muscle length and the mechanical conditions during contraction? What enzymes catalyze these reactions, and how are they controlled? These are some of the questions about the energetic aspects of contraction which can be answered experimentally. The appropriate way to evaluate theories from the point of view of energetics is to calculate both the energy output and the ATP splitting to be expected from the particular model considered; an example of this approach is the article by Yamada and Kodama in this volume.

It is well established that a large contribution to the energy output during contraction comes from the splitting of ATP and the reactions which resynthesize it, particularly the creatine kinase reaction. At one time it was thought that these reactions would account for *all* the energy output. However, a proper, quantitative comparison following the principles described by Wilkie (1960) has shown that the observed energy output in isometric tetanic contractions substantially exceeds the amount that can be explained by the known metabolic reactions. This has been confirmed in many experiments (reviews by Homsher & Kean, 1978; Curtin & Woledge, 1978). Therefore, one or more additional reactions is contributing significantly to the energy output. These reactions could be incomplete cycles of the processes involved in the splitting of ATP by the actomyosin ATPase and/or the sarcoplasmic reticulum ATPase.

We have recently followed the time course of the energy output and of ATP splitting during isometric contractions of frog muscle at $0°C$. We find that the rate of unexplained energy output is greater early in the tetanus (first 5 sec) than it is later (up to 15 sec). The quantity of unexplained energy is about the same as Aubert's "labile" part of the energy output. Homsher and Kean (personal communication) have made similar observations and have found that there is no unexplained energy output between 8 and 15 sec in a tetanus.

It is not possible to use measurements of ATP splitting, or of energy output, to infer the rate of turnover of cross-bridges. There are at least three reasons for this: (1) ATP is split, and energy produced, by non-myofibrillar systems within the muscle; (2) cross-bridges may undergo a complete cycle involving attachment and detachment but not involving ATP splitting or net energy production; (3) cross-bridges may undergo incomplete cycles, and the unexplained energy found in energy balance studies may be derived from these processes.

The third of these possibilities is suggested by the work of Rall et al. (1976). They found that during rapid shortening there is actually less ATP splitting than during isometric contraction, and that the unexplained part of the energy output is greater. Another way of influencing cross-bridge reactions is to stretch the active muscle. This gives very high tensions but greatly reduces the amount of ATP that is split during stretching compared to isometric controls (Curtin & Davies, 1975). When stimulation is continued after the stretch, the tension can remain above the level in an isometric contraction for several seconds (Abbott & Aubert, 1952a). The energy changes during this period of enhanced tension development have not been investigated previously. We have, therefore, measured the energy production and ATP splitting during this time.

DESIGN OF THE EXPERIMENT

The experiments were made on sartorius muscles of *Rana temporaria* at $0°C$. The techniques used were similar those described by Curtin & Woledge (1977). In one series of observations heat production and tension development were observed during 6.5 sec of stimulation. The work stored in the series elasticity was calculated and added to the heat produced to obtain the quantity which will be referred to as $h + w$. Note that the work done *on* the muscle during stretches has not been subtracted from this quantity; the fate of this work is discussed below. In a parallel series of experiments pairs of muscles were mounted on a hammer-freezing apparatus. Both muscles were stimulated and they were frozen when one (E) had been tetanized for 6.5 sec and the other (C) for 1.5 sec. The difference in the free creatine levels, $E - C$, in the extracts of the muscles thus measures the change during 5 sec of tetanus, and the extent of the creatine kinase reaction during this time. No other metabolite changes are reported here, but on the basis of our previous work we expect that all the ATP split would be rephosphorylated by the creatine kinase reaction. The change in free creatine level thus also measures the extent of ATP splitting.

In both series of experiments the muscle was either isometric at a muscle length of 1.2 l_0, or it was stretched 3 mm (at 38 mm/sec) to reach a final length of 1.2 l_0 at 1.1 sec after the start of the tetanus. When the tetanus was to be isometric the muscle was stretched to 1.2 l_0 at 1 sec before the start of the tetanus. Tension changes during stretches of resting muscle were also recorded and this passive tension has been subtracted from the total tension during and after stretch of active muscle to give the active tension values reported here.

RESULTS

As expected, the active tension maintained after the stretch was considerably greater than in the isometric tetanus (Figs. 1 and 2), and this persisted to the end of stimulation. In the heat experiments the mean excess tension was 37% of the isometric value at 1.5 sec and 18% at 6.5 sec. In the chemical experiments the excess tension-time integral (mechanical impulse) between 1.5 sec and 6.5 sec was 44% of the isometric value. This big increase in developed tension is not accompanied by commensurate changes in the energy production. As is shown in Fig. 1 there is no significant difference in the amount of ATP splitting. Thus the mechanical impulse for each ATP split is considerably greater after stretch.

The observed $h + w$ production is shown in Fig. 2. In the time interval which includes the stretch, the rate of $h + w$ production is much higher than it is in the isometric control, presumably because much of the work done on the muscle during the stretch appears immediately as heat in the muscle. During the next one-sec interval the rate of $h + w$ production remains higher than the isometric control value. This extra $h + w$ might also be derived from work done on the muscle during the stretch if part of the work had been mechani-

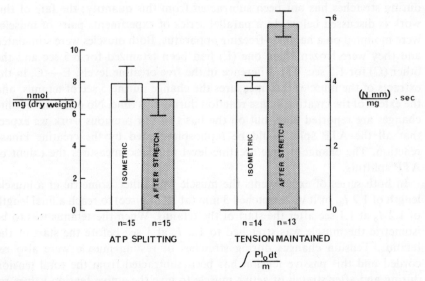

FIG. 1. Active tension and ATP splitting after stretch compared with normal isometric contraction. The measurements are for the period 1.5 sec to 6.5 sec in a tetanus at a muscle length of $1.2 l_0$. For the observations labelled "after stretch" the active muscles were stretched 3 mm between 1 sec and 1.1 sec in the tetanus. The blocks show the means and the bars ± one standard error for n observations.

cally stored within the muscle since the time of the stretch. During the remainder of the tetanus, and during relaxation, the $h + w$ output is almost the same after stretch as in the isometric control, although the tension remains significantly greater.

For the isometric results it is possible to compare the observed $h + w$ during the period from 1.5 to 6.5 sec with that explained by the splitting of ATP and the creatine kinase reaction. As usual, the observed $h + w$ (0.390 ± 0.017 mJ/mg dry weight) significantly exceeds the explained energy (0.268 ± 0.038 mJ/mg). This indicates that part of the energy output during this time comes from some additional process. Similarly, after stretching, the observed $h + w$ exceeds the explained $h + w$. However the interpretation of this result is difficult for the following reason. There may be an extra source of heat, besides chemical reactions, during this period—namely, the dissipation of energy mechanically stored in the muscle since the stretch. The data now available are not sufficient to assess the extent of this process.

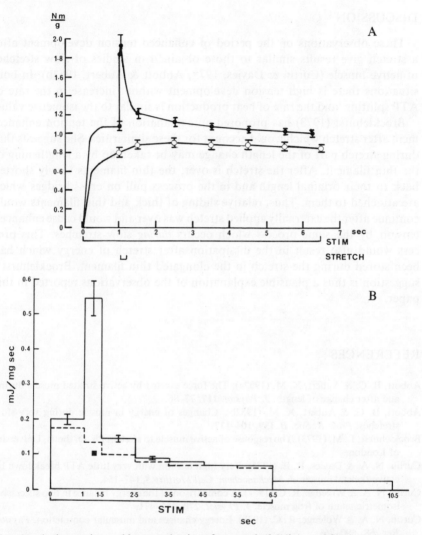

FIG. 2. Active tension and heat production after stretch (full lines and filled symbols) compared with normal isometric contraction (broken lines and open symbols). The quantities plotted are: in A, [(active tension) × (*in situ* muscle length, l_0)] ÷ (muscle dry weight, m); in B, [(observed heat production, h) + (work stored in the series elasticity, w)] ÷ (muscle dry weight, m). During the stretch, work is done on the muscle. If this work is subtracted from the observed $h + w$ during the 0.5 sec interval in which the stretch occurs, the result is shown by the symbol, ■. Symbols show means and bars ± one standard error from 15 observations of each quantity.

DISCUSSION

These observations on the period of enhanced tension development after a stretch give results similar to those obtained in studies of slow stretches of active muscle (Curtin & Davies, 1975; Abbott & Aubert, 1952b). In both situations there is high tension development without increase in the rate of ATP splitting, and the rate of heat production is similar to the isometric value.

Brocklehurst (1973) has proposed an explanation of the tension enhancement after stretch which would account for these similarities. She suggests that during stretch part of the length change may be taken up by a lengthening of the thin filament. After the stretch is over, the thin filaments slowly shorten back to their original length and in the process pull on cross-bridges which are attached to them. Thus, relative sliding of thick and thin filaments would continue after the externally applied stretch was over and would cause enhanced tension by the same process which occurs *during* slow stretches. This process would also result in the dissipation after stretch of energy which had been stored during the stretch in the elongated thin filament. Brocklehurst's suggestion is thus a plausible explanation of the observations reported in this paper.

REFERENCES

Abbott, B. C. & Aubert, X. M. (1952a). The force exerted by active striated muscle during and after change of length. *J. Physiol.* **117**, 77–86.

Abbott, B. C. & Aubert, X. M. (1952b). Changes of energy in muscle during very slow stretches. *Proc. R. Soc. B.* **139**, 104–117.

Brocklehurst, L. M. (1973). The response of active muscle to stretch. Ph. D. thesis, University of London.

Curtin, N. A. & Davies, R. E. (1975). Very high tension with very little ATP breakdown by active skeletal muscle. *J. Mechanochem. Cell Motility* **3**, 147–154.

Curtin, N. A. & Woledge, R. C. (1977). A comparison of the energy balance in two successive isometric tetani of frog muscle. *J. Physiol.* **270**, 455–471.

Curtin, N. A. & Woledge, R. C. (1978). Energy changes and muscular contraction. *Physiol. Rev.* **58**, 690–761.

Homsher, E. & Kean, C. J. (1978). Skeletal muscle energetics and metabolism. *Ann. Rev. Physiol.* **40**, 93–131.

Rall, J. A., Homsher, C., Wallner, E. & Mommaerts, W. F. H. M. (1976). A temporal dissociation of energy liberation and high energy phosphate splitting during shortening in frog skeletal muscle. *J. gen. Physiol.* **68**, 13–27.

Wilkie, D. R. (1960). Thermodynamics and the interpretation of biological heat measurements. *Prog. Biophys. biophys. Chem.* **10**, 259–289.

Discussion

Rall: Roger, you are looking at incomplete thermodynamic cycles. The suggestion might be that when the cycles become complete, you would no longer have an energy imbalance. Is that true or not?

Woledge: Well, it is certainly true from the theoretical point of view. The question is when these cycles are completed. In the case of this sort of model, one would expect them to be completed fairly soon, for instance during relaxation, or shortly thereafter. We do know that the unexplained energy is still there after relaxation is over, so that is another factor which rather argues against this. We do know, for example from Saul Winegrad's measurement (Winegrad, 1968), that the return of calcium to its original site may take quite a long time after relaxation, and so it is possible that the energy imbalance is going to be restored on that time scale. I will not say any more about what the observations are on that, because that is something Marty Kushmerick will be discussing later.

Sugi: I would like to make two comments. First, Dr. Amemiya, my colleague, has shown in this meeting that the increase in tension may not be due to increasing the number of cross-bridges. Is this consistent with your results? And second, my colleague Dr. Tanaka examined the rate of ATP splitting during and after stretch of glycerinated rabbit psoas fibers (Tanaka, Tanaka & Sugi, 1978). In that case, the rate of ATP splitting increases, the higher the tension after stretch. So, my impression is that the glycerinated system differs very much from the intact system.

Woledge: With respect to your first comment, I do not think our results have anything to say about how many cross-bridges are acting. We only see the product of the number that are there, and the rate at which they are going through this particular step of ATP splitting. Concerning your second comment, I saw those results you are referring to, and I agree with you, it looks different. I do not know why, but it certainly looks different.

Winegrad: Does the unexplained heat in a second tetanus that occurs after a very short period of rest following the first tetanus still bear the same relationship to the labile heat?

Woledge: Yes.

Winegrad: The second question is, have you examined the unexplained heat occurring during shortening when you allow shortening to occur from an initially stretched muscle out in the range where Dr. Edman showed us in this meeting that there is an increase in V_{max}?

Woledge: The answer to that is no.

REFERENCES

Tanaka, H., Tanaka, M. & Sugi, H. (1978). Effect of sarcomere length and stretch on the tension and the rate of ATP splitting in glycerinated rabbit psoas muscle fibres. *6th Int. Biophys. Congr. Kyoto*, p. 383.

Winegrad, S. (1968). Intracellular calcium movements of frog skeletal muscle during recovery from tetanus. *J. gen. Physiol.* **51**, 65–83.

An Explanation of the Shortening Heat Based on the Enthalpy Profile of the Myosin ATPase Reaction

Takao KODAMA* and Kazuhiro YAMADA**

* Department of Pharmacology, Juntendo University School of Medicine, Tokyo, Japan
** Department of Physiology, Oita Medical College, Oita, Japan

ABSTRACT

A model is proposed which could account for the time course of total energy output (heat + work) and the high-energy phosphate splitting by shortening muscle. The model is based on results from recent kinetic and calorimetric studies of myosin and actomyosin ATP spilitting. The major elements are: (1) ATP is split by the cross-bridge cycle, the predominant intermediate states of which are $A \sim M \cdot ADP \cdot Pi$ (attached state), $M \cdot ATP$ and $M \cdot ADP \cdot Pi$ (A denotes an actin monomer and M a myosin active site); (2) the transition from $M \cdot ATP$ to $M \cdot ADP \cdot Pi$ is endothermic and is the rate-limiting step of the overall cycle (the rate constant remains the same whatever the mechanical conditions of muscle); (3) the rate constants for the cross-bridge attaching ($M \cdot ADP \cdot Pi \rightarrow A \sim M \cdot ADP \cdot Pi$) and detaching ($A \sim M \cdot ADP \cdot Pi \rightarrow M \cdot ATP$) steps increase during shortening; and (4) the attaching step is endothermic but the detaching step is strongly exothermic. The model predicts that a significant amount of heat is produced by an accumulation of $M \cdot ATP$, in addition to that from ATP splitting, as muscle goes from isometric to isotonic contraction.

INTRODUCTION

When a stimulated muscle is allowed to shorten it produces heat at a faster rate than in an isometric contraction (Hill, 1938; Irving, Woledge & Yamada, 1979). The main difficulty in understanding this phenomenon is that the rate of total energy output (heat + work) during shortening increases without sufficient ATP or creatine phosphate splitting to account for it. The high-energy phosphate splitting takes place after shortening is over. This was unequivocally shown by Rall, Homsher, Wallner & Mommaerts (1976). Here we propose a model which could account for such separation of the energy output from the high-energy phosphate splitting during shortening.

It is presumed that the cross-bridge cycle in active muscle follows the actomyosin ATPase cycle, and that as soon as ADP is dissociated from actomyosin it is rephosphorylated to ATP by creatine kinase activity (this reaction produces little heat; cf. Woledge, 1972). The present model is based on the following kinetic and energetic considerations and assumptions.

KINETICS

Figure 1 shows a Lymn-Taylor scheme for the splitting of ATP by myosin and actomyosin. It seems that there are three slow steps in actomyosin ATP splitting: reaction 3, the ATP cleavage step; reaction 4, the association of actin with myosin-product complex, which probably corresponds to the cross-bridge attaching step; and reaction 5, phosphate dissociation from actomyosin. In short, taking the ATP and actin concentrations to be constant, the second-order reactions can be treated as first-order by including the concentration of ATP or actin in the rate constant. Then it turns out that the rate constants for these three reactions are significantly smaller than those for other reactions (Table 1). Thus the rate constants for reactions 5, 6, 1 and 2 can be lumped together, resulting in the simplified scheme for actomyosin ATP splitting shown in Fig. 2. Note that the scheme consists of two detached states of cross-bridge and a single attached state, but the second attached state used in the usual model for mechanical performance of muscle has been omitted because this state exists only transiently and is not necessary for the present argument.

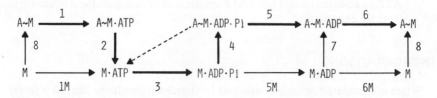

FIG. 1. Kinetic scheme for myosin and actomyosin ATP splitting. The latter is shown by the heavy arrows. The backward arrows are deleted for simplicity. The dotted line indicates the simplification used (see text).

The proportion of total cross-bridges in the different states is determined by the following set of simultaneous differential equations:

TABLE 1. Rate constants of myosin and actomyosin ATPase reactions

Reactions*	k_f	$k_{f'}(s^{-1})$	$k(s^{-1})$
1	$10^6 M^{-1}s^{-1}$	5×10^3	***
2	$\geq 10^3 s^{-1}$	$\geq 10^3$	***
3	$10^2 s^{-1}$	10^2	3
4	$\geq 10^4 M^{-1}s^{-1}$	≥ 10	0–50
5	$\geq 10 s^{-1}$	≥ 10	0–50
6	$\geq 10^3 s^{-1}$	$\geq 10^3$	***
1M	$2 \times 10^6 M^{-1}s^{-1}$	10^4	***
5M	$0.05 s^{-1}$	0.05	0.001
6M	$2.0 s^{-1}$	2	0.5
7	$5 \times 10^5 M^{-1}s^{-1}$	5×10^2	***
8	$10^6 M^{-1}s^{-1}$	10^3	***

* Reaction numbers refer to the scheme shown in Fig. 1.
k_f: forward rate constants in 0.1M KCl, 5mM MgCl$_2$, pH 7–8, 20°C. From Dr. H. D. White (personal communication at the Workshop/Conference on Muscle Energetics held in Burlington, Vermont, 1977).
$k_{f'}$: pseudo-first-order rate constants calculated for ATP = 5mM and actin = 1mM.
k: rate constants assumed for conditions pertaining to muscle cells at temperatures near 0°C.
***: very large compared to other k's.

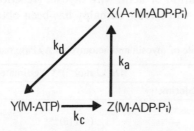

FIG. 2. Simplified scheme for actomyosin ATP splitting used in the model. X gives the proportional amount of cross-bridges in the attached state, and Y and Z give the proportional amounts of the detached cross-bridges in either the M·ATP or M·ADP·Pi state. The k's are rate constants for state transitions.

$$\frac{dX}{dt} = -k_d X + k_a Z$$
$$\frac{dY}{dt} = -k_c Y + k_d X \qquad (1)$$
$$\frac{dZ}{dt} = -k_a Z + k_c Y,$$

where X, Y and Z represent the amount of A $\sim$ M·ADP·Pi, M·ATP and M·ADP·Pi, respectively, and $X + Y + Z = 1$. In a steady state such as prolonged isometric contraction, X, Y and Z are constant. But during the transient phase, for instance, immediately after muscle has gone from rest to contraction or an isometric-isotonic transition has taken place, the amounts of these three myosin species vary with time and the values can be calculated by computer analysis (see below).

ENERGETICS

The reaction heats for many of the steps of the myosin and actomyosin ATP splitting are now known (Table 2). Using a newly designed calorimeter with an improved time resolution, Kodama & Woledge (1978a, b) measured the heat produced when myosin subfragment 1 in the M state or the M·ADP state is mixed with ATP or its analogues. The results indicate that reaction heats, large compared with the overall reaction heat of the enzyme turnover cycle, accompany intermediate steps of the myosin ATP splitting cycle shown in Fig. 1. Of particular importance is that ATP cleavage on a myosin head is endothermic in contrast to the heat of ATP splitting in free solution, which is of course exothermic ($\Delta H_{ATP} < 0$). Calorimetric information is also available about the association of actin with myosin (reaction 8) (White & Woledge, 1976), and the van't Hoff enthalpy has been obtained for reaction 6

TABLE 2. Enthalpy profile of myosin and actomyosin ATPase reactions

Reactions*	ΔH(kJ mol^{-1})** estimated by	
	Calorimetry	Kinetics ($d\ln K/dT$)
1	—	—
2	—	(-110)***
3	$+ 85$	$+ 60$
4	—	(-45)***
5	—	—
6	—	$\simeq 0$
1M	$- 90$	—
5M	$- 135$	—
6M	$+ 70$	$+ 75$
7	—	($+90$)***
8	$+ 20$	—

* Reaction numbers refer to the scheme shown in Fig. 1.
** In 0.1M KCl, 5mM MgCl$_2$, pH 8, $\simeq 20°$C.
*** Calculated from $\Sigma \Delta H = 0$.
  Calorimetric data are from Kodama and Woledge (1978a,b); kinetic data from Dr. H. D. White (personal communication, see Table 1).

(H. D. White, personal communication). The reaction heats for some steps can be calculated from $\sum \Delta H = 0$ for closed loop (see Table 2).

Returning to the simplified scheme for actomyosin ATP splitting (Fig. 2), the cleavage step is thus endothermic ($\Delta H_c > 0$). Although the reaction heats for the other two steps have not been estimated yet, a reasonable guess is that the heat for the association of actin with myosin-product complex or the cross-bridge attaching step could be endothermic ($\Delta H_a > 0$) on the analogy with reaction 7 in Fig. 1, the heat for which is calculated to be $+ 90$ kJ mol^{-1} (Table 2). Then the transition from the attached state to the detached state (M·ATP) has to be strongly exothermic, since the heat for this reaction (ΔH_d) is equal to $\Delta H_{ATP} - (\Delta H_c + \Delta H_a)$.

ENERGY OUTPUT BY ACTIVE MUSCLE

The total energy output (ΔE) per mole of cross-bridges in time t after the onset of transition between the relaxed and activated state or between isometric and isotonic contraction is given by

$$\Delta E = \Delta H_d \int_0^t k_d X dt + \Delta H_c \int_0^t k_c Y dt + \Delta H_a \int_0^t k_a Z dt. \qquad (2)$$

In a steady state this equation is reduced to

$$\Delta E = \Delta H_{ATP} \cdot k_c Y \cdot r, \qquad (3)$$

since $k_d X = k_c Y = k_a Z$ and $\Delta H_d + \Delta H_c + \Delta H_a = \Delta H_{ATP}$. Rearranging Eq. (2), we get

$$\Delta E = \Delta H_{ATP} \int_0^t k_c Y dt + \Delta H_a \Delta X + (\Delta H_{ATP} - \Delta H_c) \Delta Y, \qquad (4)$$

where $\Delta X = \int_0^t k_a Z dt - \int_0^t k_d X dt = X - X_0$ and $\Delta Y = \int_0^t k_d X dt - \int_0^t k_c Y dt = Y - Y_0$ (X_0 and Y_0 are the initial values of X and Y at $r = 0$).

The first term on the right-hand side of Eq. (4) is equivalent to the "explained" heat obtained by chemical analysis (cf. Rall et al., 1976). Thus it is clear that, during the transient phase, the time course of energy output by cross-bridge activity consists of two components: one is explained by ATP splitting and the other is due to the change in relative amounts of the different species of myosin (the latter two terms in Eq. (4)).

Shortening heat. We can now describe a model which could account for the time course of energy output and ATP splitting by active muscle. First of all we assume that the ATP cleavage step is rate-limiting in the overall cycle. This assumption is plausible as shown by recent kinetic studies (Taylor, 1977; White 1977). The rate constant is taken to be 3 sec^{-1}, which is close to the value found by Ferenczi, Homsher, Simmons & Trentham (1978). It is also assumed that this rate constant remains the same whatever the mechanical conditions. On the other hand, the rate constants for the cross-bridge attaching and detaching reactions, k_a and k_d, are assumed to be dependent on the mechanical conditions of the muscle. We assume that the attaching rate increases during shortening because each myosin site comes close to an actin site more frequently during shortening than in isometric contraction. The detaching rate should also increase because during shortening a cross-bridge remains attached only over a limited range. Thus in the present model the high-energy phosphate splitting and energy-liberating steps are intrinsically separate.

We now consider the following transitions: muscle, which has been in an isometric state for 1 sec, is allowed to shorten rapidly for 100–200 msec and then

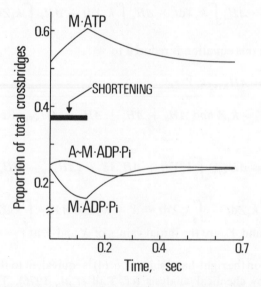

FIG. 3. Change in the proportion of cross-bridges in different states during shortening followed by isometric contraction. Shortening period is 133 msec with rate constants of k_a (9 sec^{-1}) and k_d (12 sec^{-1}). After the shortening both rate constants are reduced to 6 sec^{-1} again.

goes back to isometric contraction again. The rate constants for both attaching and detaching of a cross-bridge are taken to be 6 sec^{-1} in the isometric state. The corresponding values are assumed to increase during shortening to 9 and 12 sec^{-1}, respectively. Figure 3 shows the time course of the change in the proportion of total cross-bridges in different states. It is evident that the shortening elicits a rapid increase in M·ATP, accompanied by a rapid decrease in myosin-product complex. Such a shift in the state of the cross-bridges would produce a significant amount of heat without net ATP splitting. Note that the recovery process after the shortening is rather slow. Figure 4 shows the time course of energy output calculated by Eq. (4) taking ΔH_{ATP} to be -50 kJ mol^{-1} (Woledge, 1972) and both ΔH_a and ΔH_c to be + 40 kJ mol^{-1}. It is obvious that during shortening the rate of energy output increases markedly but without a measurable equivalent of ATP splitting. The ATP splitting "catches up" with the total energy output rather slowly after the shortening has ended.

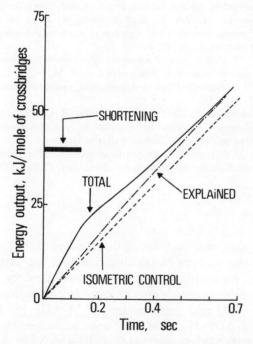

FIG. 4. Time course of energy output by muscle which is allowed to shorten for 133 msec and then back to isometric contraction again as in the case of Fig. 3. The calculation was made by using Eq. (4). The isometric heat production was calculated by Eq. (3).

Acknowledgements

We thank Dr. R.C. Woledge for his valuable comments and Prof. Y. Ogawa for his interest and support.

REFERENCES

Ferenczi, M. A., Homsher, E., Simmons, R. M. & Trentham, D. R. (1978). The reaction mechanism of the Mg^{2+}-dependent adenosine triphosphatase of frog myosin and subfragment 1. *Biochem. J.* **171**, 165–175.

Hill, A. V. (1938). The heat of shortening and the dynamic constants of muscle. *Proc. R. Soc. B.*, **126**, 136–195.

Irving, M., Woledge, R. C. & Yamada, K. (1979). The heat produced by frog muscle in a series of contractions with shortening. *J. Physiol* (in the press).

Kodama, T., & Woledge, R. C. (1978). Enthalpy profile of myosin ATPase reaction. *Abstracts of Sixth International Biophysics Congress*, p. 376.

Kodama, T., & Woledge, R. C. (1979). Enthalpy changes for intermediate steps of the ATP hydrolysis catalyzed by myosin subfragment-1. *J. biol. Chem.* (in the press).

Rall, J. A., Homsher, E., Wallner, A., & Mommaerts, W. F. H. M. (1976). A temporal dissociation of energy liberation and high energy phosphate splitting during shortening in frog skeletal muscles. *J. gen. Physiol.* **68**, 13–27.

Taylor, E. W. (1977). Transient phase of adenosine triphosphate hydrolysis by myosin, heavy meromyosin, and subfragment 1. *Biochemistry, N. Y.* **16**, 732–740.

White, H. D. (1977). Contractile filament organization, mechanics, and biochemistry. *Nature, Lond.* **267**, 754–755.

White, H. D., & Woledge, R. C. (1976), cited in Woledge, R. C. (1977). Calorimetric studies of muscle and muscle proteins. In *Applications of Calorimetry in Life Sciences*, ed. Lamprecht, I., pp. 183–197. Berlin: Walter de Gruyter.

Woledge, R. C. (1972). In vitro calorimetric studies relating to the interpretation of muscle heat experiments. *Cold Spring Harb. Symp. quant. Biol.* **37**, 629–634.

Discussion

Rall: Would you expect that, during the transient of the isometric contraction, you would also see extra energy liberation from your actomyosin system?

Kodama: Yes. We could expect to see some extra energy output from the actomyosin system when muscle goes from rest to contraction, since in resting muscle myosin would exist predominantly as $M \cdot ADP \cdot Pi$, some of which is converted to $AM \cdot ADP \cdot Pi$ and $M \cdot ATP$ upon activation. These two reactions ($M \cdot ADP \cdot Pi \to AM \cdot ADP \cdot Pi$ and $M \cdot ADP \cdot Pi \to M \cdot ATP$), however, may be compensatory with respect to the reaction heat as we have discussed here. Thus how much heat is produced would depend on the heat (ΔH_z and $-\Delta H_c$) and the extent (ΔX and ΔY) of these reactions.

Myosin Isoenzymes in Skeletal and Cardiac Muscles

J. F. Y. HOH
Department of Physiology, University of Sydney, Sydney, Australia

ABSTRACT
Pyrophosphate gel electrophoresis separates cardiac and skeletal myosins into isoenzymes. Three isoenzymes which differ in ATPase activity are present in the rat ventricle. Their structures differ in the heavy chains but not in the light chains. The profile of ventricular myosin isoenzymes is modulated by thyroid hormone. Myosin from fast-twitch skeletal muscle or single fibers also shows three isoenzymes. These differ in light chain distribution. The profile of isoenzymes in fast-twitch and slow-twitch muscles can be reversed by nerve cross-union. In both cardiac and skeletal muscles, changes in myosin isoenzyme distribution are correlated with changes in force-velocity characteristics.

Myosin is probably the most important determinant of the dynamic properties of muscle. Muscle scientists, when considering force-velocity properties, muscle energetics, the kinetics of cross-bridges and models of muscle contraction, have generally assumed tacitly that there is only one population of myosin, at least at the level of a single muscle fiber. The purpose of this paper is to show (i) that both skeletal and cardiac myosins exist in several isoenzymic forms which differ in subunit structure and enzymic properties, and (ii) that myosin isoenzyme distribution affects the contractile properties of the muscle.

Isoenzymes usually differ from each other only slightly in charge and size. Gel electrophoresis, the usual method employed for the separation of isoenzymes, has been applied to the separation of native myosin only recently (Hoh, 1975). This is because myosin is insoluble in the low ionic strength of the buffers usually employed for gel electrophoresis. This fact, together with the thermal lability and large size of myosin, posed considerable problems to the application of gel electrophoresis to native myosin. These problems were solved by the addition of ATP to a low ionic strength buffer (Hoh, 1975) or the use of pyrophosphate as a buffer (Hoh, McGrath & White, 1976). It has previously been shown that ATP, pyrophosphate and other polyanions can dissolve myosin (Brahms & Brezner, 1961) and destabilize myosin filaments (Harrington & Himmelfarb, 1972) at low ionic strength. The application of pyrophosphate gel electrophoresis to myosins from skeletal (Hoh et al., 1976), cardiac (Hoh, McGrath & Hale, 1978) and smooth (Hoh & Yeoh, 1977) muscles has revealed hitherto unsuspected heterogeneity.

Heterogeneity of the ventricular myosin of the rat myocardium is shown in Fig. 1. Three isoenzymes, termed V_1, V_2 and V_3 in the order of decreasing electrophoretic mobility, can be seen when stained for Ca^{++}- activated ATPase activity or for protein. A comparison of the densitometer profiles for ATPase and protein for the same gel shows that V_1 has the highest Ca^{++}-activated ATPase activity while V_3 has the lowest.

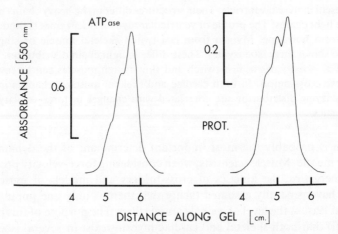

FIG. 1. Densitometer scans of ventricular myosin isoenzymes from an adult rat after separation in 4% acrylamide gel in pyrophosphate buffer. The scan on the left was done after staining for Ca^{++}-activated ATPase activity; that on the right shows the same gel after protein staining. The isoenzymes are named, from right to left, V_1, V_2 and V_3. Note the difference in these profiles illustrating the different Ca^{++}-ATPase activities of the isoenzymes.

Since ventricular myosin isoenzymes differ in enzyme activity, the mean myosin ATPase activity of a ventricle is a function of isoenzyme distribution. No significant regional variation in isoenzyme distribution within the same ventricle has been detected. On the other hand, marked variations in isoenzyme

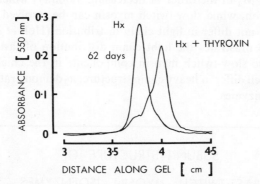

FIG. 2. Effect of the thyroid state on rat ventricular myosin isoenzyme distribution. Superimposed isoenzyme profiles of ventricular myosins from chronically hypophysectomized rat (Hx, left) and chronically hypophysectomized rat after daily treatment with 5 µg of thyroxine for 16 days (right). Note the predominance of V_3 in the hypophysectomized animal and the shift in isoenzyme distribution towards V_1 with thyroxine replacement.

distribution occur as a function of age or the thyroid state of the animal (Hoh et al., 1978). Juvenile rats are rich in V_1. When these rats are rendered hypothyroid by hypophysectomy, the isoenzyme distribution shifts towards V_3, with reduction in mean myosin ATPase activity so that, in a chronically hypophysectomized animal, the ventricular myosin is principally V_3 (Fig. 2). Thyroxine replacement at physiological doses rapidly returns the distribution towards V_1 (Fig. 2), accompanied by the restoration of mean ATPase activity.

Myosin is now known to contain 2 heavy chains and 2 pairs of light chains (Lowey & Risby, 1971). Analysis of light chains of myosin containing predominantly V_1 or V_3 isoenzymes reveal no difference in molecular size or stoichiometry (Hoh et al., 1978). However, cyanogen bromide peptide maps of heavy chains of V_1 and V_3 isoenzymes differ significantly (Hoh & Yeoh, 1979). Maps obtained from myosin containing V_1, V_2 and V_3 do not appear to differ significantly from those obtained by mixing peptides from V_1 and V_3 heavy chains. It is therefore likely that the heavy chain composition of isoenzymes V_1, V_2 and V_3 are, respectively, $\alpha\alpha$, $\alpha\beta$ and $\beta\beta$, where α and β are two structurally distinct heavy chains.

Skeletal myosins were known to exist in structurally distinct fast-twitch and slow-twitch forms (Lowey & Risby, 1971) before the advent of pyrophosphate gel electrophoresis. Fast-twitch myosin contains 3 types of light chains (LC_1^f, LC_2^f & LC_3^f) while slow-twitch myosin contains 2 types of light chains (LC_1^s &

LC_2^s) which differ in molecular size from those of fast-twitch myosin. In pyrophosphate gels, chicken fast-twitch myosin has been resolved into 3 isoenzymes (FM_1, FM_2 & FM_3 in the order of decreasing mobility) which differ in light chain distribution, while slow-twitch myosin can be resolved into 2 components which do not differ in light chain distribution (Hoh et al., 1976; Hoh, 1978). Figure 3 illustrates the light chain distribution of fast-twitch myosin isoenzymes. The slow-twitch myosin components in the chick are probably isoenzymes which differ in heavy chain structure, as demonstrated for ventricular myosin isoenzymes.

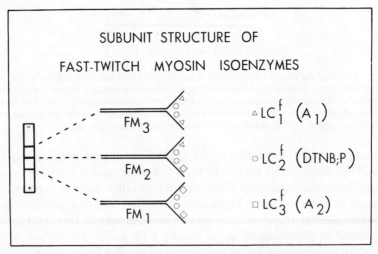

FIG. 3. Diagrammatic representation of the distribution of light chains in the three fast-twitch skeletal myosin isoenzymes. The gel on the left shows the order of electrophoretic mobility of these isoenzymes. Synonyms of the light chains are given in parentheses on the right.

The presence of three myosin isoenzymes in fast-twitch muscle has been verified in fish, amphibians, reptiles and mammals. Figure 4 shows the isoenzymes from the rabbit back. Analysis of the light chains of rabbit fast-twitch myosin isoenzymes reveals a distribution of light chains identical to that found for homologous chicken isoenzymes (Fig. 3).

Subfragment 1 prepared from fast-twitch myosin can be fractionated into LC_1^f-containing and LC_3^f-containing isoenzymes, the latter having an actin-activated ATPase activity about twice that of the former (Weeds & Taylor, 1975). It can therefore be expected that FM_1, having two identical heads containing LC_3^f, should have the highest actin-activated ATPase activity while

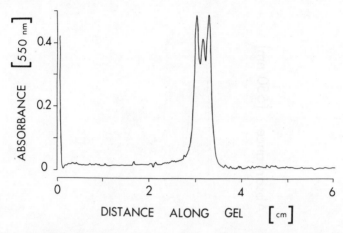

FIG. 4. Densitometer scan of fast-twitch myosin isoenzymes in rabbit longissimus dorsi muscle after staining for protein. The light chain distribution of these isoenzymes is the same as that for homologous chicken fast-twitch myosin isoenzymes and is illustrated in Fig. 3.

FM_3, with two identical heads containing LC_1^f, should have the lowest actin-activated ATPase activity. As in the case of ventricular myosin isoenzymes, the distribution of fast-twitch isoenzymes within a muscle would determine the mean enzyme activity of its myosin. The distribution of fast-twitch isoenzymes between muscles of the same animal varies widely in chicks (Hoh, 1978) and mammals.

The presence of three isoenzymes in fast-twitch muscle raises the question whether these isoenzymes occur in three populations of muscle fibers each having a single isoenzyme or in single muscle fibers containing three types of isoenzymes. This question has been resolved by pyrophosphate gel electrophoresis of myosin extracted from single rabbit psoas muscle fibers (Krieger &Hoh, 1975). Every fiber examined showed the presence of all three fast-twitch myosin isoenzymes.

The distribution of skeletal myosin isoenzymes is under neural regulation. The profiles of myosin isoenzymes in normal rat fast-twitch and slow-twitch are shown in Fig. 5. Two isoenzymes are present in the slow-twitch soleus muscle. The fast-twitch extensor digitorum longus (EDL) muscle contains the three fast-twitch myosin isoenzymes, but in addition also contains smaller proportions of the two isoenzymes found in the soleus muscle. When the nerves to these muscles are cut and reunited cross-wise, so that the nerve to the soleus is allowed to reinnervate EDL muscle and *vice versa*, the distributions of myosin isoenzymes in these muscles are altered. This is illustrated by Fig.

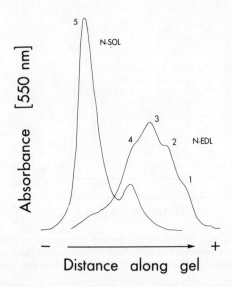

FIG. 5. Myosin isoenzyme profile in normal fast-twitch extensor digitorum longus (N-EDL) and slow-twitch soleus (N-SOL) muscles of the rat. Peaks 1, 2 & 3 correspond to FM_1, FM_2 & FM_3 in other vertebrates. Peak 4 appears to have the same light chains as FM_3. Peak 5 is slow-twitch myosin.

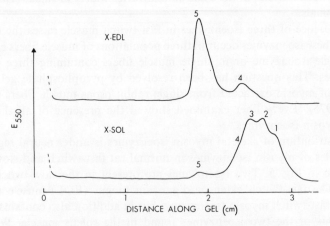

FIG. 6. Myosin isoenzyme profiles of cross-reinnervated rat skeletal muscles. The profile for cross-reinnervated extensor digitorum longus (X-EDL) resembles that of normal soleus while that of cross-reinnervated soleus (X-SOL) is like that of normal extensor digitorum longus (compare with Fig. 5).

6. The cross-reinnervated EDL now shows a distribution of isoenzymes similar to that of normal soleus, while that of cross-reinnervated soleus is similar to that of normal EDL. These muscles represent the cases in which the change brought about by nerve cross-union is most complete. Other cross-reinnervated muscles of both kinds show a wide spectrum of intermediate isoenzyme distributions.

The functional significance of isoenzymes of myosin hinges on the fact that enzymatic activity of myosin limits the rate of energy transduction in muscle. Fast muscles are intrinsically more powerful than slow ones. Bárány (1967) has shown that myosin ATPase activity is directly proportional to muscle speed. It is therefore expected that changes in myosin ATPase activity resulting from shifts in isoenzyme distribution should be accompanied by changes in contractile properties. This has indeed been shown to be the case for both cardiac and skeletal muscles. The maximum speed of shortening of rat papillary muscle is significantly reduced by hypophysectomy; this property can be restored to control level by thyroxine treatment (Korecky & Beznak, 1971). Nerve cross-union transforms the force-velocity characteristics of rat EDL in such a way as to resemble that of soleus muscle, and *vice versa* (Close, 1969).

Since changes in isoenzyme distribution are correlated with changes in force-velocity properties, the heterogeneity of isoenzymes within a muscle or single fiber may influence the shape of the force-velocity curve. Consider a muscle with two fibers, each containing a myosin isoenzyme with a different ATPase activity. Above the maximal velocity for the slower fiber, only the faster fiber with the high ATPase cross-bridges will be bearing the external load. Below the maximal velocity for the slower fiber, both types of cross-bridges will bear the load. The force-velocity curve of this two-fiber muscle would then be expected to show two distinct segments, a steeper portion due to the high ATPase cross-bridges alone and a less steep portion due to both types of cross-bridges (Fig. 7). Extrapolation of the lower segment of the curve to zero load, assuming that a hyperbolic relation holds, would give a maximal velocity below that for the faster fiber. If both types of myosin are present in the same fiber, the above arguments would be similar though allowance may have to be made for the possibility that the slower cross-bridges may contribute an internal load to the faster ones at high velocities.

A non-hyperbolic relation between force and velocity has been shown in muscles known to be heterogeneous with respect to myosin. Both rat EDL and rat extraoccular muscles show myosin heterogeneity. Close & Luff (1974) showed that the force-velocity curve for rat inferior rectus and EDL muscles deviated from Hill's equation (Hill, 1938). They found that Hill's equation could be made to fit their results reasonably well for loads greater than about

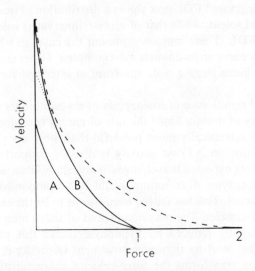

FIG. 7. Effect of myosin heterogeneity on the force-velocity curve. A & B represent curves for two muscle fibers of equal size containing myosins of different ATPase activities (hence different speeds of shortening). C represents the force-velocity curve of the two fibers arranged in parallel shortening under a common load. It is obtained by adding the loads borne by each fiber at the various velocities of shortening. Note that curve C shows two segments. Above the maximel velocity for A, curve C is coincident with B. Hyperbolic extrapolation to zero velocity (dotted line) of the portion of curve C below the maximal velocity for A would yield a value below the maximal velocity for B.

20% of isometric tetanic tension, but when this was done the observed velocities for loads less than that value were considerably higher than the predicted values. The force-velocity relation of these muscles could be adequately described by an equation having 6 constants instead of 2 as in Hill's equation. Toad muscles—and, presumably, frog single fast-twitch fibers—show 3 myosin isoenzymes. Using single frog muscle fibers, Edman (1979) found that the speed of shortening at zero load, obtained indirectly from quick release experiments, was about 10% higher than predicted by hyperbolic extrapolation from velocities recorded at finite loads.

In summary, myosin exists in skeletal and cardiac muscles as isoenzymes which differ in structure and enzymatic activity. This heterogeneity is present even in single muscle fibers. The distribution of myosin isoenzymes determines the contractile properties of the muscle. Myosin heterogeneity should therefore be taken into account in detailed considerations of muscle mechanics, energetics, cross-bridge kinetics or models of muscle contraction.

Acknowledgement

This work was supported by the National Health and Medical Research Council and the National Heart Foundation of Australia.

REFERENCES

Bárány, M. (1967). ATPase activity of myosin correlated with speed of muscle shortening. *J. gen. Physiol.* **50** (Suppl., part 2), 197–218.

Brahms, J. & Brezner, J. (1961). Interaction of myosin A with ions. *Arch. biochem. biophys.* **95**, 219–228.

Close, R. (1969). Dynamic properties of fast and slow skeletal muscles of the rat after nerve cross-union. *J. Physiol.* **204**, 331–346.

Close, R. I. & Luff, A. R. (1974). Dynamic properties of inferior rectus muscle of the rat. *J. Physiol.* **236**, 259–270.

Edman, K.A.P. (1979). The velocity of shortening at zero load: its relation to sarcomere length and degree of activation of vertebrate muscle fibers. This volume, pp. 347–356.

Harrington, W. F. & Himmelfarb, S. (1972). Effect of adenosine di- and triphosphates on the stability of synthetic myosin filaments. *Biochemistry* **11**, 2945–2952.

Hill, A. V. (1938). The heat of shortening and the dynamic constants of muscle. *Proc. R. Soc. B.* **126**, 136–195.

Hoh, J. F. Y. (1975). Neural regulation of mammalian fast and slow muscle myosins: an electrophoretic analysis. *Biochemistry, N. Y.* **14**, 742–747.

Hoh, J. F. Y. (1978). Light chain distribution of chicken skeletal muscle myosin isoenzymes. *FEBS Lett.* **90**, 297–300.

Hoh, J. F. Y., McGrath, P. A. & White, R. I. (1976). Electrophoretic analysis of multiple forms of myosin in fast-twitch and slow-twitch muscles of the chick. *Biochem. J.* **157**, 87–95.

Hoh, J. F. Y., McGrath, P. A. & Hale, P. T. (1978). Electrophoretic analysis of multiple forms of cardiac myosin: effects of hypophysectomy of thyroxine replacement. *J. molec. Cell. Cardiol.* **10**, 1053–1076.

Hoh, J. F. Y. & Yeoh, G. P. S. (1977). Electrophoretic analysis of smooth muscle myosin of the rabbit. *Proc. Aust. Physiol. Pharmacol. Soc.* **8**, 81p.

Hoh, J. F. Y. & Yeoh, G. P. S. (1979). Differences in cyanogen bromide peptide maps of the heavy chains of rat ventricular myosin isoenzymes. Submitted to the International Conference on Fibrous Proteins, Palmerston North, New Zealand, Feb. 12–16, 1979.

Korecky, B. & Beznak, M. (1971). Effect of thyroxine on growth and function of cardiac muscle. In *Cardiac Hypertrophy*, ed. Alpert, N. R., pp. 55–64, New York and London: Academic Press.

Krieger, M. & Hoh, J. F. Y. (1975). Isoenzymes of myosin in single rabbit psoas muscle fibres. *Proc. Aust. Physiol. Pharmacol. Soc.* **6**, 75.

Lowey, S. & Risby, D. (1971). Light chains from fast and slow muscle myosins. *Nature, Lond.* **234**, 81–85.

Weeds, A. G. & Taylor, R. S. (1975). Separation of subfragment-1 isoenzymes from rabbit skeletal muscle myosin. *Nature, Lond.* **257**, 54–56.

Discussion

Kushmerick: How does the time-course of the change in the isozyme pattern correlate with the time-course in the change in the mechanical properties? Do they correlate well?

Hoh: I have not systematically examined this, but I would expect so.

Drabikowski: Did you measure the ratio of light chains in the case of fast muscle? There is a possibility that you have some degradation of the LC_2 light chain, and the product of degradation moves on SDS gel exactly on the same level as the LC_3 light chain.

Hoh: I have not done it on the rabbit muscle, but I presented a paper at the Kyoto Biophysics meeting (Hoh, 1978) in which I have done it on chick muscle.

Drabikowski: You should measure the ratio of the light chains, e.g. the sum of LC_1 and LC_3 to LC_2.

Hoh: It is one to one in that particular situation.

Rall: The question I have is, it looks like in the cardiac muscle the ATPase is controlled by the heavy chain, possibly, and in skeletal muscle the ATPase is controlled by light chains. Is my impression correct, and could you comment on that?

Hoh: I think both chains might contribute, depending on the type of myosin. In the case of ventricular myosin, it would seem that the heavy chains do regulate the ATPase, whereas, although the distribution of light chains are different in these skeletal myosins, one cannot rule out the differences in heavy chains.

Tregear: On the fast skeletal muscle, how do your results fit with Susan Lowey's (Lowey & Risby, 1971)?

Hoh: Very well, I think.

REFERENCES

Hoh, J. F. Y. (1978) Chicken myosin isoenzymes: Light chain structure and developmental changes. *6th Int. Biophys. Congr. Kyoto*, p. 157.

Lowey, S. & Risby, D. (1971) Light chains from fast and slow muscle myosins. *Nature, Lond.* **234**, 81-85.

Relation between Initial and Recovery Reactions in Skeletal Muscle

M. J. KUSHMERICK

Department of Physiology, Harvard Medical School, Boston, Massachusetts, U. S. A.

ABSTRACT

We tested the hypotheses (1) that there is a fixed stoichiometric relation between the high-energy phosphate utilization during contractile activity ($\Delta \sim P$) and the extent of subsequent recovery metabolic reactions (ΔO_2 and Δ lactate) and (2) that the stoichiometry is predicted by known biochemical pathways. Our results in frog sartorius at 0° and at 20° yielded a $\Delta \sim P/\Delta O_2$ ratio of 4 rather than the value of 6.2 predicted by the stoichiometry of established biochemical pathways. In anaerobic frog sartorius at 20° we obtained a $\Delta \sim P/\Delta$lactate ratio of 1.1 rather than the value of 1.5 predicted by the stoichiometry of the Embden-Meyerhof pathway. While these results might be compatible with the reversal of a spontaneous reaction which causes the unexplained enthalpy observed during an isometric tetanus, they could equally well be explained by the occurrence of additional reactions splitting energy phosphate compounds during recovery. We therefore set out to identify additional energy-utilizing processes during contractile activity and subsequent recovery. We show that the operation of a substrate cycle involving the cyclic conversion of F6P → FDP → F6P (which uses one high energy phosphate per turn of the cycle) can occur in aerobic frog sartorius muscle. The extent of this cycling is not yet measurable. In collaboration with Barany and Gillis we have recently found a reversible phosphorylation-dephosphorylation of the P-light chain in frog skeletal muscle. The latter could account for the utilization of as much as 0.6 μmole of high energy phosphate per gram muscle during the first few sec of contraction. The occurrence of two additional types of reactions resolves at least some of the quantitative problems of initial and recovery reactions and of initial unexplained enthalpy.

In the study of muscle energetics one is interested in such questions as: What is the apparent efficiency of muscular work? Does the efficiency depend on muscle length or velocity of shortening? What are the quantitative differences in efficiency and in other energetic parameters in different muscle types, such as mammalian "fast" and "slow" muscles? How do measured rates of ATP turnover compare with predictions from models of cross-bridge interactions? These are examples of questions that were thought to be in the grasp of students of muscle energetics a decade or so ago (Kushmerick & Davies, 1969; Wilkie, 1968). However, various experimental results soon pushed such goals beyond our reach and made necessary a careful re-evaluation of the very foundations of the field. Do reactions, known to occur during contraction, account for the energy output? Measurements of creatine phosphate breakdown (or of ATP splitting in suitably poisoned muscles) did not fully explain the measured enthalpy change in a single contraction (for review, see Homsher & Kean, 1978; Curtin & Woledge, 1978). Thus there was doubt about whether we could measure all of the reactions that occurred and whether we knew the stoichiometries of the relevant reactions and side reactions.

My interpretation of this state of affairs was that a quantitative study of the overall energetic and metabolic inter-relationships was necessary. The goal was to study initial chemical reactions of ATP turnover and PCr splitting in relation to the terminal reactions of recovery, lactate production and oxygen consumption (Kushmerick, 1977). Over the whole cycle, no net change in the internal composition of a muscle is expected, except for the utilization of some endogenous substrate if not provided exogenously. An operating principle is that muscle contraction produces a rapid change in the chemical potential of the cells, for example, by transforming the chemical potential energy of creatine phosphate into work and heat. These initial reactions occur in seconds and are followed by much slower relaxation processes which restore the initial composition of resting muscle by the operation of glycolytic and oxidative "recovery" metabolism. Two uses of this information are, in principle, possible. First, the experiments would provide the basis for a biochemical or metabolic balance sheet. Secondly, the time course of the recovery metabolic events contains information concerning the kinetics and control of those reactions and is thus important for integrating cellular energy metabolism. In this paper I will be concerned only with the first aspect of our work.

The basic experimental designs are well known. Paired muscles from one animal are used since there is more variation in muscle composition among different animals than there is in paired muscles from the same animal, at least in frogs. One muscle is stimulated and the other is not. At the appropriate time both muscles are rapidly frozen to stop mechanical and chemical events. Stable extracts of the muscle are prepared and analyzed. The extent of initial reactions is deduced from changes in the composition between experimental and controlled muscle. Methods and procedures for measuring these initial chemical reactions have been described in the literature (Gilbert, Kretzschmar, Wilkie & Woledge, 1971; Kushmerick & Paul, 1976a, b). We made major efforts in improving the techniques for measuring recovery metabolism. We used chambers closed to the atmosphere and of sufficiently small volume to measure accurately and precisely the amount of oxygen consumed during recovery from a single isometric tetanus. The extent of anaerobic recovery was measured by collecting the total lactate produced by the muscle into an external solution.

Experimental results concerning aerobic recovery at 0° and anaerobic recovery at 20° from isometric tetani in frog sartorius have been published (Kushmerick and Paul, 1976a, b; DeFuria and Kushmerick, 1977). Measurements were also made of recovery oxygen consumption and of phosphorylcreatine breakdown using unpoisoned oxygenated frog muscles at 20°. Table 1 shows that such preparations were stable and reproducible. The major finding (Table 2) is that the observed extent of recovery oxygen consumption (Line D) was always greater than the predicted recovery oxygen consumption (Line E). This prediction was made by dividing the observed breakdown of high energy phosphates (Line C) by the stoichiometric coupling factor ($\Delta \sim P/\Delta O_2 = 6.2$) derived from oxidative metabolism of glycogen. The expected stoichiometry of high energy phosphate synthesis per unit recovery oxygen consumption ($\Delta \sim P/\Delta O_2$) is 6.2, whereas the observed quantities (Line F) were always

TABLE 1. The recovery oxygen consumption during a complete contraction-recovery cycle of an isolated frog sartorius subjected to isometric tetani at 20°C

Tetanus duration (sec) (in serial order)	Average tension $N \cdot cm / g\ dry\ wt$	Recovery oxygen consumption, ΔO_2 (μmoles)	Baseline O_2 consumption (nanomoles/min)	
			Pre-tetanic	Post-tetanic
→3	11.9	0.23	4.0	4.4
→5	11.5	0.31	3.8	4.1
1	12.2	0.12	4.9	4.2
→5	11.3	0.29	5.0	5.5
1	11.9	0.09	5.1	4.8
→5	10.6	0.36	4.8	5.0
3	10.6	0.22	5.0	5.1
→1	10.5	0.10	6.0	5.5
3	9.9	0.20	5.5	5.9

The dry weight of the muscle was 23 mg. The P_{O_2} of the Ringer was periodically brought to $P_{O_2} = 300$ torr at the time designated by the arrows. The P_{O_2} of the Ringer never dropped below 100 torr. The duration of this experiment was approximately 7 hr.

TABLE 2. Initial and recovery reactions: isometric contraction (aerobic frog sartorius, 20°)

Tetanus duration (sec)	1	3	5
A. ΔCr	15.9 ± 0.90	35.3 ± 1.00	50.4 ± 1.50
	$(N = 7)$	$(N = 9)$	$(N = 6)$
B. ΔCr $\sim$ P	-17.0 ± 1.80	-36.5 ± 1.50	-48.9 ± 4.90
	$(N = 7)$	$(N = 9)$	$(N = 6$
C. $\Delta \sim$ P	16.5 ± 1.40	36.4 ± 1.20	49.7 ± 3.30
D. Recovery O_2 Consumption	3.5 ± 0.10 $(N = 13)$	8.3 ± 0.22 $(N = 23)$	11.8 ± 0.31 $(N = 11)$
E. Predicted O_2 Consumption	2.7	5.9	8.0
F. Observed $\dfrac{\Delta \sim P}{\Delta O_2}$	4.7	4.4	4.2

Units: μmole/g dry weight $\pm$ S.E.M.

lower. Thus the results at 20° and 0° were the same. These aerobic experiments were many months apart from the previously published anaerobic experiments (DeFuria & Kushmerick, 1977) and were made with different batches of frogs. There is remarkably good agreement with the observed initial chemical events (Table 3). This last finding suggests that the extent of resynthesis of high energy phosphate compounds by oxidative recovery reactions during the tetanus itself is negligible. This point was extensively checked experimentally (Kushmerick & Paul, 1976a; DeFuria & Kushmerick, unpublished experiments). The same conclusion is obtained by calculating the extent of oxidative metabolism during a tetanus. This quantity is obtained by multiplying the maximal rate of recovery oxygen consumption (ΔO_2/recovery oxygen time constant) by the tetanus duration. For a typical 5-sec isometric tetanus, recovery oxygen consumption was 11.8 μmole/g dry weight. The recovery time constant is approximately 5 min at 20°. During the 5-sec tetanus

TABLE 3. Initial and recovery reactions: isometric contractions (frog sartorius, 20°)

Tetanus duration (sec)	0.5	1	2	3
ANAEROBIC RESULTS				
ATP resynthesized	11.7 ± 0.9	17.2 ± 1.2	39.0 ± 2.4	—
Recovery lactate production	10.7 ± 0.4	14.2 ± 0.6	31.8 ± 0.4	—
Recovery lactate (predicted)	7.8	11.5	26.0	—
AEROBIC RESULTS				
ATP resynthesized	—	16.5 ± 1.4	36.4 ± 1.2	49.7 ± 3.3
Recovery oxygen consumption	—	3.5 ± 0.1	8.3 ± 0.2	11.8 ± 0.3
Recovery oxygen (predicted)	—	2.8	6.1	8.2

Units: μmole/g dry weight $\pm$ S.E.M.

not more than 0.2 μmoles/g dry weight of oxygen consumption could have occurred ($< 2\%$ of the total). Thus much less than 2% of observed creatine phosphate breakdown could have been synthesized during the tetanus. Finally, net lactate production indicated that less than 3% of the aerobic resynthesis of high energy phosphate could be attributed to glycolysis.

Thus we arrive at an apparently general and consistent set of data which shows that there is a greater extent of recovery metabolism (glycolytic or oxidative) than can be explained by the measured extent of high energy phosphate splitting during contraction of frog muscle.

Our analysis of this data and of the problem of the relation between initial chemical breakdown and extent of recovery metabolism is given by the following equations. For aerobic muscles:

$$(\Delta \sim P_{contraction} + \Delta \sim P_{recovery}) = K(\Delta O_2). \tag{1}$$

For anaerobic muscles:

$$(\Delta \sim P_{contraction} + \Delta \sim P_{recovery}) = L(\Delta \text{ lactate}). \tag{2}$$

Based on control experiments, we are quite certain that the measured recovery oxygen consumption (ΔO_2) and recovery lactate production (Δlactate) are not subject to major systematic errors and that there is a definable recovery period at the end of which the initial contents of Pi, PCr, Cr and ATP are restored (Kushmerick & Paul, 1976b; DeFuria & Kushmerick, 1977; Kushmerick, 1977).

It is possible that the proportionality constants, K and L, are not $K = 6.2$ and $L = 1.5$, as expected from the stoichiometry of metabolic pathways well established *in vitro*. However, uncoupling of high energy phosphate synthesis to lactate production was not observed in a reconstituted Embden-Meyerhof pathway *in vitro* (Scopes, 1973). He found $\Delta \sim P/\Delta$lactate $= 1.5$. Although such uncoupling is possible by spontaneous hydrolysis of the enzyme—substrate complex of glyceraldehyde-phosphate dehydrogenase (especially when certain sulfhydryl groups are oxidized), it does not happen in normal muscles (Patnode, Bartle, Hill, LeQuire & Park, 1976). Mitochondria isolated from frog leg muscle showed respiratory control ratios of 4–6 and the expected $\sim P/O_2$ ratios of 6 at 0° (Kushmerick, 1977) and at 20° (DeFuria, unpublished observations) using standard assay techniques. We conclude that true uncoupling of high energy phosphate synthesis in glycolysis and in oxidative phosphorylation is unlikely. Nevertheless direct evidence for this statement in intact muscle cells is needed.

The possibility that the term ($\Delta \sim P_{contraction}$) does not measure all the high-energy phosphates broken down during contraction is a specific mechanism which could minimize or solve the problem of the initial unexplained enthalpy and the observed $\Delta \sim P/\Delta O_2$ and $\Delta \sim P/\Delta$lactate ratios. These matters have been extensively discussed and reviewed (Curtin & Woledge, 1978; Homsher & Kean, 1978; Kushmerick, 1977). Our results are thus compatible with the hypothesis that recovery metabolic reactions reverse not only the observed breakdown of creatine phosphate but also the unknown reaction(s) associated with contraction and giving rise to the unexplained enthalpy. We should not accept this hypothesis hastily because the observed quantities, $\Delta \sim P/\Delta O_2$ and $\Delta \sim P/\Delta$lactate, are independent of the extent of initial PCr breakdown. This means that the extent of the reaction(s) giving rise to the excess unexplained enthalpy production would have to be proportional to the PCr splitting at 0° and at 20°, in anaerobic and aerobic muscles. Studies of the time course and magnitude of the unexplained enthalpy will be very informative. Because a discrepancy between the work done by contracting muscle and the extent of initial high energy phosphate splitting has never been observed (Curtin, Gilbert, Kretzschmar & Wilkie, 1974; Gilbert & Kushmerick, 1970; Kushmerick & Davies, 1969), it is possible that the reaction(s) giving rise to the unexplained enthalpy might be only indirectly related to the contractile mechanism.

Another approach to this problem is to ask whether there is a significant amount of high-energy phosphate utilization during the recovery period itself. This hypothesis is a plausible explanation for our studies of initial and recovery metabolic processes. Specifically, we postulate that both $\sim$ P terms in Eqs. (1) and (2) are quantitatively significant. We recognize that past attempts to detect high energy phosphate utilization after relaxation gave negative results. A possible mechanism for the proposed high-energy phosphate utilization during recovery would be related to the structure and/or regulation of the metabolic pathways themselves and so could be proportional to the extent of those reactions. An interesting mechanism of this type is substrate cycling (Katz & Rognstad, 1976; Newsholme & Crabtree, 1976; Scrutton & Utter, 1968). Except as a mechanism of thermogenesis in bumblebee flight muscle (Clark, Bloxham, Holland & Lardy, 1973), substrate cycles have received little attention in the field of muscle energetics. We studied the possibility that substrate cycles can occur in frog sartorius muscle. Our argument is that if a substrate cycle can occur to a quantitatively significant extent, then the term $\Delta \sim P_{recovery}$ in Eqs. (1) and (2) must be added to the overall description of the energy cost of muscular activity. If this hypothesis were correct, then measurements of the extent of recovery processes would indeed provide a measure of the total energy cost of contraction, but not necessarily the initial energy cost which occurs during mechanical activity. The initial chemical reactions measured by rapid freezing techniques do not include a metabolically related energy cost and could indeed measure the energy cost directly associated with the contractile processes.

Specifically, we tested the occurrence of the F6P substrate cycle in frog sartorius at 20°. Phosphofructokinase (PFK) and fructose diphosphatase (FDPase) are obligate enzymes of this cycle:

$$\begin{array}{r} \text{F6P} + \text{ATP} \longrightarrow \text{FDP} + \text{ADP} \\ \text{FDP} \longrightarrow \text{F6P} + \text{Pi} \\ \hline \text{net:} \quad \text{ATP} \longrightarrow \text{ADP} + \text{Pi} \end{array}.$$

There is no question of the presence and activity of PFK in frog muscle; glycolysis is well known and PFK has been isolated from this source (Trivedi & Danforth, 1966). Homogenates of frog leg muscle contain FDPase activity (Krebs & Woodford, 1966), and we purified a specific FDPase from frog muscle (DeFuria & Kushmerick, unpublished observations). However, important questions concerning the degree of activity of FDPase *in vivo* and the possible significance of its activity remain. Thus our experiments made a detailed study of the activity of FDPase in frog sartorius as well as tests of the simultaneous activity of FDPase and PFK. The latter condition, by definition, is the occurrence of the F6P substrate cycle.

The activity of FDPase *in vivo* was studied by measuring the rate of glycogen synthesis from lactate supplied in the Ringer. This experimental approach gives an estimate of the *minimal* chemical flux through FDPase, since this enzyme catalyzes one of the required reactions. We designed these experiments to confirm and extend the original observations of Meyerhof and the results of Bendall & Taylor (1970) (Fig. 1 and Table 4). Clearly there was a net synthesis

TABLE 4. Glycogen synthesis (frog sartorius paired, 20°, 16 hr)

Conditions	Glycogen (μmoles C_6/mg protein)
No substrate	−0.05
30 mM Lactate, Na	0.12[a]
30 mM Lactate + 10 mM Glucose	0.12[a]
30 mM Lactate, HCO_3^- Ringers	0.17
30 mM L(+) Lactate, Li	0.12[b]
30 mM D(−) Lactate, Li	−0.03[b]

[a,b] Experiments on one day, same batch

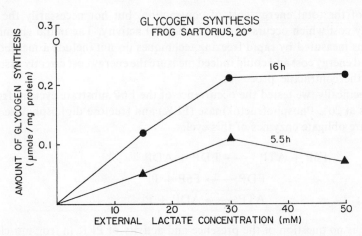

FIG. 1. Paired frog sartorius muscles were used at 20°. Control muscles were not incubated. Their glycogen content is given by the open circle. Glycogen assays were made as described (DeFuria and Kushmerick, 1977). Experimental muscles were incubated for either 5.5 hr or 16 hr in phosphate-Ringers gassed with 100% O_2 and containing the concentration of lactate indicated, before measurement of their glycogen content. The closed symbols show the increase in glycogen content in experimental muscle minus its paired control.

of glycogen from lactate. Control levels of glycogen were 0.2 μmole/mg protein. Therefore the glycogen content more than doubled. Synthesis increased with time and with lactate concentration. Other experiments indicate the half maximal rate of synthesis was obtained at a concentration of 10–15 mM lactate. Glycogen synthesis is specific for the natural isomer ($L(+)$lactate) and not $D(-)$lactate, a finding which excludes nonspecific chemical artifacts (Table 4). Gluconeogenesis from lactate was found in bicarbonate ringer as well as in phosphate ringer. This observation suggests that CO_2-fixing reactions and inappropriate cytoplasmic pH are unlikely. The maximal rate of gluconeogenesis (0.1–0.3 nmoles/min-mg protein) is about equal to the steady-state flux through the Embden-Meyerhof glycolytic pathway in resting and anaerobic muscles and is somewhat greater than the glycolytic flux in aerobic and resting muscle (Table 5). Plasma lactate concentrations of 10–30 mM have been measured in frogs and are elevated for at least 30 min after a period of hopping (J. Leff & C.R. Taylor, unpublished experiments). We conclude that FDPase can have physiologically significant activity in frog sartorius muscles. The conditions we studied are compatible with the metabolic status during recovery.

The question of the simultaneous activity of PFK and FDPase is more difficult to answer. An experiment suggesting simultaneous activity is summarized

TABLE 5. Carbon fluxes (frog sartorius, 20°)

	Nanomole C_6
	Min·mg muscle protein
Aerobic	
Glycogen synthesis (lactate → C_6 in glycogen)	0.10–0.30
Glycogen breakdown ($C_6 + 6O_2 \rightarrow 6CO_2 + 6H_2O$)	0.02–0.04
Anaerobic glycolysis	
$C_6 \rightarrow$ Lactate: Resting	0.10–0.20
contracting	$\geqslant 6$

TABLE 6. Simultaneous glycogen synthesis from lactate, and CO_2 production from glucose in isolated frog sartorius at 20°C metabolizing both lactate and glucose

	Net glycogen synthesis	^{14}C Incorporation into glycogen	$^{14}CO_2$ Production
	nanomoles glucose units/min/mg protein		
5 mM [U-^{14}C]Glucose and	0.27 ± 0.03	0.007 ± 0.001	0.010 ± 0.002
35 mM Lactate	$N = 8$	$N = 8$	$N = 8$
5 mM Glucose and	0.31 ± 0.02	0.290 ± 0.030	0.260 ± 0.060
35 mM [U-^{14}C]Lactate	$N = 5$	$N = 5$	$N = 5$

Isotopic data are reported in glucose units using the specific activity of the appropriate substrate. Net glycogen synthesis was measured chemically. Incubations were for 18 hr. The data are normalized to total muscle protein and incubation time and are given in the format, mean ± S.E.M. Each N represents an experiment with a single muscle. Rates are calculated by dividing the total changes by the incubation time.

in Table 6. Oxygenated muscles were incubated in Ringer containing both lactate and glucose as substrates. Either lactate or glucose was labelled uniformly with ^{14}C. In both types of experiments there was a similar net synthesis of glycogen measured chemically. The predominant source of the carbons for this synthesis was lactate, as the middle column of data shows. Simultaneously there was a clearly measurable oxidation of glucose to CO_2, 0.01 nmoles glucose/mg protein-min. This experiment shows simultaneous synthesis of glycogen (FDPase activity) and oxidation of glucose (PFK activity). This experiment is not entirely satisfactory because the rate of oxidation of glucose was low, an interesting finding we have repeatedly made.

A better test of the simultaneous activity of PFK and FDPase is provided by the method of Clark et al. (1973). The results of our experiments following this approach, to be published in detail elsewhere (DeFuria & Kushmerick, 1979), require (1) incubation of sartorius muscles in U-^{14}C and 5-^{3}H-glucose, and (2) measurement of the ^{3}H and ^{14}C radioactivity in F6P and G6P extracted from muscles. The principle of the method is based on the equilibration of fructose-

1,6-diphosphate, dihydroxyacetone phosphate and glyceraldehyde-3-phosphate, catalyzed by aldolase and triose-isomerase. The isomerase catalyzes a rapid exchange between the trioses, a reaction which leads to equilibration of the tritium on position 5 of the original hexose with intracellular water (cytoplasmic H_2O becomes labeled). Thus any F6P generated from fructose-1,6-diphosphate, which is depleted of tritium by the equilibration described, would lead to a decrease in the $^3H/^{14}C$ ratio of the F6P-G6P pool extracted from muscle. The experimental results are given in Table 7. A group of muscles was incubated in a flask containing 5 mM glucose (5-^3H- and U- ^{14}C-glucose with a $^3H/^{14}C$ ratio equal to 10) and 100 mU insulin/ml in oxygenated Ringer. At the times indicated sartorii were removed and extracted. Fructose-6-phosphate and glucose-6-phosphate were separated electrophoretically from other ^3H- and ^{14}C-containing compounds in the extract, combined and counted. The upper part of Table 7 shows that after an hour of incubation a steady state labeling of F6P and G6P was approached. Then 35 mM lactate (*not* labeled

TABLE 7. Test of F6P substrate cycling

Incubation conditions	^3H (dpm)	^{14}C (dpm)	$^3H/^{14}C$
5 mM glucose and 100 mU insulin/ml; 5-^3H, U-^{14}C glucose			
1 min	1266	129	9.8
30 min	10744	1080	9.9
90 min	13974	1389	10.1
Addition of 35 mM lactate, unlabelled			
5 min	10561	1157	9.1
20 min	5385	770	7.0
60 min	2487	416	5.9
120 min	1154	219	5.3

TABLE 8. Phosphorylation of 18,000 dalton light chain of myosin

		Extent of phosphorylation	
Experimental conditions	N	Multiple of control	Mole phosphate incorporation mole of light chain
0.2 sec isometric tetanus	3	1.35	0.14
0.5 sec isometric tetanus	5	1.56	0.22
5.0 sec isometric tetanus	1	1.60	0.24
5.0 sec isometric tetanus plus 3 sec relaxation	3	1.25	0.10

with ^{3}H or ^{14}C) was added. Additional muscles were sequentially removed and similarly analyzed. The unlabelled lactate clearly diluted the intracellular pool of the hexose monophosphates. The most important observation is that the ratio, ^{3}H/^{14}C, clearly decreased (Table 7). This finding indicates that F6P substrate cycling was induced by the addition of lactate. The ^{3}H/^{14}C ratio of the glucose and the bathing solution was sampled and was found to remain constant at 10 throughout the experiment. Present data do not allow an unambiguous calculation of the rate of cycling because all the requisite intracellular specific activities were not measured (see Katz & Rognstad, 1976, for a critical review). The first part of this data shows that we could not detect F6P substrate cycling in resting muscles. The conclusion drawn from the results presented in Tables 6 and 7 is that the F6P substrate cycle can occur in frog muscle. These results therefore identify F6P substrate cycling as a plausible and specific mechanism contributing to the term $\Delta P_{recovery}$ discussed earlier.

The physiological role of FDPase activity in amphibian muscles is not fully defined. Since mammalian muscles do not have a significant glyconeogenic capability from 3-carbon sources, there may be evolutionary reasons why that function was relegated predominantly to the liver and kidney in mammals. The lack of evidence for F6P substrate cycling in resting muscles makes unlikely a significant regulatory role for it in the control of glycolysis as postulated by Newsholme. The F6P substrate cycle may thus represent an "inefficiency" in metabolism as a consequence of the presence of oppositely directed metabolic pathways in a cell.

The final experiments I will discuss concern an initial chemical reaction not heretofore described during a single isometric tetanus of frog sartorius muscle. Frearson & Perry (1975) described the phosphorylation of a specific light chain of skeletal muscle myosin, the P-light chain which has a molecular mass of about 18,000 daltons. Phosphorylation of the analogous light chains of smooth muscle myosin and of myosin of non-motile cells may be required for actomyosin interaction and ATPase activity (see Adelstein, 1978, for a review). The role of phosphorylation of skeletal and cardiac myosin is not established. Barany & Barany (1977) showed that phosphorylation of P-light chain could be detected in caffeine contractures or prolonged tetanic of frog sartorii, and that the extent of phosphorylation of tropomyosin remained unchanged. We collaborated in testing the hypothesis that phosphorylation of P-light chain is an early contractile event in oxygenated frog sartorius and semitendinosus muscles.

Frogs were injected with ^{32}P-inorganic phosphate, and after three days the intracellular specific activity of Pi, PCr and ATP epuilibrated. Paired muscles were used. Experimental muscles were stimulated tetanically under isometric conditions for 0.2, 0.5 and 5 sec. The proteins were extracted, washed and separated on urea-SDS polyacrylamide gels. Incorporation of ^{32}P into the P-light chain was measured. The constant labelling of tropomyosin provided an internal reference. The results obtained for contractions at 0° are summarized in Table 8. Control experiments indicated that an unexplained extent of phosphorylation of the P-light chain occurred in unstimulated muscles which were frozen on the rapid-freezing device; this amounted to 0.3 moles of phosphate per mole light chain. Increased phosphorylation amounting to 0.2 mole P per mole light chain was detected during a single isometric tetanus at 0°. The reversibility of this phosphorylation was shown by the finding that the extent of phosphorylation decreased with relaxation. Similar results were obtained at higher temperatures. The maximal total incorporation of phosphate during tetanic stimulation thus amounted to about 0.5 mole of P per mole light chain. An increase of 0.2 mole P per mole light chain corresponds to a net splitting of PCr of about 0.06–0.08 μmole/g which is just at the limit of resolution of current techniques for detecting initial chemical changes. Such a change would be manifested by a decrease in PCr without a corresponding increase in Pi.

Muscle activation with increased cytoplasmic Ca^{++} levels but not actomyosin interactions are required for P-light chain phosphorylation, since experiments with highly stretched semitendinosus muscles which gave no or insignificant active isometric force gave results similar to the data in Table 8. Complete phosphorylation of P-light chains (which we have not yet observed) requires a turnover of high energy phosphates of about 0.6 μmole/g muscle [(2 moles P-light chain/mole myosin) × (0.3 μmole myosin/g muscle)]. The enthalpy of this process has not been measured. It may be large if phosphorylation induces a configurational or a hydration change in the myosin molecule. Thus P-light chain phosphorylation is a candidate for the unexplained enthalpy early in a tetanus. The physiological role of light chain phosphorylation remains speculative in striated muscle.

In summary, we have made a consistent set of observations in frog sartorius muscles, aerobic and anaerobic at 0° and at 20°, that the extent of glycolysis or oxidative metabolism is greater than predicted from the measured initial PCr splitting. The coupling coefficients in whole cells defined in Eqs. (1) and (2), $K = \Delta \sim P/\Delta O_2$ and $L = \Delta \sim P/\Delta$lactate, are likely to be the stoichiometries established by *in vitro* biochemical techniques: $K = 6.2$ and $L = 1.5$. Direct experimental tests *in vivo* are needed, however. We have evidence for the existence of a new class of energy-dissipating processes which is associated with the metabolic recovery processes themselves. Specifically, we have shown that the operation of the F6P substrate cycle can occur in frog sartorius muscles. The possibility of other substrate cycles (see Katz & Rognstad, 1976) has not been tested. We have not been able to measure unambiguously the rate of extent of F6P substrate cycling. Thus we cannot assess whether the magnitude is sufficient to account for the observed $\Delta \sim P/\Delta O_2$ and $\Delta O_2/\Delta$ lactate ratios. Phosphorylation-dephosphorylation of the P-light chain of myosin has been identified during a single brief tetanus and thus should be included in a list of reactions in an energy balance sheet. There are insufficient data to assess what fraction of the unexplained enthalpy is due to this process. Finally, an obvious conclusion is that the identification of the two specific reactions as described in this paper does not exclude the existence of others. Thus it is quite a realistic possibility that further work will identify other initial and recovery reactions of energetic importance.

Acknowledgements

The work described was supported by a grant from N.I.H. (AM 14485) and from the Muscular Dystrophy Associations of America. M.J.K. is the recipient of a Research Career Development Award (IKO4 AM00178). The anaerobic recovery experiments and the studies on substrate cycling were made by R.R. DeFuria and comprise part of his Ph.D. thesis.

REFERENCES

Adelstein, R. S. (1978). Myosin phosphorylation, cell motility and smooth muscle contraction. *Trends in Biochem. Sci.* **3**, 27–30.

Bárány K. & Bárány, M. (1977). Phosphorylation of the 18,000 dalton light chain of myosin during a single tetanus of frog muscle. *J. Biol. Chem.* **252**, 4752–4754.

Bendall, J. R. & Taylor, A. A. (1970). The Meyerhof quotient and the synthesis of glycogen from lactate in frog and rabbit muscle. *Biochem. J.* **118**, 887–893.

Clark, M. G., Bloxham, D. P., Holland, P. C. & Lardy, H. A. (1973). Estimation of the fructose diphosphatase-phosphofructokinase substrate cycle in the flight muscle of *Bombus affinis*. *Biochem. J.* **134**, 589–597.

Curtin, N. A., Gilbert, C., Kretzschmar, K. M. & Wilkie, D. R. (1974). The effect of the performance of work on total energy output and metabolism during muscular contraction. *J. Physiol.* **238**, 455–472.

Curtin, N. A. & Woledge, R. C. (1978). Energy changes and muscular contraction. *Physiol. Revs.* **58**, 690–761.

DeFuria, R. R. & Kushmerick, M. J. (1977). ATP utilization associated with recovery metabolism in anaerobic frog muscle. *Am. J. Physiol.* **232**, C30–C36.

Frearson, N. & Perry, S. V. (1975). Phosphorylation of the light-chain component of myosin from cardiac and red skeletal muscles. *Biochem. J.* **151**, 99–107.

Gilbert, C., Kretzschmar, K. M., Wilkie, D. R. & Woledge, R. C. (1971). Chemical change and energy output during muscular contraction. *J. Physiol.* **218**, 163–193.

Gilbert, C. & Kushmerick, M. J. (1970). Energy balance during working contractions of frog muscle. *J. Physiol.* **210**, 146–147P.

Homsher, E. & Kean, C. J. (1978). Skeletal muscle energetics and metabolism. *Ann. Rev. Physiol.* **40**, 93–131.
Katz, J. & Rognstad, R. (1976). Futile cycles in the metabolism of glucose. *Curr. Topics in Cell Reg.* **10**, 237–289.
Krebs, H. A. & Woodford, M. (1965). Fructose 1,6 diphosphatase in striated muscle. *Biochem. J.* **94**, 436–445.
Kushmerick, M. J. (1977). Energetics of muscle contraction: a biochemical approach. A review requested for *Current Topics in Bioenergetics* **6**, 1–37.
Kushmerick, M. J. & Davies, R. E. (1969). The chemical energetics of muscle contraction. 2. The chemistry, efficiency and power of maximally working sartorius muscle. *Proc. R. Soc.* **B174**, 315–353.
Kushmerick, M. J. & Paul, R. J. (1976a). Aerobic recovery metabolism following a single isometric tetanus in frog sartorius muscle at 0°C. *J. Physiol.* **254**, 693–709.
Kushmerick, M. J. & Paul, R. J. (1976b). Relation of initial chemical reactions and oxidative recovery metabolism for isometric contraction of frog sartorius at 0°. *J. Physiol.* **254**, 711–727.
Newsholme, E. A. & Crabtree, B. (1976). Substrate cycles in metabolic regulation and in heat generation. *Biochem. Soc. Symp.* **41**, 61–109.
Patnode, R., Bartle, E., Hill, E. T., LeQuire, V. & Park, J. H. (1976). Enzymological studies on hereditary avian muscular dystrophy. *J. biol. Chem.* **251**, 4468–4475.
Scopes, R. K. (1973). Studies with a reconstituted glycolytic system. *Biochem. J.* **134**, 197–208.
Scrutton, M. C. & Utter, M. F. (1968). The regulation of glycolysis and gluconeogenesis in animal tissues. *Ann. Rev. Biochem.* **37**, 249–302.
Trivedi, B. and Danforth, W. H. (1966). Effect of pH on the kinetics of frog muscle phosphofructokinase. *J. biol. Chem.* **241**, 4110–4114.
Wilkie, D. R. (1968). Heat, work and phosphorylcreatine breakdown in muscle. *J. Physiol.* **195**, 157–183.

Discussion

Pollack: Do you think there is any possibility that the rate of phosphorylation has anything to do with the rate of rise of tension? For example, at the long sarcomere lengths could the slow creep of tension be related to slow phosphorylation?

Kushmerick: I can only say that speculation is possible. If I may speculate maybe even more wildly, one can say phosphorylation could conceivably alter the rate of ATP splitting and therefore change the apparent... let's call it "efficacy." That is absolute, wild speculation, but I will say it anyway.

Wilkie: Actually, the business about muscles resynthesizing glycogen; it is one of these amazing stories about fashion in physiology, because after all, Myerhof showed in 1920 that you could resynthesize glycogen over that sort of time scale, 25 hr.

Kushmerick: There is no question about that, and, in fact, the reason we got into this is that there is a long-standing argument, I understand, between Hans Krebs and Myerhof about this, and of course Krebs is still arguing but Myerhof isn't available anymore. Basically, he does not believe those results at all, and we have gone to some lengths to do some experiments, many along the lines that he suggested, to prove that in fact it can occur.

Wilkie: Well, it was shown by Bendall and Taylor actually in 1970 (Bendall & Taylor, 1970).

Kushmerick: He does not believe the significance of Bendall either for different reasons. The first reason that Krebs does not find the apparent synthesis of glycogen significant is the argument that the concentration of lactate that is necessary is unphysiologically high, and the second reason is that the rates are much too slow to be significant. I already said something about the rates, and the only other thing I can say is that there are some people at Harvard University who do a lot of exercising of a great number of animals, including frogs, and they have measured frog plasma lactates on the order of 30 or 40 mM after a bout of hopping on a treadmill, so indeed they can get that high.

Rall: With regard to phosphorylation of myofibrillar proteins, you saw a change in the DTNB, but also we know that there can be a change in the level of phosphorylation of other myofibrillar proteins. Was any of that apparent with stimulation?

Kushmerick: I went through those pretty quickly, but I hope you were able to see that, in fact, there was phosphorylation, but not a contraction-dependent phosphorylation, so all I can say is that, in skeletal muscle for sure, the role of phosphorylation is unclear. Completely. And the only one that seemed to be dependent in a reversible fashion with contraction was the DTNB light chain phosphorylation.

Woledge: I have a comment on one of the points that you made. You have shown that there is a very low stoichiometry of rephosphorylations by lactate in glycolysis. The question then is, what happens to the energy which is undoubtedly produced when the lactate is produced? It does not seem to go into rephosphorylating ADP, not as much as we thought. Does it become heat, or does something else happen to it? Now, if it became heat, I just calculated on the back of an envelope that the heat you would see in recovery from an anaerobic contraction would be rather bigger than the heat that you see during the initial processes. Now the classical observations on this point show that in fact the heat during recovery is only about a quarter of what is produced in the initial processes. This seems to suggest that the energy is not being liberated as heat, for instance by the substrate cycling that you are talking about, but is actually being used to do something else. There must be some heat-absorbing process going on in this recovery period. Would you like to comment on that?

Kushmerick: Well, I would first of all say that to the extent that these substrate cycles operate, it would simply produce heat. The second thing is that we do not have an estimate, except very approximate, of the amount of the substrate cycling. So I cannot really say. I would have thought that the time during which this process can occur is very large, so your quoting of the amount—a quarter of the initial heat—is that the total heat that is produced?

Woledge: Yes. That is the heat that has produced during anaerobic recovery in a period of 20 min or so.

Kushmerick: O.K. Well, as far as we know, the anaerobic recovery that we have measured is much longer, and the full time course and amount of anaerobic recovery heat are not fully known. Nonetheless, at face value it appears my results would imply more heat than appears to have been observed.

Woledge: Well, the question of time-course complicates it because I had not realized how slow your recoveries were. So I think we need to make a direct comparison of that point.

REFERENCE

Bendall, J. R. & Taylor, A. A. (1970) The Meyerhof quotient and the synthesis of glycogen from lactate in frog and rabbit muscle. *Biochem. J.* **118**, 887–893.

Studies of the Biochemistry of Contracting Muscle Using ^{31}P Nuclear Magnetic Resonance (^{31}P NMR)

M. Joan DAWSON,* D. R. WILKIE* and D. GADIAN**

* Department of Physiology, University College London, London, England
** Department of Biochemistry, University of Oxford, Oxford, England

ABSTRACT

We have recently devised methods for applying ^{31}P nuclear magnetic resonance (^{31}P NMR) to the study of living muscle; it is now possible to maintain muscles in a physiological state within the spectrometer, to stimulate them electrically and to record their force development at the same time that ^{31}P NMR spectra are being obtained.

We are now using ^{31}P NMR in a study of fatiguing anaerobic frog muscle in which we relate isometric force development and the rate constant for the exponential phase of relaxation ($1/\tau$) to each of the following: 1) the concentrations of PCr, ATP, free ADP, Pi, H$^+$ and Cr; 2) the free-energy change for ATP hydrolysis; 3) the rate of glycolysis; and 4) the rate of phosphorus utilization. Of these only the concentrations of H$^+$ and of phosphorus compounds (with the exception of free ADP) can be determined directly by ^{31}P NMR. The other quantities are estimated on the basis of current knowledge of muscle biochemistry, together with the ^{31}P NMR results.

We find that both force development and $1/\tau$ are closely correlated with metabolite levels rather than with any *independent* changes in excitatory conduction or excitation-contraction coupling.

INTRODUCTION

We have already described (Dawson, Gadian & Wilkie, 1977a, b; 1978) how the non-injurious radio-frequency technique of ^{31}P NMR can be applied to isolated muscles during rest, contraction and recovery. However, it seems that the physical principles underlying the method are not universally understood and that it would be helpful to explain this aspect briefly so that the potentialities and the limitations of NMR may be more readily appreciated.

ELEMENTARY NMR

The use of NMR to measure tissue metabolites imposes constraints which are quite different from those encountered in conventional physiological or NMR experiments. In order to appreciate these constraints it is necessary to know something about the physical principles of NMR. All atomic nuclei have positive charge and mass but only some of them, including ^{1}H, ^{2}H and ^{31}P, possess spin. The rotating charge causes the nucleus to behave like a tiny magnet and the spinning mass endows it with the properties of a gyroscope. Thus when an external magnetic field is applied, the nuclear magnets do not line up with it in the same way that non-spinning magnets would. Instead, they precess like a child's top around the direction of the field at a characteristic frequency, the Larmor frequency ν_L, which is directly proportional to the magnetic flux density B_0. For ^{31}P nuclei, $\nu_L = 1.723 \times 10^7 B_0$.

Thus in the (vertical) 7.5 Tesla field of the superconducting Oxford instrument, designed by Hoult & Richards (1975), the nuclei precess at 129.2 MHz. If a horizontal magnetic field oscillating at this frequency is applied, the nuclei absorb energy and will continue to radiate a signal after the exciting field is switched off at a frequency proportional to the local magnetic field experienced by the nucleus. This differs very slightly from the applied magnetic field B_0 because of the effects of local circulating electrons, neighboring magnetic nuclei, etc., which are characteristic of different types of compound. The resulting frequency shift, the chemical shift, can be detected and used to identify the different phosphorus compounds present. The chemical shift is usually expressed in parts per million (ppm).

Unfortunately NMR is a relatively insensitive technique. Recent technical advances have made it possible to identify the principal compounds in a single signal, but it is usually necessary to average many signals or "scans" in order to obtain useful spectra. Although the signal decays within about 20 msec ($T_2 \simeq$ 10msec), the pulse can be repeated only every 2sec because it takes a relatively long time (determined by T_1, the spin-lattice relaxation time) for the ^{31}P nuclei to revert to their unexcited state. As a result, the accumulation of a spectrum can take from a few minutes to several hours. In studying muscular contraction and recovery, the inherent slowness of the method can be overcome by stimulating repeatedly and averaging the signals during preset time intervals phased with the stimulus (see Fig. 2).

PHYSIOLOGICAL TECHNIQUES

The physiological problem is the counterpart of the physical situation. The choice of experimental preparation thus involves a compromise; thick muscles would give large signals but only thin ones can be adequately oxygenated by diffusion. They should, if possible, function at low temperature. Accordingly, most of our experiments were performed on frog muscles from *Rana temporaria* at 4°C.

Our first studies were of sartorius muscles undergoing various patterns of stimulation. We used ^{31}P NMR to determine concentrations of phosphorus metabolites and intracellular pH at rest, during contraction and during the following metabolic recovery period. I do not propose to say a great deal about this work. since it was published over a year ago. I shall simply note that, where comparison is possible, the results of these studies agreed well with information derived from other techniques. We feel that this verifies the accuracy of both the ^{31}P NMR and the conventional analytical methods, since the potential artifacts are so very different in each case.

The spectra from resting, oxygenated frog and toad muscles are shown in Figs. 1A and B. The positions of the peaks serve to identify each compound, and the area under the peak is determined by the amount of compound present. Note the small amount of inorganic phosphate (Pi) present in these intact living muscles. The region marked "Sugar P" indicates the zone occupied by the hexose and triose phosphates of the glycolytic pathway, and the peaks marked "?" represent three compounds whose functional significance is at present unknown, although two of them have recently been chemically identified (Burt, Glonek & Bárány, 1976; Chalovich, Burt, Cohen, Glonek & Bárány, 1977).

The position of the Pi peak shifts according to the pH; the observed position thus provides an excellent measure of the intracellular pH. The positions of the ATP peaks, especially those arising from the terminal β and γ phosphorus atoms, are changed when Mg^{2+} is bound; we have taken advantage of this to determine by a method which is independent of binding constants that the free $[Mg^{2+}] > 2.5$ mmol kg^{-1}. At this free $[Mg^{2+}]$ almost all of the ATP must be Mg^{2+}-bound (see Dawson *et al.*, 1978).

STUDIES OF MUSCULAR FATIGUE USING ^{31}P NMR

We should like now to describe in some detail a recently completed study of muscular fatigue. In these experiments we deliberately made the muscles anaerobic with N_2 and NaCN, so as to be able to work with large frog gastrocnemii and thus obtain adequate spectra (see Fig. 2) in a reasonably short period of time. ^{31}P NMR was used to monitor the metabolic changes the muscles underwent as a result of repeated fatiguing isometric contractions at

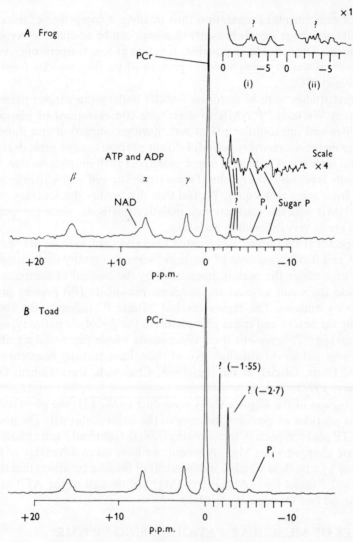

FIG. 1. Spectra from resting muscles at 4°C. A. Four frog sartorii, May 19, 1976. Lowest line: Average of 10,000 scans at 2sec intervals. The multiple peak centered at −3.0 ppm that is marked"?" represents three compounds of unknown function, two of which have been chemically identified (see text). The inset Figs. (i) and (ii) show how these peaks vary in size, though not in position (i) November 16, 1975; (ii) April 29, 1976. The peak labeled "Sugar P" at −7.5 ppm represents combined resonances from several hexose and triose phosphates; AMP and IMP appear in the same general region. The short length of record on the right-hand side of the PCr peak has been enlarged 4 × vertically to show fine detail. B. Two toad gastrocnemii, May 20, 1976. Average of 6,000 scans at 4 sec intervals.

4°C. Our analysis of the relation between force development and metabolic change has recently appeared in *Nature* (Dawson et al., 1978), and a manuscript concerning the relation between the rate constant for the exponential phase of relaxation and metabolic change is now in preparation.

A typical experiment is illustrated in Fig. 2. The relationships between decline in force development, the decrease in relaxation rate and changes in all of the following were determined: 1) concentrations of phosphorus compounds, creatine and H^+; 2) free-energy change for ATP hydrolysis; 3) rate of lactic acid (LA) production; and 4) total ATP hydrolysis. Of these, the concentrations of H^+ and of the phosphorus compounds can be determined directly by ^{31}P NMR as explained above. The other quantities were estimated on the basis of current knowledge of muscle biochemistry together with the ^{31}P NMR results. After 2 hrs poisoning with 2 mM NaCN, muscles were tetanized for 1sec/20sec, 1sec/60sec or 5sec/300sec. We performed two experiments of each type, lasting from 18 to 92 min.

The different patterns of stimulation used in these experiments ensured that there were differences in all of the following: 1) the total duration of the experiment, 2) the total number of contractions, and 3) the total period for which the muscles were stimulated (*i.e.*, the total number of contractions × duration of each contraction). When force development in individual experiments was plotted against these three parameters, the results separated themselves into groups according to the different patterns of stimulation. In contrast, the relationship between either force development or relaxation rate and concentration of each of the measured chemical substances was completely independent of the pattern of stimulation. We conclude that the changes in these mechanical manifestations of fatigue are linked to the biochemical status of the muscle rather than to any *independent* changes in excitatory conduction or excitation-contraction coupling. Changes in conduction or excitation-contraction coupling could be responsible for the mechanical manifestations of fatigue only if they themselves are somehow linked to biochemical changes.

Since there was no discernible effect of the pattern of stimulation on the biochemical results we have averaged all six experiments after grouping them for statistical purposes by linear interpolation. I shall discuss first the results concerning force development and then, briefly, our conclusions concerning the decline in relaxation rate.

The decline in force development

Metabolite and H^+ levels. In addition to measuring directly [ATP] [PCr], [Pi] and [H^+], we calculated [creatine] from the observed changes in

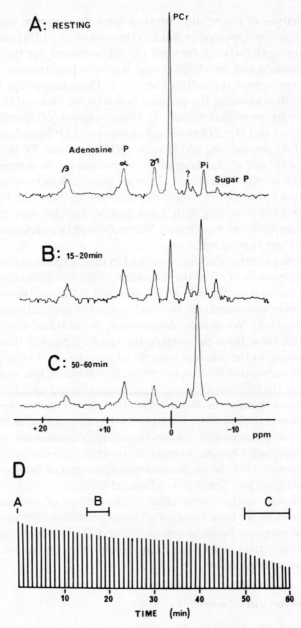

FIG. 2. Results from a single experiment in which a pair of anaerobic frog gastrocnemii (total wt = 773 mg) were stimulated 1sec/60sec. Panels A, B and C are examples of the ^{31}P

[PCr] and [free ADP] from the creatine phosphotransferase equilibrium. A strong correlation was found between force development and the levels of each of these metabolites; these relationships were completely independent of the pattern of stimulation, as shown in Figs. 3 and 4A. The [ATP] dropped by about 1mM and [PCr] dropped nearly to zero while force declined to 20% of its initial value. The relations between force and [PCr], [Pi] or [Cr] were clearly nonlinear, thus running counter to the results reported by Spande & Schottelius (1970), but the relations between force and [free ADP] or [H$^+$] may be linear.

Free-energy change for ATP hydrolysis. The most direct measure of the chemical energy available for muscle contraction is not the levels of substrates, PCr and/or ATP, but the free-energy change for ATP hydrolysis. Determination of the free-energy change, or affinity for ATP hydrolysis requires estimates of [ATP], [ADP], [Mg^{2+}] and [H$^+$]; it is only using ^{31}P NMR that all of these quantities can be determined simultaneously and *in vivo*. Affinity ($A = -\mathrm{d}G/\mathrm{d}\xi$), often miscalled $-\Delta G$ (The Royal Society, 1975), for ATP hydrolysis was calculated by the method of Alberty (1969). The decline in A was similar in each experiment, and its relation to force development appears to contain two linear components (see Fig. 4B). The significance of A is that it measures the maximum work theoretically available per mole of ATP hydrolyzed.

Glycolysis and ATP turnover. We have so far been concerned with relating force development to *levels* of affinity or of metabolites. However, the situation in muscle is dynamic, the metabolite levels reflecting the balance between the rate at which ATP is being hydrolyzed and the rate at which it is being resynthesized. Comparison of the changes in rates of these reactions with the decline in force development yields additional information concerning the mechanism of fatigue.

Under anaerobic conditions, resynthesis of ATP occurs mainly from glycolysis, and from the breakdown of PCr. The quantity of LA produced can be estimated from the pH changes and the buffering capacity of frog muscle. Other reactions known to be occurring in contracting muscle have much smaller effects on [H$^+$] and have therefore been neglected. It is known that LA diffuses only very slowly out of frog muscle; the results of Mainwood and co-workers (Mainwood, Worsley-Brown & Paterson, 1972) on fatigued frog

---

NMR spectra obtained, with peaks identified in spectrum A. We found no consistent or substantial changes in the ? or Sugar P peaks (see legend to Fig. 1), so they are neglected in the body of the text. Spectrum A was accumulated for 10 min before the first stimulation and C is the sum of two spectra accumulated in separate 5 min bins. Panel D is a drawing indicating the average value of each force response; the initial force was 238 mN mm^{-2}.

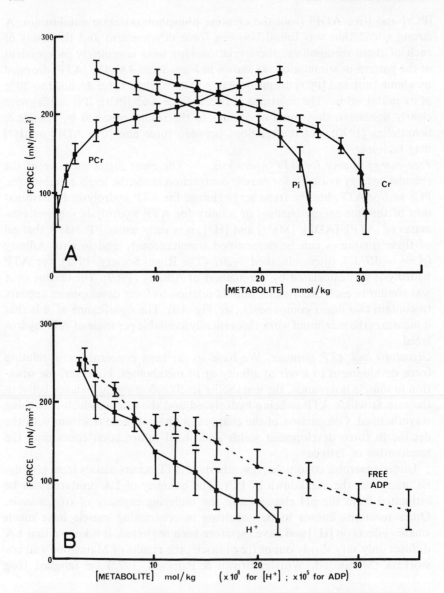

FIG. 3. Force developed as a function of metabolite levels; averaged results of six experiments. Spectra were accumulated for 5-min periods if the muscles were stimulated for 1sec/60sec or 5sec/5 min, and for 2-min periods if the stimulation was for 1sec/20sec. In the 5sec/5 min experiments the average force developed by each contraction was related to the metabolite levels indicated by the spectrum preceding that contraction. For the 1sec stimulation

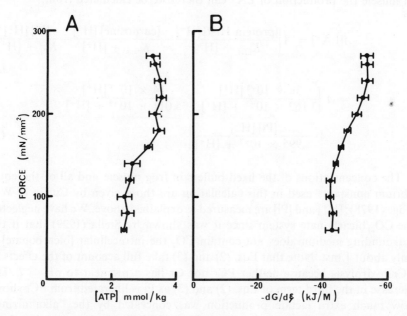

FIG. 4. Force as a function of [ATP] (A) and affinity (B). The data were obtained as explained in the legend to Fig. 2. The horizontal bars represent standard errors.

muscle indicate maximum rates of LA diffusion into the medium of only a few mmol kg^{-1}h^{-1}. We have accordingly neglected LA diffusion from the muscle in the following calculations.

Over the range of pH covered in these experiments (6.5 — 7.5), the concentrations of undissociated LA (pK = 3.9) and of H$^+$ are both negligible compared with the concentrations (> 1 mM) that concern us here. LA production can thus be estimated from the increase in the BH$^+$ forms of the buffers present. For one buffer (B, dissociatation constant K) whose total concentration is:

$$([BH^+] + [B]) = C$$
$$[BH^+] = C[H^+]/(K + [H^+]).$$
(1)

---

experiments the force developed by the contraction nearest the middle of the spectral-averaging period was related to the metabolite levels. The force developed at particular metabolite levels was obtained by linear interpolation between the data points, so as to obtain the standard errors shown by vertical bars.

In muscle the production of LA^- can therefore be calculated from:

$$\Delta[LA^-] = \Delta\left\{\frac{[\text{protein Hist}][H^+]}{K_{\text{hist}} + [H^+]} + \frac{[\text{carnosine}][H^+]}{K_{\text{carn}} + [H^+]} + \frac{[Pi][H^+]}{K_{P_i} + [H^+]}\right\} \quad (2)$$

$$= \Delta\left\{\frac{36 \times 10^{-3}[H^+]}{3.162 \times 10^{-7} + [H^+]} + \frac{14 \times 10^{-3}[H^+]}{5.011 \times 10^{-8} + [H^+]}\right.$$
$$\left. + \frac{[Pi][H^+]}{1.995 \times 10^{-7} + [H^+]}\right\}. \quad (3)$$

The concentrations of the fixed buffers in frog muscle and all of the equilibrium constants used in this calculation are those given by Curtin & Woledge (1978); $[H^+]$ and $[Pi]$ are measured as explained above. We have neglected the CO_2/bicarbonate system since it was shown by Stella (1929) that if the surrounding medium does not contain CO_2 the intracellular [bicarbonate] is only about 1 mM. Note that Eqs. (2) and (3) take full account of the effects of PCr hydrolysis because neither PCr nor Cr has a pK near to pH = 7. The increase in the third term of Eqs. (2) and (3) as Pi is liberated from PCr shows how much extra lactate production was concealed by the "alkalinizing" effect of PCr splitting. Note that Eqs. (1) – (3) do not involve any approximations or assumptions that restrict accuracy unless only small changes in $[H^+]$ have occurred. Incidentally, it has now become possible to estimate LA production *directly* by proton NMR (Brown, Campbell, Kuchel & Rabenstein, 1977) rather than indirectly from the change of pH interacting with the intracellular buffers present, and we are currently beginning studies using phosphorus and proton NMR simultaneously.

Total LA^- produced and total phosphorus utilized (P_{util}; the amount of ATP breakdown that would have been required in the absence of all recovery processes) are shown in Fig. 5 as a function of time for each of the six experiments. P_{util} was calculated according to the following equation:

$$\Delta P_{\text{util}} = (3/2)\Delta[LA] - \Delta[PCr] - \Delta[ATP]. \quad (4)$$

(There is some doubt as to the amount of ATP resynthesized per mole of LA produced (De Furia & Kushmerick, 1977); however, until this question is resolved we think it best to use the commonly accepted stoichiometry.) Both LA production and phosphorus utilization depend strongly upon the pattern of stimulation. Even in those cases in which the muscles were stimulated for the same total amount of time per hour (5sec/5 min and 1sec/60sec) there is

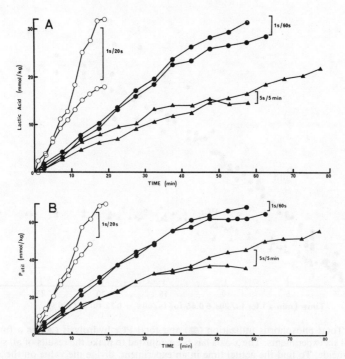

FIG. 5. Total lactic acid produced (A) and total phosphorus utilized (B) as a function of time for each of the six experiments.

a twofold difference in the total phosphorus utilized and in the total LA produced. These results show that the factor(s) determining the rate of glycolysis under the conditions of our experiments are closely linked to the rate at which phosphorus is being utilized.

Total phosphorus utilization and the contribution to it from PCr hydrolysis are replotted in Fig. 6. Here the results of all six experiments have been made to coincide by appropriate adjustment of the x-axis, the time scale. The difference between the two plotted curves represents the contribution from glycolysis. Figure 6 shows that, for a given amount of phosphorus utilization, a fixed proportion of the recovery is from LA formation and a fixed proportion is from PCr hydrolysis, regardless of the pattern of stimulation. It also shows that the muscles rely heavily upon PCr breakdown for metabolic recovery until the PCr is virtually exhausted, even though, at least in the two less demanding patterns of stimulation, the rate of glycolysis must be submaximal. From the viewpoint of animal design it is surprising that glycolysis

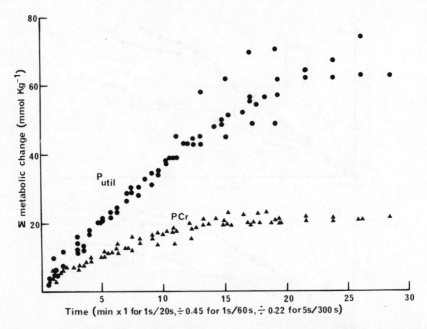

FIG. 6. Total phosphorus utilization (●) and total PCr hydrolysis (▲) as a function of time in all six experiments. The x-axis has been adjusted to make the results of all six experiments coincide. To find the actual time in an experiment, divide the value on the x-axis by 0.45 for the 1sec/60sec stimulation pattern and by 0.22 for the 5sec/300sec pattern. The difference between the two plotted curves represents recovery via glycolysis.

is not turned on more fully by the milder patterns of stimulation, for this would seem to provide a better stratagem for conserving PCr to use in emergencies.

The amount of PCr available to contribute to recovery becomes negligible after 22 repeated 1sec contractions. This change from two sources to a single source of rephosphorylation of ADP occurs when force development has fallen to about 160 mN mm^{-2}. This is the same point at which the relation between force development and relaxation rate constant changes suddenly in slope (see Fig. 8B) and is the point at which all the curvilinear relations between biochemical parameters and either force development or rate constant show their major curvature.

Force development as a function of phosphorus utilization. In order to relate our experimental results to theoretical schemes of contractility, we have examined the relation between force development and phosphorus utilization in several different ways. In Fig. 7, P_{util} is plotted against the cumulative sum of the force development in each contraction (F) times the duration of stimula-

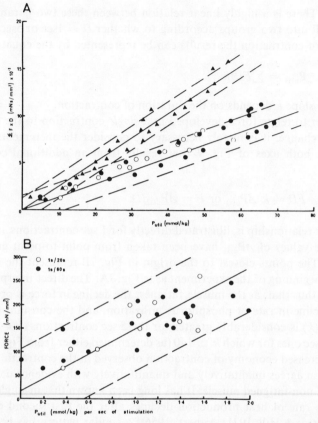

FIG. 7. A. The sum of the force developed in each contraction × the duration of stimulation [$\sum(FD)$] as a function of phosphorus utilized (P_{util}). FD was obtained by multiplying the force developed at the midpoint of each stimulation by the duration of stimulation. This method was chosen both for its simplicity and because it does not assume that ATP is being hydrolyzed during the relaxation period. ▲ = 5 sec/5min ● = 1 sec/60 sec and 0 = 1 sec/20 sec. The solid lines were obtained by least square regression analysis, assuming that all the experimental error is associated with P_{util}. In both cases the intercepts were negligible. For 5 sec stimulation $\sum FD$ (mN mm^{-2}sec) = 261 P_{util} (mmol kg^{-1}); $r = 0.99$; $n = 25$. For 1 sec stimulation $\sum FD$ (mN mm^{-2}sec) = 142 P_{util} (mmol kg^{-1}); $r = 0.97$; $n = 49$. The dashed lines are 95% confidence limits.

B. Force as a function of P_{util} per second of stimulation; P_{util}/D was estimated directly from the slopes in Fig. 5B. For clarity only the data from stimulations of 1 sec duration are included. For the 1sec/20sec experiments, the results were averaged over 4- rather than the usual 2-min intervals. The solid line was obtained by least-square regression analysis, assuming that all the experimental error is in P_{util}; slope = 134 mN mm^{-2} mmol^{-1} kg sec; $r = 0.80$; $n = 34$; $t = 8.0$; $P \simeq 0$. The dashed lines are 95% confidence limits and the intercept is not significant.

tion (D). There is a highly linear relation between these two parameters; the results fall into two groups according to whether $D = 1\text{sec}$ or 5sec. For each duration of contraction the results can be represented by the equation:

$$P_{util} = \sum FD/k \tag{5}$$

where the slope k depends on the duration of contraction.

In order to relate force developed in a *single* contraction to the associated chemical change, it is merely necessary to consider the increments (FD and ΔP_{util}) in both axes of Fig. 7 that results from an additional contraction cycle:

$$FD = k \ \Delta P_{util} \text{ or } F = \Delta P_{util}/D. \tag{6}$$

The latter relationship is illustrated directly for 1 sec contractions in Fig. 7B, where the values of ΔP_{util} have been taken from point-to-point gradients in Fig. 5B. The points closest to the origin in Fig. 7B represent the end rather than the beginning of the experiment as in Fig. 7A. The direct interpretation of Eq. (6) is thus that, as the muscle fatigues, the decline in force is proportional to the decline in rate of phosphorus utilization, and the constant of proportionality (k) is considerably greater for the 5 sec contractions ($k = 260$) than for the 1 sec ones for which $k \simeq 140$ (as determined either from Fig. 7A or 7B).

The increased economy of contraction observed in the contractions of longer duration agrees qualitatively and quantitatively with evidence derived from studies of non-fatigued muscle. It has long been known that in single isometric tetani the rate of heat production declines after the first second of stimulation (Hartree & Hill, 1921; Aubert, 1956); a similar pattern has been demonstrated for PCr hydrolysis (Marechal & Mommaerts, 1963; Homsher, Rall, Wallner & Ricchiuti, 1975). Marechal and Mommaerts' results on single and repeated tetani of frog sartorii at 0°C fitted the equation.

$$\Delta PCr = 0.33 + 0.28 \ D \text{ mmol kg}^{-1} \tag{7}$$

for values of D up to 60sec. In absolute terms Eq. (7) accords reasonably well with other measurements on frog sartorii by chemical methods at 0°C (Homsher et al., 1975; Gilbert, Kretzschmar, Wilkie & Woledge, 1971; Curtin, Gilbert, Kretzschmar & Wilkie, 1974) or by ^{31}P NMR at 4°C (Dawson et al., 1977a, b).

Our present results for $D = 1$ sec and 5 sec can be analyzed in a similar way:

$$P_{util} = F(4.01 + 3.03D) \times 10^{-3} \text{ mmol kg}^{-1}. \tag{8}$$

This equation, and the ratio between its constants, closely resembles Eq. (7) and furnishes the additional information that both terms in the bracket are directly proportional to force development as it declines during fatigue. We cannot distinguish in the present experiments between phosphorus utilization *during* contraction and utilization *between* contractions, but we have reason from other experiments for thinking that the latter contribution is small. Studies of non-fatigued frog (see Marechal & Mommaerts, 1963, p. 62) and toad (Dawson *et al.*, 1977a, b) gastrocnemii at 4°C support the view that these muscles break down 1.5–1.7 mmol kg^{-1} of PCr during a 1 sec contraction; this is in quantitative agreement with the present results and is roughly three times more than the values obtained on sartorii at 0–4°C.

In trying to relate our results to the fundamental processes thought to be occurring in contracting muscle, Eqs. (7) and (8) are particularly attractive since it seems natural to associate the first bracketed term with the chemical and mechanical events required to set up isometric force; while the second, duration-dependent term could be associated with the actual cross-bridge cycling that maintains the force.

Certainly there is good evidence that a quantity of PCr is hydrolyzed at the onset of contraction even when actomyosin interaction has been prevented by stretching (Homsher, Mommaerts, Ricchiuti & Wallner, 1972; Smith, 1972) If the duration-dependent term in Eq. (8) can indeed be equated with cross-bridge cycling, then our results show that the economy of maintenance of force remains constant in fatiguing muscle at about 300 Nmm^{-2}sec per mol ATP/kg split, and that the force declines because the rate of ATP utilization falls progressively. If so, anything that reduced the rate of ATP hydrolysis (*e.g.* product inhibition of cross-bridge cycling or decreased Ca^{+2} release) would automatically reduce the force proportionately and result in fatigue.

The decrease in relaxation rate

Skeletal muscle shows a characteristic pattern of relaxation following a tetanic contraction. After the last stimulus there is a delay, during which isometric force falls only slightly, followed by an exponential return to resting level. It has been shown that the exponential phase of relaxation does not arise from the characteristics of passive elements in the muscle, but reflects the progress of some underlying biochemical reaction (Jewell & Wilkie, 1960). We now turn to the question of how the rate constant ($1/\tau$) for this final phase of relaxation is related to biochemical and energetic changes.

It has been postulated that the rate constant for the exponential phase of relaxation is equal to that for cross-bridge detachment and thus to the steady rate of cross-bridge cycling during the preceding contraction (Edwards, Hill & Jones, 1975a, b). Alternatively, relaxation rate may be determined by the rate of Ca^{+2} removal into the sarcoplasmic reticulum (Sandow, 1965; Connolly, Gough & Winegrad, 1971; Briggs, Poland & Solaro, 1977; Blinks, Rudel & Taylor, 1978), and thus by the mechanism that deactivates cross-bridge cycling.

Since the concentrations of various metabolites change considerably during the course of fatigue, it is of interest to ask whether the decline in relaxation rate may be actually *caused* by any of the biochemical changes. Knowledge of such causation could help to rule out particular candidates for the step or steps whose change in time-course accounts for the prolongation of relaxation.

Like force development, $1/\tau$ is closely correlated with metabolite levels independently of the pattern of stimulation (within the limits that we have studied). However, since $1/\tau$ is not proportional to force (see Fig. 8), the relations between $1/\tau$ and metabolite levels are quite different from those between force and metabolite levels. We found that $1/\tau$ is linearly related to [PCr], [Cr] and [Pi], but very nonlinearly related to [H⁺] and [ADP].

Relation of $1/\tau$ to [ATP]. The relation between $1/\tau$ and [ATP] is of particular interest. Edwards and co-workers (1975a) have proposed a model of fatigue in which the decrease in relaxation rate is postulated to represent the slowing of

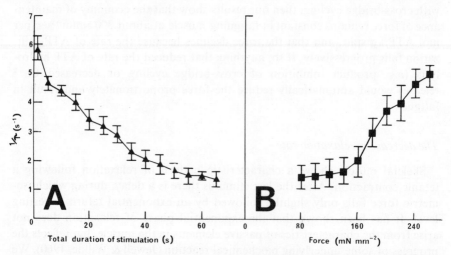

FIG. 8. $1/\tau$ as a function of (A) total duration of stimulation and (B) force development. The points represent the combined results of all six experiments. In each experiment the value of Y for a given value of X was obtained by linear interpolation between the original data points. Limits are standard errors.

cross-bridge cycling due to a diminished [ATP]. Our results are inconsistent with this model in several respects. *Firstly*, the model predicts (Edwards *et al.*, 1975b) that the economy of contraction must increase as $1/\tau$ increases, whereas we find that the economy of contraction remains unchanged as fatigue progresses (see Fig. 7B). *Secondly*, the model is based on a linear relation between [ATP] and $1/\tau$ in mouse soleus muscle (Edwards *et al.*, 1975a). We find a very different pattern in these experiments; during the first part of the experiment, over which $1/\tau$ changes by about five-fold, there is no change in [ATP], while the small changes in [ATP] that do occur are towards the end of the experiment, when $1/\tau$ remains fairly constant. This is shown by the two regression lines in Fig. 9A. Each experiment was divided into two parts, the division being made at the point where the relation between force and $1/\tau$ changes suddenly in slope; linear regression analysis was then done separately on the two parts.

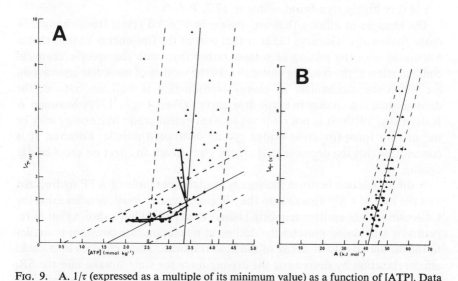

FIG. 9. A. $1/\tau$ (expressed as a multiple of its minimum value) as a function of [ATP]. Data represent results of all six experiments. Each experiment was divided into two parts, initial (represented by ▲) and final (●). Linear regression analyses were done separately on the two parts, yielding the following linear equations for the regression of X upon Y:
initial: [ATP] mmol kg^{-1} = 0.043 $1/\tau_{rel}$ + 3.39 ($r = 0.203$; $n = 38$)
final: $1/\tau_{rel}$ = $-1.834 + 1.11$ [ATP] mmol kg^{-1} ($r = 0.56$; $n = 36$; $P = 8 \times 10^{-5}$).
The correlation between $1/\tau_{rel}$ and [ATP] during the first part of the experiment was not significant; this is to be expected when one of the parameters is approximately constant. The dashed lines are 95% confidence limits. The solid curve represents results of a single experiment in which the muscles were stimulated at 1 sec/60 sec. B. $1/\tau$ as a function of affinity or free-energy change for ATP hydrolysis. Data represent results of all six experiments. Solid line is regression of X upon Y (see text); dashed lines are 95% confidence limits.

The difference in slope of the two regression lines was tested (Brownlee, 1957, p. 67) and found to be very highly significant ($t = 495$; $P \simeq 0$). It is thus clear that the best fit for these data is *not* a single linear relation between $1/\tau$ and [ATP].

Affinity for ATP hydrolysis. Although the changes in affinity are small (total change approximately 15kJ M^{-1}), they do have the same time-course as the changes in relaxation rate constant, both occurring mainly during the first half of the experiment. Figure 9B shows a single highly linear relationship between these two variables, and linear regression analysis yields the following equation of best fit, assuming that all of the experimental error is in the estimate of affinity:

$$1/\tau \,(\text{sec}^{-1}) = -16.21 + 0.397\, A(\text{kJ M}^{-1})\, (r = 0.85; n = 74); \qquad (9)$$

τ is very highly significant with $t = 13.7$, $P \simeq 0$.

The changes in affinity that are shown in Fig. 9B result from changes in *cratic free-energy* (Gurney, 1953) or that part of the free-energy change that is associated with the mixing of solutes rather than with the specific chemical configuration of the reacting molecules. In the models of muscular contraction for which the mechanism of energy transduction is well understood, the driving force is a change in cratic free-energy (Weis-Fogh, 1975; Sussman & Katchalsky, 1970). It is not clear yet to what extent cratic free energy may be the driving force for cross-bridge cycling in striated muscle; however, it is conceivable that the decrease in A affects $1/\tau$ through an effect on cross-bridge cycling.

A simple relation between the energy available per mole of ATP hydrolyzed and the rate of Ca^{+2} uptake into the SR also suggests itself. Studies using the Ca-sensitive light emitter aequorin (Blinks *et al.*, 1978) have shown that in repetitively stimulated muscles the decline in relaxation rate constant is similar to the decline in rate of decay of the light response. Thus, changes in A could affect relaxation by decreasing the driving force for Ca^{2+} uptake into the SR. This suggestion is testable by determining the effect of changing affinity for ATP hydrolysis on the rate of Ca^{2+} uptake by isolated SR.

Conclusions. In this study we have used ^{31}P NMR together with other techniques to relate both force development and rate constant for the exponential phase of relaxation to biochemical and energetic changes in fatiguing frog muscle. We find that both of these mechanical manifestations of fatigue are closely correlated with metabolite levels rather than with any independent changes in excitatory conduction or excitation-contraction coupling. Isometric force development (but not $1/\tau$) is proportional to the decrease in rate of phosphorus utilization. These results suggest that the hydrolysis of ATP in each crossbridge cycle produces a fixed mechanical impulse (force × time) and that the economy of ATP hydrolysis remains unchanged as the muscle fatigues. We found no evidence that $1/\tau$ is related in a simple way to the rate of crossbridge cycling during the preceding contraction, but whatever process causes the decline in $1/\tau$ is nevertheless dependent upon metabolic factors.

REFERENCES

Alberty, R. A. (1969). Standard Gibbs free energy, enthalpy and entropy changes as a function of pH and pMg for several reactions involving adenosine phosphates. *J. biol. Chem.* **244**, 3290–3302.
Aubert, X. (1956). In *Le Couplage Energetique de la Contraction Musculaire*. Brussels: Editions Arscia.
Blinks, J. R., Rudel, R. & Taylor, S. R. (1978). Calcium transients in isolated amphibian skeletal muscle fibres: Detection with aequorin. *J. Physiol.* **277**, 291–323.
Briggs, F. N., Poland, J. L. & Solaro, J. H. (1977). Relative capabilities of sarcoplasmic reticulum in fast and slow mammalian skeletal muscles. *J. Physiol.* **266**, 587–594.
Brown, F. F., Campbell, I. D., Kuchel, P. W. & Rabenstein, D. C. (1977). Human erythrocyte metabolism studies by '¹H spin echo NMR. *FEBS Lett.* **82**, 12–16.
Brownlee, K. A. (1957). In *Industrial Experimentation*. Ch. 9. 4th Ed. London: Her Majesty's Stationery Office.
Burt, C. T., Glonek, T. & Bárány, M. (1976). Phosphorus-31 nuclear magnetic resonance detection of unexpected phosphodiesters in muscle. *Biochemistry, N. Y.* **15**, 4850–4853.
Chalovich, J. M., Burt, C. T., Cohen, S. M., Glonek, T. & Bárány, M. (1977). Identification of an unknown ^{31}P nuclear magnetic resonance from dystrophic chicken as L-serine ethanolamine phosphodiester. *Arch. Biochem. Biophys.* **182**, 683–689.
Connolly, R., Gough, W. & Winegrad, S. (1971). Characteristics of the isometric twitch of skeletal muscle immediately after a tetanus. *J. gen. Physiol.* **57**, 697–709.
Curtin, N. A., Gilbert, C., Kretzschmar, K. M. & Wilkie, D. R. (1974). The effect of the performance of work on total energy output and metabolism during muscular contraction. *J. Physiol.* **238**, 455–472.
Curtin, N. A. & Woledge, R. C. (1978). Energy changes and muscular contraction. *Physiol. Rev.* **58**, 690–761.
Dawson, M. J., Gadian, D. C. & Wilkie, D. R. (1977a). Contraction and recovery of living muscles studied by ^{31}P nuclear magnetic resonance. *J. Physiol.* **267**, 703–735.
Dawson, M. J. Gadian, D. C. & Wilkie, D. R. (1977b). Studies of living contracting muscle by ^{31}P nuclear magnetic resonance. In *NMR in Biology*. eds. Dwek, R. A., Campbell, I. D., Richards, R. E., & Williams, R. J. P., pp. 289–322. Oxford: Oxford University Press.
Dawson, M. J., Gadian, D. G. & Wilkie, D. R. (1978). The energetics of muscular fatigue investigated by phosphorus nuclear magnetic resonance *Nature, Lond.* **274**, 861–866.
DeFuria, R. R. & Kushmerick, M. J. (1977). ATP utilization associated with recovery metabolism in anaerobic frog muscle. *Am. J. Physiol.* **232**; *Cell Physiol.* **1**, C30–C36.
Edwards, R. H. T., Hill, D. K. & Jones, D. A. (1975a). Metabolic changes associated with the slowing of relaxation in fatigued mouse muscle. *J. Physiol.* **251**, 287–301.

Edwards, R. H. T., Hill, D. K. & Jones, D. A. (1975b). Heat production and chemical changes during isometric contractions of the human quadriceps muscle. *J. Physiol.* **251**, 303–315.
Gilbert, C., Kretzschmar, K. M., Wilkie, D. R. & Woledge, R. C. (1971). Chemical change and energy output during muscular contraction. *J. Physiol.* **218**, 163–193.
Gurney, R. W. (1953). In *Ionic processes in solution*. New York: McGraw-Hill.
Hartree, W. & Hill, A. V. (1921). Regulation of the supply of energy in muscular contraction. *J. Physiol.* **55**, 133–158.
Homsher, E., Mommaerts, W. F. H. M., Ricchiuti, N. V. & Wallner, A. (1972). Activation heat, activation metabolism and tension-related heat in frog semitendinosus muscles. *J. Physiol.* **220**, 601–625.
Homsher, E., Rall, J. A., Wallner, A. & Ricchiuti, N. V. (1975). Energy liberation and chemical change in frog skeletal muscle during single isometric tetanic contractions. *J. gen. Physiol.* **65**, 1–21.
Hoult, D. I. & Richards, R. E. (1975). Critical factors in the design of sensitive high resolution nuclear magnetic resonance spectrometers. *Proc. R. Soc. A*, **344**, 311–340.
Jewell, B. R. & Wilkie, D. R. (1960). The mechanical properties of relaxing muscle. *J. Physiol.* **152**, 30–47.
Mainwood, G. W. Worsley-Brown, P. & Paterson, R. A. (1972). The metabolic changes in frog sartorius muscles during recovery from fatigue at different external bicarbonate concentrations. *Can. J. Physiol. Pharmacol.* **50**, 143–155.
Marechal, G. & Mommaerts, W. F. H. M. (1963). The metabolism of phosphocreatine during an isometric tetanus in the frog sartorius muscle. *Biochim. biophys. Acta* **70**, 53–67.
Sandow, A. (1965). Excitation-contraction coupling in skeletal muscle. *Pharmac. Rev.* **17**, 265–320.
Smith, I. C. H. (1972). Energetics of activation in frog and toad muscle. *J. Physiol.* **220**, 583–599.
Spande, J. I. & Schottelius, B. A. (1970). Chemical basis of fatigue in isolated mouse soleus muscle. *Am. J. Physiol.* **219**, 1490–1495.
Stella, G. (1929). The combination of carbon dioxide with muscle: its heat of neutralization and its dissociation curve. *J. Physiol.* **68**, 49–66.
Sussman, M. V. & Katchalsky, A. (1970). Mechanochemical turbine: A new power cycle. *Science, N. Y.* **167**, 45–47.
The Royal Society, London (1975). *Quantities Units and Symbols.*
Weis-Fogh, T. (1975). Spasmonemes, a new contractile system. In *Ciba Foundation Symposium* 31 (new series), Amsterdam: Elsevier.

Discussion

Rall: You mentioned cause and effect. Can you talk a little more about cause and effect, and speculate in terms of what you see in your data as to why muscle fatigues?

Wilkie: One example of cause and effect arises from our experiments on the decline of force and of phosphorus turnover during fatigue. We have considered two general hypotheses.

First, we have shown by NMR that the products of ATP hydrolysis—ADP, Pi and H^+—increase markedly as fatigue progresses. One obvious mechanism would be that the accumulation of products reduces the rate of ATP splitting by the actomyosin ATPase system. A second possibility is that the products interfere with "activation" at some point, *e.g.* H^+ might compete with Ca^{2+} at the troponin sites.

The first possibility is being tested by Susan Smith in our laboratory. She is measuring the net rate of ATP hydrolysis by frog actomyosin with reagent concentrations imitating those that we observe by NMR at the beginning and at the end of our experiments. If the rate of ATP hydrolysis is substantially reduced we shall incline towards the first possibility; if not, to the second possibility.

Mohan: Is it the total free energy you are talking about, or just one part of the cycle?

Wilkie: It is the total free energy for going completely around the cycle and hydrolyzing one mole of ATP.

Mohan: Is this linear with the force? Is it linear to the rate constant?

Wilkie: It is linear to the rate constant, but as you saw, not really strictly linear to the force.

Winegrad: This is a question directed towards people doing energy balancing. I notice that very little, or simply nothing, was said about the energy involved in the restoration of the ion gradients that are altered by the depolarization, by the action potential, during tetanus. I have not done the calculations, but is this sufficiently small so as to be negligible? Otherwise, it would present a phase lag in that ATP would be broken down after the contraction in order to restore the ion gradients, and might influence the measurements of net ATP synthesis that you do see after the contraction.

Wilkie: Well, as far as the action potential is concerned, and sodium and potassium gradients, it is certainly completely negligible. Now calcium might be another story, and it is much harder to get at; but as far as sodium and potassium are concerned, if you take a squid axon and you poison it, it goes on and gives you 10,000 action potentials. The actual energy cost of them is very small.

Kushmerick: But I think as regards calcium, it certainly would be included in this post-relaxation or recovery ATP utilization. Actually, I should have mentioned it; you are right, but having mentioned it is about all one can do, because there is not any direct way of measuring those quantities.

Winegrad: I do not think the evidence from the squid axon is entirely satisfying because you can get action potentials that are sufficient to trigger a contraction that occur with smaller ion gradients than normally exist; but the muscle, obviously, over a certain period of time, must restore the original ion gradients, so the fact that you can get many action potentials without any energy source does not indicate that the energy involved in the restoration is small; and then, as you point ous, the question of calcium movements across the surface membrane must be taken into consideration as well.

Wilkie: Well, perhaps the squid axon was an unfortunate example to choose. Lüttgau wrote two very good papers about fatigue in single skeletal muscle fibers (Lüttgau, 1965; Grabowski, Lobsiger & Lüttgau, 1972) in which he showed that when the metabolism was severely run down—he describes it in those very words—the muscle fiber behaved like a squid axon in giving thousands of action potentials, and you can show by calculation, too, that the energy cost of it is small. But over calcium, one does not know.

Kushmerick: I think the most direct answer is that it is very difficult to estimate these quantities, and all you can do is play around with paper. I think most of us would rather do experiments.

Wilkie: I do not think that is true of action potentials, absolutely not. It is small, very small.

Dawson: I have a question that is similar to Dr. Winegrad's in that it involves fluxes, and I would like to direct it primarily to Roger Woledge, although there are other people in this room who might like to comment. I know you are aware of the work of Blinks and co-workers (Blinks, Rüdel & Taylor, 1978) showing that the aequorin light response is variable with the length of the muscle, and also the work of David Allen in our department (Allen, 1978) showing after a period of isometric contraction that length changes are associated with changes in the aequorin light response. I should think that these results would complicate interpretation of experiments such as you have mentioned, your own experiments on stretched muscle, and also the paper of Rall et al. (Rall, Homsher, Wallner & Mommaerto, 1976). I would like to know your opinions about this.

Woledge: I quite agree that interpreting an experiment when you try to eliminate actomyosin interaction by progressively stretching the muscle is difficult. If you do that, you make the assumption that the activation processes continue unaltered. Now, the evidence on that is equivocal. As I understand Dr. Allen's work, he sometimes finds the calcium transients to be more or less the same when he stretches muscles; he sometimes finds them going down, and sometimes up, so there could be changes there, and one would like to be able to follow them, but we cannot. Now, another problem in interpreting them is the very fact that you get a lot of tension developed at long muscle lengths due to this "creep" phenomenon, which we have heard quite a bit about at this meeting, and that also complicates the question of interpreting an observation made with a muscle with long sarcomere length. The problem then is that there are these difficulties, but we do not have a better way of dissociating these two possible sources of energy output, and so I think we will just have to do the experiment and make the best of it that we can.

REFERENCES

Allen, D. G. (1978) Shortening of tetanized skeletal muscle causes a fall of intracellular calcium concentration. *J. Physiol.* **275**, 63 *P.*

Blinks, J. R., Rüdel, R. & Taylor, S. R. (1978) Calcium transients in isolated amphibian skeletal muscle fibres. *J. Physiol.* **227**, 291–323.

Grabowski, W., Lobsiger, E. A. & Luttgau, H. C. (1972) The effect of repetitive stimulation at low frequencies upon the electrical and mechanical activity of single muscle fibres. *Pflugers Arch. ges. Physiol.* **334**, 222–239.

Lüttgau, H. C. (1965) The effect of metabolic inhibitors on the fatigue of the action potential in single muscle fibres. *J. Physiol.* **178**, 45–67.

Rall, J. A., Homsher, E., Wallner, A. & Mommaerts, W. F. H. M. (1976) A temporal dissociation of energy liberation and high energy phosphate splitting during shortening in frog skeletal muscle. *J. gen. Physiol.* **68**, 13–27.

VIII. THEORIES OF CONTRACTION

Simplified Theory of the Huxley-Simmons T_0, T_1 and T_2 in Muscle Models with Two Attached States

Terrell L. HILL* and Evan EISENBERG**

* Laboratory of Molecular Biology, National Institute of Arthritis, Metabolism and Digestive Diseases, National Institutes of Health, Bethesda, Maryland, U.S.A.
** Laboratory of Cell Biology, National Heart, Lung and Blood Institute, National Institutes of Health, Bethesda, Maryland, U.S.A.

ABSTRACT

If we assume, following Huxley and Simmons, that the quick recovery of force in an isometric transient is associated with fast transitions between attached states, then it is easy to derive relatively simple, approximate expressions for the Huxley-Simmons quantities T_0, T_1 and T_2. These results are useful in the selection of parameters for more quantitative and detailed models. We illustrate this procedure using, primarily, models with two attached states and with one, two, three or an infinite sequence of actin sites (55 Å apart) available to a given myosin cross-bridge. Models with three attached states or with "slipping" of the myosin head (in the same attached state) from one actin site to another are also discussed briefly.

1. INTRODUCTION

We have been involved in the past few years, with our colleague Yi-der Chen, in the construction and testing of a large number of self-consistent (Hill, 1968, 1974, 1975) sliding filament models of muscle contraction. We expect to continue this program. So far these models have included four biochemical states of the myosin, two of which are attached (to actin) and two of which are unattached (Eisenberg & Hill, 1978). In the future we plan to examine more complex models in view of recent biochemical information that several more weakly attached states may occur (Stein, Schwarz, Chock & Eisenberg, forthcoming). Most of our work has been on single site models (*i. e.*, a particular cross-bridge, at any given time, can attach to only one actin site). However, we have also studied, to some extent, models in which there is an indefinite array of equivalent actin sites 55 Å apart (Hill, 1975). Actually, we believe that models intermediate between these two extremes are probably most realistic; because of the twist of the actin filament, a group of three nonequivalent actin sites may be available to a given cross-bridge at any one time (Hill, 1975). We plan to examine such models in the near future.

The first phase of the above program is now complete; a quantitative treatment of four-state models with a single actin site is now being prepared for publication (Eisenberg, Hill & Chen, forthcoming). This paper will present numerical results that exemplify the more qualitative biochemical arguments in the introductory treatment of Eisenberg & Hill (1978).

The present paper occupies a position intermediate between the qualitative and the quantitative papers just mentioned. We found it indispensable in the construction of complete, quantitative models to use the simple theory to be outlined here as a guide in selecting certain parameters of the complete model. Others may find this analysis to be equally useful. In particular, from relatively elementary considerations, it is possible to anticipate the approximate behavior of detailed models with respect to the quantities Huxley & Simmons (1971) denoted by T_0, T_1 and T_2.

All of our models so far, and especially the present paper, adopt the Huxley-Simmons premise that the quick recovery of tension following a sudden length change is a consequence, primarily, of an adjustment between attached states only. Huxley & Simmons (1971) outlined the basic theory of the quick recovery using only a single value of the conventional variable x, namely x_0 (see Section 2). Hill (1974, pp. 311-320) extended this work to include the proper x averaging. The present paper gives a very much simplified version of Hill's (1974) treatment——but with x averaging retained.

There has been some recent renewed interest in the possibility of a significant passive elastic element in series with the active cross-bridge system. This complication will not be dealt with in the main text but will be considered in the Appendix.

2. SIMPLIFIED THEORY FOR A SINGLE ACTIN SITE

We use the notation and approach in Hill (1974). Many details of the general formalism will not be repeated here.

We assume, as before (Huxley & Simmons, 1971; Hill, 1974; Eisenberg & Hill, 1978), that there are two attached states, 1 and 2, with transitions between them that are fast compared with the "on-off" transitions between attached and unattached states. Consequently, under isometric conditions ($T_0 =$ isometric force or tension per overlap cross-bridge), states 1 and 2 are essentially in equilibrium with each other. These attached states are, of course, the only ones that contribute to the force or tension. A sudden release or stretch (Huxley & Simmons, 1971; Ford, Huxley & Simmons, 1977) of length y per halfsarcomere takes the 1, 2 system out of equilibrium and produces an instantaneous force $T_1(y)$. The subsequent quick recovery brings the 1, 2 system back to a new equilibrium state, with force $T_2(y)$, before (ideally) there is any on-off adjustment. We shall be concerned here with the values of T_0, T_1 and T_2, but not with the actual kinetics of the quick recovery ($T_1 \to T_2$) nor with the slow recovery from T_2 back to T_0.

The entire analysis we give here assumes that there is a clean separation (two different time domains) between the 1, 2 adjustment and the on-off adjustment. This is, of course, only an approximation——but a very useful one.

Figure 1 shows the basic free energy curves (Hill & Simmons, 1976; Hill, 1977; Eisenberg & Hill, 1978) for the two attached states, and also indicates schematically the nature of the new simplification that we introduced in Appendix 2 of Eisenberg & Hill (1978) and use in this paper. Namely, we assume the on-off rate constants (there are four of these) have values such that the probability a cross-bridge is attached (in either state 1 or 2), in the isometric state, is a constant, p, between $x = x' - l$ and $x = x'$, and zero otherwise. Thus, attachment occurs over a range l. A probability of attachement of this form is

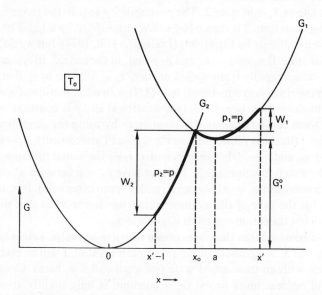

FIG. 1. Free energy curves of two attached states. Heavy portions show regions of attachment assumed, under isometric conditions. See text.

not overly idealized; in fact, our complete single-site model (Eisenberg, Hill & Chen, to be published) has this property, approximately. If the repeat distance between actin sites is d (e.g., $d = 360$ Å), the fraction of all cross-bridges in the actin-myosin overlap zone that are attached in the isometric state is then pl/d (e.g., 0.85×80 Å$/360$ Å $= 0.19$).

The free energy functions in Fig. 1 can be written as

$$G_1(x) = G_1^0 + \frac{1}{2} K(x-a)^2$$
$$G_2(x) = \frac{1}{2} Kx^2, \tag{1}$$

where, it should be noted, we use the same force constant K for both states. The two free energy minima are separated by a distance a. The corresponding force functions are

$$F_1(x) = K(x-a), \quad F_2(x) = Kx. \tag{2}$$

Let $n_2(x)$ be the (equilibrium) probability that an *attached* cross-bridge, with actin site at x, is in state 2. The probability $p_2(x)$, in the range l, that the cross-bridge is in state 2 is then p (see above) multiplied by $n_2(x)$. The function $n_2(x)$ is given explicitly by Eq. (I–99) (I refers to Hill, 1974), but we do not need it here. For state 1, $n_1 = 1 - n_2$ and $p_1 = pn_1$ in the range l. In typical cases n_2 decreases rather rapidly from $n_2 \cong 1$ at, say, $x_0 - 10$ Å to $n_2 \cong 0$ at $x_0 + 10$ Å. This change in n_2 is symmetrical: $n_2 - (1/2)$ is an odd function of $x - x_0$. It is not hard to show that, because n_2 is symmetrical and p is constant over l, the isometric force can be calculated *without error* by using the simplification implicit in Fig. 1 (the heavy lines indicate the region of attachment); $n_2 = 1$ between $x' - l$ and x_0, and $n_2 = 0$ between x_0 and x' (i.e., the actual function n_2 can be replaced by a step function at x_0). Thus we have $p_2 = p$ between $x' - l$ and x_0 (zero otherwise) and $p_1 = p$ between x_0 and x' (zero otherwise). It is, of course, necessary for the sake of this argument that we choose x' so that the range l of p covers the transition region in n_2 around x_0.

It should be noted that there are certain restrictions on the values chosen for p in Fig. 1. For $x > x_0$, under isometric conditions, state 1 will be essentially in equilibrium with an unattached state (we shall call it u, here). Consequently $p_1 = p$ will be near unity unless $G_u = $ constant is only slightly above G_1^0. In this case, the choice $p_1 = p$ would be somewhat flexible but then $x' - a$ would have to be small. There is not a similar problem with $p_2 = p$ (i.e., for $x < x_0$)

because a true steady state, with adjustable rate constants, is involved here (not an equilibrium).

The above discussion is by way of justifying, *as an approximation*, the simple probability distribution shown schematically (by the heavy lines) in Fig. 1. We now proceed to use this distribution to calculate T_0, T_1 and T_2. This can be done using either force functions (Eq. 2) or free energy functions (Eq. 1). We choose the latter method (except in Fig. 6).

Isometric force (T_0). In a very slow pass of an actin site (Hill, 1974, 1977) through a distance d (the repeat distance), past a particular cross-bridge, the work done by the cross-bridge on the site is $T_0 d$, where T_0 is the mean force exerted by the cross-bridge in the interval d (Hill, 1974). But this work is also equal to the drop in free energy of the cross-bridge, while attached. This quantity is $(W_1 + W_2)p$ (Fig. 1). Thus we can calculate T_0 very simply from (see Eq. 1):

$$T_0 = (W_1 + W_2)p/d \tag{3}$$

$$W_1 = (K/2)[(x' - a)^2 - (x_0 - a)^2] \tag{4}$$

$$W_2 = (K/2)[x_0^2 - (x' - l)^2]. \tag{5}$$

This gives

$$T_0 = (Kp/2d)[2x'(l - a) + 2x_0 a - l^2]. \tag{6}$$

(The same result is found, after a more complicated calculation, from Eqs. (I-97) to (I-100), taking p constant over l.)

Instantaneous force after length change (T_1). Figures 2a (stretch) and 2b (release) show the probability distributions for the attached states 1 and 2 immediately following a sudden length change y per half-sarcomere. States 1 and 2 are not in equilibrium here. The distributions in Fig. 1 are merely shifted along the x axis a distance y without any readjustment between states 1 and 2. The instantaneous force per cross-bridge, $T_1(y)$, can be calculated from Figs. 2a and 2b using the same method as in Eqs. (3–5). But even more simply, we can go directly to Eq. (6) and replace x_0 by $x_0 + y$ and x' by $x' + y$ (compare Figs. 1 and 2). The result is

$$T_1(y) = T_0 + (Kpl/d)y, \tag{7}$$

with T_0 given by Eq. (6). This is easily seen to be a special case (p constant over

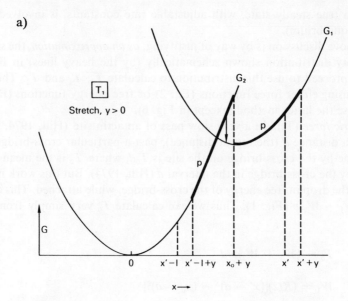

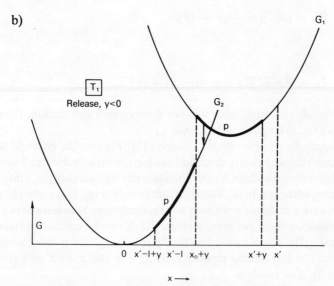

FIG. 2a. In a sudden stretch y, regions of attachment are shifted to the right along x axis without transitions between the two states. This figure allows calculation of T_1. The arrow shows the transitions that occur in the subsequent quick recovery, leading to Fig. 4. See text.

FIG. 2b. Same as (a) except that y is negative (release).

l) of the more general Eq. (I–44). There is no error from use of the n_2 step function. T_1 is linear in y because the free energy functions are quadratic in x. After suitable corrections, Ford *et al.* (1977) find the experimental T_1 to be approximately linear in y. Let us denote by y' the value of y that gives $T_1 = 0$. Ford *et al.* (1977) report that $-y' \cong 40$ Å. From Eq. (7), we have

$$-y' = T_0 d/Kpl. \tag{8}$$

$T_1(y)$ is shown schematically in Fig. 3.

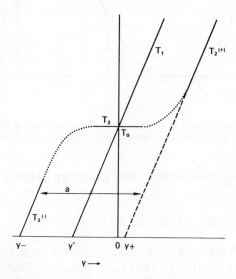

FIG. 3. This figure shows, schematically, the relations between T_0, T_1 and the three branches of T_2 that can be calculated by the present methods. See text.

Force after quick recovery (T_2). The arrows (on the free energy curves) in Figs. 2a and 2b indicate the process that occurs during the quick recovery. At the end of the quick recovery (using the stretch in Fig. 2a as an example), the attached state distributions appear as in Fig. 4. States 1 and 2 are again in equilibrium with each other. We are assuming here that $|y|$ is small enough so that the transition region in n_2 around x_0 is covered by the new range in p (*i.e.*, after the step y). The subsequent slow recovery ($T_2 \to T_0$) involves detachment from state 1 in region A of Fig. 4 and attachment into state 2 in region B of Fig. 4. The final state is again the same as that shown in Fig. 1 (isometric).

On comparing Figs. 1 and 4, we see that $T_2(y)$, for $|y|$ not too large (see

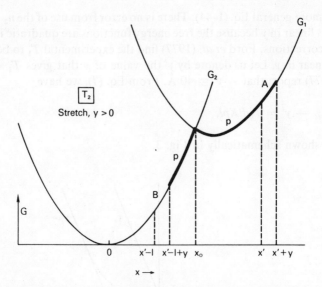

FIG. 4. Regions of attachment in states 1 and 2 after completion of the fast transitions indicated by the arrow in Fig. 2a. This figure allows calculation of T_2 for $|y|$ not too large. See text.

above), may be calculated directly from Eq. (6) by replacing x' by $x' + y$. We obtain

$$T_2(y) = T_0 + [Kp(l - a)/d]y. \qquad (9)$$

This same result is obtained, for p constant over the range l, from the constant and linear terms in Eq. (I-113). Thus again there is no error from use of the n_2 step function. Experimentally, Huxley & Simmons (1971) and Ford et al. (1977) have found that the slope of $T_2(y)$ around $y = 0$ is essentially zero. Thus the on-off rate constants should be chosen (in a more complete model) to make the range l of p equal to a. Equation (9) then reduces to $T_2(y) = T_0$. This is shown in Fig. 3 as the horizontal branch of T_2 around $y = 0$. Also, it should be noted that, if $l = a$, the two regions A and B in Fig. 4 are in equivalent positions on the two free energy curves. Hence, shifting p in region B (T_0) to p in region A (T_2) has no effect on the work or force. Thus, it is obvious why $T_2 = T_0$ (Eisenberg & Hill, 1978).

If the step y is positive (stretch) and large enough so that all of the new range of p in Fig. 4 falls in state 1 (i.e., $x' - l + y > x_0$), then we can easily derive the equation of the branch of T_2 marked $T_2^{(+)}$ in Fig. 3. Using the method of Eqs. (3)–(5), we have

$$T_2^{(+)}d = (Kp/2)[(x' + y - a)^2 - (x' - l + y - a)^2] \qquad (10)$$

or

$$T_2^{(+)}(y) = (Kpl/d)[x' - a - (l/2) + y].$$

This is a straight line, with the same slope as $T_1(y)$ which, when extrapolated, intersects the abscissa (Fig. 3) at

$$y_+ = a + (l/2) - x'. \qquad (11)$$

Similarly, if the step y is sufficiently negative (large release), all of the new range of p falls in state 2. In this case,

$$T_2^{(-)}d = (Kp/2)[(x' + y)^2 - (x' - l + y)^2]$$

or

$$T_2^{(-)}(y) = (Kpl/d)[x' - (l/2) + y]. \qquad (12)$$

This straight line also has the same slope as $T_1(y)$, although the experimental $T_2^{(-)}$ curve (Ford et al., 1977) has a smaller slope than T_1. The line intersects the abscissa (Fig. 3) at

$$y_- = (l/2) - x'. \qquad (13)$$

The "width" of the complete $T_2(y)$ curve is $y_+ - y_- = a$ (Fig. 3). As expected this is just the interval on the x-axis between the two free energy or force curves (Eq. (1) or (2)). The dotted "corners" of T_2, shown schematically in Fig. 3, cannot be deduced by the present simple argument.

It is easy to verify that the $T_2^{(+)}$ and $T_2^{(-)}$ branches of T_2 are equidistant from $T_1(y)$, i.e., $y' = (y_+ + y_-)/2$, if $x' = x_0 + (l/2)$, i.e., if the original range l of p (Fig. 1) is symmetrical about x_0. As might be anticipated, this same condition, $x' = x_0 + (l/2)$, makes the y^2 term in Eq. (I-113) equal to zero when p is constant over l. That is, $T_2(y)$ has an inflection point at $y = 0$ in this case. The inflection point obtained experimentally, as shown in Fig. 13 of Ford et al. (1977), also occurs at $y \cong 0$.

In Fig. 5 we include the free energy level G_u of the unattached state u that precedes state 1 in the cycle (u → 1 is the attachment step). There may be other unattached states, not shown. Because of the close relation (Hill, 1974) be-

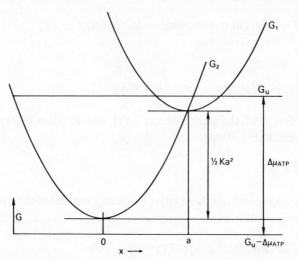

FIG. 5. Thermodynamic relations between states, including one unattached state, u. See text.

tween $G_u - G_1$ and the rate constants for u → 1 and 1 → u, it is more likely that the range l of p would be symmetrical about $x = a$ than about $x = x_0$. (Recall, however, the discussion above of restrictions on the value of p in Fig. 1.) This suggests that, in order to maintain the symmetry in T_2 discussed in the preceding paragraph, we should take $x_0 = a$.

The special case $l = a$. As already mentioned, it is necessary to choose $l = a$, in a single site model, in order that $T_2(y) = T_0$ near $y = 0$, as observed experimentally. In this case, our main results simplify as follows:

$$T_0 = Kpa(2x_0 - a)/2d \tag{14}$$

$$T_1(y) = T_0 + (Kpa/d)y \tag{15}$$

$$-y' = (2x_0 - a)/2 \tag{16}$$

$$T_2(y) = T_0 \tag{17}$$

$$T_2^{(+)}(y) = (Kpa/d)[x' - (3a/2) + y] \tag{18}$$

$$T_2^{(-)}(y) = (Kpa/d)[x' - (a/2) + y]. \tag{19}$$

It should be noted that x' is not involved in Eqs. (14)–(17).

The special case $l = a$, $x_0 = a$. We have already mentioned one reason to select $x_0 = a$. Another is the following. In a steady isotonic contraction, attachment is likely to occur near the minimum in G_1, *i.e.*, near $x = a$ (Hill, 1974; Eisenberg & Hill, 1978). To prevent negative force being generated in state 1 before the transition $1 \to 2$ occurs, x_0 should be chosen equal to or close to a. If we therefore set $x_0 = a$, Eqs. (14) and (16) become simply

$$\text{(a) } T_0 = Kpa^2/2d \qquad \text{(b) } -y' = a/2. \tag{20}$$

Further, the condition $x' = x_0 + (l/2)$ for symmetry in T_2 becomes here $x' = 3a/2$. Then Eqs. (18) and (19) simplify to

$$T_2^{(+)}(y) = (Kpa/d)y \tag{21}$$

$$T_2^{(-)}(y) = (Kpa/d)(a + y). \tag{22}$$

In Fig. 3, in this special case, $y_+ = 0$ and $y_- = -a$.

We have already mentioned that Ford *et al.* (1977) report that $-y'$ is about 40 Å. Hence, if we use this y', $a \cong 80$ Å (Eq. (20b)). The distance $y' - y_-$ in Fig. 3 (half-width of T_2) is also $a/2$ (or 40 Å). Experimentally, this distance is much larger, about 80 Å (see Fig. 13 of Ford *et al.*, 1977), because, as already mentioned, the experimental $T_2^{(-)}$ has a smaller slope than T_1. We return to this topic below and in the next section.

In Fig. 5, $\Delta\mu_{\text{ATP}}$ (free energy of hydrolysis of ATP) is about $23\ kT$. A reasonable value for the quantity $Ka^2/2$ shown in Fig. 5 is then, say, $16\ kT$, about 70% of $\Delta\mu_{\text{ATP}}$. This is adequate for a realistic thermodynamic efficiency (40–50%) and also allows acceptable kinetics (Eisenberg and Hill, 1978). From Eq. 20a, we have $T_0 d = (Ka^2/2)p$. If we take p as about 0.85, then $T_0 d \cong 13.6 kT$. (As was noted just before Eq. (1), $p = 0.85$ corresponds to 19% of all cross-bridges in the actin-myosin overlap zone attached.) Finally, if we take the repeat distance $d = 360$ Å, we get (at 3°C) $T_0 \cong 1.44 \times 10^{-7}$ erg cm^{-1}. This is just one-half of the experimental value of 2.9×10^{-7} erg cm^{-1} (Gordon, Huxley & Julian, 1966; Squire, 1977; Hill, Eisenberg, Chen & Podolsky, 1975). This result suggests that single site models cannot provide a large enough isometric force. Hence, it is likely that a cross-bridge can attach to more than one actin site per 360 Å (Hill, 1975).

Three attached states. Our object is to keep the analysis in this paper simple. Hence we are concentrating on models with two attached states only. But the general approach would obviously be valid for three or more attached states if we assume that all attached states are interconnected by transitions that are fast compared to the on-off transitions.

Figure 6 is a special case of a three-attached-state model. There is much symmetry, for simplicity. States 1 and 2 have the same K ($K_1 = K_2$), but K_3 is smaller by a factor of two. Hence F_3 has a correspondingly smaller slope. In the isometric state, we assume (Fig. 6) that the "on" distribution p is centered about the G_1, G_2 intersection point (state 3 is not occupied). We omit the proof, but it follows from the "on" distribution shown that T_0, T_1 and the three branches of T_2 are as indicated (dashed lines) in the figure. Note the location of the zero on the y axis. These results are in any case rather obvious from the symmetry of the figure. This example suggests one possible explanation of the Ford et al. (1977) observation that $T_2^{(-)}$ has a smaller slope than T_1. Another arises if detachment of the cross-bridge is rapid for $x < 0$, thus reducing negative force (note that this falls outside of our assumption, at the beginning of this section, about the "clean separation" of two time domains).

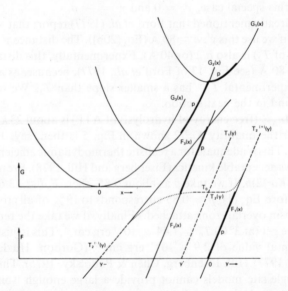

FIG. 6. Example with three attached states. The upper curves, $G_i(x)$, are free energy curves. Below are the corresponding force curves, $F_i(x)$. $F_i = 0$ where G_i has a minimum, in each case ($i = 1, 2, 3$). The assumed isometric attachment regions (heavy portions) are shown in both the free energy and force curves. Superimposed on the force curves are the dashed functions $T_1(y)$ and $T_2(y)$ (three branches), where the zero on the y axis is located as shown in the figure. The dotted portions of $T_2(y)$ are schematic only. See text.

3. SIMPLIFIED THEORY FOR TWO ACTIN SITES

In this section we extend the above results to a group of two equivalent actin sites. We use the same general procedure; hence we omit all algebraic details and give results only. The motivation here is to come one step closer to the presumably more realistic case, mentioned in Section 1, of a group of three nonequivalent actin sites (Hill, 1975).

Figure 7 shows, schematically, the two actin sites in relation to a stationary cross-bridge. The $m = 0$ site is at x; the $m = -1$ site is at $x - \delta$ (Hill, 1975), where $\delta = 55$ Å. For some values of x, the cross-bridge might attach to either site. The actin sites move to the left in a contraction (arrow). The repeat distance for pairs of sites is $d = 360$ Å. The free energy curves for states 1 and 2 are shown in Fig. 8 (Hill, 1975). To simplify the algebra, we take $x_0 = a$ at the outset. We assume, as in Section 2, that the range of attachment would be l, with probability p, for either site acting alone. But between $x = x' - l + \delta$ and $x = x'$ (cross-hatches in Fig. 7), over an interval of length $l - \delta$, both sites compete for the one cross-bridge. The possible range in l that we allow in the main treatment below is δ to 2δ. The case $l < \delta$ is much simpler (no competition); we discuss it separately.

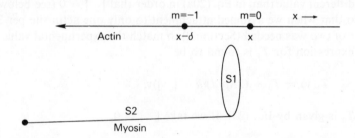

FIG. 7. Notation for a group of two actin sites in the presence of a myosin cross-bridge. See text.

We use the following simple argument to handle the competition interval $l - \delta$. With only one site interacting with the cross-bridge, the on and off rate constants are proportional to p and $1 - p$, respectively. With two sites competing, these quantities become $2p$ and $1 - p$, respectively. This gives a total probability of attachment of $2p/(1 + p)$ (the denominator being the sum of $2p$ and $1 - p$) or an attachment probability of $p/(1 + p)$ on each of the two sites.

In this treatment (accept for a brief digression, below) we assume no "slipping" between sites (Hill, 1975). That is, transitions occur between states 1 and

2 on either site, but not *between* sites, *e.g.*, state 2 ($m = 0$) → state 2 ($m = -1$).

On using weights p and $p/(1 + p)$ in the equivalent of Eq. 3, we find for the isometric force

$$T_0 = \frac{Kp}{d}\left\{a^2 + (2x' - l - a)\left[\left(\frac{l - \delta}{1 + p}\right) - (a - \delta)\right]\right\}. \tag{23}$$

For small steps y, we can find T_2 by the transformation $x' \to x' + y$ in Eq. (23). The result is

$$T_2(y) = T_0 + (2Kp/d)[\ ]y, \tag{24}$$

where [ ] is the square bracket expression in Eq. (23). To give $T_2(y)$ a zero slope at $y = 0$ (Section 2), we need here to choose parameters so that [ ] = 0. In this case, the isometric force becomes

$$T_0 = Kpa^2/d.$$

Although this appears to be double the result in Eq. (20a), p will presumably have a different value than in Eq. (20a) in order that [ ] = 0 (see below). Recall also that when we assumed attachment to only one actin site per 360 Å, a factor of two was needed (Section 2) to match the experimental value of T_0.

The expression for T_1 is found to be

$$T_1(y) = T_0 + (2Kp/d)(a + [\ ])y, \tag{26}$$

where T_0 is given by Eq. (23). If we take [ ] = 0,

$$T_1(y) = T_0 + (2Kpa/d)y, \tag{27}$$

with T_0 given by Eq. (25). We then find $-y' = a/2$ as in Eq. (20b). If, for example, $-y' = 40$ Å, then $a = 80$ Å.

We can now examine what assumptions are required for [ ] = 0. We take $\delta = 55$ Å and, as an example, $a = 80$ Å. Then l depends on p and falls between $l = 80$ Å ($p = 0$) and $l = 105$ Å ($p = 1$). This is within the allowed range of l values, δ to 2δ, for the present treatment. If, for example, $p = 0.85$, then $l = 101.25$ Å. The isometric force is correct in this case (see Section 2). The fraction of cross-bridges, in the actin-myosin overlap zone, that are attached is found to be 0.38.

For large steps $|y|$, we obtain the two other branches of T_2:

$$T_2^{(+)}(y) = (Kp/d)(a + [\ \ ])(2x' - l - 2a + 2y) \tag{28}$$

$$T_2^{(-)}(y) = (Kp/d)(a + [\ \ ])(2x' - l + 2y). \tag{29}$$

These have the same slope as $T_1(y)$ in Eq. (26). The values of y_+ and y_- are given by Eqs. (11) and (13). The "width" of the complete $T_2(y)$ curve is again a. This is because each site has two free energy or force curves separated by a, and the two sites do not interact with each other in the quick recovery. When $[\ \ ] = 0$, the two branches of T_2, above, are equidistant from $T_2(y)$ if $x' = a + (l/2)$. This is presumably also the condition for an inflection point in $T_2(y)$ at $y = 0$ (see Section 2).

Slipping. If there is fast slipping between the two sites, the width of the complete $T_2(y)$ would be increased from a to $\delta + a$ because state 2, site $m = 0$ would be used for y very negative, and state 1, site $m = -1$ would be used for y very positive. This is an improvement in T_2 width (see Section 2), but it would be at the expense of a decrease in isometric tension. The use of three or more attached states is an alternative way of widening the complete $T_2(y)$ (see Fig. 6).

With a group of three sites and fast slipping, the complete width of T_2 would be $2\delta + a$.

The case $l < \delta$. If, in Fig. 8, $l < \delta$ rather than $l > \delta$ (as drawn), there would be a gap between the p distributions on the two sites rather than an overlap (cross-hatched intervals). In this case, as the two sites move past a given cross-bridge in a slow contraction, the cross-bridge has two sites to interact with, but only one at a time. Thus the treatment in Section 2 applies except that T_0, T_1 and T_2 are all multiplied by a factor of two (there is simply a change in ordinate scale in Fig. 3). If we put $x_0 = a$ in Section 2 (to conform to our choice in the present section), and insert appropriate factors of two as just mentioned, one can verify that the Section 2 results agree, as expected, with those of the present section in the "boundary" case $l = \delta$.

The condition for a flat T_2 at small $|y|$ is again $l = a$ (Section 2). If, for example, it should turn out that $-y' \cong 20$ Å and hence $a \cong 40$ Å, then we would need $l \cong 40$ Å. Since $\delta = 55$ Å, the assumed condition $l < \delta$ is satisfied.

Three actin sites. The above treatment of two equivalent sites can be extended to three equivalent sites without difficulty. We give a few results. The condition for zero slope in $T_2(y)$ at $y = 0$ is found to be

$$\frac{l - \delta}{1 + p} - \frac{3}{4}(a - \delta) - \frac{1}{4}(l - \delta) = 0. \tag{30}$$

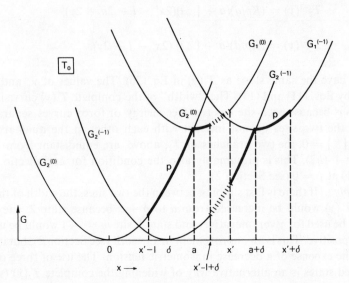

FIG. 8. Free energy curves and "on" probability distribution for a cross-bridge in the presence of two equivalent actin sites (see Fig. 7). See text for details.

When this condition is satisfied,

$$T_0 = 3Kpa^2/2d \tag{31}$$

$$T_1(y) = T_0 + (3Kpa/d)y \tag{32}$$

$$y' = a/2.$$

From Eq. (30), taking $a = 80$ Å and $\delta = 55$ Å, we find $l = 80$ Å when $p = 0$, and $l = 110$ Å when $p = 0.69$ (the acceptable range in l for this treatment is 55 Å to 110 Å). For the large values of l in the range 80Å $\leq l \leq$ 110 Å, the corresponding range $0 \leq p \leq 0.69$ is not near enough unity (for p_1) to be realistic (see the discussion of restrictions on the value of p in Fig. 1).

As a more satisfactory alternative, suppose $l < \delta$. In this case, the $T_2'(0) = 0$ condition is $l = a$ (see above). If this is satisfied, $T_0 = 3Kpa^2/2d$. If we use the example $a = 40$ Å, mentioned above, we have then $l = 40$ Å. Also, if we use the experimental value of T_0, $d = 360$ Å, and $Ka^2/2 = 16\ kT$, we deduce $p = 0.57$. But since the $p = p_1$ range is only $l/2 = 20$ Å in this case, this value of p is reasonable (i.e., G_u is not much greater than G_1^0).

4. SIMPLIFIED THEORY FOR MULTIPLE ACTIN SITES

Suppose there is an indefinite row of equivalent actin sites with spacing $\delta = 55$ Å (Hill, 1975). Figure 8 applies if we confine our attention to the interval between $x = a$ and $x = a + \delta$. The repeat distance is now δ rather than d.

The isometric tension can be shown to be

$$T_0 = \frac{Kp}{\delta} \left\{ \frac{a^2}{2} + (2x' - l - a) \left\langle \left(\frac{l-\delta}{1+p}\right) - (a - \delta) - \frac{1}{2}(l-a) \right\rangle \right\}. \tag{33}$$

For small steps y, we obtain

$$T_2(y) = T_0 + (2Kp/\delta) \langle \ \rangle y, \tag{34}$$

where T_0 and $\langle \ \rangle$ are given in Eq. (33). To make $T_2(y) = T_0$ near $y = 0$, we take $\langle \ \rangle = 0$. Then

$$T_0 = Kpa^2/2\delta. \tag{35}$$

The complete T_2 curve would again have a width a (with no slipping).

For T_1 we find

$$T_1(y) = T_0 + (2Kp/\delta)[(a/2) + \langle \ \rangle] y. \tag{36}$$

When $\langle \ \rangle = 0$,

$$T_1(y) = T_0 + (Kpa/\delta)y, \tag{37}$$

with T_0 as in Eq. (35). Equation (37) gives $-y' = a/2$ as usual.

We return to examine the conditions required for $\langle \ \rangle = 0$. We put $\delta = 55$ Å and take $a = 80$ Å, as an example. Then l is a function of p. For $p = 0$, $l = 80$ Å and for $p = 0.375$, $l = 110$ Å (upper limit). Even more so than in the discussion of three-site models above, this range in p is much too low to be realistic for p_1.

As to the magnitude of T_0, we should compare Eq. (35) with Eq. (20a). As an example, Eq. (35) with $p = 0.26$ would give a T_0 twice as large (this is needed to match experiment) as Eq. (20a) with $p = 0.85$. This follows because $d/\delta = 6.55$.

If we ignore the condition $\langle\ \rangle = 0$ and consider the special case $l = \delta$ (no cross-hatch region in Fig. 7), Eq. (34) becomes

$$T_2(y) = T_0 + (Kp/\delta)(\delta - a)y. \tag{38}$$

This $T_2(y)$ will have a significant negative slope at $y = 0$ because $a > \delta$ (still using $a = 80$Å).

The case $l < \delta$. Because there is no competition between sites for the cross-bridge in this case, the results in Section 2 (single site) apply here if we put $x_0 = a$ (special case used in Fig. 8) and $d = \delta$ (repeat distance). The condition for a flat T_2 at small $|y|$ is $l = a$.

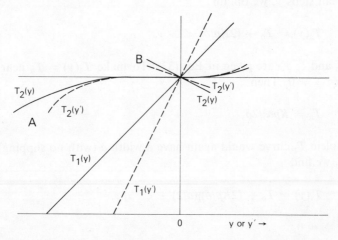

FIG. 9. Relations between observed functions $T_1(y)$ and $T_2(y)$ and the "cross-bridge functions" $T_1(y')$ and $T_2(y')$, in two hypothetical cases A and B. See text for details.

As a check, we obtain the same results in the "boundary" case $l = \delta$ either from Section 2 ($x_0 = a$, $d = \delta$) or from Eqs. (33), (34) and (36).

If we work out the same numerical example as at the end of Section 3 ($l = a = 40$ Å), using $T_0 = Kpa^2/2\delta$ and the experimental T_0, we find $p = 0.26$. With $l/2 = 20$ Å, this is rather too low for p_1 (if $G_u = G_1$, $p_1 = 0.5$ at equilibrium).

APPENDIX

If a passive series elasticity is present, the theory of the isometric transient

MODELS WITH TWO ATTACHED STATES

becomes complicated (Hill, 1975, pp. 135–138), but the modification in T_1 and T_2 is easy to handle. There is no change in T_0. In this appendix, we adopt the notation in Hill (1975) without review. This notation differs somewhat from that used above. We consider the special case that both T_1 and the passive force are linear. For the latter, we write

$$\mathscr{F}(\zeta) = a(\zeta - \zeta_R), \tag{39}$$

where ζ_R is the rest length. Then using Eq. (II-219) (Hill, 1975),

$$T_1(y') = \mathscr{N}(\bar{F}_0 + Cy') = a[\zeta(y') - \zeta_R]. \tag{40}$$

The isometric case is $y' = 0$, with $\zeta(0) \equiv \zeta_0$. If we solve Eq. (40) for $\zeta(y')$ and substitute into Eq. (II-216),

$$y = y'[1 + (\mathscr{N}C/a)]. \tag{41}$$

The y scale is expanded compared to the y' scale. Equation (40) gives the "crossbridge" $T_1(y')$. The "observed" $T_1(y)$ is found on using Eq. (41) to replace y' by y in Eq. (40). For example, if $\mathscr{N}C/a = 1$,

$$y = 2y', \quad T_1(y) = \mathscr{N}[\bar{F}_0 + (C/2)y]. \tag{42}$$

$T_1(y)$ has half the slope of $T_1(y')$.

We can treat T_2 for small y' just as we did T_1. For this purpose, we replace C by D in Eq. (40). Then Eq. (41) follows with D in place of C. As already noted, the observed slope of $T_2(y)$ at $y = 0$ is zero. Since this slope is $\mathscr{N}D/[1 + (\mathscr{N}D/a)]$ here, we must have $D = 0$. Hence the "cross-bridge" slope $\mathscr{N}D$ of $T_2(y')$ at $y' = 0$ is also zero.

For the complete T_2 curves, we write

$$\begin{aligned} T_2(y') &= \mathscr{N}[\bar{F}_0 + f(y')] \\ T_2(y) &= \mathscr{N}[\bar{F}_0 + g(y)] \end{aligned} \tag{43}$$

where $f(y') = g(y)$. From Eq. (II-216), we find

$$y = y' + (\mathscr{N}/a)f(y')$$

or

$$y' = y - (\mathcal{N}/a)g(y). \tag{44}$$

Equation (44) shows how to calculate y' from each y, to obtain the function $T_2(y')$, given the observed $g(y)$. In the special case $\mathcal{N}C/a = 1$ (above),

$$Cy' = Cy - g(y). \tag{45}$$

The y' scale is compressed somewhat compared to the y scale (except near $y = 0$, where $y' = y$).

Figure 9 illustrates a hypothetical but self-consistent calculation of $T_2(y')$, given $T_1(y)$, $T_1(y')$ and $T_2(y)$. In case A, $T_2(y)$ has zero slope at $y = 0$. In case B, $T_2(y)$ has a negative slope at $y = 0$.

REFERENCES

Eisenberg, E. & Hill, T. L. (1978). A cross-bridge model of muscle contraction. *Prog. Biophys. molec. Biol.* **33**, 55–82.
Ford, L. E., Huxley, A. F. & Simmons, R. M. (1977). Tension responses to sudden length change in stimulated frog muscle fibers near slack length. *J. Physiol.* **269**, 441–515.
Gordon, A. M., Huxley, A. F. & Julian, F. J. (1966). Tension development in highly stretched vertebrate muscle fibers. *J. Physiol.* **184**, 170–192.
Hill, T. L. (1968). On the sliding filament model of muscular contraction. II. *Proc. natn. Acad. Sci. U.S.A.* **61**, 98–105.
Hill, T. L. (1974). Theoretical formalism for the sliding filament model of contraction of striated muscle. Part I. *Prog. Biophys. molec. Biol.* **28**, 267–340.
Hill, T. L. (1975). Theoretical formalism for the sliding filament model of contraction of striated muscle. Part II. *Prog. Biophys. molec. Biol.* **29**, 105–159.
Hill, T. L. (1977). *Free Energy Transduction in Biology.* New York: Academic Press.
Hill, T. L., Eisenberg, E., Chen, Y. & Podolsky, R. J. (1975). Some self-consistent two-state sliding filament models of muscle contraction. *Biophys. J.* **15**, 335–372.
Hill, T. L. & Simmons, R. M. (1976). Free energy levels and entropy production in muscle contraction and in related solution systems. *Proc. natn. Acad. Sci. U.S.A.* **73**, 336–340.
Huxley, A. F. & Simmons, R. M. (1971). Proposed mechanism of force generation in striated muscle. *Nature, Lond.* **233**, 533–538.
Squire, J. M. (1973). General model of myosin filament structure. III. Molecular packing arrangements in myosin filaments. *J. molec. Biol.* **77**, 291–310.
Stein, L., Schwarz, R. P., Chock, P. B. & Eisenberg, E. (1979), in preparation.

Discussion

Tregear: Can you at all distinguish the likelihood of the two or three site hypotheses, or is that too naïve a question at this point?

Hill: Well, I think that if we extend this to two sites we get just the right isometric force. A factor of two then disappears, and you get the right isometric force, but I do not take that too seriously. I feel just on structural grounds, given the helical structure of the actin filament, that the most obvious simple guess is to use a three-site model, with the central site being a "nicer" site, so to speak, than the other two. That is to say, it would have a lower free energy than either of the other two. That is the kind of thing, I think, we will put in our model. The main problem with the one-site calculation is just the T_0 business. Everything else seems pretty nice. T_0 is just too small. Two sites take care of that problem—two equivalent sites—and I think, clearly, three nonequivalent sites will take care of it just as well.

Tregear: I think then it is of interest that the various methods of modeling on the arthropod muscle have all come up with numbers between two and three sevenths for the fractional number of sites occupied in rigor.

Hill: Splendid.

Mohan: Dr. Hill, in your potential energy diagrams you ignore the rotation of the head. Is that true?

Hill: No. I did not ignore anything in that respect, that is, no commitment is made about the origin of the elasticity. When you draw these free energy curves, it is a little bit quasi-thermodynamic in the sense that if you just look at the free energy curves, and even look at a set of rate constants, you can go through the formalities of the calculation of the properties of the model without deciding what the molecular origin of the free energy curves were. So I did not say anything about whether the free energy curves are related to the bending of the actin-S-1 angle, or whether, on the other hand, they might be due to the elasticity of the S-2. Either one could give you exactly the same looking free energy curves.

Gergely: You have this parameter. I noticed alongside it you put 360 Å; is that the range of the movement of one cross-bridge?

Hill: Well, I have been laboring under a misapprehension for a few years. I just talked to Hugh Huxley about this. The repeat distance in the actin filament, I thought I had read a long time ago, was 360 Å, using "repeat" in quotes, but there is more to it than that. Probably it is more appropriate for the purposes of these calculations to use 7 x 55 Å, namely 385 Å, 55 Å being the distance between actin sites, and taking 7 in a group, even though that is not a true repeat. I think in our numerical work—in the paper I mentioned earlier that Evan's working on—I think that we will switch from 360 Å to 385 A. I have been using 360 Å for several years, and I guess I should not have.

Active Potential and Dynamic Cooperativity in the Chemo-Mechanical Conversion in Active Streaming and Muscle Contraction

H. SHIMIZU, M. YANO, K. NISHIYAMA, K. KOMETANI, S. CHAEN and T. YAMADA

Faculty of Pharmaceutical Sciences, University of Tokyo, Tokyo, Japan

ABSTRACT

A streaming system was reconstituted from acto-HMM of rabbit skeletal muscle; under physiological conditions, an active streaming of velocity of 20 µm/sec occurs for about 90 minutes in the direction specifically determined by the polarity of F-actins fixed in the system. The streaming system has a second-order phase transition, which can be explained well in terms of our three-state theory for the chemo-mechanical cycle in muscle contraction. In the disordered phase appearing below T_c, streamings are not observed and the ATPase activity is substantially the same as that in isotropic systems. Steady and uniform streamings occur only when the molecular dynamics of the elementary cycle has an order in the phase above T_c. The order is caused by a new mechanism, dynamic cooperativity, which, on the other hand, breaks the thermodynamic detailed balance in the ATPase activity and results in a kind of Fenn's effect. The dynamic cooperativity also exists in muscles and works to significantly promote the velocity of shortening and the efficiency of the chemo-mechanical conversion. It is probable that tension is generated by a single head of myosin, while the dynamic cooperativity requires two heads on a myosin molecule.

1. INTRODUCTION

A large gap still exists between the biochemistry of acto-heavy meromyosins (acto-HMM) system in a test tube and the physiology of contracting muscles. For instance, from experiences obtained *in vitro*, we know that an enzymatic reaction such as

$$E + S \underset{k_{-1}}{\overset{k_1}{\rightleftharpoons}} ES \underset{k_{-2}}{\overset{k_2}{\rightleftharpoons}} E + P \tag{1}$$

is nothing but a special case of ordinary chemical reactions; therefore, the ratios k_1/k_{-1} and k_2/k_{-2} are given in terms of the free energy differences ΔG between E + S and ES and between ES and E + P, respectively. Such *thermodynamic detailed balances* have been assumed by Bagshaw & Trentham (1973), Arata, Inoue & Tonomura (1975) and White & Taylor (1976) in the biochemistry of acto-HMM. Generally, isolated molecules in a medium are strongly perturbed by thermal agitations from their surroundings. As a result, molecular dynamics have the random and stochastic character as Brownian movements of small and independent particles. This random nature of the molecular dynamics gives an assembly of molecules a macroscopic dynamics in which changes occur in the direction of increasing randomness (entropy)—more exactly, in the direction of decreasing free energy G. At the same time, the molecular dynamics satisfy the thermodynamic detailed balance when we observe them from a statistical point of view. Therefore, the validity of the thermodynamic detailed balance in a dynamical process such as (1) means that the process can be discussed in terms of thermodynamics (for an equilibrium system).

The validity of the thermodynamic detailed balance may be present in muscle, provided that the elementary cycles of ATP hydrolysis are independent from each other as they are in acto-HMM in a test tube. Gordon, Huxley & Julian (1966) concluded that the force-generators work independently in muscle contraction. Therefore, it seems natural that Hill (1974) argued that any model for the elementary cycle must satisfy the thermodynamic detailed balance. We think that the concept of the independent force-generator is not only an acceptable one as the zeroth order approximation but also the most reasonable one as a working hypothesis for the molecular mechanism of muscle contraction. The "independent" force-generator was hypothesized only from measurements on mechanical properties of contracting muscles. Hence, it can say nothing about details of the chemical reaction such as the thermodynamic detailed balance of the elementary cycle. Or we may say that the "independent" force-generator does not always mean an independent elementary cycle (Hill, 1975). This is by no means a matter of minor correction but is very important for the thermodynamics of the chemo-mechanical conversion in muscle contraction.

According to Schrödinger (1944), life phenomena can be regarded as a kind of ordered dynamics of a macroscopic system, and the appearance of a life phenomenon requires a decrease of randomness in a specific part of the system, even continuously. Consequently, he asserts that life phenomena can appear in the system when and only when molecules acquire an order instead of randomness in their dynamics. If his view is correct, the anisotropic and regular arrangments of component molecules in sarcomeres is not a sufficient condition for muscle contraction, a kind of life phenomenon, but is only a necessary condition. For life phenomena, the molecular dynamics must change their random character, which breaks the thermodynamic detailed balance. Therefore, thermodynamics (for equilibrium systems) cannot be applied to the chemo-mechanical conversion in muscle contraction. (Rigorously speaking, there is no proof which supports the validity of the thermodynamic detailed balance in acto-HMM ATPase activity. However, we will accept this as a reasonable hypothesis and will discuss the detailed balance in the actomyosin ATPase activity in muscles.)

Schrödinger could succeed neither in finding a reasonable mechanism for carrying an order to the molecular dynamics nor in showing evidence that molecular dynamics have an order when a life phenomenon appears in a living system. Even today, these two points are still left unsolved as fundamental problems in the dynamic aspect of life phenomena. This is because living systems are so complex that quantitative studies on the molecular dynamics are accompanied by technical difficulties. Therefore, it is highly desirable to obtain a system which is between acto-HMM in a homogeneous solution, where no life phenomenon appears, and muscle, which is a living system having a complex structure, and to link the biochemistry of acto-HMM *in vitro* to the physiology of muscle contraction. This was the motivation for our reconstituting an artificial streaming system from acto-HMM of rabbit skeletal muscle. The structure of the system must be simple enough and the steady state must continue long enough that quantitative observations on the relation between the streaming velocity and the ATPase activity are possible. In the following discussions we will show evidence that the molecular dynamics of acto-HMM has an order as argued by Schrödinger when the active streaming is produced in our streaming system. A new mechanism, *dynamic cooperativity*, will be proposed as the molecular mechanism to yield order in the molecular dynamics. It will be argued that dynamic cooperativity is also present in muscle contraction and promotes the efficiency of the chemo-mechanical conversion.

2. STREAM CELLS

Figure 1 sketches an outline of our stream cell (Yano, 1978), in which active streaming was caused by acto-HMM from rabbit skeletal muscle, utilizing the chemical energy of ATP. The cell is composed of two concentric cylinders, A and B, tightly attached to a base-plate C. There is a narrow slit, called the circular slit, between the two cylinders. Both cylinders are not

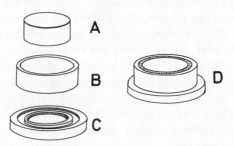

FIG. 1. A schematic representation of the stream cell; the inner cylinder (A), the outer cylinder (B), and the bottom plate (C) on which the two cylinders can be attached mechanically (D) are shown.

only rotatable around the center but also can be removed from the base-plate.

On both walls of the circular slit we wanted to fix F-actins and to fill the slit with a solution containing HMM, ATP and suitable amounts of proper ions. This solution will be named the HMM solution. According to Curie's symmetry principle, no directed flow can be produced in an isotropic system by chemical reactions (Katchalsky & Curran, 1965). Consequently, the most crucial point in our study was to realize a polarity or anisotropy in the arrangement of F-actins in the circular slit. Our basic idea was as follows (Yano, 1978; Yano, Yamada & Shimizu, 1978). F-actin itself has a polarity; only one of the two terminals of an F-actin filament can show a high affinity to certain material under some suitable conditions, while the other terminal moves relatively freely. Therefore, in an externally driven flow the latter terminal will be moved in the direction where the friction between the fluid and the filament is minimized, resulting in filamental reorientation in the direction of flow. Then, this orientation of the F-actin filament will be fixed on the wall of the slit by means of chemical bonds between the filament and the wall (Yano, 1978; Yano, Yamada & Shimizu, 1978).

We found that Millipore filter had a reasonable ability to select the specific terminal under suitable concentrations of magnesium ions. A sheet of Millipore filter was cut to cover almost the whole surface of the walls and was attached to the walls by means of double-sided Scotch tape. Then the Millipore filter was activated by cyanogen bromide for the chemical fixing of F-actin filaments. The slit was filled with an F-actin solution, and was incubated for three hours at 4°C. In the initial 10 min of the incubation, only the inner cylinders A_1 and A_2 were rotated in the opposite directions between paired stream cells 1 and 2, giving rise to a flow in the direction of the rotation as shown in

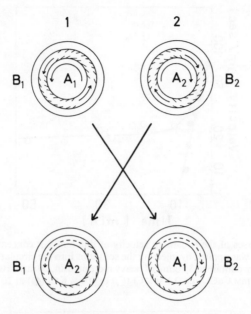

FIG. 2. Preparation of the stream cells. Inner cylinders A_1 and A_2, outer cylinders B_1 and B_2, and active streamings (--→).

Fig. 2. Resulting polarity in the F-actin system will be different between the inner and outer cylinders of each cell. Hence, after the incubation, the inner cylinders were exchanged between the paired cells to make the directions of the polarity of F-actins the same on both walls of each cell, as shown in Fig. 2. Steady and uniform streaming was observed in the directions indicated by dotted arrows, which are those of the externally driven flows, after the F-actin solution was replaced by an HMM solution. The streaming directions were opposite between the paired cells (Yano, Yamada & Shimizu, 1978). When an initial velocity was given to the HMM solution in the counter-specific direction, the initial flow came to a stop within 10 min and a new streaming appeared spontaneously in the specific direction that continued steadily for more than 1hr.

The steady and uniform streaming appeared only under physiological conditions for muscle contraction (Yano, Yamada & Shimizu, 1978). In other words, it required the presence of Mg-ATP in the HMM solution. The addition of EDTA or the replacement of Mg-ATP by $Mg-PP_i$ stopped the streaming within 2–5 min. As shown in Fig. 3, the initial velocity which was given ex-

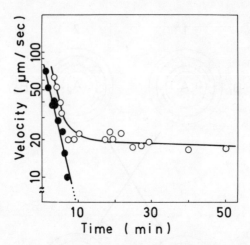

Fig. 3. Time courses of the streaming velocity at different salt concentrations; the initial velocity of the fluid was applied by injecting the solutions into the circular slit at a rate of 100 μm/sec. The solution contained 3 mg/ml of heavy meromyosin, 20 mM Tris-maleate (pH 7.0), 4 mM MgATP, 0.1 mM $CaCl_2$ and 0.1 M KCl_2(○) or 0.6 M KCl (●) at 20°C.

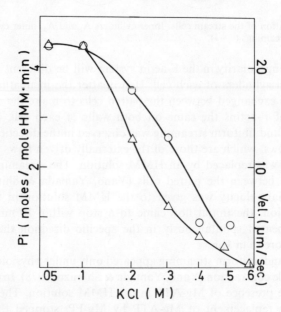

Fig. 4. Plots of the ATPase activity (○) and streaming velocity (△) as a function of salt concentration with 3 mg/ml of heavy meromyosin, 0.25 mg/ml of native tropomyosin, 40 mM Tris-maleate (pH 7.0), 10 mM ATP, 4 mM $MgCl_2$ and 0.1 mM $CaCl_2$ at 20°C.

ternally to the HMM solution in the cell decayed within 10 min at 20°C when the salt concentration in the solution was 0.6 M, but when the concentration was 0.1 M, a steady and uniform streaming continued for about 90 min after the decay of the initial velocity. The maximum velocity of the steady and uniform streaming was about 20 µm/sec, which was determined by measuring the time needed for small particles in the HMM solution to move a constant distance on the micrometer scale of an optical microscope.

When the salt concentration was varied, a strong correlation was observed (as shown in Fig. 4) between the ATPase activity, indicated by open circles, and the streaming velocity, indicated by triangles. Furthermore, the streaming was controlled by calcium through the troponin and tropomyosin system as muscle contraction (Ebashi, Endo & Ohtsuki, 1969). Native tropomyosin was bound to F-actin in a stream cell. As shown in Fig. 5, there was a strong correlation between the ATPase activity, denoted by open circles, and the streaming velocity, denoted by triangles, as the Ca concentration was varied. The addition of EGTA to the HMM solution brought the streaming to a complete stop within 2–4 min. From the above evidence we may conclude that active streaming is produced in our artificial system by acto-HMM due to a molecular mechanism which is substantially the same as that of muscle contraction.

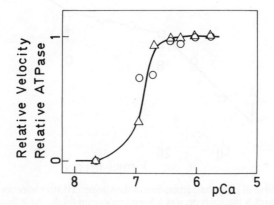

FIG. 5. Plots of the ATPase activity (○) and streaming velocity (△) at 20°C as a function of pCa with 3 mg/ml of heavy meromyosin, 0.25 mg/ml of native tropomyosin, 50 mM KCl, 40 mM Tris-maleate (pH 6.8), 10 mM ATP, 4 mM $MgCl_2$ and 2 mM EGTA. The values of the ATPase activity plotted are normalized as [(activity)-(the minimum activity)]/[(the maximum activity)-(the minimum activity)]. The values of the streaming velocity plotted are normalized to the value of the maximum velocity. The ATPase activities at pCa = 6.07 and 7.70 were 5.17 and 0.73 mole P_i/mole HMM·min, respectively.

3. THE ACTO-HMM ATPASE ACTIVITY IN THE STREAMING SYSTEM

In our streaming system a steady state lasts for a long time. Therefore, it is particularly suitable for a study of quantitative relations between the mechanical motion and the chemical reaction in a steady state. The relation between the streaming velocity and the ATPase activity is shown in Fig. 6 (Yano & Shimizu, 1978); during times A and B the HMM solution was allowed to stream freely with the velocity about 20 μm/sec. Then at time B, four resistances were placed in the circular slit to reduce the streaming velocity. Streaming was observed only very vicinity of the walls of the slit, with velocity of the order of 1 μm/sec or less. And the resistances were removed again at time C. The gradient of the lines in Fig. 6 represents the size of the ATPase activity. Evidently the ATPase activity was increased in the active streaming, which means that a kind of Fenn's effect is present in our streaming system.

Figure 7 shows the dependence of the streaming velocity and the ATPase activity on the substrate concentration. Open circles indicate the ATPase ac-

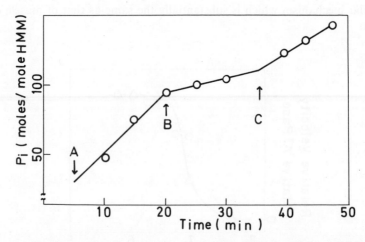

FIG. 6. Time course of phosphate liberation in the stream cell after injection of the solution. Between times A and B the solution was allowed to stream freely. At B four pieces of chromium-coated wire were inserted in the circular slit and at the time C they were removed, giving rise to a steady and uniform streaming again. The solution consisted of 3 mg/ml of heavy meromyosin, 50 mM KCl, 40 mM Tris-maleate (pH 7.0), 10 mM ATP, 4 mM MgCl$_2$ and 0.1 mM CaCl$_2$ at 20°C.

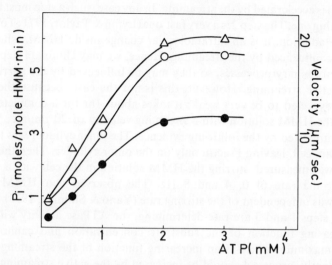

FIG. 7. Representation of the ATPase activity and the streaming velocity at 20°C as a function of MgATP concentration. Open triangles denote the streaming velocities. Open circles and closed circles show the ATPase activities of free streaming and resisted streaming, respectively. The solutions consisted of 3 mg/ml of heavy meromyosin, 0.25 mg/ml of native tropomyosin, 10 mM ATP, 50 mM KCl, 40 mM Tris-maleate (pH 7.0), 0.1 mM $CaCl_2$ and various amounts of $MgCl_2$.

tivity in the case of free streaming, the velocity of which is denoted by triangles. Closed circles represent the ATPase activity when the four resistances were inserted in the circular slit. It is remarkable that there is a good correlation between the ATPase activity and the streaming velocity, and that the ATPase activity at high concentrations of substrate has a plateau which is an increasing function of the streaming velocity.

The elementary processes in the acto-HMM ATPase activity may be simplified as

$$A + M + S \underset{k_{-1}}{\overset{k_1}{\rightleftharpoons}} A + MS \underset{k_{-2}}{\overset{k_2}{\rightleftharpoons}} A + M^*P \underset{k_{-3}}{\overset{k_3}{\rightleftharpoons}} AM^*P \underset{k_{-4}}{\overset{k_4}{\rightleftharpoons}}$$
$$A + M + P, \qquad (2)$$

where A and M respectively stand for actin and HMM, and the symbol * on M indicates that HMM has a nonequilibrum conformation storing excess energy (Yamada, Shimizu & Suga, 1973). Now we want to determine which step in (2) is accelerated by the streaming. In any case, such a step must be a rate-determining one. The step 2 is a very fast one (Lymn & Taylor, 1971; Tonomura, 1972). In addition, it is an intramolecular change inside HMM, which could hardly be influenced by the streaming. Hence, we may eliminate step 2. Steps 1 and 3 are binary processes, so they may be influenced by the stirring effect due to active streaming. However, this is not the case, because the stirring effect is expected to be very weak; it takes about 1hr for a complete circulation of the HMM solution with a streaming velocity of 20 μm/sec. This was further supported by the following evidence. The inner cylinder was taken out of a stream cell, leaving F-actin only on the outer cylinder. Then the ATPase activity was measured, stirring the HMM solution in the cell with a magnetic stirrer at the rate of 0, 4 and 8 Hz. The observed acto-HMM ATPase activity was independent of the stirring rate (Yano & Shimizu, 1978). Furthermore, if steps 1 and 3 are rate-determining, the ATPase activity will become an increasing function of the substrate concentration and cannot have a plateau maximum which is an increasing function of the streaming velocity. Consequently, process 4 should be increased by the active streaming.

Figure 8 shows the temperature dependence of the streaming velocity (Yano & Shimizu, 1978) by triangles and of the ATPase activity by open circles. Remarkably, our streaming system has a biphasic behavior with respect to temperature change. In the phase below the critical temperature T_c, which is about 10°C in this example ($1/T_c$ is about 3.53×10^{-3} deg^{-1} in Fig. 8.), no uniform streaming appeared, and the temperature dependence of the ATPase activity was essentially the same as that in isotropic systems, which is indicated by a dotted line. On the other side, the active streaming was observable only in the phase appearing above the critical temperature T_c, and the ATPase activity there showed a nonlinear change, *i.e.*, a remarkable increase with respect to a negative change of the inverse of the temperature. The critical temperature T_c was shifted to a higher temperature region when resistances were placed in the circular slit, and was shifted to a lower temperature region when the concentration of F-actins on the walls of the slit was increased. Consequently, the phase transition of the streaming system at the critical temperature T_c is by no means attributable to an intramolecular phase change inside the AM*P complex; it must be caused by some intermolecular interactions.

Generally, the state of the thermodynamic system is determined under a competition of counterworking forces——namely, interacting forces among the elements of the system and random forces due to thermal agitations. When the effect of the former forces exceeds that of the latter forces, the elements show a cooperativity in their behavior and we have an ordered phase in the

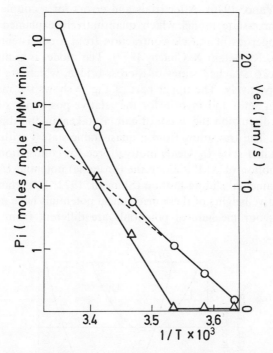

FIG. 8. Temperature dependence of the streaming velocity and ATPase activity. Abscissa: the inverse of the absolute temperature. Ordinate: the streaming velocity (△) and the logarithm of the ATPase activity (○). The solution consisted of 3 mg/ml of heavy meromyosin, 10 mM ATP, 4 mM $MgCl_2$, 50 mM KCl and 40 mM Tris-maleate (pH 7.0).

system. Clearly, in our streaming system the phase appearing above T_c is an ordered phase. What kind of interaction is involved among the elements?

4. METACHRONAL ROTATION MODEL OF MYOSIN HEADS

To explain the cause of the cooperativity in our streaming system, we must change the subject momentarily. It has been gradually accepted that the AM*P complex in actomyosin ATPase activity in muscle contraction has at least two different states. Important contributions have been made to this idea by Huxley & Simmons (1971), White (1972), Huxley (1973), Julian, Sollins & Sollins (1974) and Hill (1974). We have also come to this conclusion in recent years in our study of the shortening and the ATPase activity in isolated sarcomeres and in muscle (Shimizu & Yamada, 1975; Shimizu, Yamada,

Nishiyama & Yano, 1976). After trials and errors for a couple of years, we arrived at a three-state model which quantitatively explained most of the mechanical properties of muscle contraction from a consistent base (Nishiyama, Shimizu, Kometani & Chaen, 1977). The model is composed of one detached and two attached states of a cross-bridge, which are named states 0, 1 and 2, respectively. The upper part of Fig. 9 shows the mechanical potentials between actin and myosin for the relative position x of the myosin. In the present discussion the x coordinate is taken in the direction of the sliding motion. The transition from a quasi-stable state 1 with potential U_1 to a quasi-unstable state U_2 yields motive force for contraction, and the intramolecular motion of AM*P along the (unstable) potential U_2 is the direct cause of the filamental sliding motion (Shimizu, 1972; Kometani & Shimizu, 1977). The relative heights of these mechanical potentials have no significance in this model; our mechanical potentials are different from Hill's (1974)

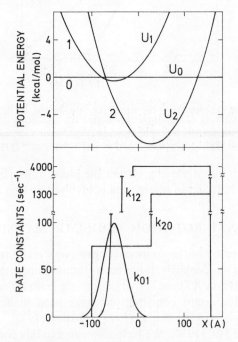

FIG. 9. The upper curves: potential energies for the three states as a function of x. U_0, U_1 and U_2 denote, respectively, the potential energies of the states 0, 1 and 2. The lower curves: important transition probabilities among three states as a function of x.
The value of x in this figure differs from that in the text by a constant factor.

thermodynamic potentials. Some important transition probabilities among these states are shown in the lower part of Fig. 9. The most remarkable point in our three-state model is that the rate constant k_{12} for the transition from state 1 to state 2 is a very steeply increasing function of x as $k_{12}(x) = k_{12}(0)$ exp (Kx/kT), with the Boltzmann constant k and a positive constant K.

Now, going back to our streaming system. As shown schematically in Fig. 10, the active motion of HMM (e.g., the rotation of the S–1 part of HMM [Huxley & Simmons, 1971]) along the unstable potential U_2 on F-actin will push fluid around the molecule in the x-direction with velocity v_f. This active motion will produce a microscopic flow which will be significantly weaker at a certain relaxation distance d^* because of friction inside the fluid. (d^* is an increasing function of the temperature). At low temperatures the mean distance d between neighboring AM*P complexes (in state 1) is larger than d^*. Consequently, each AM*P will decompose independently without mutual interference, as schematically illustrated in the upper part of Fig. 10. The greater part of the mechanical energy of the rotational motion of HMM will be dissipated to thermal energy through friction. Macroscopic streaming will appear in the system only when the production rate of the microscopic flows is larger than their decay rate. In the case of $d^* < d$, we have only a low efficiency of chemomechanical conversion because the rate of the dissipation is high. The condition for the production of macroscopic streaming is specifically satisfied, as will be discussed below, provided that d^* is larger than d, which is obtained at high temperatures. In such a case the decompositions of AM*P complexes become cooperative, and microscopic flows produced at the molecular level will expand with a high efficiency to a uniform (ordered) streaming in the stream cell. Nonuniform (disordered) streaming may appear in the case of $d^* < d$. However, the efficiency of the chemo-mechanical conversion will become small in this case. Therefore, in the first-order approximation, we may neglect such nonuniform streamings.

When AM*P complexes are placed under perturbations of microscopic flows from other AM*P complexes, as shown in the lower part of Fig. 10, the the microscopic flows can induce the active motion (head rotation) to HMM by the perturbed AM*P by the mechanism illustrated in Fig. 11 (Shimizu & Yano, 1978). The key mechanism of the induction of the active motion is that the microscopic flow pushes or tilts the perturbed HMM on the potential U_1 in the x direction by means of a frictional force between the flows and the perturbed HMM, and the force F_s is balanced with the potential force F_1 from U_1 at a certain position $x°$ on the x coordinate. The resulting increase in the x coordinate of the HMM will significantly promote the probability of the transition from state 1 to state 2, because this probability is a steeply in-

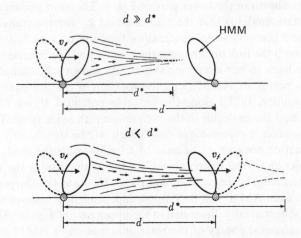

FIG. 10. Schematic representations of the active motion of heavy meromyosin on F-actin. (a) In the case of $d^* < d$, each heavy meromyosin moves independently, and (b) in the case of $d^* \geq d$, heavy meromyosin molecules move cooperatively through a dynamic interaction mediated by the microscopic flow, the initial velocity of which is v_f.

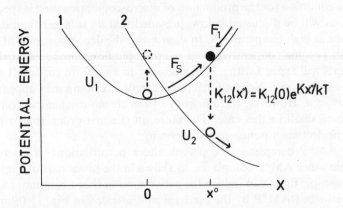

FIG. 11. The transition from state 1 to state 2 is induced by the local streaming which moves the mean position of heavy meromyosin from $x = 0$ to $x°$, where the force F_S from the streaming is balanced by the potential force F_1. In the independent process the transition $1 \rightarrow 2$ occurs around $x = 0$ with a much smaller probability than in the cooperative processes.

creasing function of x as $\exp(Kx/kT)$. The microscopic flow from an AM*P complex will induce active motion on the perturbed AM*P complex before the flow is appreciably decayed by friction. The second AM*P will add a new microscopic flow to the flow from the first AM*P. In other words, the microscopic flow will be amplified while the flow is relayed from complex to complex successively in a metachronal manner. Such an amplification may be compared with the amplification of an electric current by membrane excitation in conduction through a nerve fiber. Such an amplification is necessary for the growth of the microscopic flows to a macroscopic streaming. A microscopic flow induces a cooperativity, *dynamic cooperativity*, in the molecular dynamics of AM*P complexes and gives rise to, for instance, *metachronal rotations of myosin heads* on F-actin filaments. This metachronal motion may be compared to the motions of toothed wheels in a watch mechanism by Schrödinger. The metachronal motions having dynamic cooperativity will increase the acto-HMM ATPase activity in an autocatalytic manner, because the transition $1 \to 2$ is a rate-determing step for the processes (2). This increase in ATPase activity is considered as the cause of the Fenn effect observed in our streaming system.

The above concept of the self-organization of the macroscopic streaming was tested by comparing our data on the temperature dependence of the streaming velocity and of the ATPase activity with theoretical results which were calculated from assumptions based on the concept (Shimizu & Yano, 1978). We assumed that n AM*P complexes decompose per unit time through the independent and the cooperative processes. It was shown that the number n_{coop} decomposing per unit time through the latter route could be written in the first-order approximation as $n_{\text{coop}} = ncv$, where v represents the streaming velocity and c is a proportionality constant which is independent of temperature T. The fluid is moved in the circular slit by HMM moving actively with a velocity v_f in the x-direction. Hence, the fluid is given a motive force F_m which approaches the velocity v to v_f. This means that F_m is proportional to $v_f - v$, at least in the first-order approximation. Neglecting the contribution from independent motions of HMM, we may write

$$F_m = f(v_f - v)n_{\text{coop}}, \qquad (3)$$

where f is a force constant which is an increasing function of the adhesion of the fluid. In addition to the motive force, the streaming solution is given a resistive force $F_r = -\gamma v$ by the friction inside the fluid, whose friction constant is γ. Therefore, the equation of motion for the streaming fluid may be given as

$$m dv/dt = F_m + F_r = f(v_f - v)n_{\text{coop}} - \gamma v = cnf(v_f - v)v - \gamma v, \quad (4)$$

which is a nonlinear equation with respective to v, where m denotes the mass of the solution. In Eq. (4) the friction constant γ is inversely proportional to T according to the fluctuation-dissipation theorem (Kubo, 1966), and the velocity v_f of the active motion of HMM can be regarded as a constant in the first-order approximation because it depends only weakly on the temperature (Kometani & Shimizu, 1977). The streaming velocity v in the steady state can be calculated from Eq. (4) as

$$v = \begin{cases} 0 & \text{for } T < T_c \\ a(1/T_c - 1/T) & \text{for } T > T_c, \end{cases} \qquad (5)$$

where $a/T \equiv \gamma/cnf$ and $a/T_c \equiv v_f$ and the values of a and T_c are to be determined from comparisons with experimental data.

The calculation of the ATPase activity is simple (Shimizu & Yano, 1978). We may assume that the rate of the ATPase activity is proportional to that of the rate determining step $1 \to 2$, and that the transition occurs at $x = 0$ for independent routes, and at the position $x°$ in which $F_s(v) = F_1(x°)$ in Fig. 11 for cooperative routes. On the other hand, we may write $F_s = \beta v$ with a friction constant β and $F_1(x) = \alpha x$. Hence, the value $x°$ is proportional to v such as $x° = (\beta/\alpha)v$ and the rate constant for ATP hydrolysis through the cooperative process may be written as

$$k_{\text{coop}}(v) = k° \exp(Kx°/kT) = k° \exp(\beta Kv/\alpha kT), \qquad (6)$$

where $k°$ represents the rate constant for the independent process. The observed activity is the sum of the rates of the ATP decompositions through the independent and the cooperative processes:

$$J(v) = k°(n - n_{\text{coop}}) + k_{\text{coop}} n_{\text{coop}}$$
$$= nk°[(1 - cv) + cv \cdot \exp(\beta Kv/\alpha kT)];$$

or, if we define the effective rate constant $k(v)$ by $J(v) = nk(v)$, we obtain

$$k(v) = k° + ck°[\exp(\beta Kv/\alpha kT) - 1]v. \qquad (7)$$

Substituting (5) into (7) and taking the logarithm of the resulting equation, we obtain

$$\log k(v) = \log k° + \log[1 + ac(1/T_c - 1/T)(\exp\{(Kab/\alpha kT^2) \\ \times (1/T_c - 1/T)\} - 1)], \qquad (5)$$

where $\beta = b/T$ and $\log k°$ is a linearly decreasing function of $1/T$, and three unknown quantities α, b and $\log k°$ are to be determined from comparisons with experimental data. Comparisons between the theory and the experiment are given in Fig. 12, where theoretical values are indicated by lines. Excellent agreement is seen, from which we may conclude that our concept of the self organization of the macroscopic streaming in terms of the dynamic cooperativity of the chemo-mechanical processes is a reasonable one.

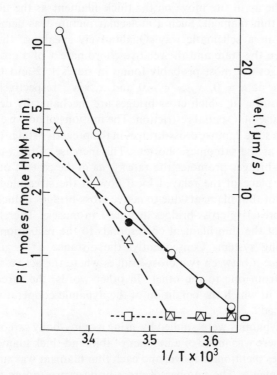

FIG. 12. Temperature dependence of the streaming velocity and the ATPase activity. Abscissa: the inverse of the absolute temperature. Ordinate: the streaming velocity and the logarithm of the ATPase activity. Open circles (○) and closed circles (●) denote experimental values of the ATPase activity for free and resisted streaming, respectively, and the solid line (– –) and dotted-dashed line (– - –) indicate the theoretical values of the ATPase activity for free and resisted streamings, respectively. The triangles (△) and squares (□) represent the experimental values for the streaming velocity in free and resisted streamings, respectively, and the dashed line (- - -) and dotted line (. . . .) show the theoretical values of the streaming velocity of free and resisted streamings, respectively. The solution consisted of 3 mg/ml of heavy meromyosin, 10 mM ATP, 4 mM $MgCl_2$, 50 mM KCl and 40 mM Tris-maleate (pH 7.0).

5. THE DYNAMIC COOPERATIVITY OF CHEMO-MECHANICAL PROCESSES IN MUSCLE CONTRACTION

In the sarcomere, the spatial pitch of the helical structure of the thick filament, 429 Å, is different from the corresponding one of the filament, 370 Å. Consequently, the x-coordinate relative to the interacting actin site is in general different between cross-bridges which are located at different positions on the same thick filament. Hence, the "good" positions where cross-bridges can interact with the actin site move on the thick filament as the filament slides relative to the thin filament. Such a molecular picture was demonstrated by Huxley (1969) in a schematic way. Qualitatively speaking, there is a correlation between the state and the relative coordinate x of a cross-bridge, so that cross-bridges are most probably found in states 1, 2 and 0 when their coordinates are at $x \simeq 0$, $x_d > x \gg 0$ and $x > x_d$, respectively, where x_d denotes the position at which cross-bridges are mechanically detached from actin sites because of a spatial restriction. The motions of active cross-bridges, cross-bridges in state 2, move cross-bridges in the preactive state, the state 1, in the x direction as the sarcomere shortens. This increase of the x-coordinate in the latter cross-bridges promotes the rate k_{12} as in the case of our streaming system. In the place of the relayed local flow in the streaming system, the sliding motion of thin filaments due to active cross-bridges induces the transition $1 \rightarrow 2$ in preactive cross-bridges in state 1 in muscle. Therefore, in muscle the length of the thin filament corresponds to the relaxation distance d^* in the streaming system. Consequently, the distance d^* is always larger than the distance d between two cross-bridges where the active state, state 2, is transferred from one to the other. In other words, the sarcomere has a steric structure in which the condition of the dynamic cooperativity, $d < d^*$, is always satisfied.

Sarcomere dynamics were studied by using a many-body sarcomere model, where a sarcomere was made of a number of thin and thick filaments with the helical structures mentioned above, and each thin filament was surrounded by three thick filaments. The number of cross-bridges protruding from a thick filament to a thin filament was assumed to be 25, which means that a thin filament is surrounded by 75 cross-bridges. The mechanical potentials and the transition probabilities which were used in our three-state model (Nishiyama et al., 1977) were assumed for the molecular dynamics for each cross-bridge. One of the two Z-bands of the sarcomere was spatially fixed, and a load was imposed on the other Z-band.

Figure 13 shows one of the results obtained from our sarcomere model; it shows the variations in the state of the cross-bridges in a half-sarcomere

DYNAMIC COOPERATIVITY IN CHEMO-MECHANICAL CONVERSION 581

during a steady shortening under conditions of maximum efficiency. The tiny squares in sets of three lines represent 75 cross-bridges protruding from three thick filaments to a thin filament, and there are 25 squares in each line. These 75 cross-bridges surround the thin filament in a helical way. The dark squares in the left figure in Fig. 13 represent those cross-bridges which are in the preactive state, state 1, with a probability of 15–25% (25% is the maximum probability). The closed squares in the right figure indicate cross-bridges in the active state with a probability greater than 35%. Qualitatively speaking, it is clear that cross-bridges which are in the preactive state change to the active and unstable state 10 msec later. These changes become more clear than those in Fig. 13 if a time-interval shorter than 10 msec is chosen as the time scale. Furthermore, our sarcomere model showed that the maximum velocity of shortening of the sarcomere is independent of the overlapping length between thick and thin filaments. This means that the "independent" force generator by Gordon, Huxley & Julian (1966) is essentially compatible with the cross-bridge motions in the presence of dynamic cooperativity.

Effects of dynamic cooperativity on the efficiency of the chemo-mechanical conversion in muscle contraction was studied as follows. Three kinds of $k_{12}(x)$ as shown in Fig. 14 were given for the sake of comparison, taking the maximum value of $k_{12}(x)$ as parameter. The apparent steepness of the x dependence of $k_{12}(x)$ differs in these three cases. (In our model, $k_{12}(x)$ with the largest value, the case a, corresponds to the real case.) Clearly, the

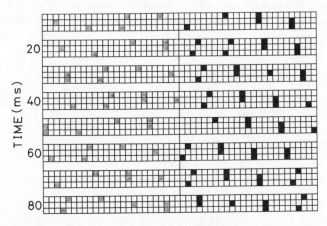

FIG. 13. The time evolution of the states of the cross-bridges surrounding a thin filament. The figure on the left-hand side represents the distribution of state 1 among cross-bridges and the figure on the right-hand side that of state 2.

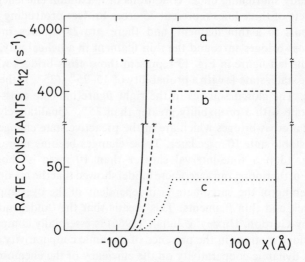

FIG. 14. An illustration of the rates $k_{12}(x)$ which are used in the discussion of the influence of $k_{12}(x)$ on muscle contraction; a, b and c, respectively, correspond to the cases where $k_{12}(x)$ has 4,000, 400 and 40 sec^{-1} for the maximum value.

magnitude of the dynamic cooperativity is an increasing function of the size of k_{12}. Normalized P-v relations calculated for these three cases are shown in Fig. 15. A. V. Hill's constant **a** (more exactly, a/P_0) was 0.25, 1.32 and $\sim \infty$ for 4,000, 400 and 40 sec^{-1} of the maximum value of $k_{12}(x)$. It was shown that the maximum velocity of shortening depends much more sharply on k_{12}

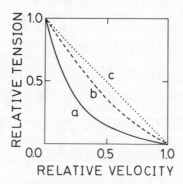

FIG. 15. Normalized tension-velocity relation. Curves denoted by a, b and c correspond to those of Fig. 14.

than the maximum tension does. Figure 16 shows the efficiency of the energy conversion in these three cases. In the calculation of the efficiency, we tentatively assumed that one ATP molecule is hydrolyzed after one elementary cycle $0 \to 1 \to 2 \to 0$. Remarkably, the efficiency depends greatly on the rate $k_{12}(x)$. Figure 16 suggests that the dynamic cooperativity is an important factor for the efficiency of the chemo-mechanical conversion.

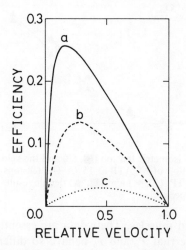

FIG. 16. Efficiency of energy conversion. Curves denoted by a, b and c correspond to those of Fig. 14.

Fibers of glycerinated rabbit psoas muscle were isolated, and each one was allowed to contract in 108 mM KCl, 4 mM $MgCl_2$, 5.3 mM $CaCl_2$, 4 mM EGTA, 1 mg/ml creatine kinase (64–69 U/ml), 16 mM creatine phosphate and various amounts of ATP. Then isometric tensions P_0, the velocity v_0 of no-load shortening and the tension change after a quick release in muscular length in an isometric contraction were observed. Experimental data for P_0 and v_0 are respectively denoted in Fig. 17 by closed squares and triangles vs. the concentration of Mg-ATP, [Mg-ATP]. The isometric tension was at maximum at about 30–50 μM and then gradually decreased as the substrate concentration was increased. The velocity v_0 was almost inappreciable at 10 μM and still had a very small value at 30 μM Mg-ATP. Then it increased gradually to a plateau which appeared at concentrations higher than 1 mM. P_0 and v_0 were also measured by Arata, Mukohata & Tonomura (1977) for myofibrils in various concentrations of Mg-ATP up to 100 μM, and by Cooke & Bialek

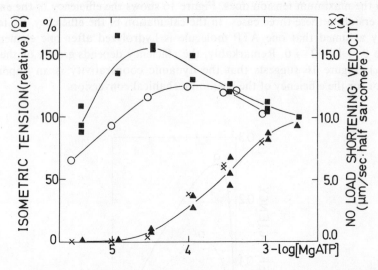

FIG. 17. Dependence of isometric tension (■, ○) and the velocity of no-load shortening (▲, ×) on the concentration of MgATP at 15°C. The tensions were measured as relative values to that at 4 mM ATP. (○, ×) and (■, ▲) denote results obtained with and without addition of 1 mM PPi, respectively.

(1978) for glycerinated muscles. Their data are essentially the same as ours. It is quite remarkable that P_0 and v_0 behave so differently in a wide concentration range of Mg-ATP. It seems to suggest that the molecular mechanism for the generation of the tension differs from that of the shortening velocity, because no rigor complex gives the principal cause for the difference in the behavior of P_0 and v_0, as will be shown in the following discussions.

First, the amount of rigor complex was evaluated from the stiffness of glycerinated fibers in a resting solution containing 4 mM EGTA. It was found that the amount of rigor complex was about 20% (of that in the rigor solution) at 10 μM Mg-ATP, and, if decreased to about 10%, at 30 μM. No rigor complex was appreciable at 100 μM. Then, 1 mM of PP_i was added to the resting solutions, which led to a significant reduction in the amount of rigor complex, which was about 5% at 10 μM Mg-ATP and gradually decreased to 0% at about 100 μM Mg-ATP. As shown by the open circles in Fig. 17, isometric tension P_0 in the concentration range 30–50 μM significantly decreased by the addition of 1 mM PP_i; and P_0 varied only a little in the concentration range from 50 to 500 μM Mg-ATP (with the maximum around at 100 μM Mg-ATP), where a small reduction occurs in P_0, probably due to a decrease in the concentration of AM*P complex caused by competition between ATP and PP_i in binding to myosin. In concentrations higher than 200 μM Mg-ATP, PP_i caused no appreciable change in P_0. On the other hand, v_0 was influenced by the addition of PP_i hardly, as denoted by crosses. v_0 is still an increasing function of [Mg-ATP] in concentrations higher than 100 μM where no appreciable amount of rigor complex is present in the fiber. Therefore, the apparent dependence of v_0 on the substrate concentration is not caused by rigor complex but by Mg-ATP itself. Perhaps an increase of the rate k_{20} for the transition from the attached state 2 to the detached state 0 may contribute to the dependence of v_0 on the substrate concentration. However, it cannot be the principal cause because the addition of 1 mM PP_i (which surely increases the rate k_{20}) gives very little change in the dependence of v_0 on [Mg-ATP]. There is likely to be another process which is influenced by the substrate concentration. To confirm this point, the following work was performed.

As described by Huxley & Simmons (1971), when muscular length is suddenly shortened in an isometric contraction of muscle fiber, the original tension simultaneously falls to a value T_1 and is followed by a quick recovery to another value T_2. Then tension returns slowly to the original value. The quantity $T_2 - T_1$, the amount of the quick recovery of tension, is roughly proportional to the size of the transition probability k_{12}. Similar measurements were performed for our glycerinated fibers: T_1 and T_2 were observed for a length change of 3 nm/half-sarcomere in various concentrations of Mg-ATP. As shown in Fig. 18, T_1 depends only weakly on the concentration of substrate, while T_2 shows a fairly sharp increase for an increase of the substrate concentration. Remarkably, the dependence of $T_2 - T_1$ on the concentration of Mg-ATP resembles that of v_0, as shown by the lowest curve in Fig. 18. Addition of 1 mM PP_i to the contracting solutions did not cause an appreciable change in the behavior of $T_2 - T_1$. This is consistent with the effect of PP_i on the velocity v_0. As discussed earlier in this section, a change in the size of k_{12} influences v_0 much more strongly than it does P_0. Therefore, P_0 may not change significantly in the concentration range where k_{12} clearly depends on the substrate concentration. This evidence indicates the existence of a strong correlation between v_0 and k_{12}. Hence, we may conclude that the dependence of v_0 on the substrate concentration is caused primarily by the change in the size of the probability k_{12}.

The simplest explanation for the dependence of the size of k_{12} on the substrate concentration is that the size (including the steepness on the x coordinate) of $k_{12}(x)$ becomes much larger when the two heads of the cross-bridge are simultaneously bound to substrate than when only one head is bound. There are two possibilities concerning this mechanism: (a) One head of myosin

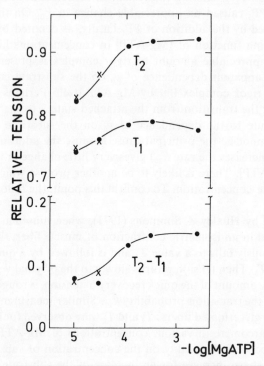

Fig. 18. Dependence of T_1, T_2 and $T_1 - T_2$ for a length change of 3 nm per half-sarcomere on the concentration of MgATP at 20°C. Symbol (×) denotes a result obtained in the presence of 1 mM PP_i.

originally has a large k_{12}. However, the unbound head works to prevent the bound head from making the transition $1 \to 2$ with a large probability. Simultaneous binding of the two heads to substrate releases the block. (b) The two heads work cooperatively to increase the probability k_{12} when they are bound to substrate. Possibility (a) is inconsistent with the evidence that the addition of 1 mM PP_i causes no appreciable change in $T_2 - T_1$ nor in v_0, though it reduces the amount of rigor complex. Furthermore, it is hard to imagine the reason why myosin would have a "dangerous" structure such that one head can easily disturb the function of the other head. Therefore, it would be helpful to recall here a hypothesis which has been proposed by us (Shimizu & Yamada, 1975; Shimizu et al., 1976) in analyses of the dependence of the velocity of shortening and of the ATPase activity of isolated sarcomeres (Takahashi et al., 1965). That is, the transition probability $k_{12}(x)$ is increased by utilizing the chemical energy of ATP under a cooperative action of the two heads of myosin. (This cooperative action of the two heads at a cross-bridge produces a sliding motion of a thin filament, which, on the other hand, induces cooperative action and the transition 1 → 2 at other cross-bridges in the preactive state, forming a feedback loop of action.) The two-headed structure of myosin is necessary for dynamic cooperativity, provided that the hypothesis is correct.

Based on the hypothesis of the cooperative action of the two heads, we may propose the following mechanism: tension is generated by cross-bridges when at least one of the two heads of myosin binds substrate, and the double bindings of substrate to cross-bridges gives no significant change in the tension but increases the probability k_{12} significantly, which promotes the dynamic cooperativity in muscle contraction. In other words, rapid shortening of sarcomeres is produced only by cross-bridges which bind two substrate molecules. This mechanism can explain our observations as follows. As the substrate concentration is gradually increased, one of the two heads of myosin begins to bind substrate, and the resulting increase in the number of cross-bridges occupied by a single substrate molecule gives an increase in P_0 until all the possible cross-bridges are occupied by at least one substrate molecule. Observed P_0 in the presence of PP_i indicates that this concentration for the saturation of the cross-bridges with one substrate molecule appears at about 100 μM. This is consistent with the evidence that rigor complex disappears when the substrate concentration is increased to 100 μM, because HMM produces no rigor complex when at least one of its two heads binds substrate in the form of MS or M*P (Inoue, Shigekawa & Tonomura, 1973). The species doubly occupied by substrate begins to increase from concentrations which are a little below 100 μM, say from 30—50 μM, and it continues to increase until both heads of all the possible cross-bridges are bound to substrate. Associated with this increase in the number of the doubly occupied species, the average size of k_{12} is increased. Consequently, the velocity v_0 will be increased. The rate k_{20} is a linearly increasing function of the substrate concentration, if the dissociation of AMP by substrate is a second-order reaction. Therefore, the effect of the increase of k_{20} becomes appreciable at high concentrations of substrate, which leads to a gradual decrease in P_0 and may give an increase in v_0 as a second-order effect.

6. CONCLUDING REMARKS

The present work shows that the molecular mechanism of actomyosin ATPase activity is generally different between an isotropic and homogeneous system, such as a solution in a test tube, and a chemo-mechanical system, such as our streaming system or muscle where actin and myosin distribute in an anisotropic and heterogeneous way. The cause of the difference is the presence of *dynamic cooperativity* in the decomposition of AM* P in the chemo-mechanical system. (Actomyosin threads studied by Cooke and Franks (1978) seem to indicate that k_{12} is so small that dynamic cooperativity is essentially absent there.)

Dynamic cooperativity brings order to the molecular dynamics of actomyosin in chemo-mechanical systems and produces ordered motions of molecules such as those proposed by Schrödinger. It is possible to show that there is a close correspondence between the thermodynamic mechanism of the production of active streamings by acto-HMM and that of the generation of laser lights (Shimizu, 1979). Both are typical (nonlinear) phenomena appearing in a system *far from thermodynamic equilibrium*. Therefore, ordinary thermodynamics (for equilibrium systems) is not applicable to the streaming system or to the laser. This will also be the case for muscle. The high efficiency of the chemo-mechanical conversion in muscle contraction is related to this thermodynamic property of muscle.

Dynamic cooperativity introduces new aspects to the molecular dynamics of muscle contraction. First of all, cross-bridges should *not* be regarded as *independent* force generators in the strict sense of the word. Secondly, attention to the interaction between the cross-bridge dynamics inevitably leads to a study of the structure and function of sarcomeres and to a study of the relation between the cross-bridge dynamics and the sarcomere dynamics. It is not surprising that evidences which are incompatible with the independence of cross-bridge dynamics have been found. Finally, it is probable that the presence of two heads on a myosin molecule is related to the function of producing dynamic cooperativity. (Active streamings were also produced in our stream cell by the use of S-1 instead of HMM. In contrast with the case of HMM, they seem to suggest the absence of dynamic cooperativity. However, we should be careful in drawing a final conclusion.)

Dynamic cooperativity is different from the static cooperativity between subunits of protein molecule in the following ways. First, it is caused by an indirect and long-range interaction between protein molecules transmitted by an ordered dynamics of the chemo-mechanical system. Second, it is a kind of induced cooperativity whose appearance requires free energy.

REFERENCES

Arata, T., Inoue, A. & Tonomura, Y. (1975). Thermodynamic and kinetic parameters of elementary steps in the reaction of H-meromyosin adenosinetriphosphatase. *J. Biochem. Tokyo* **78**, 277–286.

Arata, T., Mukohata, Y. & Tonomura, Y. (1977). Structure and function of the two heads of the myosin molecule. VI. ATP hydrolysis, shortening, and tension development of myofibrils. *J. Biochem. Tokyo* **82**, 801–812.

Bagshaw, C. R. & Trentham, D. R. (1973). The reversibility of adenosine triphosphate cleavage by myosin. *Biochem. J.* **133**, 323–328.

Cooke, R. & Bialek, W. (1978). The contraction of glycerinated skeletal muscle fibers as a function of ATP concentration. *Biophys. J.* **21**, 104a.

Cooke, R. & Franks, K. E. (1978). Generation of force by single-headed myosin. *J. molec. Biol.* **120**, 361–373.

Ebashi, S., Endo, M. & Ohtsuki, I. (1969). Control of muscle contraction. *Quart. Rev. Biophys.* **2**, 351–384.

Gordon, A. M., Huxley, A. F. & Julian, F. J. (1966). The variation in isometric tension with sarcomere length in vertebrate muscle fibres. *J. Physiol.* **184**, 170–192.

Hill, T. L. (1974). Theoretical formalism for the sliding filament model of contraction of striated muscle. Part I. *Prog. Biophys. molec. Biol.* **28**, 267–340.

Hill, T. L. (1975). Theoretical formalism for the sliding filament model of contraction of striated muscle. Part II. *Prog. Biophys. molec. Biol.* **29**, 105–159.

Huxley, A. F. (1973). A note suggesting that the cross-bridge attachment during muscle contraction may take place in two stages. *Proc. R. Soc. Lond. B.* **183**, 83–86.

Huxley, A. F. & Simmons, R. M. (1971). Proposed mechanism of force generation in striated muscle. *Nature*, Lond. **233**, 533–538.

Huxley, H. E. (1969). The mechanism of muscle contraction. *Science, N. Y.* **164**, 1356–1366.

Inoue, A., Shigekawa, M. & Tonomura, Y. (1973). Direct evidence for the two route mechanism of the acto-H-meromyosin-ATPase reaction. *J. Biochem. Tokyo* **74**, 923–934.

Julian, F. J., Sollins, K.R. & Sollins, M. R. (1974). A model for the transient and steady-state mechanical behavior of contracting muscle. *Biophys. J.* **14**, 546–562.

Katchalsky, A. & Curran, P. F. (1965). In *Nonequilibrium Thermodynamics in Biophysics.* Cambridge, Massachusetts: Harvard Univ. Press.

Kometani, K. & Shimizu, H. (1977). On the molecular dynamics of the chemo-mechanical conversion in muscle contraction. *J. theor. Biol.* **69**, 415–428.

Kubo, R. (1966). The fluctuation-dissipation theorem. *Rept. Progr. Phys.* **29**, 255–284.

Lymn, R. W. & Taylor, E. W. (1971). Mechanism of adenosine triphosphate hydrolysis by actomyosin. *Biochemistry, N. Y.* **10**, 4617–4624.

Nishiyama, K., Shimizu, H., Kometani, K. & Chaen, S. (1977). The three-state model for the elementary process of energy conversion in muscle. *Biochim. biophys. Acta* **460**, 523–536.

Schrödinger, E. (1944). *What is Life?* Cambridge Univ. Press.

Shimizu, H. (1972). Instability theory of biological motility. I. On the statistical mechanical nature of biological motility. *J. Phys. Soc. Japan* **32**, 1323–1330.

Shimizu, H. (1979). *Adv. Biophys.* **13**, in press.

Shimizu, H. & Yamada, T. (1975). The synergetic enzyme theory of muscular contraction: A two-headed myosin model. *J. theor. Biol.* **49**, 89–109.

Shimizu, H., Yamada, T., Nishiyama, K. & Yano, M. (1976). The synergetic enzyme theory of muscular contraction. II. Relation between Hill's equations and functions of two-headed myosin. *J. theor. Biol.* **63**, 165–189.

Shimizu, H. & Yano, M. (1978). Studies of the chemo-mechanical conversion in artificially

produced streamings III. Dynamic cooperativity–A new cooperativity in actomyosin systems with a polarized arrangement of F-actin. *J. Biochem. Tokyo* **84**, 1093–1102.
Takahashi, K., Mori, T., Nakamura, H. & Tonomura, Y. (1965). ATP-induced contraction of sarcomeres. *J. Biochem. Tokyo* **57**, 637–649.
Tonomura, Y. (1972). *Muscle Proteins, Muscle Contraction and Cation Transport.* Univ. Tokyo Press and Univ. Park Press, Tokyo and Baltimore.
White, D. C. S. (1972). Dynamics of contraction in insect flight muscle. *Cold Spring Harb. Symp. quant. Biol.* **37**, 201–213.
White, H. D. & Taylor, E. W. (1976). Energetics and mechanism of actomyosin triphosphatase. *Biochemistry, N. Y.* **15**, 5818–5826.
Yamada, T., Shimizu, H. & Suga, H. (1973). A kinetic study of the energy storing enzyme-product complex in the hydrolysis of ATP by heavy meromyosin. *Biochim. biophys. Acta* **305**, 642–653.
Yano, M. (1978). Observation of steady streamings in solution of Mg-ATP and acto-heavy meromyosin from rabbit skeletal muscle. *J. Biochem. Tokyo* **83**, 1203–1204.
Yano, M. & Shimizu, H. (1978). Studies of the chemo-mechanical conversion in artificially produced streamings. II. An order-disorder phase transition in the chemo-mechanical conversion. *J. Biochem. Tokyo* **84**, 1087–1092.
Yano, M., Yamada, T. & Shimizu, H. (1978). Studies of the chemo-mechanical conversion in artificially produced streamings. I. Reconstruction of a chemo-mechanical system from acto-HMM of rabbit skeletal muscle. *J. Biochem. Tokyo* **84**, 277–283.

Discussion

Hill: I would like to ask just a technical question. When you showed that two free-energy curves of states 1 and 2, it appeared that perhaps you had your x axis going in the opposite direction than I have. It is just a matter of convention, right?

Shimizu: Right. Because I want to have consistency between our streaming velocity and the coordinate.

Rüegg: I would like to confirm the results that you got in the quick phase. Yamamoto also found that the quick phase was very small, or nearly abolished, at very low MgATP concentrations in skinned fibers of the frog muscle. And the question is, can you, by increasing the viscosity of the medium, slow down the movement of streaming, and does this influence the ATPase?

Shimizu: I have not yet done such an experiment, but we put resistances in the circular slit and measured the ATPase acitivty. In Fig.12, the closed circles are the ATPase activity, which was obtained by putting four resistances in the circular slit, while the open circles indicate that no resistances were placed in the circular slit. As you see in Fig.12, the transition point is shifted to a higher temperature region.

Gergely: I have a question that perhaps Dr. Shimizu or Dr. Hill might answer. In this kind of model, the movements are supposed to occur extremely rapidly when the system changes along those parabolae, and the steps that are associated with the rates are the transitions between the two curves where there is no motion; so how does one relate this hydrodynamic type of motion to this kind of diagram that you showed?

Shimizu: Are you talking about the rate constant from state one to state two?

Gergely: As I understand Dr. Hill's models, there is no motion associated with the transitions which are governed by the rate constants.

Shimizu: In our model, the transition takes place first, between the two states, and then myosin moves along the potential—that is our assumption.

Mohan: I have two questions to ask. First, if you have a streaming flow where the walls are charged, you must get a potential difference between one end and the other end of the muscle fiber, because your flow takes place where the walls are charged. It is not just a simple streaming hydrodynamic flow and an inert wall. You have a wall which is fully charged, so these charges at the end would create kind of an electric potential difference at the ends.

Second, your streaming is not unidirectional; how do you explain the unidirectional flow? Or is it flowing in both directions? Is the streaming flow unidirectional?

Shimizu: I think so. I do not have enough time to explain how we make and measure the streaming, but as far as we observe, the steady stream was unidirectional, and the direction of the stream was determined by the condition under which we prepare the streaming cell.

A Hydrodynamic Mechanism for Muscular Contraction

R. TIROSH,[*] N. LIRON [**] and A. OPLATKA[*]

[*] Department of Polymer Research, The Weizmann Institute of Science, Rehovot, Israel
[**] Department of Applied Mathematics, The Weizmann Institute of Science, Rehovot, Israel

ABSTRACT

It is proposed that in a contracting muscle chemical energy is vectorially transformed into kinetic energy of protons which are ejected during the hydrolysis of ATP. Energy transfer to the fluid may cause active streaming. The role of the muscle proteins is to ensure the vectorial nature of the mechano-chemical transduction, allowing streaming of the fluid in a well-defined direction. In striated muscle, streaming is assumed to occur along the overlap region toward the center of the sarcomeres. External resistance to shortening will create a sub-pressure in the I-band region. The pressure differences developed on the envelope of the sarcomere will cause hydraulic compression which is revealed as tension. The hydraulic force may cause shrinkage of the I-band and expansion of the central region. This can explain the following phenomena: 1) variation in the value of the isometric tension in the range of 1–4 kg/cm^2 for muscles differing in their enzymatic activity; 2) the possibility that the tension-length relationship may assume different profiles, not necessarily related to the degree of overlap; 3) the time dependence of tension development and of relaxation which are considered to be affected by the rate at which fluid transport within the sarcomeres takes place. Since the hydraulic forces also operate laterally, transverse elastic elements should connect the sarcomeres and thus contribute to stored elastic energy. A reversible baroentropic effect, which may contribute to reversible heat exchange, is expected to accompany tension variations. The vectorial movement of protons may also cause movement of other ions in the same direction, especially under isometric conditions. The electric field produced may affect the distribution of Ca^{2+} ions and therefore also the regulatory process. The mechanism suggested offers a consistent explanation for the capability of non-filamentous myosin species to induce contraction in various systems and for various mechanical and heat phenomena. The proposed relationship between structure and function in muscular contraction is fundamentally different from that suggested by the cycling cross-bridge theory.

It is well established that muscular contraction is the outcome of a mechano-chemical process in which ATP is enzymatically hydrolyzed by myosin and actin. The enzymatic cycle is composed of several stages, including the formation of a complex between the two proteins and its dissociation by Mg-ATP. It is generally believed that the proteins are also physically involved in the generation and/or transmission of force. We have considered the possibility that the interaction between ATP and water alone (*i.e.,* in the absence of the proteins) constitutes the elementary mechano-chemical transduction event, in which chemical energy is transformed into kinetic energy of the surrounding fluid. The function of the enzymatic proteins is then to impose directionality on the otherwise random and dissipative mechano-chemical process so as to enable massive streaming of the fluid in a well-defined direction. Tension generation is thus the outcome of active streaming of the fluid, rather than of the "contraction" of the so-called contractile proteins, either through conformational changes or as a result of electrical interaction.

THE ELEMENTARY MECHANO-CHEMICAL EVENT

In an exothermic reaction, taking place in solution or in the gaseous phase, chemical energy is transformed into heat which is equivalent to an increase in the kinetic energy of the molecules. The energy liberated during such a reaction is thus supposed to be distributed as kinetic energy among the products. If we assume conservation of mechanical momenta, then the smaller the particle, the larger the energy acquired by it. The excess kinetic energy will eventually be transferred to surrounding fluid molecules. In order to generate local streaming in a definite direction, posessing the maximal portion of the total energy, it is preferable that a single, very light particle be produced during the reaction. Scattering can be minimized if the particle is electrically charged, thus enabling continuity of the transfer through long range interaction, provided it possesses a minimal cross-sectional area for head-on collision. In aqueous media, protons thus appear to be the best candidates for our purpose. It is essential that the kinetic energy of the ejected proton be high enough to overcome the cohesive energy of the water phase which is equivalent to the latent heat of evaporation — *i.e.,* about 11 kcal/mole. It is interesting to note that at neutral pH the potential energy difference ΔE between a hydronium (H_3O^+) complex and a water molecule, given by the Boltzmann relation

$$\frac{(H_3O^+)}{(H_2O)} = \frac{10^{-7}}{55} = e^{-\Delta E/RT},$$

also equals 11 kcal/mole. Hydration of a proton thus appears to be energetically comparable to the removal of a water molecule from the water lattice by evaporation. The dissociation of the hydronium ion will therefore be accompanied by the liberation of 11 kcal/mole, most of which should be possessed by the light bare proton. This process thus seems to be ideal for the efficient conversion of chemical energy into vectorial kinetic energy. So far we have discussed the discrete elementary molecular event of mechano-chemical transformation. In order to obtain massive streaming, it is necessary to organize many H_3O^+ molecules in such a manner that all protons will be ejected in the same direction. For the process to be continuous, a cyclic mechanism should be operating. These goals seem to be achieved in actomyosin systems in the following fashion:

1. attachment of the hydronium ion to Mg-ATP, which in turn forms a complex with a myosin head, the orientation of which can be directed by a neighboring actin filament;
2. interaction of many myosin molecules with the same actin filament;
3. the parallel assembly of many actin filaments.

Let us now examine to what extent do the molecular features of ATP and of the proteins indeed fit the proposed functions. The Mg-ATP molecule and the complex it forms with the active site of myosin are both characterized by multiple ring structures involving the Mg^+ ions, -SH groups of myosin, and purine and ribose rings (*cf.* Bendall, 1969). Binding to actin of the two heads of a myosin molecule will give rise to a macromolecular ring structure. The overall construction of the actomyosin complex should thus be capable of holding the H_3O^+ ion in a definite position with respect to the actin filament at the moment of ejection of the energetic proton.

For actin and myosin to be able to impose vectoriality in a *continuous* process, cyclic enzymatic activity is obviously essential. The question then arises: at which step of the enzymatic cycle does the mechano-chemical transformation occur? We suggest the following scheme:

$$A\text{-}M + ATP\text{-}H_3O^+ \underset{}{\overset{(a)}{\rightleftharpoons}} A\text{-}M\text{-}ATP\text{-}H_3O^+ \overset{(b)}{\longrightarrow}$$

$$\uparrow H^+ + A + M\text{-}ATP^+ \underset{}{\overset{(c)}{\rightleftharpoons}} A + M\text{-}P_r \underset{}{\overset{(d)}{\rightleftharpoons}} A\text{-}M + P_r$$

where A, M, P_r and $\uparrow H^+$ denote actin, myosin, products and the energetic proton. Step (b), similar to step (b) in the scheme proposed by Sleep & Taylor (1976), is essentially irreversible. Furthermore, if MATP$^+$ is identical to M* ATP (cf. Gutfreund & Trentham, 1975; Sleep & Taylor, 1976), then step (b) should be accompanied by a free-energy change of about 11 kcal/mole, whereas all other steps in the cycle are considered to be reversible and to involve smaller changes in energy.

The movement of energetic protons along the actin filaments may cause transport of other ions in the same direction creating dynamic electrical dipoles and electrical polarization. Coupling between the mechanical and the electrical effects has to be anticipated.

Application of the proposed electro-mechano-chemical coupling mechanism to activate streaming phenomena in various non-muscle cells is presented in another communication (Tirosh, Liron & Oplatka, 1978).

TENSION GENERATION IN STRIATED MUSCLE

a) *The sarcomere as a hydraulic machine*

We shall now consider the contraction of a sarcomere in a striated muscle fiber as the outcome of an active streaming process. The latter is supposed to take place along the enzymatically active region of overlap between the actin and the myosin filaments. Since the sarcomere contains two sets of actin filaments which are polarized in opposite directions, streaming in the two halves of the sarcomere will also take place in opposite directions. During shortening, the volume of the I-band region is diminished while that of the A-band region increases; streaming should therefore proceed towards the center of the sarcomere. When an external force opposing shortening is applied to the sarcomere ends, sub-pressure will develop in the I-band region which would otherwise be evacuated by the active streaming process. Under isometric conditions, the mechano-chemical engine in the overlap region will operate against passive streaming towards the I-bands resulting from the pressure difference ΔP between the central and the I-band regions. Such passive streaming would lead to the extension of the sarcomere under the influence of the external force. The same pressure difference should exist on the envelope of the sarcomere (which includes Z-bands) and give rise to longitudinal compression of the sarcomere as well as to transverse compression of the I-band region.

The tensile force will thus be given by

$$F = A \cdot \Delta P \tag{1}$$

where A is the effective cross-sectional area which is equal to projection of the envelope on a plane perpendicular to the direction of the tensile force (Tirosh, Oplatka & Liron, submitted for publication). In a myofibril, the tensile force should be transmitted through elastic elements connecting sarcomeres in series. Transverse elastic elements are required in order to prevent total lateral shrinkage due to the lateral compression effect. The latter may, however, give rise to a limited transport of fluid from the I-band region towards the center of the sarcomere. As a result, activation of a relaxed sarcomere with an homogenous cross-sectional area may lead to shrinkage in the I-band and to expansion of the central region. The effective area A defined above is practically equal to the cross-sectional area of the expanded central region.

Since the lowest value that the sub-pressure may reach is obviously zero, we assume that under normal atmospheric pressure, the maximal value that ΔP can attain is 1 kg/cm². In the following we shall attempt to justify this assumption.

b) *Force-length relationship for isometric contraction*

The value of the isometric tension is generally represented by the ratio

$$T_0^* = \frac{F_0}{A_r^*} \tag{2}$$

where A_r^* is the average cross-sectional area of the relaxed muscle at rest length. From Eqs. (1) and (2):

$$T_0^* = \frac{A}{A_r^*} \cdot \Delta P_0. \tag{3}$$

The lower part of Fig. 1 represents in a schematic manner a sarcomere at 4 different lengths which define 5 ranges of lengthening a to e (also given in the upper part of the figure). For the sake of illustration it is assumed that, in the myosin filament, the bare zone occupies 1/3 of the total length. The lengths of the actin and the myosin filaments are taken to be 1.0 and 1.5 μm, respectively. The maximal length of the sarcomere at which there is still full overlap is therefore 2.5 μm. This length, l^*, is approximately equal to the rest length. The variable $\bar{l}$ is the length in units of l^*. From Eq. (3):

$$T_0^* = \frac{A}{A_r^*} \cdot \Delta P_0 = \frac{A}{A_r} \cdot \frac{A_r}{A_r^*} \cdot \Delta P_0 = e \cdot h \cdot \Delta P_0, \tag{4}$$

where A_r is the cross-sectional area of the relaxed muscle at the variable length $\bar{l}$. The ratio $e = A/A_r \geq 1$ is a measure of the expansion effect at a given length. For isovolumic contraction,

$$h = \frac{A_r}{A_r^*} = \frac{l^*}{l} = \frac{1}{\bar{l}}. \tag{5}$$

The isometric tension T_0^* is thus the product of three factors: ΔP_0, h, and e. Curve (1) in Fig. 1 represents the hypothetical case in which $e = h = 1$, with $\Delta P_0 = 1$ kg/cm² as we have assumed above. The geometrical factor h should, according to Eq. (5), vary hyperbolically with length, therefore, for $e = 1$, T_0^* will also decrease hyperbolically with length (curve (2)). The third factor, $e \geq 1$, is expected to cause elevation of the profile represented by curve (2). In the following we shall discuss the variation with $\bar{l}$ of e (and therefore also of T_0^*) in the limiting case of maximal expansion in which all the fluid is accumulated in the central region in which there is no overlap between the actin and the myosin filaments. The maximal value of e will then be given by

$$e_m = \frac{A_m}{A_r} = \frac{l}{l_c} \tag{6}$$

where A_m and l_c denote, respectively, the maximal cross-sectional area and the width of the inactive region. Therefore, from Eqs. (4), (5) and (6) one gets

$$\frac{T_{0,m}^*}{\Delta P_0} = e_m h = \frac{l}{l_c}\frac{l^*}{l} = \frac{l^*}{l_c}. \tag{7}$$

When the sarcomere length changes in the (a) region, (i.e., between 2.0 and 2.5 μm), $l_c = 0.5$ μm, and therefore

$$T_{0,m}^* = \frac{2.5}{0.5} \Delta P_0 = 5 \text{ kg/cm}^2.$$

However, in the (b) region, $l_c = l - 2.0$ μm, and therefore

$$T_0^* = \frac{2.5}{l - 2.0} = \frac{1}{\bar{l} - 0.8}$$

Curve (3) demonstrates the constancy of $T_{0,m}^*$ in the (a) region and its hyperbolic

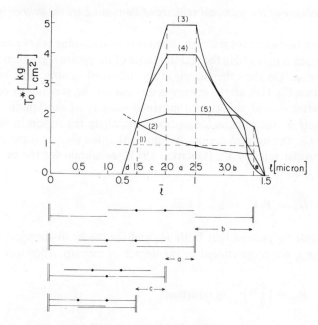

FIG. 1. Tension-length relationships for a sarcomere.

decrease in the (b) region (*i.e.*, 2.5 − 3.5 μm). In the (c) region, overlap of the two sets of actin filaments might interfere with the expansion effect. We assume that at $l = 1.5$ μm *i.e.,* when the ends of the myosin filaments just hit the Z-bands, expansion cannot occur, and therefore curves (2) and (3) should intersect at that length. As a first approximation, we assume linear change of T_0^* in the c region (1.5 − 2.0 μm). Resistance to further decrease in length will cause diminution in the value of T_0^*. Above 3.5 μm, a steep decrease in tension should occur since there is no overlap. Nonhomogeneity of length as well as mechano-chemical activity of non-filamentous myosin may be responsible for residual tension in this region (e).

As indicated above, curves (2) and (3) represent contractions with no expansion and with maximal expansion, respectively. Since expansion might be opposed by internal elastic elements, the actual force-length relationship will probably assume an intermediate profile such as those represented by curves (4) and (5) in Fig. 1. Comparison of curves (3)–(5) shows that when the extent of expansion is decreased, the maximal value as well as the variation with length of the isometric tension will also decrease.

c) *Independence of the maximal velocity of shortening on the degree of overlap*

During an isometric contraction, when no net streaming takes place, all the chemical input is invested in the maintainance of the pressure gradient along the overlap region. On the other hand, in an unloaded contraction, when $F = A\Delta P = 0$ (see Eq. (1)), all the energy is utilized for the generation of streaming. The latter should give rise to maximal velocity of shortening since it is opposed only by the viscous forces operating along the region in which it is produced, *i.e.*, the overlap region. Assuming that the viscous force F_η, is proportional to the velocity V_m (Oplatka, 1972), we obtain for the rate, H_m, of energy dissipation

$$H_m = F_\eta V_m = \eta_{mc} \cdot V_m^2. \tag{8}$$

It is plausible to assume that both H_m and the mechano-chemical frictional coefficient η_{mc} are proportional to the degree of overlap. Therefore

$$V_m = \left(\frac{H_m}{\eta_{mc}}\right)^{1/2} = \text{constant}. \tag{9}$$

d) *Latency relaxation and elongation*

During the initial phase of mechano-chemical reactivity, when shortening or tension has not yet developed, maximal streaming takes place momentarily. According to Bernoulli's (hydrodynamic) theorem, the hydrostatic pressure in the streaming fluid should drop. The pressure difference developed will cause lateral compression, which, for constant volume behavior, may lead to elongation. If a tensile force has been applied to the muscle prior to stimulation, this elongation will be revealed as a momentary relaxation which is known as latency relaxation.

e) *Tension development in an isometrically contracting muscle*

As we have shown above, the value of the tensile force is determined by the pressure difference and maximal cross-sectional area at a given time. In the following we shall consider possible variations with time of these two factors and their contribution to the time-dependency of the tensile force.

Since the expansion effect is an outcome of the pressure difference, expansion may still be taking place when ΔP has already attained its maximal value. In an isometric contraction, one should therefore expect discontinuity in the time-dependence of the tension development at about 1 kg/cm², and the latest phase should be associated with the expansion effect (see Fig. 2). Such an abrupt change in slope will probably become more pronounced when the degree of overlap is decreased.

The pressure difference is an outcome of the prevention of shortening, and therefore of net streaming, by the external world. The compressive force generated, acting on the envelope, will be sustained by elastic elements which consume mechanical energy. The expansion effect involves transport of fluid against the pressure difference and will therefore also require expenditure of energy. The energy source for both processes must be the hydrolysis of ATP, the rate of which should affect their time duration. Since muscles varying in their chemical input develop about the same isometric tension, it is plausible to assume that they spend the same amount of energy in order to charge their elastic elements and to transport fluid. The duration of maximal tension development should therefore be inversely proportional to the rate of their chemical input.

f) *The response of contracting muscle to stretch and release*

When a relaxed sarcomere is stretched, fluid must move from the central region to the I-band region. This passive streaming requires the creation of a pressure difference. The generation of sub-pressure in the I-band region must therefore be anticipated also in a relaxed sarcomere during stretch. Upon increasing the velocity of stretching, the value of the sub-pressure will decline down to zero so that ΔP, and therefore also the tensile force, may attain the maximal value of 1 kg/cm² at a certain velocity. Further increase in velocity will not cause further rise in tension.

Let us now consider the effect of stretching during tension development in an activated sarcomere. In this case, just as with a relaxed sarcomere, stretching can charge the elastic elements, thus enabling the mechanochemical engine to spend a larger portion of its power for the expansion effect. One should then expect a higher rate of expansion and therefore also of tension development. Upon increasing the velocity of stretching, the rate of tension development above 1 kg/cm² will be higher and reach a maximal value. The latter will be determined by the maximal possible velocity of streaming, which, in turn, will be limited by the mechanochemical power of the engine. Up to this velocity of stretching, there was a net streaming of the fluid towards the central region of the sarcomere. At still higher vleocities, passive streaming will take over, causing diminution of the expansion effect and therefore also a drop in tension ("give effect").

When an isometrically contracting muscle is quickly released over a short distance so that the tension drops down to zero, energy stored in the elastic elements will be dissipated as heat. The muscle will now redevelop tension isometrically. The time course of tension development will be governed by the mechano-chemical power, the elastic elements and the expansion effect, as discussed in paragraph (e) above for the initial phase of an isometric contraction. However, in the present case, one should take into account the fact that expansion has already increased up to a certain value prior to release and that its passive decrease during the quick release is probably small. Therefore, the duration of tension development will be shorter.

g) *Memory effects associated with the expansion effect*

Passive reversal of the expansion effect should take place after switching off the mechano-chemical engine (*i.e.,* during and after relaxation). In all probability, this process will be much slower than the active expansion at the initial phase. If the muscle is re-activated before full reversal took place, it is to be anticipated that tension development will now be faster. The expansion effect may thus be responsible for a variety of memory effects.

h) *Volume changes due to sub-pressure*

Sub-pressure in the I-band region should be associated with an increase Δv in the volume of the muscle, which is given by

$$\Delta v = - \gamma \Delta P \cdot v_I \tag{10}$$

where γ is the compressibility of the fluid and v_I is the volume of the I-region. The time course of Δv should thus be affected by changes in both ΔP and v_I. As we have shown above, there is a fast increase in ΔP during the initial phase of an isometric contraction. This should, according to Eq. (10), be accompanied by a proportional change in Δv, up to a maximum (Fig. 2). Later on, the volume v_I will decrease due to the expansion effect. This should lead to a decrease in Δv.

i) *Baro-entropic effect*

Changes in pressure may cause changes in the entropy of a fluid and may therefore lead to reversible heat exchange with its surroundings. Sub-pressure in the I-band region should thus be accompanied by the release of heat, q_b,

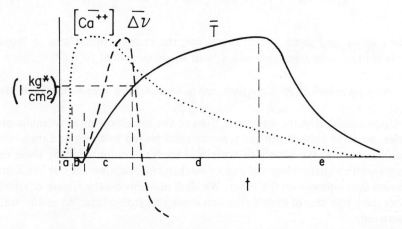

FIG. 2. Different phases (a to e) in the time-dependence of the tension T, the volume change Δv and the free Ca^{2+} ions content in an isometric twitch. All three variables are normalized with respect to their maximal value.

according to the equation

$$q_b = -\alpha \cdot T \cdot v_\mathrm{I} \cdot \Delta P \tag{11}$$

where α is the thermal expansion coefficient and T the absolute temperature. Since tension is related to ΔP, it is to be expected that tension development and relaxation will be accompanied by reversible baro-entropic heat exchanges. For water below 4°C, $\alpha < 0$, whereas above that temperature $\alpha > 0$, and therefore, according to Eq. (11), heat should be released below, and absorbed above, 4°C during the initial fast phase of tension increase and vice versa during decrease in tension. According to Eq. (11), q_b is proportional also to v_I; the magnitude of the baro-entropic effect should therefore become smaller as expansion increases. It is therefore to be anticipated that above 1 kg/cm², the higher the tension, the smaller will be the associated heat exchange. Furthermore, due to memory effects, one should not be surprised to find that the magnitude of the baro-entropic effect becomes much smaller if contraction has been preceded by another one (especially one involving stretch). Since v_I increases with increase in muscle length, heat exchange due to the baro-entropic effect should be more pronounced at higher lengths. Assuming that the sarcomeres occupy most of the volume of a muscle, the baro-entropic heat per unit volume $\bar{q}_b$ of the muscle is given, using Eqs. (1), (2) and (11), by

$$\bar{q}_b = -\alpha \cdot T \cdot \frac{v_\mathrm{I}}{v} \cdot \Delta P = -\alpha \cdot T \cdot \frac{v_\mathrm{I}}{v} \cdot \frac{A_r^*}{A} \cdot \Delta T^*. \tag{12}$$

For a quick and short stretch or release, the cross-sectional area A should remain practically constant, and $\bar{q}_b$ will be proportional to ΔT^*.

j) *Heat phenomena which are supposed to be associated with elastic elements*

Upon considering the structural basis of the sarcomere as a hydraulic machine, we have pointed out the separate roles played by series and transverse elastic elements. We have also correlated energy accumulation by these elements with mechano-chemical and/or mechanical inputs, on the one hand, and tension development on the other. We shall now discuss the release of energy from these two sets of elastic elements during tension relaxation and isotonic shortening

During a fast release, all the energy of both sets is dissipated as heat. However, during a slowed-down release, the velocity of which is still higher than the maximal velocity of shortening, V_m, the series elements do mechanical work on the releasing device, whereas all the energy of the transverse elements appears as heat.

Let us now assume that the elastic elements obey Hooke's law and that they can be analyzed in a framework of three orthogonal axes so that their effective compliances per unit volume, c, are equal. The elastic energy per unit volume, E_{el}, will then be given by

$$E_{el} = 3 \cdot \frac{1}{2} \cdot c \cdot T^{*2}; \tag{13}$$

therefore, for a fast release from T_0^* to T_f^*, heat production, $\bar{q}_{el}$, by the elastic elements will be given by

$$\bar{q}_{el} = \frac{3}{2} \cdot c \cdot (T_0^{*2} - T_f^{*2}) \approx \frac{3}{2} \cdot c \cdot T_0^* \cdot \Delta T^*. \tag{14}$$

Therefore, $\bar{q}_{el}$, like $\bar{q}_b$, is proportional to ΔT^*.

During isotonic shortening, the energy associated with the series elastic elements should remain constant; however, energy stored in the transverse elastic elements should be dissipated in proportion to the extent of shortening, since its amount is proportional to the area of the envelope of the I-region and therefore also to the sarcomere length.

The irreversible heat production by the elastic elements occurs simultaneously with the reversible baro-entropic heat exchange which may, as we have seen, be either positive or negative. It is important to note that while the charging of the elastic elements requires a mechano-chemical or mechanical energy source, the energy associated with the baro-entropic effect, being purely thermal, does not have a chemical or a mechanical equivalent. During tension release or isotonic shortening, the heat liberated by the elastic elements together with the heat exchange through the baro-entropic effect cannot, therefore, be accounted for by a chemical change occurring at the same time. However, in considering the energetics of a closed cycle comprising contraction and relaxation, the total amount of heat liberated will include the energy accumulated and released by the elastic elements. The baro-entropic effect, being reversible, will not, however, contribute to the net energy balance.

k) *Implication of the electro-mechano-chemical coupling in the movement and distribution of Ca^{2+} ions*

As we have proposed above, mechano-chemical activity is associated with an electromotive force along the overlap region, directed from the I-band region towards the center of the sarcomere. Under isometric conditions, significant electric polarization will be established as soon as tension reaches its maximal value, since then there is no streaming which interferes with polarization. As the concentration of Ca^{2+} ions, which are involved in the regulatory process, is relatively low, the movement and spatial distribution of these ions will be strongly affected by the electromotive force and by the electrical polarization. The latter will be responsible for a counter-current of the ions in the periphery of the sarcomere, thus creating closed loops of Ca^{2+} currents the intensity of which will be proportional to the rate of the ATPase activity. Polarization will also cause an uneven distribution of free Ca^{2+} ions. The moving Ca^{2+} ions will continuously be absorbed by the sarcoplasmic reticulum. However, repetitive stimulation can maintain a reservoir of free Ca^{2+} ions around the sarcomere. Cessation of stimulation will therefore be followed by depletion of the reservoir at a rate which will be proportional to the Ca^{2+} ions current and therefore also to the rate of the ATPase activity. Enzymatic activity will continue until the level of Ca^{2+} ions in the reservoir falls to a certain value. Only then will relaxation start (Fig. 2). The time interval between the cessation of stimulation and the onset of relaxation is expected to be inversely proportional to the rate of ATP hydrolysis. We consider the possibility that in a relaxed muscle there exists a very low level of mechano-chemical activity which, by maintaining a minimal electrical polarization, will accelerate the first Ca^{2+} ions to be released on stimulation from the sarcoplasmic reticulum towards the overlap region. Such a "stand-by" mechanism will ensure an immediate and autoaccelerated response of the stimulated muscle. The above arguments relating to the accelerated initiation and to the delayed relaxation lead to the expectation that the initial phase of tension development should be the same irrespective of the rate of stimulation, including the case of a single pulse. Since stretching should accelerate tension generation, and as depletion of the Ca^{2+} ions reservoir is supposed to start when tension reaches its maximal value, the duration of the active phase in a twitch should become shorter upon stretching (and longer as a result of release).

Following stimulation, the Ca^{2+} ions which have arrived at the overlap region will not be pushed out of the sarcomere until the tension reaches a high enough value. At higher tensions, there is a possibility that the electro-mechano-chemical engine will remove Ca^{2+} ions faster than their entrance from the reservoir. When this happens, the activity of the engine will be slowed down, enabling a repetition of the course and giving rise to oscillatory behavior.

DISCUSSION

The central idea of the model proposed above is that, in a contracting muscle, chemical energy is directly converted into kinetic energy which is transferred to the surrounding fluid molecules. In an internal combustion engine, such a process is utilized in order to obtain macroscopic movement and/or force in a well-defined direction by transferring part of the molecular energy and momenta to a movable piston. In the cylinder, the molecules move at random, and therefore the excess kinetic energy acquired by them as a result of the chemical reaction is actually heat energy. Such an engine cannot be efficient under isothermal conditions. If indeed, mechano-chemical coupling in muscle involves direct conversion of chemical into kinetic energy, then for muscle to be able to function isothermally and efficiently, it must avoid the intermediate step of heat energy. This can be achieved if transduction is vectorial at the level of the reactive molecules (*i.e.*, ATP and its products). This, in turn, necessitates the existence of highly ordered, catalytic and linear structural elements. Such a task can be fulfilled by the actin filaments in combination with the "fuel"-carrying myosin molecules. We do not suggest an additional role for the proteins such as force generation and/or transmission as a result of conformational changes in a continuous, inflexible network as assumed by others. Our picture can explain force generation in systems in which the proteins do not form a continuous three-dimensional network *e.g.*, glycerinated muscle fibers, in which the filamentous myosin has been inactivated by poisoning, which can develop tension after irrigation with heavy meromyson (HMM) (Borejdo & Oplatka, 1976). On the other hand, we have demonstrated active streaming in solutions containing HMM and a complex of F-actin with natural tropomyosin (Tirosh & Oplatka, submitted for publication).

We have considered the possibility that in the elementary mechano-chemical event, energetic protons possessing 11 kcal/mole are ejected. In a secondary event, this energy is transferred to water molecules of the muscle. Release and absorption of energetic protons have been claimed to be implicated in the hydrolysis and the synthesis of ATP in membranous systems (Mitchell, 1976). Hydrolysis may lead to the development of a pH gradient and/or of an electric potential difference of up to 0.5 volts across the membrane as a result of the vectorial movement of bare, energetic protons. Electro-chemical potential differences may thus reach 0.5 proton-volts, which correspond to about 11 kcal/mole. The full conversion of the kinetic energy of the protons into electrical energy in the membrane suggests that no energy is dissipated in this process. When the bare protons hit the aqueous phase they are probably captured as hydronium ions. A pare proton, accelerated by the extremely high electric field inside the membrane may acquire enough kinetic energy to enable the synthesis of ATP. In an aqueous solution of actomyosin, contrary to the three-phase phospholipid membraneous systems, one should not expect measurable pH changes as a result of the ejection of the protons, since the release of each proton should be preceded by complex formation of H_3O^+ with ATP according to our scheme.

The hydraulic picture we have presented for muscular contraction is in line with the following structural features: The need for hydrostatic continuity of the surroundings of each sarcomere is realized by membraneous invaginations which reach each sarcomere and by the open structure of the Z-region (Ulrick, Toselli, Saide & Phear, 1977). For higher efficiency of active streaming in the densely packed filamentous structure of the sarcomere, it is preferable that the myosin heads be in the close vicinity of the actin filaments and that they be dissociated from the actin during the larger part of the enzymatic cycle. The last requirement seems to be fulfilled in the "refractory state" (Chock, Chock & Eisenberg, 1976).

Our model is capable of accounting for a large variety of phenomena which have been reported in the literature. Analysis of experimental data on the basis of the model will be presented in another communication. A mathematical formulation of the energetics of muscular contraction, based on our model, has been submitted for publication (Tirosh, Oplatka & Liron).

Acknowledgement

This work was supported in part by a grant from the Muscular Dystrophy Associations of America.

REFERENCES

Bendall, J. R. (1969). In *Muscles, Molecules and Movement*. London: Heinemann.
Borejdo, J. & Oplatka, A. (1976). Tension development in skinned glycerinated rabbit psoas fiber segments irrigated with soluble myosin fragments. *Biochim. Biophys. Acta* **440**, 241–258.
Chock, S. P., Chock, P. B. & Eisenberg, E. (1976). Pre-steady state kinetic evidence for a cyclic interaction of myosin subfragment-1 with actin during the hydrolysis of adenosine 5'-triphosphate. *Biochemistry, N. Y.* **15**, 3244–3253.
Gutfreund, H. & Trentham, D. R. (1975). Energy changes during the formation and interconversion of enzyme-substrate complexes. In *Energy Transformation in Biological Systems. Ciba Foundation Symposium* **31**, 69–86.
Mitchell, P. (1976). Possible molecular mechanisms of the proton motive function of cytochrome systems. *J. theor. Biol.* **12**, 327–368.
Oplatka, A. (1972). On the mechanochemistry of muscular contraction. *J. theor. Biol.* **34**, 379–403.
Sleep, J. A. & Taylor, E. W. (1976). Intermediate state of actomyosin adenosine triphosphatase. *Biochemistry, N. Y.* **15**, 5813–5817.
Tirosh, R., Liron, N. & Oplatka, A. (1979). A proposal of the molecular basis of cytoplasmic streaming. In *Cell Motility: Molecules and Organization*, eds. Hatano, S., Ishikawa, J., and Sato, H., Tokyo: University of Tokyo Press.
Ulrick, U. W., Toselli, P. A., Saide, J. D. & Phear, W. P. C. (1977). Fine structure of the vertebrate Z-disc. *J. molec. Biol.* **115**, 61–74.

Discussion

Woledge: In looking for some predictions that your theory makes that can be tested, I am interested in the baroentropic effect. When there are sudden tension changes in muscle, there are, of course, heat effects. So, I am interested to know how large you would predict those effects to be?

Oplatka: I do not recall numbers, but we have calculated a value which fits experimental data nicely.

Pollack: It was not clear to me where the location of the elastic elements was in your model.

Oplatka: I did not get into details. All I said was that it makes sense to believe that such elements do exist.

ter Keurs: What are the passive viscous properties of muscle in your model?

Oplatka: I did not get into this at all.

Sagawa: I wonder how your model can explain shortening or force development in skinned fibers, glycerinated fibers, or extracted fiber proteins?

Oplatka: This is a good point. I have been talking of pressure differences, and I did not actually say what I meant by those presure differences. Are these differences between the atmospheric and the inside pressures or anything else? Recently we did an experiment which was quite trivial. We checked for the possibility that reduction of external pressure might affect the mechanical performance of a muscle. There was no effect, as maybe one could have anticipated, so that we are now inclined to believe that the pressure we are talking about is not the atmospheric pressure but rather the so-called internal pressure. This is related very simply to the cohesive energy of the medium.

So, it appears that—though it may sound crazy—that one should go one step further and not only forget a lot about the proteins, but in a sense ascribe the very generation of force, and the value of force as we measure it, somehow to a property intimately associated with water; meaning to say that streaming water differs in internal pressure from stationary water. Maybe one should not think of water only as a component of the ATPase hydrolysis reaction. I was quite encouraged to hear about Dr. Endo's paper in this meeting, though one should be very careful about them since they may be at least partly interpreted as being due to reduction in the affinity of binding of calcium to troponin. So maybe one should repeat those experiments with a system which is calcium insensitive just to be on the safe side. Actually we had in mind to measure isometric tension not as a function of viscosity of various liquids, but rather of the cohesive energy density of the liquids. And as I mentioned before, it is somewhat striking to notice that the enthalpy change accompanying the hydrolysis of ATP is exactly equal to the cohesive energy density of the water, which is about 10 kcal/mole.

A New Field Theory of Muscle Contraction

Tatsuo IWAZUMI

Department of Anesthesiology and Division of Bioengineering, University of Washington, Seattle, Washington, U. S. A.

PROLOGUE

I feel a little awkward about the title of my presentation because this theory was developed almost ten years ago. However, I think the title is justified because the majority of the distinguished audience of this Symposium will hear about my theory for the first time. I am sincerely grateful to the organizers of this symposium for providing an opportunity for the theory to emerge from ten years of obscurity.

1. DEVELOPMENT OF THEORY

Traditionally, biological scientists have treated the biological systems with a phenomenological approach. The theory I have developed is a radical departure from this tradition. It may be called a synthetic approach. Perhaps because of my engineering background, the first question I asked when I faced the problems of muscle contraction was, "If I were Nature, how would I go about designing a basic contractile machine which can be used for any motor application?" Of course, I am bound by all physical laws and only a handful of raw materials. I assumed that the physical laws when Nature first created muscle millions of years ago were the same as those we know today. The biggest problem was that I did not know what kind of principle Nature employed to generate contractile forces, and what kind of raw materials were available at that time. So, I went out to seek tips from Nature.

While trying to digest a mountainous volume of discoveries about muscles made by great scientists, one thing that struck me most was the beautiful regularity of the ultrastructure of striated muscle. Particularly, I was impressed by many micrographs showing that, when sarcomere length was varied, the A-band remained at the center of the sarcomere and the thin filaments slid in parallel to the thick filaments (Huxley & Niedergerke, 1954; Huxley & Hanson, 1964). I thought there ought to be strong stabilizing forces acting between these two kinds of filaments, otherwise the thick filaments would eventually evenly spread out in the sarcomere and the thin filaments would tangle up with the thick filaments like spaghetti. I took these outstanding features of the ultrastructure of the sarcomere as tips from Nature; in other words, my contractile machine must be designed based upon the sliding filament mechanism, and the design must be made such that the thick filaments are stabilized at the center of the sarcomere and the thin filaments are freely suspended between thick filaments for friction-free sliding motion. Figure 1 shows a schematic drawing of the machine. A contraction takes place by generating a force between the thick and thin filaments, which results in a sliding motion between them while maintaining the thick filament at the center and suspending the thin filament between the thick filaments. I shall call the centering of the thick filament "longitudinal stability" and the free suspension of the thin filament "lateral stability."

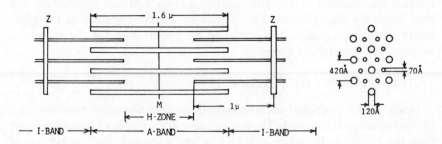

FIG. 1. A schematic sarcomere, showing spatial relationships and sizes. The dimensions are approximate values found in typical skeletal and cardiac muscle.

Of course, this drawing alone would not indicate how force is generated between the thick and thin filaments. But, whatever the principle of force generation, I knew that it must be compatible with the longitudinal and lateral stabilities. In other words, the longitudinal and lateral stabilities impose absolute constraints on the choice of force-generating mechanisms. Let us consider the constraints arising from the longitudinal stability. Figure 2 illustrates a simple way to derive the longitudinal stability condition (see Iwazumi, 1970, for a more rigorous method). In order that the thick filaments tend to remain at the center of the sarcomere, any perturbation from the center,

A NEW FIELD THEORY OF MUSCLE CONTRACTION

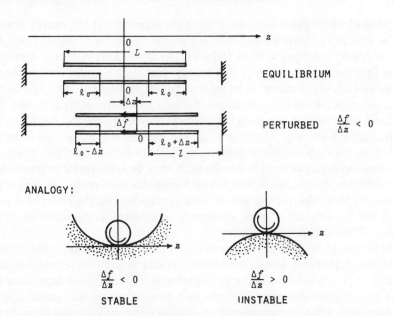

FIG. 2. Longitudinal stability condition of the thick filament. In order for the thick filament to be stable at the center, the restoring force Δf must always point in the direction opposite to the direction of perturbation, Δz.

Δz, must be met with a restoring force, Δf, which must always act in the direction opposite to that of the perturbation; therefore, $\Delta f/\Delta z$ must be negative. Since Δf is the difference between the contractile forces on the right and left side of the thick filament, a greater force must be generated at the side with less overlap. So the longitudinal stability condition requires that the force-generating mechanism must produce more force with decreasing overlap length. Now the question is: what kind of force-generating mechanism would do that? I considered many possible mechanisms, and found one which stood out above all others: an electrostatic force that acts between a dipole field generator and a dielectric filament. This is analogous to the force between a magnet and a piece of iron needle. Schematically, it is convenient to explain it by a capacitor model as shown in Fig. 3. The energy source is a battery which delivers positive and negative charges to the capacitor plates and establishes an electrostatic field. In muscle the battery is ATPase in myosin. When a filament with high dielectric constant is introduced into the field, induced charges appear on the surface, thus resulting in an axial force which is unidirectional and tends to pull the filament into the field. One outstanding property of this force-generating mechanism is that the magnitude of force is proportional to the product of the cross-sectional area of the filament and the energy density of the field in the surrounding medium on the tip of the filament. Here, the energy density is defined as $W = (1/2)\varepsilon E^2$. The reason I prefer energy density to more familiar field strength is that energy density is a scaler (rather than a vector as E is), which proves to be most useful for dealing with dielectric bodies in an electrostatic field. Nevertheless, it may appear strange by intuition that the axial force acts on the dielectric filament only at the *tip* and not along the shaft. Figure 4 explains why that is so. The induced charges arising from the field along the shaft distribute symmetrically about the center of the capacitor; therefore, no net axial force is developed. It may be appropriate to mention at this point that the maximum field strength needed to account for the maximum contractile force observed in actual skeletal muscle is on the order of 1 volt/ 100 Å (or 10^6 volt/cm), a value commonly encountered at the solid-liquid interface of electrochemical systems.

Let us now go back to the longitudinal stability problem. With the electrostatic force acting on a dielectric filament, it is easy to fulfill the stability requirement. Figure 5 shows how it can be implemented. A number of capacitors are lined up along the dielectric filament, and they are connected individually to batteries with different voltages the magnitude of which increases towards the

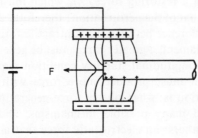

$F = AW_{(tip)}$

F: AXIAL FORCE ACTING ON THE THIN FILAMENT

A: CROSSSECTIONAL AREA OF THE THIN FILAMENT

W(tip): ENERGY DENSITY AT THE TIP OF THE THIN FILAMENT

FIG. 3. Capacitor model of a contractile force generation. The dielectric cylinder induces surface charges which result in a net axial force by Coulomb's law. The field energy is supplied by a battery which is the myosin-ATP complex in muscle.

A NEW FIELD THEORY OF MUSCLE CONTRACTION

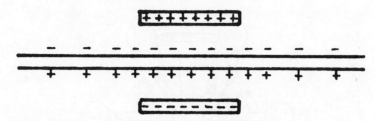

FIG. 4. Induced charge distribution on a dielectric cylinder with its tip extending far outside of the capacitor. No net axial force acts on the cylinder because of the symmetry of the charge distribution.

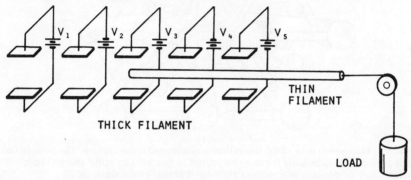

FIG. 5. A bank of capacitors along the thin filament. The thick and thin filament in this figure represent one half of a sarcomere. The voltage of each myosin-ATP battery is independently controlled by the local Ca^{++} concentration at the battery. The longitudinal stability condition is fulfilled by setting $V_1 < V_2 < V_3 < V_4 < V_5$.

right, *i.e.*, $V_1 < V_2 < V_3 < V_4 < V_5$. This arrangement will give us an electrostatic field whose strength increases towards the right end; therefore, the axial force acting on the tip of the filament increases with decreasing overlap length; hence, longitudinal stability is attained.

Since the force acting on the dielectric filament is determined only by the field which is generated by a capacitor near the tip of the filament, and since no other capacitors generating fields along the shaft contribute to the axial force, it may appear that this is a very wasteful design. Actually, these fields along the shaft are necessary to secure the lateral stability of the filament. In order to suspend the filament freely in space by an electrostatic field, it proved necessary that the field vector be perpendicular to and helically rotating along the axis of the filament; this also satisfies some conditions on the gradient and curvature of the energy density. The field that satisfies all these requirements would surround the filament with a spiral wall of potential energy well, somewhat similar to a threaded hole which supports a screw.

Having fulfilled all the requirements for an electrostatic field, I wanted to

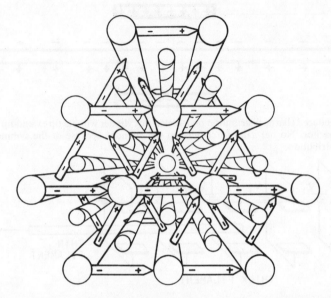

FIG. 6. Transversal view of the theoretically synthesized ultrastructure. The pointed end of each projection is to indicate the orientation, not to be meant its actual shape. The left-hand helical array of projections generates right-hand helical electrostatic fields.

find a spatial charge distribution that would generate the required field. This is a field synthesis problem which proved to be the toughest part of this theory. I regret I am not able to explain the very important and interesting field synthesis methods here due to limited space. I shall describe only the result of the synthesis.

The three-dimensional distribution of charges is shown in Fig. 6, in a transverse cross-section of the synthesized structure. Each cross-projection of the thick filament has clusters of positive and negative charges. The pointed end of the projection is meant to indicate that it does not connect two thick filaments together. The projections point to the adjacent thick filaments and are distributed along the shaft of the thick filament helically with a 60° turn per advance. The direction of the helix is left-turn (counterclockwise), and the resulting field vector rotates around the thin filament in a right-turn (clockwise) direction. It is emphasized here that this is the one and only configuration that can generate a field which satisfies the lateral stability condition.

There are many interesting properties of this synthesized structure. However, due to limited space I must reluctantly forego important questions such as the connection between the biochemistry of myosin and the field generator, the dielectric properties of the thin filament, ions in the aqueous medium, energetics, and the role of Ca^{++}, Mg^{++} and troponin in the control of electrostatic field generation. I shall limit the scope of the discussion to the mechanics which are directly associated with the synthesized structure. Let me emphasize here that the synthesized structure is an artificially designed machine which is physically realizable and guaranteed to be stable both statically and dynamically. I would like to show that the synthesized structure is of the same design as that which Nature made millions of years ago, by showing that the structural and mechanical properties of the synthesized structure are identical to those of real striated muscles.

2. JERKY MOTION OF THE THIN FILAMENT

It would be appropriate to begin the kinetic analysis of the synthesized structure from a single field generator (dipole) as shown in Fig. 7. The charge separation distance of about 100 Å was derived from the lateral stability requirement; if the distance is much greater the lateral stability cannot be achieved, and if it is much smaller the field does not extend far enough to do the job. The electrostatic field around the field generator is highly nonuniform and similar to the dipole field; the field strength rapidly diminishes with distance from the generator. Let us then consider the consequence of such a nonuniform

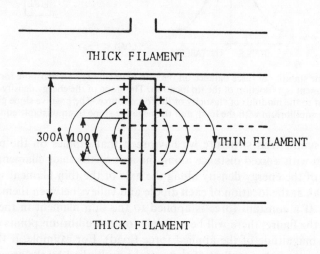

FIG. 7. Theoretical charge distribution on a projection. It can be shown that a charge separation distance much greater than 100 Å will result in lateral instability. If the distance is much less, the field does not extend far enough to do the job.

field on the force acting on a thin filament. Figure 8 illustrates the situation schematically. Since the energy density is maximal at the center of the capacitor and diminishes in the fringe region, the axial force acting on the tip varies from zero to the maximum depending upon the location of the tip. Suppose that the thin filament is axially translated from the far right to the far left, passing through the center of the capacitor. As the tip traverses the capacitor, the force acting on the filament increases from zero to a maximum, F_{max}, which occurs when the tip is at the center, and then decreases again to zero. If we apply a constant force to the thin filament in the direction opposite to and less than F_{max}, the force equilibrium is always reached at the location where the tip lies ahead of the center of the capacitor. The force balance cannot be maintained in the region before the center because the force balance here is unstable.

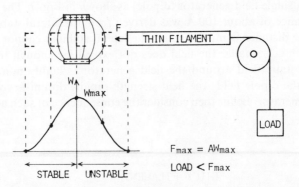

FIG. 8. The stability of force balance on the thin filament. The electrostatic force acting on the thin filament is a function of the tip location. The slope of the energy density distribution is equivalent to the modulus of elasticity of the force source. The positive slope gives rise to a stable force equilibrium with the load, and the negative slope to an unstable equilibrium.

In the synthesized structure we have seen that dipoles on the projections are spaced with a fixed distance along the axis of the thick filament. The distribution of the energy density along the axis of the thin filament will clearly have a peak at the location of each dipole and valleys between them, as shown in Fig. 9. If a constant force is applied to the thin filament in the direction shown in the figure, there will be multiple force equilibrium points depending upon the magnitude of the applied force (load). For example, if the applied force is less than the smallest of these peaks multiplied by the cross-sectional area of the thin filament, then a force equilibrium could be achieved at any dipole. Now, suppose that the applied force is gradually increased and exceeds

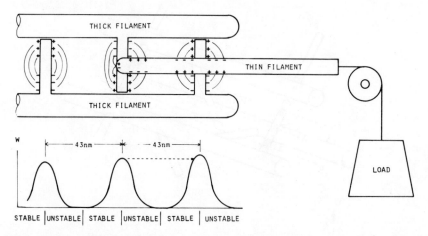

FIG. 9. The energy density distribution along the axis of the thin filament. Due to alternating stable and unstable regions along the axis, the thin filament tip can find multiple equilibrium points for a given load. If the load becomes greater than the field can support the thin filament slips backward until the tip enters a strong enough field.

the maximum that the dipole near the tip can support. The thin filament then jumps to the next dipole, which is generating a stronger field as required by the longitudinal stability. As the applied force is increased further, the thin filament tip jumps many successive dipoles.

Let us then consider what would happen when the applied force is reduced. Clearly, the maximum distance that the thin filament can move is the width of the stable region, which is about 200 Å, when the applied force is varied from the maximum to zero. This might appear contrary to what we observe in real muscle, which can contract to a great extent. The synthesized structure does shorten quite differently from a single thin filament because of helically arranged dipoles which generate energy density distributions with a spatial three-phase relationship to one another. The top part of Fig. 10 shows three thick filaments cut from the synthesized structure for clarity's sake. The distance between succeeding projections is depicted disproportionately longer to avoid cluttering. In the lower part of Fig. 10, the distributions of the energy density along the three lines A, B and C are indicated. Let us suppose that we are on line A walking to the right starting from the projection which points upward. Evidently, at this location we experience the greatest field strength; hence, the energy density is greatest as shown in the figure. As we walk to the right, the field strength rapidly diminishes, and at the same time the field vector rotates clockwise. Note that the direction of projection rotates counterclock-

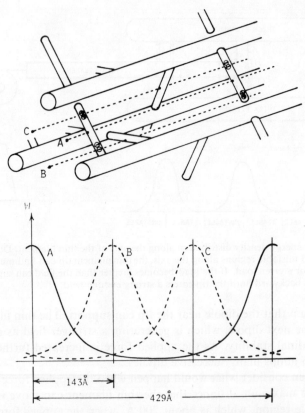

FIG. 10. Top: Three dimensional illustration of the orientation of projections. The projection spacing is shown disproportionately long so as to avoid cluttering. Lines A, B and C pass through the center of dipole on each projection. Bottom: Three-phase distribution of the energy density along the lines A, B and C. The fringe field extends about 150 Å.

wise. At the midpoint to the next projection, the field strength is nearly zero. As we walk further ahead, the field strength begins to increase, and at the next projection it is again greatest. The direction of the field is now 180° from that of the starting point. If we repeat the same process for lines B and C, the energy density distribution along each line will be the same as that of line A, except that the distribution curve is shifted axially by 143 Å, as indicated by dotted lines in the lower figure. Because of this spatial three-phase arrangement of the energy density, the thin filaments whose tips are in the low-energy density region are pushed forward (through Z-line connections) by other thin filaments whose tips are in the high-energy density region. As soon as those pushed tips enter the high-energy density region, which is the unstable region of the dipole, they immediately jump forward to carry the load, thus relieving the load from those thin filaments whose tips were previously in the high-energy density region. This explanation still does not include a full account of the shortening process since the ATPase activation by the thin filament is not taken into consideration. Nevertheless, it is clear that the thin filament motion is jerky in either the shortening or the lengthening direction due to alternating slow and fast movement. It may seem that this jerky motion of the individual thin filament is not too important so far as the dynamics of the sarcomere are concerned, because one might expect that such a movement would be averaged out by the thousands of thin filaments in the sarcomere, and remain unobservable as a macroscopic behavior of the sarcomere. That this is not the case will be explained next.

3. THIN FILAMENT TIP ALIGNMENT WITH DIPOLE

To keep the argument simple let us assume that we are looking at one sarcomere at rest. Initially the tips of thin filaments are not well aligned with respect to the projections. This is true even if both A-bands and Z-lines are perfectly straight and no length variations of filament exist because the projections are helically arranged. Upon activation of dipoles of the projections by ATP and Ca^{++}, a force acts on each thin filament with a magnitude dependent upon the position of its tip with respect to the nearest dipole. Those filaments with tips close to the dipoles feel a great force and are pulled forward, while those with tips in the zero field feel no force. These forces are transmitted to the Z-line, and of course the reaction forces are applied to the projections and transmitted to the M-line. These forces strain their respective structures, but the greatest strain will appear in the Z-line (or Z-disc) because of its network-like structure. The result is the relative movement of tip positions: those in high fields move forward to lesser fields and those in low fields are pulled back to higher fields. Figure 11 shows the relative movement of filaments and resulting strains in the Z-line and M-line. Besides these strains there will be small strains in the thick and thin filaments. All these strains allow a greater and greater proportion of the thin filament tips to closely align with the projections as the energy density increases.

There are many consequences of this tip alignment process. Of the greatest importance to the mechanics of the sarcomere is that the alignment effect makes the dynamics of the sarcomere very close to that of a single thin filament. I shall deal with this subject again when discussing the quick stretch and release responses of the sarcomere. Other effects of the tip alignment pro-

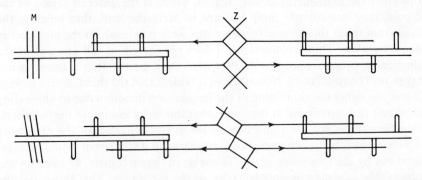

FIG. 11. Thin filament tip alignment with projection. Shear strains at the Z-line and M-line allow the tips of thin filaments to align with the projection. The projection itself also suffers a bending strain which is not shown here.

cess include vigorous jumping motion between the thick and thin filaments during the alignment, and the discreteness of the sarcomere length distribution. The former gives rise to fast fluctuations of the scattered light from muscle (Bonner & Carlson, 1975) which subside as filaments settle down at most favorable positions in tetanic contractions. The latter can be observed as sharp parallel lines in the diffraction patterns of muscle in extremely steady tetanic contractions (Iwazumi, ter Keurs & Pollack, 1977).

4. ELECTROSTATIC INDUCTION

When a dielectric body is moved with respect to an electrostatic field, a reactive force acts on the body in the direction opposing the motion. This is analogous to magnetic induction. More specifically, when the thin filament is quickly pulled out of the capacitor, the capacitance drops stepwise, as shown in Fig. 12. Due to the rate-limiting chemical process in the battery (myosin-ATP complex), immediately after the step change of capacitance, the charge Q remains constant. From the relationship, $Q = CV_0$, where V_0 is the potential difference between the capacitor plates, a step decrease of C must be counteracted by a step increase of V_0. Afterwards, a finite current flows which removes the excess charges from the plates, thus causing V_0 to diminish. The force f acts on the thin filament changes in proportion to $(V_0)^2$. Conversely, when the thin filament moves into the capacitor, the capacitance increases, and the reactive force appears to decrease the force acting on the thin filament. I shall summarize the peculiarity of the electrostatic induction in the following.

1. The induced force is a strongly nonlinear function of velocity. It is ap-

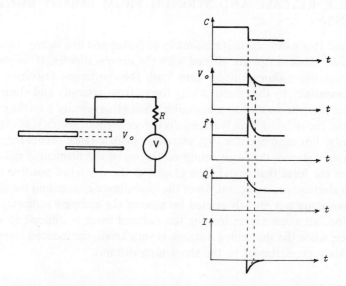

FIG. 12. Electrostatic induction associated with a sudden pull-out of a dielectric body. R represents a rate-limiting process, not a resistive element; C = capacitance V_0 = voltage across capacitor; f = force; Q = charge; I = current; and V = voltage of an ideal source (EMF).

proximately linear with velocity up to 0.1 μm/sec (relative velocity of the thick and thin filament), but becomes virtually constant above 10 μm/sec.

2. For a fixed velocity the magnitude of induced force is proportional to the force that existed before the motion. In other words, the normalized induced force appears to have the same time course.

3. The induced force has a relaxation time constant which is determined by the chemical process of the battery and by geometrical factors such as the number of thin filaments in the capacitor. Although the relaxation time is definable, the time course is not a single exponential function because of the rate-limiting process. It is also clear that the shortening and lengthening relaxation times are different due to different rates of forward and backward chemical reaction.

4. The induced force is a reactive force; therefore, it is not associated with an energy loss (except for the secondary effect of increased charge loss due to greater field strength).

5. The length of travel over which the induction effect is clearly observable is limited. It is about 150 Å per half-sarcomere. Beyond this range thin filaments begin jumping either forward or backward, thus obscuring the induction effect.

5. QUICK RELEASE AND STRETCH FROM STEADY CONTRACTIONS

Suppose that a sarcomere is maximally activated and in a steady state. Most of the thin filament tips are aligned with the nearest dipoles. If we suddenly let the sarcomere shorten under zero load, the shortening velocity is determined essentially by the counteracting forces from viscosity and electrostatic induction (*i.e.*, the mass effect is negligible). Since this velocity is on the order of 10μ m/sec, the induced force is essentially independent of velocity, as explained previously. For example, if a 10% increase of capacitance occurs for a given shortening distance, the force acting on the tip of the filament is reduced to 82.6% of the force that would have existed at the shortened position if there were no electrostatic induction. Since the capacitance change and the shortening distance are not linearly related because of the complex geometry of the projection, an exact expression for the induced force is difficult to obtain. However, when the shortening distance is very small, the induced force is approximately proportional to the shortening distance.

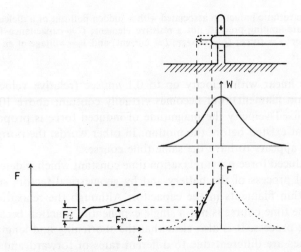

FIG. 13. Quick release from a steady isometric contraction. The average position of the tip of the thin filament is slightly ahead of the projection at a maximum steady force. Upon quick release over a small distance, the tip moves forward at the velocity limited by viscosity and electrostatic induction. The capacitance increase due to forward movement of the thin filament brings about a momentary reduction of the field strength by the electrostatic induction. The induced force F_i decays and the force approaches F_r which is the value determined by the fringe energy density at the new tip location. The tip does not remain at this location because the reduced strain of the structure forces thin filaments to settle at different locations. This redistribution process takes time, and the force gradually recovers toward the previous level.

The time course of force after a quick release is shown in Fig. 13. F_i is the magnitude of an induced force (negative), and F_r is the force reached after shortening. The force will not remain at F_r since the tips of thin filaments begin to redistribute again to settle down at the most favorable positions to bear the maximum force. Therefore, during this recovery phase, there should be a vigorous jumping motion of filaments.

Since the synchronous movement of thin filament tips is possible only at high force levels which produce large enough strains in the synthesized structure, the quick release from low force levels will not result in responses in proportion to those obtained at high forces. At a low force, F_r should approach the initial force since the tip locations are not well aligned with the dipoles.

It is evident from Fig. 13 that the quick release response depends upon the spatial distribution of the energy density. For example, the distribution changes with ionic strength and temperature; therefore, the quick release responses change accordingly. The relaxation time is also a function of ATPase activity, ionic strength, temperature and geometrical factors (such as the number of thin filaments per unit cell).

Let us now deal with the quick stretch responses of the synthesized structure. As before, we begin with a sarcomere which is fully activated and the

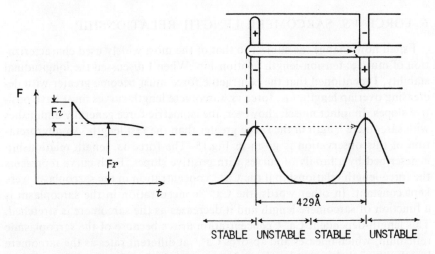

FIG. 14. Quick stretch from a steady isometric contraction. The average position of the thin filament tip is slightly ahead of the projection at a maximum steady force. When the thin filament is pulled quickly with a force that overcomes the electrostatic induction at the left projection, the tip jumps backward to the right projection. The new tip locations are determined by the strains formed in the structure after stretch.

majority of thin filament tips are aligned with dipoles. Figure 14 shows the time course of force when the sarcomere is quickly stretched. The average tip position of the thin filament is always ahead of the projection. This is because tip positions have some distribution around the projection, and tips cannot be in the unstable region. When the sarcomere is stretched slightly, some of the thin filaments which were supporting a maximum possible force at exactly the center of the projection must jump back to the next projection. As the stretch distance is increased, more and more thin filaments make jumps until a distance is reached at which all thin filaments have completed jumps. Since each thin filament produces electrostatic induction when it makes a jump, the stretch response of the sarcomere is a summation of all the inductions which occur at slightly different times. In passing it should be noted that filament jumping does not necessarily imply sarcomere length jump because sarcomere length jump can only be observed only if all thin filaments in the sarcomere make jumps simultaneously.

The time course of force, after the relaxation of electrostatic induction has ended, reflects the readjustment of tip positions, which involves elastic properties of the structure, Ca^{++} distribution changes at the new length and other factors such as stretch velocity, stretch distance, uniformity of sarcomere length along muscle and the periodicity of troponin sites on the thin filament.

6. FORCE *VS.* SARCOMERE LENGTH RELATIONSHIP

I shall now change the subject to that of the most widely used characterization of muscle: tension-length relationship. When I discussed the longitudinal stability, I mentioned that the contractile force must become greater with decreasing overlap length, *i.e.*, force *vs.* sarcomere length curves must have positive slopes. In intact muscle, however, the isometric force generally diminishes with sarcomere length in the range greater than the rest length. An interpretation of this observation is given in Fig.15. The force *vs.* length relationship is described by a family of curves with positive slopes. Each curve represents the force-length relationship if the Ca^{++} concentration in the sarcoplasm were kept constant. In other words, the Ca^{++} concentration in the sarcoplasm is a function of sarcomere length and it decreases as the sarcomere is stretched. This length-dependent Ca^{++} concentration arises because of the sarcoplasmic reticulum, which releases and absorbs Ca^{++} at different rates as the sarcomere length is varied. As a result, a large force is generated at short sarcomere length (l_s) since the energy density (which increases with Ca^{++} concentration) is very high at the tip of the thin filament despite the concave distribution curve. On the other hand, the contractile force becomes small at long sarcomere length

A NEW FIELD THEORY OF MUSCLE CONTRACTION

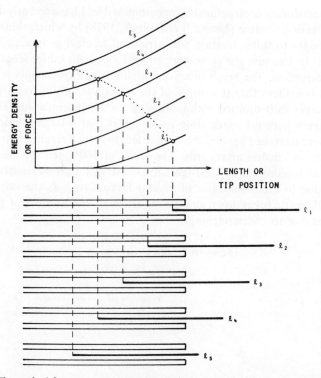

FIG. 15. Theoretical force *vs.* sarcomere length in intact striated muscle. The relationship is described as a family of curves of the energy density distribution because of the length-dependent Ca^{++} concentration in the sarcoplasm. High forces at short sarcomere lengths are due to high energy density resulting from high Ca^{++} concentration of the sarcoplasm, not due to large overlap length.

since much less Ca^{++} is available. So, an oversimplified conclusion is that the force decrease at longer muscle length is due to decreased level of Ca^{++} concentration in the sarcoplasm, not due to the decrease of overlap length. If this explanation were correct, then we would expect to see an increasing force with sarcomere length if it were possible to keep the Ca^{++} concentration in the sarcoplasm constant. This could be done by removing the membrane of a muscle cell, thus subjecting the intracellular environment to outside control. This "skinned fiber preparation" has been used by a large number of investigators, but there has been a serious problem with this preparation. It is nonuniform activation of sarcomeres; upon exposure to Ca^{++} containing solution ($[Ca^{++}]$ = 10^{-6}M or higher), the skinned fiber promptly loses its regular striations, an indication of highly nonuniform contraction, and the sarcomere length measurement during contraction becomes impossible. I have recently developed a slow activation scheme (Iwazumi & Pollack, 1978) by which skinned fibers can be brought to full activation with striations as clear as intact fibers, thus allowing us to measure the sarcomere lengths along the entire length of fiber during contraction. The result of several preliminary experiments is shown in Fig. 16. It is evident that the slopes of the tension-length curves are positive and the curves shift upward with increasing Ca^{++} concentration. The reason measurements were not made close to 3.6 μm is that a large fraction of sarcomeres were stretched beyond the overlap length, which resulted in irreversible degradation of tension afterwards. The stress at 100% tension was about 3 kg/cm². The leveling off of the high Ca^{2+} curve at longer sarcomere lengths could be due to a combination of several factors such as the saturation of ATPase, dielectric breakdown and the reduction of the number of functional sarcomeres due to overstretching.

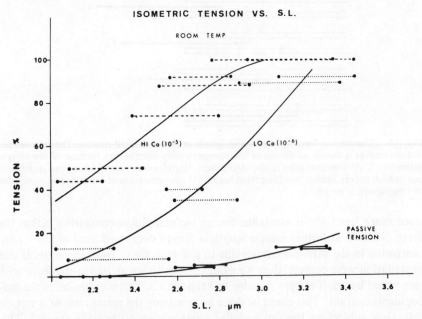

FIG. 16. Experimental isometric tension *vs.* sarcomere length in skinned rabbit soleus fibers. At constant Ca^{++} concentration the slopes of the curves are always positive as predicted by the theory. Note the upward shift of curves at higher Ca^{++} concentration.

7. CONCLUSION

I have introduced a field theory of muscle contraction and discussed a limited number of topics in sarcomere mechanics. This theory is still under development to integrate all aspects of muscle biochemistry, ultrastructure and mechanics into a coherent point of view. The outlook is quite promising, and I invite more people to join me in exploring the extremely interesting properties of the synthesized structure which I believe holds the key to understanding real muscle.

REFERENCES

Bonner, R. F. & Carlson, F. D. (1975). Structural dynamics of frog muscle during isometric contraction. *J. gen. Physiol.* **65**, 555–581.
Huxley, H. E. & Hanson, J. (1954). Changes in the cross-striations of muscle during contraction and stretch and their structural interpretation. *Nature, Lond.* **173**, 973–976.
Huxley, A. F. & Niedergerke, R. (1954). Structural changes in muscle during contraction. *Nature, Lond.* **173**, 971–973.
Iwazumi, T. (1970). A new field theory of muscle contraction. Ph. D. Thesis, University of Pennsylvania. *Univ. Microfilms, Inc.*
Iwazumi, T. & Pollack, G. H. (1978). Stretch responses of maximally activated skinned rabbit soleus fibers. *Int. Biophys. Congress*, VIII-8-(J).
Iwazumi, T., ter Keurs, H. E. D. J. & Pollack, G. H. (1977). Do sarcomeres assume discrete lengths? *Biophys. Soc. Abstr.* **17**, 199a.

Discussion

Edman: You would have difficulty, I think, when the tips of the I filaments come to the M region, and I think you could overcome that by assuming that you have a certain amount of staggering of the I filaments which would take care of a lightly loaded shortening. But I would think that you should expect a decrease of the maximum force rather than a plateau. Wouldn't you expect a trough there because you would have a smaller number of tips in that region?

Iwazumi: Well, that is possible, assuming you have good regularity of sarcomere length.

Edman: When you pass that region you would expect to have a smaller number of tips which experience fields; therefore, you would have a trough maybe down, say, to 50% of the maximum force.

Iwazumi: Depending upon non-uniformity, yes.

Tregear: A rather good way of testing theories is by energetics, and I have been trying to think of a critical test of your hypothesis. Obviously a very interesting point about your hypothesis is that just the tip is working, and I have come to one from my own experience; I am sure that the people who work on vertebrate muscle could come up with equally good ones from theirs, but this is from the insect flight muscle. Now, in that muscle, as in all other muscles, the ATPase sites are distributed along the thick filament, on the myosin, evenly and this muscle oscillates, and when you oscillate it at the right frequency and at the right amplitude, which is about one of your 43 nm moves, you get the following characteristics: You get the usage of ATP which is equivalent to—if you divide it up—about one ATP per myosin head per oscillation. And you also get an efficiency of about 50%, a chemomechanical transduction of about 50% efficiency. Now, that is easy enough to explain; it is possible enough to explain on the cross-bridge theory or indeed any theory which invokes interaction of the sites along the length, but I would be very interested to hear if you have got any explanation of it on end motion.

Iwazumi: (Ed.: This response is expanded from the original at the request of R. Tregear.) I am really happy to have that question, because this theory turns out to be the most beautiful to explain how the oscillation in the insect flight muscle occurs.

If you look at the electron micrograph of the relaxed insect flight muscle very carefully, the tip of the thin filament is found well beyond the end projection and into the bare zone as shown in Fig. VIII-1. When the field is generated upon activation of the ATPase of all projections, the tip of the thin filament is still outside of the field; therefore, it is necessary to pull the thin filament back

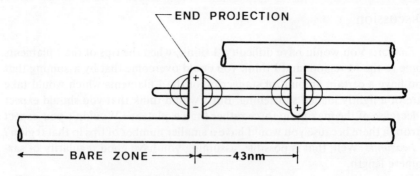

FIG. VIII-1. The location of the tip of the thin filament in insect flight muscle. At rest length, the tip lies beyond the end projection and into the bare zone. Oscillation becomes possible when the tip is pulled back before the end projection, *i.e.*, into the unstable region resulting from the negative slope of the energy density (or negative modulus of elasticity). Any other unstable regions are unsuitable to sustain the synchronous oscillations of all the thin filaments in the sarcomere.

a little in order to observe a tension (so-called stretch activation). If you pull the thin filament a little further back, it jumps to the next projection. As I have explained (Fig.14), this jumping occurs because of the unstable region, *i.e.*, the negative modulus of elasticity (the slope of the energy density distribution curve). Now, after the thin filament jump, what happens to the ATPase activity of the end projection is that the rate of ATPase stops momentarily, due to the electrostatic induction as I explained in the text, then diminishes rapidly because the Ca^{++} that stimulated the ATPase of the end projection is carried away with the thin filament, The energy associated with the jump, *i.e.*, force times distance of jump, is transferred from the negative impedance source to the springs and the mass, M, which is the wing of the insect. Consequently the mass begins to move to the right. However, the parallel elastic element of the insect muscle is highly nonlinear (stiffens with stretch); thus the mass bounces back to the left because the energy stored in the parallel elastic element is now transferred to the mass. This motion pushes the thin filament forward, and the tip approaches the end projection. The Ca^{++} carried by the thin filament stimulates the ATPase activity of the end projection which rapidly increases the field strength as the tip is sucked into the field. This proximity effect is a regenerative process (*i.e.*, positive feedback) and this is really the driving source of the oscillation since the negative modulus of elasticity alone will not produce oscillations. The tip of the thin filament keeps moving forward, then reaches the original position where this oscillatory motion started, and the whole process repeats itself. The frequency of the oscillation is, therefore, determined primarily by the mass and the spring constant of the series elastic element. The optimum frequency and amplitude of the oscillation depend upon the ATPase rate of the end projection; the faster the rate, the higher the optimum frequency.

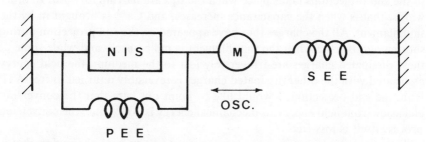

FIG. VIII–2. A model representation of the oscillation dynamics of insect flight muscle. NIS = negative impedance source; PEE = parallel elastic element; M = mass (the wing of the insect): and SEE = series elastic element. NIS and PEE represent the flight muscle, and M and SEE are the external load.

The key to make this oscillation mechanism workable is the synchronization of all the thin filaments in each sarcomere and of all the sarcomeres in the muscle. The synchronization is accomplished by the end projection, since those thin filament tips that have moved too far ahead of the end projections are out of the field; therefore, they do not bear tension any longer, and thus stop shortening. If there were more projections ahead, the thin filaments would keep moving forward; hence, no synchronization is possible.

Tregear: I am sure you can come up with the oscillation mechanism, but what I am challenging you to do is to produce a thing which produces an enormous energetic throughput from sites which appear to be distributed all the way along the thing, and at a high efficiency, and I just do not see how it can do that.

Iwazumi: The efficiency is high because all the projections along the shaft of the thin filament are not splitting much ATP even though they are generating strong fields to suspend the thin filament in space. The reason why this is possible is that the field is generated when ATP is adsorbed by myosin, and the process that presumably occurs is: adsorption energy of ATP = field energy + molecular strain energy in myosin, and this is a reversible process. The splitting of ATP occurs when the charge is lost to the medium. One thing I want to add to this is that the rate of ATPase is not uniform among the projections in this region along the shaft, because the troponin sites on the thin filament move in and out with respect to the projections during oscillatory motion, consequently creating local Ca fluctuations which modify the rate of ATPase.

Since there is no capacitance change, *i.e.*, transduction from field energy to mechanical energy, in these projections, all the ATP split in this region is entirely lost as heat. On the other hand, the greatest rate of ATP adsorption to the end projections takes place when the tips are moving forward. In other words, that is when the capacitance increases, and Ca^{++} is brought in by the thin filament. All the charges that have appeared on the end projection during shortening dissipate when the thin filament is pulled back, and while some of these dissipated charges are irreversibly lost to the medium, the field energy associated with the other dissipated charges is reversibly returned to free ATP from the end projection. I would like to point out here that the conversion efficiency from field energy to mechanical energy is 100%, *i.e.*, the conversion process itself is loss-free.

Tregear: I think that in that case you are going to have the region that is like the tip of a red hot poker. It is undergoing a vast frequency of hydrolysis. That is a very interesting hypothesis.

General Discussion

Problems of the current sliding filament/cross-bridge model

Sugi: In order to make further progress in the field of the mechanism of muscle contraction, it seems essential to throughly discuss the validity of the current sliding filament/cross-bridge concept. To begin with, I would like to summarize basic questions concerning muscle mechanics, though I notice that the following comments are very biased and cover only a limited area of research work. It is indeed a pity that we could not have Professor Andrew Huxley and Dr. Simmons here with us.

The first question is, what is the meaning of the stiffness? Though the stiffness is generally taken as a measure of the number of attached cross-bridges, there seems to be no concrete evidence that the instantaneous elasticity resides mostly in the cross-bridges; it seems possible that the compliance of myofilaments and the other structures is also involved in the instantaneous elasticity. Moreover, a quick length change applied to a muscle or muscle fiber should be distributed uniformly along its entire length in order that the resulting tension changes can be interpreted in terms of cross-bridges: but our high-speed cinematographic studies indicate (Sugi *et al.*, Session II) that the above condition may not be fulfilled in the case of extremely quick length changes.

The second question is related to the site of force-generators. It is of course generally believed that the force generator is the cross-bridge. But in *Limulus* striated muscle, it seems possible that the thick filament *per se* posesses the tendency to shorten in some conditions (Dewey, Session I). According to Iwazumi's field theory (Iwazumi, Session VIII), the contractile force may be generated only at the tips of thin filaments.

The third question is, whether the force generators work independently or cooperatively? The stepwise sarcomere shortening (Pollack *et al.*, Session I) may be taken as evidence for the cooperativity of the cross-bridges. The marked oscillatory length changes following step increases in load (Tsuchiya *et al.*, Session III) might also result from synchronized detachment and attachment of the cross-links.

The fourth point is the length-tension relation. Dr. Pollack and his colleagues (ter Keurs *et al.*, Session IV) presented some data against the well-known length-tension relation in which the isometric tension decreases linearly with decreasing amount of overlap between the filaments. This topic is closely related to the problem of so-called longitudinal stability; why, after stretch, can sarcomeres generate much greater force than their normal isometric tension?

Wilkie: It has been rightly said that only artists love their models more than scientists do. I think we should start by discussing experimental points where there seems to be real disagreement about some very important issues. As an ex-mechanical properties person, I have not forgotten all I once knew, so over your points one and four, the question of stiffness, I noticed a great laxity in the people who use the expression, in that any relationship between length change and tension change is called stiffness. Well, the most elementary mechanics says that is not true, even for a one-element situation, and stiffness should really be strictly applied to situations where you are dealing with a compliant element, where there is a time-independent relationship between force change and length change. And people have not done that; I am appalled. They have taken any sort of length step, measured the consequent tension step or *vice versa*, and called the result a stiffness. It is rubbish of course, even for one-element situations, and mostly they are not one-element situations. So over this I think there has been too much laxity.

Moving to point number four, which is more fundamental as far as sliding filament theories are concerned, I think most of us would now need a lot of convincing that the original elegance of the 1966 view of the length-tension diagram can be sustained in its original simple form. It is easy to invent various excuses why it does not any longer apply, but it certainly does no longer apply. I think we have to start from that viewpoint. That has been made very clear by several of the experimental demonstrations here. I think it would be much more profitable to start with the experimental questions of disagreement than to start with, what is at the moment, as far as I can judge from conversations with Andrew Huxley and Bob Simons, an incomplete theory—that is, an incomplete model.

So I am making two absolutely definite criticisms. On your point number one I am saying that people are being very sloppy when they use the word stiffness, and they should not be. And on number four, I think, as a meeting we are forced to the conclusion that one can no longer accept the textbook view that force is proportional to filament overlap.

Length-tention relation

Rall: I have a question along the lines of point number four (length-tension

LENGTH-TENSION RELATION

relation). I gather from Dr. Edman's results that his length-tension diagram looks more textbook-like than Dr. Pollack's. Possibly Dr. Edman could comment. Did you extrapolate back, or did you take a steady force level, and how close is your length-tension diagram to what was seen in 1966 by Gordon et al. (Gordon, Huxley & Julian, 1966)?

Edman: Those plottings refer to the plateau tension during tetanus, and the data I published in 1966 (Edman, 1966) also refer to tetanus plateau. They show rather similar diagrams as produced by Gordon et al. (1966), although slightly curved, but as I said yesterday, if you accept the curvature, then it extrapolates down rather close to 3.65 μm or so. I did not follow down to that region though. If you take longer tetani, say 5 to 10 sec, the descending limb is slightly lifted up, but not very much. It departs from the plateau very close to 2.2 μm in our experiments.

Winegrad: I wonder whether we cannot follow up on Dr. Wilkie's comments on the fourth point, the length-tension relationship, by asking the question much more specifically and pointedly, and that is, what is the length-tension relationship of sliding filaments *per se*, and the second is what is the length-tension relation for an intact cell? I would agree with what Dr. Wilkie says, that there is no question that the length-tension relationship of the intact cell does not follow the classical description of the 60's. We also have some idea about what other factors are involved—activation, internal forces, etc. For instance, you just simply take a myofibril and activate it, and remove the activation and it returns to a certain length. Something more is involved there than just filaments sliding past each other in the simplest idea. However, I do not believe that the data have been presented here to evaluate one way or another whether the isolated overlapping of thick and thin filaments does or does not behave consistently with the sort of model that has been proposed by the sliding filament hypothesis.

We must look at muscle contraction in a cell as a series of steps, only one of which is the interaction between two sets of filaments.

Pollack: One of the most straightforward predictions of the sliding filament/cross-bridge model is that the amount of force should vary in a specific way with the amount of overlap, and there are several experiments that have been presented at this meeting which have shown that such a prediction is not satisfied. We have discussed these. There is one other kind of experiment that has not been discussed fully, and that is the data from skinned fibers at high pCa levels. In that case, there is agreement among various workers that the tension actually *increases* as overlap *decreases*, and I wonder whether or not that observation is consistent with the prediction of the cross-bridge/sliding filament hypothesis?

Rüegg: I do not think there is agreement about this, because the results which I showed show clearly that there is a decrease in tension and stiffness as you increase the sarcomere length in skinned frog fibers, and moreover, these tensions were steady-state tensions obtained after full development of tension. It was not the sort of extrapolated thing, but it was a plateau tension we plotted.

Pollack: I apologize. I was referring to Dr. Sakai's data and Dr. Endo's data and Dr. Iwazumi's data, but let us suppose there is some difference in different preparations. Wouldn't one expect, on the basis of the sliding filament/cross-bridge theory, that one could not get increased force with decreased overlap under some conditions? Is that not a violation?

Iwazumi: May I comment about the skinned frog fiber preparation? I have gone through all the problems with activation, and what I have found was that unless you have very clean striations during contractions, the data are not worthwhile. The reason is whenever I see smeared-out striations I get less tension at long sarcomere lengths than when the striations remain absolutely beautiful during contraction. It is really essential that the specimen be examined from end to end to make sure that the sarcomere length is uniform; also, that the width of the first order, if you use a diffraction pattern, is just as good as that of the resting fiber, which is an extremely tough experiment. I have gone through hundreds of fibers, and the results, in my view, which show a decreasing tension at longer sarcomere lengths with constant calcium, are essentially due to non-uniform contraction; because at longer fiber lengths with a highly non-uniform sarcomere length, what you end up with is extremely shortened sarcomeres stretching the other sarcomeres way out of overlap. That, I believe, is the reason why you get less tension.

Rüegg: Excuse me. I must say something again. We checked, of course, in our experiments the sarcomere length very carefully, and in many, although not in all, experiments we were able to obtain diffraction patterns during contraction, and were careful enough to check various portions of the fiber; not all through, but at the end, at the middle, and at the other end, so we have control of sarcomere length.

Iwazumi: Have you checked the sharpness of the diffraction pattern? If you can see a diffraction pattern but the intensity goes way down and the width increases, this is an indication of non-uniformity.

Rüegg: No. That we have not done.

Sugi: I would like to make some comment. The largest problem may be the non-uniformity of sarcomere length, and it is normally very difficult to assess the extent of non-uniformity produced under various conditions. In this connection, Dr. Edman told me that he could succeed in clamping sarcomere lengths by using laser beam diffraction patterns, so I think the sarcomere length clamp may be very helpful in future work.

Edman: Yes, I agree, and I would comment on what Dr. Iwazumi said about non-uniformity, that this would be a reason for the decrease of tension. I think just the opposite is true. Dr. Godt, now at the Mayo Clinic, visited Lund, and we looked at the length-tension curve of skinned fibers and found that at a higher temperature, we got a very flat length-tension curve, but when we fixed these fibers and looked at them in the microscope, we found a whole spectrum of sarcomere lengths from very stretched down to very short. And under those conditions I believe that the tension produced in that preparation is determined by the strongest sarcomeres, which happen to be down to 2.0 or 2.2 μm. Because of the stability of the descending limb, those sarcomeres which are being stretched are probably holding the tension which is produced by the strongest sarcomeres, so I would not agree with that.

Iwazumi: Could you tell me about the stress developed? How much tension did you get from the fiber?

Edman: I do not remember now, exactly, but it was about what you can get maximally, I think, in living fibers—so it was a maximal contraction.

Winegrad: We have done a few experiments where we have quickly frozen skinned fibers just after the addition of calcium, and what you see is that the superficial myofibrils are contracted, and the ones inside are not. Even the contraction along the surface is not uniform, presumably because the resistance to diffusion may not be exactly the same, and I think that this may be one of the major problems. I think if people using skinned fibers would at least think along the lines Dr. Moisescu (1976) has suggested for trying to initiate uniform activation across the whole cross-sectional area of a skinned fiber at the same time, that these problems of the production of non-uniformity can be decreased markedly.

Longitudinal stability

Sugi: Now I would appreciate hearing opinions on other topics.

Pollack: I would like to discuss the idea of stability. It has been demonstrated quite adequately that the muscle is stable on the descending limb of the length-tension curve. In particular, I would like to refer to the experiments that Dr. Noble presented, where if you stretch the muscle, the amount of tension that you get after stretch exceeds the amount of tension before stretch for a period of many seconds. I wonder if somebody, who is familiar enough with the current sliding filament/cross-bridge theory, can interpret this observation in those terms.

White: Well, I do not think there is any problem in explaining the dynamic part. Do you mean the long-term thing?

Pollack: Yes. Why is there more force persisting after stretch?

Woledge: I will say something about that. I am just covering an idea which comes from Lydia Hill (1977); I do not know if she originated it but it seems plausible, and her idea is that part of the length change goes into the thin filaments, so that they are extended, and that the sliding motion of thick and thin filaments in the overlap region continues after the stretch is over, and that the long-lasting enhancement of tension development is due to this continuing sliding motion. Now, that theory has a merit that can be tested; it can be tested by measuring the filament lengths, and that, as I understand it, is the experiment which she is doing now.

Noble: Could I comment on that? Of course, it should be tested as you say, but it would seem to me that it is still necessary for the force generators to support that force, and they have to do that having remembered that there was a previous stretch, so I cannot see, even if there was a slight strain in the thin filament, that you can get away from the fact that those force generators have to generate that force as well as the increased velocity of shortening and the increased mechanical performance.

Huxley: Well, I believe I was once said to be responsible for a suggestion about how that might occur, and it simply was the notion of "locked-on" bridges: that when a bridge was bent backwards it not only might generate more tension than it would in its normal configuration, but it might be pulled into a configuration where its dissociation from actin became extremely slow at that particular length, and it just stayed on generating a lot of tension. And, if you pulled the muscle some more, you eventually broke those attachments, but there were then some other cross-bridges which were now pulled into this locked-on state. I do not see that there is anything very unlikely about that.

Noble: If I could just reply to that; I am not, of course, challenging what you say, but I am merely pointing out that you would need to explain why these "locked-on" bridges can generate an extra velocity, and why the ATP consumption is the same, why the X-ray diffraction pattern $(I_{1,0}/I_{1,1})$ is the same. These are not easy to explain. One should also point out that the delivery of calcium apparently is the same according to Allen and Blinks. So these are not easy to explain, I think.

Edman: I think there would be a difficulty in using that explanation to explain the length-dependency of that remainder, *i.e.,* of the extra tension after stretch. If you perform this stretch at shorter lengths, then you do not get any remainder. At longer sarcomere lengths you do. It seems as if we are recruiting some parallel force. For some reason it behaves like that. Where it comes from, that is of course a question mark. Perhaps some structure is reorganized, or formed, during activation.

Huxley: Could I ask Dr. Noble? What was this evidence from Blinks? Did they do that sort of stretch experiment?

Noble: There is no difference in the light signal compared with an isometric control.

Huxley: Even when they extend the muscle?

Noble: It is stretched during activity, yes; whereas if they shorten during activity, the signal drops (Allen, 1978).

Sugi: So, I think the length-tension curve appears to be very complicated. Are there any experimental means to clearly determine the relations between sarcomere length and force? I think it is also related to the question of whether force generators work independently or not.

Tregear: I am not going to try to attempt an answer to that question, Dr. Sugi. I was very impressed by Dr. Kushmerick's phosphorylation, and how fast it occurred. It does seem to me that perhaps we are being a bit naïve in our logic generally, in assuming that we have always got the same thing there after we have given it, say, this pull. Perhaps the whole state of the muscle has been changed, in which case where is the logic? How can one apply logic? I suppose it means that we should be measuring, or attempting to measure, phosphorylation as a routine thing, at any rate, to know under what circumstances it occurs in these experiments and in order to ask whether it is a possible cause of the phenomenon.

Winegrad: We have also been operating under the assumption that there is a force that a force generator can produce that is essentially solely length-dependent, or dependent upon its geography, let us say, with respect to the backbone of the filament. But there are other observations suggesting that it may not be quite as simple as this. I think the first was made by Annemarie Weber (Bremel & Weber, 1972), when she found that the combination of low ATP and calcium produced a higher rate of ATP splitting in the test tube than existed with normal ATP concentrations and maximal calcium activation, and analogous experiments have been done on both skinned skeletal and skinned cardiac fibers (Godt, 1974; Fabiato & Fabiato, 1975) and show the same thing: that, if you drop the ATP concentration down in the range of 20–50 μM, and then add calcium, you can get somewhere between 50 and 100% greater force production than you can by maximal calcium activation with 5 μM ATP. I think that these are observations that have not yet really been incorporated, and we should bear in mind the possibility that the force generator is capable of developing different amounts of force depending on specific conditions.

Sugi: Thank you very much. I think it is a very important point.

Pollack: I was intrigued by Dr. Dewey's finding, and I wonder if Dr. Dewey would care to comment on the possible significance of thick filament shortening in terms of force generation.

Dewey: Unfortunately, I think we are in a state where we have no information on that whatsoever. With *Limulus* muscle, we are really back in the early 60's. But it is clear that there is a sliding of thick and thin filaments from long sarcomeres to rest length, and that from rest sarcomeres to shortest sarcomere length there is the filament thickening, which I would assume would have to be force-generating as well as involve actin-myosin interaction. It is interesting that, at very short sarcomeres, you can get 50% of maximum tension development, and in the length-tension curve you get 50% of tension also at a sarcomere length of 12 μm, which is at a point where there should be non-overlap, except for the skewing of the filaments.

Vectors of the cross-bridge force

Noble: What I would like to ask is the question of how, if one has a cross-bridge model (Fig. IX-1), the lateral vector of cross-bridge force is opposed, because since we have lateral stability of the filament lattice, it follows that any force recorded in the longitudinal direction is a vector, a longitudinal vector of a cross-bridge force which is exerted at an angle. We can argue about what that angle is, but there has to be an angle of some sort, and therefore there has to be a lateral vector. And that lateral vector has to be opposed by an opposing force, equal and opposite with the same time course, if the thin filaments are going to stay stationary, and they presumably are according to the X-ray diffraction data as I understand it. So what I would like to ask is, what is the mechanism of the lateral opposing force?

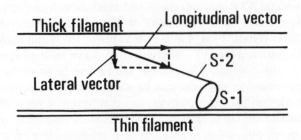

FIG. IX-1. Longitudinal and lateral vectors of the force exerted by a cross-bridge.

White: Well, I suppose one answer to that is that you will have vectors which have the same longitudinal component, but different sideways components. Because of the number of cross-bridges involved the ones which are perpendicular to the filaments will average out.

Noble: Isn't the lateral vector always going to be tending to produce a movement from the thin filament towards the thick filaments?

White: Yes, but you have got thick filaments surrounded by thin filaments, and the bridges will go to those presumably, roughly speaking, equally, and therefore the vectors will be going out from one thick filament to several different thin filaments, and the average will come to be zero.

Noble: In other words, you depend on a rigid cell membrane?

White: No.

Tregear: David, there is a residual at the surface.

Huxley: I think that what Dr. Noble is getting at is that at certain filament separations, it is at least likely that the S-2 link is not directed purely longitudinally, and therefore there is bound to be a component of sideways force between the filaments. I had always imagined that question had been answered by the calculation that Elliott and other people have done about the balance of forces between the thin and thick filaments (Elliott, 1968; Elliott, Rome & Spencer, 1970), which are quite large forces, and which, it is plausible, balance out at about the distances that you find the filaments apart in a muscle. Within a muscle contained within a membrane, the filaments may be compressed slightly closer together than they would be in free solution, and that this balance of forces holds the filaments at an approximately fixed distance, for a particular overlap of thin and thick filaments.

Pollack: Certainly, those forces are there. I think what Dr. Noble is referring to is the additional force that occurs when cross-bridges attach and generate force. The longitudinal component of force then increases and so must the lateral component of force. So when the cross-bridges generate longitudinal force, some additional mechanism has to generate equal and opposite lateral force to oppose the lateral contribution of cross-bridge force.

Huxley: Yes. And the mechanism is that the force generated between the filaments is in equilibrium at their resting separation, and it is at the bottom of a rather sharp well, so that you only need a very small difference in the spacing to increase the force to balance the longitudinal force.

Winegrad: I think, actually, there are additional experimental observations to support that notion. If you do serial sections transversely through a sarcomere, and look at what happens when double overlap of thin filaments occurs, you have just what Dr. Huxley's information would predict, and that is that instead of two thin filaments which are opposite each other coming close together, the entire rosette rearranges so that those filaments are equally spaced, and the bend that takes place in the filament that is moving to the contralateral side takes place something like 500 to 1,000 Å before it ever comes to the tip of the ipsilateral thin filament. So I think this is reasonable evidence that there is considerable resistance against thin filaments coming closer together, and that would answer the question you have raised.

Kawai: Further experimental evidence to answer that question is that, if you look at the diameter of the skinned fiber and then activate it, it shrinks. I take this to occur because of the extra concentric force produced. In this respect I wonder if somebody has done X-ray analysis to measure the filament spacing. I suppose it is a little less than in the relaxed state, since I observe the shrinkage of the full diameter. Dr. Huxley, do you have any evidence whether the lattice spacing also shrinks in the equatorial diffraction pattern?

Huxley: Not on skinned fibers.

Kawai: How about on intact ones? I presume that the same thing happens in the intact preparation.

Huxley: There is very little change of spacing in intact muscles, but of course one is getting a certain amount of shortening at the same time, so it could be that one is seeing a few percent change in the spacing.

White: Perhaps I can mention some experiments that David Maughan (Maughan & Godt, 1978) has done on skinned fibers, where he has looked at the diameter of fibers as a function of the concentration of PVP in the external solution, and has developed a very elegant theory to explain the changes he sees in terms of the osmotic pressure difference between the inside of the fibers and the outside. I think that would go along with answering your question as well.

O'Brien: Dr. Matsubara has done the experiment with the skinned fiber and showed that the filaments do get closer together from the X-ray diffraction pattern.

Kawai: To what extent?

O'Brien: I do not know. Also, when the muscle goes into rigor, the filaments move closer together.

Extra tension produced by stretch

Huxley: Perhaps we can return to the effect of stretch and the augmented tension. Perhaps Dr. Noble was referring to some later work of Blinks, but what is published in their paper (Blinks, Rüde & Taylor, 1978) is that they always find an increase in the amount of the aequorin response in a stretched fiber, and that the only thing that remains constant when there is a stretch during contraction is the duration of the aequorin response.

Noble: Well, I should not speak to this since it is not my work, and there are people who are more familiar with the authors, but as I understand their presentation to the Physiological Society in London, there is, of course, a creep upwards of the light response during a contraction; they have isometric controls showing this (Allen, 1978). Then, when they did a stretch during contraction, this was unaltered. I am open to correction there, if anybody from University College can correct me.

Dawson: Well, I do not know if any of us can speak for it. David Allen has done some experiments on stretched muscles since the time of his presentation to the Physiological Society, but since he has not yet published them, I do not know how much you could say about them.

Pollack: Dr. Huxley, does your interpretation of the effect of the extra tension after stretch depend on the fact that the sarcomere lengths throughout the preparation remain precisely constant after stretch? It would seem that if you are arguing that after stretch the cross-bridges all "hang on," and that is responsible for the extra tension, one would need to demand that no sarcomeres shorten; otherwise the cross-bridges in the region generating the most tension would no longer be hanging on. The situation would be quite different.

Huxley: Well, if the whole muscle is shortened, yes.

Pollack: No, I mean a redistribution of sarcomere lengths, where some sarcomeres along the preparation might shorten a bit, and some might lengthen.

Huxley: Well, I suppose it depended how much they shorten and lengthen. If they shorten down to mere overlap and other ones were stretched more, then I imagine you can get any behavior. Is there an experiment that you were referring to?

Pollack: I wonder if someone has done the stretch experiment, and looked to determine whether during the period after stretch there is some redistribution of sarcomere lengths taking place. Certainly if some sarcomeres along the fiber tended to shorten by more than a cross-bridge stroke, then the argument that the cross-bridges are all "hanging on" and generating that extra tension would be invalid, would it not? I do not know if anybody has done that experiment.

Wilkie: Well, surely Lydia Hill (Hill, 1977) has done that experiment very elegantly by stretching, and then found that the sarcomere lengths remained uniform for about 20sec stimulations. Then she said she was able to predict when the next one was going to look bad because she could see the sarcomere lengths become uneven. And she has done that experiment very beautifully.

Noble: We have looked at this problem with laser diffraction. I think Dr. Edman can explain.

Edman: Well, as Dr. Noble showed in this meeting, the laser diffraction pattern wat quite constant.

Noble: Yes. We have actually done detailed analysis of the first order with densitometry, and found that it is extraordinarily uniform after stretch.

Filament length change

White: Can I make one other point? Insect muscle shows strain activation; maybe vertebrate muscle does too. One suggestion is that, in insect muscle, this is dependent upon the strain of the A filaments, and if you were able to strain—and obviously you do strain the A filament in vertebrate muscle—maybe that is sufficient to maintain the long-lasting tension changes which you see.

Pollack: At this meeting we have been considering the posidility that some length changes could occur by filament strain, both A and I. I wanted to mention that there are some who take exception to the view that the major changes of muscle length occur exclusively as a result of sliding between the filaments. Some of the studies supporting this view are not well known, so I have taken the liberty of constructing a histogram (Fig. IX-2) showing a sampling of papers published during the past fifteen or twenty years, some of which claimed that there can be a substantial amount of A-band or thick filament shortening. I include one of Dr. Dewey's papers. The reason that I present this is not necessarily to argue in favor of the fact that we should abandon the sliding-filament theory, but only to bring up the point that some people have other views on the subject.

Huxley: Well, I have answered this question on a number of previous occasions to Dr. Pollack, but obviously my answer has not been very satisfactory to him. So I had better to answer it again. And, I think the thing one has to realize is that to measure a filament length in the electron microscope is not an absolutely straightforward business. You have to fix the muscle with as little shrinkage as possible; you have to section it at the correct angle so that you do not get any compression during the cutting stroke, and so that the axis of the filaments is parallel to the plane of observation. You have to calibrate your microscope carefully, and you have to take considerable care that you make a note of the magnification of the particular picture you were taking; and, some of these are considerable pitfalls that originally many people were

FILAMENT LENGTH CHANGE

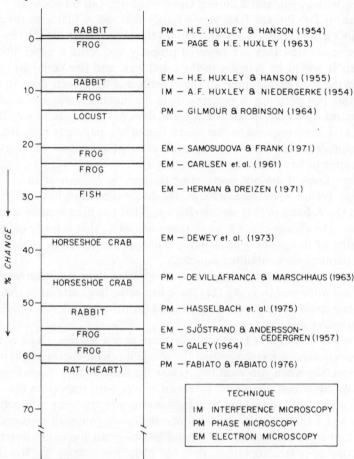

FIG. IX-2. Histogram showing papers reporting length changes of thick filaments or A-bands.

unaware of, particularly the effect of flow during sectioning on the sarcomere lengths. It has been my experience, and that of a number of other people, that when you control all of these factors very carefully, then you find that the A filaments are approximately constant in length, and the I filaments are approximately constant in length. And there have been a number of cases where apparent filament length changes were seen and where it became apparent what the artifact was in that case; there are still things in the literature which have not been withdrawn, and I do not know what one can do about it.

I mean, if Dr. Pollack feels very strongly that the A filaments do change in certain circumstances, I wish him to do the experiments himself: cut some sections, make sure they are oriented properly and write a paper about it, and then it would be referred in the usual way, and one could go through it and make sure exactly how all the experiments have been done in detail. But I think just pointing to experiments in the literature, where one has very little control over what happened in them, does not really get us very far.

Dewey: I must respond to that, since one of my papers is included. I feel rather confident that I have had enough years of experience as an electron microscopist to have met all of your criteria.

Huxley: Look, I am not questioning the ones in *Limulus* at all . . .

Dewey: Well, I would like to finish. We do see something like an 8% shortening in the A-band in frog semitendinosus. That has been seen by a number of people. The change that we see in *Limulus* is 40%; that is really quite a different kind of thing, and I believe it is a unique phenomenon of slow invertebrate paramyosin-containing muscles.

White: I wish I had said stress activation rather than strain activation because I did not mean to imply that there had to be large strains in the A filament; they could be very, very small indeed for the phenomenon which could occur in insect muscle.

Winegrad: As Andrew Huxley pointed out quite a few years ago, one must be extremely careful in interpreting the image in phase contrast light microscopy, especially when you want to measure the band widths, because you can have reversal of contrast if the specimen moves with respect to the optical section, and this is something that happens very easily when you activate a muscle, and I think in view of that, one really—if you wish to support the notion that there are changes in A-band width—must use either interference microscopy or polarizing microscopy, but phase microscopy has treacherous potential.

White: Dr. Huxley, why couldn't you use the 145Å spacing on the A filament and do optical diffraction studies on it?

Huxley: Well, if you have got good enough fixed specimens to show that, sure you could, but that is not normally what is done under these circumstances.

Dewey: But that has been done. We have done that in *Limulus*, and we have done that in frog semitendinosus and it does not change, because they were sectioned as Dr. Huxley has described.

Huxley: Yes, but I think there is some confusion here. What Dr. Pollack is talking about is large-scale shortening of the A-band that certain people have reported. I think some of Samosudova's was supposed to be down to about half the normal length, but what Dr. Dewey is talking about are these extensive shortenings of the *Limulus* muscle, which are not in question. I think, at present; and the approximate 10% shortening or apparent shortening, that you can see sometimes in the semitendinosus, is easy to account for on the basis of the changes of the cross-bridge orientation at the ends of the filament.

O'Brien: I think that the X-ray diffraction technique—if you have a suitable specimen, *i.e.,* a live specimen—may be helpful on two counts. One is, as Dr. Dewey mentioned, that you could check the 145 Å spacing and see if that changes or not, but if there is a profound difference in the overall size of the A-band, then the line width in the axial direction of that reflection should change.

Dewey: No. I said precisely the opposite of that. That was the criticism of the work early on, that in shortened thick filaments in *Limulus* you should see a change of the 145Å. The two X-ray diffraction studies that were done did not see it, so, therefore, it was difficult to accept the fact that they shortened. In our optical diffraction studies of both long and shortened thick filaments we see no change in the 145Å. We see a change in the layer line going from 1450Å to 2200Å, and we see both layer lines in intermediate length filaments. So in the case of *Limulus*, the filaments shorten, we believe, without a change in the 145Å.

O'Brien: Well, I accepted that; I said that you observed no change in the 145Å spacing, but what I am adding is there is another feature of this reflection, and that is how broad it is in the axial direction. The breadth of that reflection is a feature of the axial dimension of the whole A-band, like a line width from a diffraction grating. The larger you have a diffraction grating the smaller the line width will be. So if you have a suitably good diffraction pattern, and there is a profound shortening, I would have thought it would be easy to pick up this in line width. For example, in the normal X-ray pattern from, say, frog sartorius, you can measure the line width of the myosin reflections and show that it is commensurate with the 1.6 μm thick filament length.

Dewey: I agree.

Other problems

Huxley: I would like to make sort of a general point here. It seems perfectly well to look at particular problems of some mechanisms and say, "Well, you know it is puzzling how it accounts for this. Can we figure out how it does it?" But, if you are going to use that to say, we think we cannot immediately explain it, so we had better throw the mechanism over and go to an alternate mechanism—and it seems to me that is what Dr. Pollack is suggesting—you then are under some obligation to put forward an alternate mechanism which explains that observation much better, and explains all the other observations, too, and I would like to hear what exactly his idea is of what this other mechanism is.

Pollack: I thought I stated that the sole reason for presenting this was to suggest that a number of people disagreed with the accepted view. That was the reason for presenting the histogram. I am not suggesting an alternative model. I just want to bring up the fact that there were some conflicting opinions in the literature, and I think these need to be aired.

Huxley: No, I was making a more general point, not just about the A-band shortening, but about the conflict between the two models that, as I see it, are being opposed to each other at the moment: the model in which tension is developed by a large number of cross-bridges acting in parallel, and the model in which tension is being generated by, if I understand it right, one or two cross-bridges which are acting on the tip of the thin filaments.

Pollack: That is Iwazumi's theory, yes.

Huxley: Is that your theory also?

Pollack: No, it is his theory. I think it has some attractive features, though.

Huxley: My point was, if you want to discard a cross-bridge theory, I think you are under some obligation to hoist some colors of your own, and see what it is you are suggesting that explains the phenomenon more adequately.

Pollack: I must disagree. One does not necessarily require a new hypothesis to be critical of an old one. I agree with Dr. Noble's view that we must put up a hypothesis and go on testing it.

Huxley: That is what I am asking you to do.

ter Keurs: With respect to the number of cross-bridges which are participating in generating force, I would like to know your opinion, Dr. Huxley, as to how we could get information about it? We have heard that the X-ray diffraction data do not necessarily indicate that there are cross-bridges attached and generating force, so I am still in a little bit of trouble trying to figure out how many cross-bridges are attached and generating force.

Huxley: I agree. It is not possible to arrive at a number from the X-ray pattern. But I think there is a general point about this which was raised by Dr. Tregear earlier, that if you want to develop force just at the ends of the thin filaments, then the only myosin molecules that are going to be involved in generating that force are at that particular level of cross-bridges, and since that number must be something like one-fiftieth or one-sixtieth of the total number of cross-bridges, this implies that their activity must be about sixty times that of all the other ones, and therefore, since this phenomenon can apparently happen at any sarcomere length, then, you would expect that the ATPase activity of the whole system to be fifty or a hundred times higher than is found in practice. I mean in practice the actin activated ATPase activity that you can measure is approximately right, and the numbers of myosin molecules approximately right for accounting for the amount of energy that is liberated in the muscle on the basis that all the cross-bridges, or a high proportion of them, are involved in generating tension, and converting energy into mechanical work. But if you are only allowed to use one level of cross-bridges, then I think you find yourself in real trouble over this.

Pollack: Unfortunately we are running out of time. One final comment... Dr. Oplatka.

Oplatka: I come back to my original stupid question, and that is, what is the message of the sliding filament theory? A theory has a value only if it is general and not limited to rabbit psoas or a few other muscles. Is the message that the length of the filaments does not chenge? Is the message that myosin and actin must be aggregated in the form of filaments? We have heard of cases in which heavy meromyosin and subfragment-1 are mechanochemically competent. In many non-muscle cells, in all probability myosin either does not form filaments or does so very slowly. If there are no myosin filaments then there is no room for speaking about sliding of filaments. So, is the point that tension generation and movement are the outcome of interactions between actin and myosin, involving the splitting of ATP? Well, if this is all, this kind of statement was made by Albert Szent-Györgyi around 1961. So, I ask myself a question. What does the sliding filament theory teach us? What are its predictions? I am sorry, but it is not at all clear to me.

Pollack: Maybe we can now turn to Dr. Huxley who is going to present some concluding remarks. Perhaps he will answer your question.

REFERENCES

Allen, D. G. (1978). Shortening of tetanized skeletal muscle causes a fall of intracellular calcium concentration. *J. Physiol.* **275**, 63*P*.

Blinks, J. R., Rüdel, R. & Taylor, S. R. (1978). Calcium transients in isolated amphibian skeletal muscle fibres: detection with aequorin. *J. Physiol.* **277**, 291–323.

Bremel, R. D. & Weber, A. (1972). Cooperation within actin filament in vertebrate skeletal muscle. *Nature, Lond.* **238**, 97–101.

Edman, K. A. P. (1966). The relation between length and active tension in isolated semitendimosus fibres of the frog. *J. Physiol.* **183**, 407–417.

Elliott, G. F. (1968). Force-balances and stability in hexagonally-packed polyelectrolyte systems. *J. theoret. Biol.* **21**, 71–87.

Elliott, G. F. Rome, E., & Spencer, M. (1970). A type of contraction hypothesis applicable to all muscles. *Nature, Lond.* **226**, 417–420.

Fabiato, A. & Fabiato, F. (1975). Effects of magnesium on contractile activation of skinned cardiac cells. *J. Physiol.* **249**, 497–517.

Godt, R. E. (1974). Calcium-activated tension of skinned muscle fibers of the frog. *J. gen. Physiol.* **63**, 722–739.

Gordon, A. M., Huxley, A. F. & Julian, F. J. (1966). The variation in isometric tension with sarcomere length in vertebrate muscle fibres. *J. Physiol.* **184**, 170–192.

Hill, L. (1977). A-band length, striation spacing and tension change on stretch of active muscle. *J. Physiol.* **266**, 677–685.

Manghan, D. W. & Godt, R. E. (1978). Interfilament pressure opposing radial compression of relaxed skinned skeletal muscle cells. *Biophys. J.* **21**, 62a.

Moisescu, D. G. (1976). Kinetics of reaction in calcium-activated skinned muscle fibres. *Nature, Lond.* **262**, 610–613.

Concluding Remarks

H. E. HUXLEY

MRC Laboratory of Molecular Biology, Cambridge, England

First, I would like to express all our thanks to Dr. Okinaga, the President of Teikyo University, and to Professor Sugi, for making this meeting possible and for organizing it so well.

I feel very flattered to be asked to sum up; and perhaps the implication is that I should summarize everything that has been said in this meeting. I think it is becoming a tradition in Europe and in the United States that whenever there is any muscle meeting, Andrew Huxley is invited to come along and sum up. He sits and takes notes quietly throughout the meeting, and then at the end he explains to everyone what really was said at the meeting, and politely but firmly, indicates which part of it he believed.

Well, to begin with, I am not quite sure whether I have his qualities of encyclopedic knowledge and universality, but, in any case, I think it would be rather difficult to do this on this occasion, since there has been no time during the meeting to really think about what has gone before, and since I only have about 20 min or so in which to do it, even if I was going to use all that time.

I think I can make a few general points, however. The title of the meeting was "Current Problems of the Sliding Filament Model and Muscle Mechanics", and I think "Problems" really has a double meaning; there may be some real problem that is difficult or impossible to solve, or there may be a problem which is sort of an exercise, like a problem in multiplication, where it is a question of putting together certain facts and making the right deductions from them.

It seems to me that problems with the sliding filament mechanism, or indeed with any sort of mechanism, fall into a number of different categories. First of all, there may be problems where, to put it rather clumsily, we know that we do not know what is going on; that is, a problem where we can recognize our ignorance and where we would like to know the answer. In terms of the cross-bridge mechanism, for example, we do not know what the exact manner of movement of the cross-bridges is. We do not really have any idea of the exact configuration during the pulling movement, and would like to think of ways of getting at it. Also, I think we do not know exactly what is in the rate-limiting step in the biochemical cycle *in vivo*, and we can think of ways of getting at that too.

Then, there are other types of problems where we know, or rather where we think we know, how something should be explained, but where it may well turn out that we have been mistaken in some way. For example, it may be that the rapid tension recovery— the most rapid tension recovery that can be seen after a quick release— is really not measuring bridge rotation at all as Huxley and Simmons supposed, but is really measuring some rapid attachment or detachment. Also, it may be that the simplest idea, namely that the cross-bridges are independent tension generators, which is obviously one's starting point for a theory of that kind, because it simplifies the calculations very much and makes it much neater and easier to understand—it may be that that is an over-simplification under some conditions. It may turn out, as Annemarie Weber has suggested, that there are some important types of cooperativity between the interaction of the cross-bridges with the thin filaments, and that these have to be taken into account in certain circumstances during contraction.

Then, there may be other types of problems where some altogether new phenomenon is observed, something that has not been recognized before, and which one would not at first glance think automatically followed from one's own particular model. I think the effects of stretch on the tension of an active muscle, and the exact shape of the length-tension diagram come into that category. Then, either you solve it by finding some additional characteristic that you have to give your existing mechanism; or you satisfy it by the discovery that there is some external factors which have not been appreciated previously, possibly some calcium release mechanisms and so on; or you may find that you cannot explain it at all, and that you have to modify your original theory in some way.

I think one should really distinguish between essential and non-essential changes that one might make as one learns more about the mechanism. Dr. Oplatka has asked what is it actually that we are talking about, *i.e.*, the sliding filament mechanism. My pictures of it have generally been rather simplified ones, and then, what has happened on, I think, two occasions is that they have been brought to a much greater degree of perfection by Andrew Huxley, and given a mathematical shape which has been extremely stimulating, because it put them in a form where they can be tested much more readily. But my ideas have been simply that the myosin and actin are organized into separate filaments, and that these filaments move past each other during contraction, and that the force for this movement is produced by a projection on the myosin filament, and that this projection or cross-bridge interacts with the actin, moves, and in some way pulls the actin along.

Now, that is probably too crude a mechanism to test very easily, so it is much nicer if you can embody it in a precisely stated mechanism either like the A.F. Huxley 1957 or the Huxley and Simmons 1971 one; one that gives you a lot of numbers and a lot of precise predictions. One can test them and find out whether they are all fulfilled or not. But, if you find that one of them is not fulfilled, well, if you choose to, you can say that the whole mechanism is wrong; but I think that some people would need a little more persuading than that.

However, as an interested party, I think I would hesitate to try and go through the various things that have been discussed at the meeting and say whether I believe them or not, or, how much relative importance I ascribe to them. I think that is up to the various people at the meeting, to go away afterwards, read the publication when it comes out, think about these things, do experiments and make up their minds about them. I think that the value of a model or theory lies in whether people find it helpful in constructing experiments, and in thinking about their experiments, and I think people will go on believing in or using it—perhaps they do not have to believe in it—but they use a particular model as long as it is helpful to them. But if they find that experiments just do not make any sense in terms of that model, eventually they are going to turn to some other model.

I think, in the case of muscle, it is clear that very many experiments can be thought of in terms of the sliding filament and the Huxley-Simmons model. And so it seems to me very unlikely that that model is totally wrong. I think it is quite conceivable, indeed quite likely, that there are aspects of it which will need modifying; and as people learn to do more and more clever experiments, they will find out at what points these modifications are necessary, and the model will gradually become modified, and maybe, eventually, it will gain some other name because of some additional features that have been put into it.

What I think I should say perhaps a little bit about, is what I, myself, would take away from this meeting in terms of important issues that have been raised. Concerning the shortening of the A-filaments in *Limulus* and possibly in certain other muscles which Dr. Dewey has described, it seems to me the evidence of that is very strong. It is a very interesting finding and it will be very interesting to see whether he or somebody else can find out how it is brought about, or exactly what role it plays in the contraction of these muscles. So that is something very interesting, but which needs to be confirmed.

Then, the length-tension diagram. Well, I think it is very clear that there are unexplained differences between different observers of these diagrams, and it is something that people who do those kinds of experiments have got to sort out and come to some agreement on, and I think the meeting has brought out those differences very clearly. I look forward in the next year or two to hearing what the solution is, if it is agreed in that time.

And I think the experiments on the tension behaviour during stretch are very interesting, and it will be interesting to see if anyone can come up with an explanation of those.

Those are some of the things that stick in my mind, as well as the question of what phase of the cycle is responsible for the rapid tension recovery after quick releases. But perhaps I might finally say one or two things that have not come up the meeting very extensively but which I think are problems where it would be nice if we could get a bit more *direct* evidence.

First of all, in terms of the cross-bridge model—and this is directed towards the people who think that that is something that they should continue bearing in mind!—I think it is very important to try and get direct evidence of movement of attached cross-bridges on the actin during the time when one would suppose they were developing tension. One can get, I think, quite good evidence that cross-bridges do attach to actin, and that they attach in a very definite way to actin. Moreover, I think there is very good biochemical evidence that the cross-bridges need to go through this cycle of attachment and detachment in order to split ATP, and so produce the energy for contraction. But there is a break in the evidence in that, essentially for technical reasons, we do not have direct evidence showing actual movement of the cross-bridges while they are attached, and if someone can think of a way of doing that, then that would be a great step forward.

Secondly, we have not heard much about biochemistry during this meeting, although it has been extensively discussed at other meetings. There is the very stimulating Tonomura-Trentham-Taylor biochemical scheme for the splitting of ATP, and there are rate constants being measured for all the various steps of it; hopefully, this can be done for frog myosin as well as rabbit myosin. So, one of the things that is very important to do is to try and correlate those rate constants, measured biochemically, with the rate constants that various physiological models would need, on the cross-bridge theory, to account for the observed phenomena. That is something that is still to be done.

CONCLUDING REMARKS

And clearly, the X-ray crystallographers have got to try to crystallize the myosin heads, and to look for structural changes in them during enzymatic activity, and to try and figure out whether these structural changes could be used as the basis of a movement and a force-producing mechanism.

So, I think there are all sorts of things we can go home and get on with now. But I think it has been a most valuable meeting, in particular, in that people have challenged each other's experiments very hard, and we have had some very good, and I hope very fruitful, discussion as a result of that.

In conclusion then, I would like to thank Dr. Sugi very much indeed again, for getting us all together, stirring up such a hornet's nest; and also I would like to thank Dr. Okinaga of Teikyo University for giving the very generous financial support that has enabled us to have these three days in Tokyo to try, if not to settle these problems, at least to find out where the essential points of our disagreements are.

Contributors Index

Entries in bold type indicate papers, while other entries are contributions to discussions.

Amemiya, Y. **425**, 441
Bartels, E.M. 36
Blangé, T. **211**, 222, 223, 224, 361
Brann, L. 3
Brutsaert, D.L. 238, **241**, 256, 257, 258, 271
Chaen, S. **563**
Colflesh, D. 3
Curtin, N.A. **473**
Dawson, J. 339, **515**, 536, 643
Delay, M.J. **23, 71**
Dewey, M.M. 3, 20, 21, 343, 640, 646, 647
Drabikowski, W. 498
Ebashi, S. **1**
Edman, K.A.P. 20, 35, 100, 101, 238, 293, 294, **297**, **347**, 356, 357, 358, 359, 360, 361, 362, 374, 375, 403, 629, 635, 637, 638, 644
Eisenberg, E. **541**
Elzinga, G. **297**
Endo, M. **365**, 374, 375, 376
Fujime, S. **51**, 67
Gadian, D. **515**
Gergley, J. **561, 591**
Goody, R.S. **407**
Gott, A.H. **103**, 421
Griffiths, P.J. **125**
Güth, K. **125**
Hashizume, H. **425**, 441, 442
Herzing, J.W. **125**
Hill, T.L. 239, 358, **541**, 561, 590
Hoh, J.F.Y. **489**, 498
Holmes, K.C. **407**
Housmans, R. **241**
Huxley, H.E. 21, 22, 37, 101, 102, 166, 239, 341, 357, 358, 361, 376, **391**, 401, 402, 403, 404, 441, 442, 443, 470, 638, 639, 641, 642, 643, 644, 646, 647, 648, **651**
Iino, M. **365**
Iwazumi, T. **23, 71**, 147, **277, 611**, 629, 630, 636, 637
Jacobson, R.C. **23**
Kakuta, Y. **365**
Kawai, M. 67, 81, 148, **149**, 166, 167, 168, 189, 209, 273, 309, 328, 402, 642
Kitazawa, T. **365**
Kodama, T. **481**, 488
Kometani, K. **225**, 239, 563
Kramer, E.J.L. 39, 40, 341
Kuhn, H.J. **125**
Kushmerick, M.J. 101, 145, 166, 223, 273, 275, 342, 498, **499**, 512, 513, 514, 535, 536
Levine, R.J.C. 3
Liron, N. **593**
Loeffler, L. **171**
Maéda, Y. **457**, 469, 470
Maruyama, K. **319**, 328
Mashima, H. 83, 377, 387
McClellan, G.B. **329**
Milch, J.R. **407**
Mistui, T. **445**, 469, 470
Mohan, R. 38, 123, 309, 339, 535, 561, 591
Nakayama, K. **171**
Namba, K. **445**
Natori, Reibun **41**
Natori, Reiji **41**, 49
Nishiyama, K. **563**

Noble, M.I.M 147, 292, **297**, 309, 310, 341, 342, 362, 441, 638, 639, 640, 641, 643, 644
O'Brien, E.J. 422, 443, 469, 470, 642, 647
Oplatka, A. **593**, 608, 609, 649
Pollack, G.H. 20, 23, 35, 36, 37, 38, 39, 40, 49, 67, **71**, 82, 83, 100, 143, 222, 238, 239, 257, 277, 309, 318, 328, 343, 359, 360, 375, 387, 401, 402, 404, 512, 608, 635, 636, 637, 638, 640, 641, 643, 644, 648, 649
Rall, J.A. 256, 358, 479, 488, 498, 513, 534, 634
Rodger, C.D. **407**
Rüegg, J.C. 101, **125**, 143, 144, 145, 146, 147, 148, 167, 168, 223, 238, 309, 339, 360, 361, 402, 422, 590, 636
Saeki, Y. **171**, 190, 272
Sagawa, K. 82, 145, 146, 168, **171**, 189, 190, 258, 359, 361, 403, 442, 609
Sakai, T. **311**, 318, 341
Shimizu, H. **563**, 590, 591
Steiger, G.J. 82, 167, 189, 190, **259**, 271, 272, 273, 421, 422
Stienen, G.J.M. **211**
Sugi, H. 20, 40, **85**, 100, 101, 102, 145, 147, 224, **225**, 238, 239, 272, 309, 356, 358, 360, 361, 401, **425**, 441, 442, 479, 633, 636, 637, 639, 640
ter Keurs, H.E.D.J. 21, 38, 146, 190, **277**, 292, 293, 294, 328, 341, 608, 648
Thorson, J. **193**
Tirosh, R. **593**
Tregear, R.T. 144, 145, 166, 168, 209, 223, 272, 358, 359, 375, **407**, 421, 422, 442, 498, 560, 561, 629, 632, 639, 641
Tsuchiya, T. **225**
Umazume, Y. **41, 51**
Vassallo, D.V. **23, 71**
Wakabayashi, K. **445**
Walcott, B. **3**
White, D.C.S. 39, 67, 146, **193**, 209, 210, 239, 343, 357, 362, 638, 641, 642, 644, 646
Wilkie, E.R. 35, 292, 374, 375, 512, 513, **515**, 534, 535, 536, 634, 643
Wilson, M.G.A. **193**
Winegrad, S. 21, 36, 39, 40, 100, 222, 223, 271, 294, **329**, 339, 342, 343, 361, 441, 479, 535, 635, 637, 639, 641, 646
Woledge, R.C. 168, **473**, 479, 513, 514, 536, 608, 638
Yamada, K. **481**
Yamamoto, K. **319**
Yamamoto, T. **125**, 146
Yano, M. **563**
Yoshino, S. **51**

Subject Index

A-band
  length 3–19, 100, 342
  shortening 3–5, 633, 644–647
actin
  F- 10, 563–590
  fllament 47
  monomer 457
S1 angle 561
activation
  by shortening 256
  energy 265
  by stretch 262
active state 356, 385, 377
actomyosin
  ATP-ase 145, 166, 474, 481, 484, 535, 565
  dissociation of 267
  product complex (ATP complex) 139–140, 481–488, 563–590
ADP
  accumulation of 223
  rephosphorylation of 526
aerobic
  muscle 499–512, 517
  recovery 501
afterloaded twitch 435
AMP-PNP
  and cross-bridge configuration 140, 141, 422
  effect on rigor 205
  treated muscle 14, 200–202
anaerobic
  muscle 499–512
  recovery 499–512
Arrhenius plot 64
ATP
  analog 12, 140, 407, 414
  effect of Mg- 149–164, 563–590
  splitting after stretch 479
  turnover 139, 500
autocorrelation function 446
axial reflection 398

barium contracture 171, 172, 176
baro-entropic effect 593–608
Bessel function 447, 460
Bode plot 173, 189
Boltzmann constant 594
Brownian movement
  of cross-bridges 395, 564

caffeine rapid cooling contracture 311–318, 509
calcium
  binding constant of -EGTA complex 366, 375
  distribution within sarcomeres 626
  jump 127
  sensitivity of contractile system 330, 369, 370
  transients 638, 639, 642, 643
cardiac muscle
  dynamic stiffness of 171–188
  series elasticity of 71, 72, 78, 83, 86–90
  force-velocity relation of 377–387
  length transients in 241–254
  regulation of contractile proteins in 329–338
  tetanic contraction of 243, 378
  visco-elastic model of 171
chemical shift 516
coefficient
  elastic 182–186
  viscous 182–186, 577, 578, 600
complex stiffness
  see stiffness
compliance

SUBJECT INDEX

see also series elasticity
at damaged end in cardiac muscle 72, 80, 90
of I-filaments 67
of filaments 633
connectin 41, 44, 319–328
contraction
high-speed cinematography of 86–96
isotonic 225–237, 396–397, 432–440
isovolumic 440, 598
non-uniformity during 88–90, 301
sarcomere length changes during 23–35, 281, 298
X-ray diffraction during 391–400, 425–440
cooperativity
see dynamic cooperativity
crab muscle 445–456, 457–468
crayfish muscle 90–93
creatine
kinase 473, 475, 477, 482, 483
phosphate 149, 418, 499–512, 515–534, 583
phosphotransferase 521
phosphokinase 149
cross-bridges
and AMP–PNP 140
angle of 132, 398
as independent force generators 343, 563–590, 633
ATP-ase sites in 630–632
attached states of 125, 132, 134, 138, 633
attachment/detachment rate of 392
attachment region of 231, 357, 392
back rotation of 140, 188
distorted 238, 239, 431
effect of Mg-ATP on kinetics 149–166
elastic extension of 86, 92, 93
elasticity of 132, 148
helical arrangement of 398
in rigor 97, 98
kinetic properties of 149–164, 231, 239, 425
lateral (radial) movement of 393
locked-on 638
refractory period of 247, 607
rotation of 93, 131, 132, 138, 139
slippage of 131, 136, 238, 541, 553, 555
turnover rate of 231, 354, 474
vectors of 640–642
cyclic nucleotides (cyclic AMP and GNP) 332–338

deactivation
by shortening 239, 256, 359
delayed tension change 162, 171, 266
dielectric constant
of filaments 613
dipole moment
of filaments 45, 51, 52, 611–629
dissociation
of actomyosin-product complex 139, 267
double overlap
of filaments 271
double hexagonal lattice
of filaments 393
dynamic cooperativity
of cross-bridges 563–590
dynamic stiffness
of contracting muscle 171–188, 233–237
of resting muscle 171
relation to cross-bridges 127–138, 187, 233–237

efficiency
of muscle 551, 632
electric field
effect on thin filaments 51–66
electron microscopy
of cross-bridges 466
of rigor muscle 97–98
of thick filaments in *Limulus* muscle 3–19
electrostatic field

SUBJECT INDEX

of cross-bridges 611–629
elongation
of 143A periodicity of thick filaments 398–399, 401
of H-zone 98
of I-band 98
of Z-band 98
enhancement
of mechanical performance after loaded steps 234–237
of tension by stretch 298–310
equatorial reflection 393–397, 426–440

F-actin
see actin
Fenn effect 563, 570
fatigue 441, 515–534
FENB
treated muscle 213
fibrillar muscle
see insect flight muscle
field theory 611–629
filaments
see also thick and thin filaments
C- 169, 199, 202–207, 209
elongation of, in rigor muscle 96–98
flexural rigidity of 51–66
-lattice volume 439
length changes 3–22, 644–647
polarity of 563
shortening of 3–22
spacing 641, 642
flight muscle
see insect flight muscle
FAA (fluoroacetic acid)
treated muscle 212
force
and stiffness (*see* stiffness-force relation)
-extension curve (*see* series elasticty)
restoring 361
-step experiments 225–237, 241–256
transients (*see* isometric tension transients)

force-velocity-length relation 386
force-velocity relation
departure from a hyperbola 350–352
effect of isotonic lengthening on 234–237
effect of stretch on 302–303
effect of temperature on 111
in cardiac muscle 111, 377–387
in negative force region 352, 356–359
Fourier analysis
of equatorial reflections 441
of meridional reflections 446–456, 464–466
of optical diffraction 3, 5, 56
frequency response analysis 171–188

ghost skinned fiber 45, 54
give
of muscle (*see also* yielding) 39, 40, 233, 247, 601
glycolysis 510, 517, 521

heat
labile 474
of ATP hydrolysis 481–488
shortening 481–488
unexplained 474, 499
heat production
after stretch 473–478
high energy phosphate
aerobic resynthesis of 503
in glycolysis 503
splitting 481, 482, 486, 504
utilization during recovery 499
Hill equation 350, 352, 361, 385, 386, 495, 496
HMM (heavy meromyosin) 46, 563–590, 607
Hook's law
in cross-bridges 219, 352, 604
Huxley (1957) model 96, 104, 106, 126, 231, 237, 357, 358, 653
Huxley-Simmons model (Ford-Huxley

Simmons model) 72, 86, 93, 95, 101, 125, 126, 131–135, 149, 161, 166, 213, 222, 236, 245, 253, 265–266, 541, 634, 652, 653
hydrodynamic mechanism in contraction 593–608
hypertonic solutions
 effect on contraction 374

independent force-generators
 see cross-bridges
insect flight muscle 136, 193–207, 407–423, 450, 464, 469, 629–632
instantaneous elasticity 64, 65, 72, 93, 265, 633
intensity ratio ($I_{1,0}/I_{1,1}$), of equatorial reflections
 during isometric contraction 433–434
 during isotonic shortening 396–397, 403, 435–438
 effect of stretch on 426–432
internal load 383, 384, 386
ionic strength
 effect on contraction 374, 375
isometric tension transients
 see also quick tension recovery
 effect of metabolic inhibition on 213–218
 in insect flight muscle 202
 kinetic analysis of 262–265, 272
isotonic velocity transients
 after step increases in load 227–233, 241–254
 after step decreases in load 227

lactic acid 499–512, 521–526
laser light diffraction
 see optical diffraction
latency relaxation 36
layer lines
 of X-ray diffraction 398, 412–423, 446–452, 460
length-tension relation 277–295, 307, 311–317, 341–343, 634–637
 by Gordon, Huxley and Julian (1966) 105, 277–295, 311–317, 352, 380, 581, 634, 635
 effect of stretch on 297–308
 stability of the discending limb of 292–293
light diffraction
 see optical diffraction
L_{max} (optimum length in cardiac muscle) 176, 243, 377
load-bearing capacity 234, 237
load-step experiments
 see force-step experiments
load-stiffness relation 233, 238
load-velocity relation
 see force-velocity relation
longitudinal stability
 of sarcomeres 127, 612–617, 626, 637–640
Lymn-Taylor scheme 160, 161, 267, 482

M-line 44, 621, 629
maximum shortening velocity
 see V_{max}
meridional reflection 398, 412–423, 445–456, 457–470
metabolic inhibition
 effect on isometric tension transients 212
model
 four state 541–560
 Huxley (see Huxley model)
 Huxley-Simmons (see Huxley-Simmons model)
 Podolsky-Nolan (see Podolsky-Nolan model)
 three state 563–590
 two component 85
 two state 231
 viscoelastic 171
myokinase 408
myosin
 anti- 4
 ATP-ase 64, 373, 489–497

SUBJECT INDEX

heavy chain 16, 489, 491, 498
isoenzyme 489-497
light chain phosphorylation of 489, 491, 492, 498, 513
two heads of 587

negative force region 356, 357, 358, 359, 365
negative tension (force) 359, 361, 365, 397, 552, 625
nerve cross union 493-495
non-hydrolysable ATP analog
see ATP analog
non-uniformity
of fiber segments during quick release 88-90, 95
of sarcomere length during contraction 23-35, 279, 301, 636, 637
NMR (nuclear magnetic resonance) of ^{31}P 515-535
Nyquist plot 155, 157, 167, 189, 207

optical diffraction
of intact muscle 23-35, 71-81, 146-147, 279-291, 298, 312
of skinned fiber 51-66
of electron micrograph 468
oscillatory length changes 127, 226-231
oscillatory motion
in insect flight muscle 630-632
oscillatory tension response 128
oscillatory work 152, 162
overlap
between thick and thin filaments 8, 21, 57, 62, 98, 289, 306, 307, 341, 386, 544, 551, 554, 635
oxygen consumption 500-503

parallel elastic component 44, 86, 319, 387
paramyosin 3, 4, 9, 16, 21
pCa
effect of glucose on 368

-tension relation 368, 626-628
phosphorylation
of myosin light chain 13, 14, 499, 509, 510
of tropomyosin 509, 510
of troponin 337, 339
Podolsky-Nolan model 231, 233
position-sensitive counter 393, 410, 425, 433

quick release
high-speed cinematography during 86-96
isometric tension transients after 82, 138
localized shortening during 88, 95
quick stretch
high-speed cinematography during 96
of rigor muscle 141, 148
quick tension recovery
after quck release 82, 95, 100, 132, 133, 138-139, 163, 213-218

rapid elasticity
see instantaneous elasticity
rate constants
of cross-bridge attachment (f) 231, 351, 357, 392
of cross-bridge cycling 155, 347
of cross-bridge detachment (g) 161, 231, 239, 351, 357, 358
of isometric tension transients 262-269
rigor
cross-bridges in 97-98, 150, 416
effect of calcium on 199
sliding inhibition in 97
stiffness in 199
stretch of muscle in 97-98
X-ray diffraction of muscle in 407-423, 445-456, 457-470

sarcomere(s)

see also striation spacing
length changes during contraction 23–35, 281–291, 298
length changes during quick stretch and release 78–79
overstretched 41–49
stepwise shortening of 23–40, 50, 633
-tension relation 78, 284–286, 626–628
series elasticity (series elastic component)
  anatomical origin of 85–96
  at damaged end in cardiac muscle 72, 80, 90
  distribution along the fiber 85–96
  extension of 86, 91
  force-extension curve of 86, 87, 91, 133, 136–137, 213–215, 218
  of non-cross-bridge origin 85–96, 98, 401
shortening
  deactivation by 239, 256
  heat of (see heat)
  isotonic (see isotonic shortening)
  maximum velocity of (see V_{max})
  velocity at zero load (V_0) 347–356, 357, 359–362
sinusoidal analysis 149–164, 171–188
skinned fiber
  mechanically 41–49, 51–66, 125–141, 365–374, 627
  chemically 149–164, 330
stiffness
  complex 152
  dynamic (see dynamic stiffness)
  effect of temperature on 125, 138, 180
  -force relation 125, 127–131, 146, 176, 181, 185, 195–197, 233, 238
  -frequency relation 155, 172, 179, 180
stochastic process 34, 564
strain activation 644
stretch

activation 260, 262
  effect on X-ray diffraction 638, 642–644
  force enhancement by 638, 642–644
  heat production after 473–478
striation spacing
  irregularity of 146–147, 287–289, 314–315, 627–628 (see also sarcomeres)
S_1 (subfragment-1) head 2, 399, 445, 453, 454, 484, 492, 575
synchrotron radiation 393, 409

T_1, T_2 courves
  in Huxley-Simmons model 125, 127, 129–130, 138, 148, 231–216, 220, 222, 541–560, 585, 586
thick filaments
  elongation of 85, 98
  shortening of 3–19, 644–647
  X-ray diffraction of 398–405, 412–423, 440, 647
  lengthening of 8, 12, 14
  misalignment of 342
thin filaments
  dipole moment of 45, 51, 52
  elongation of 85, 98
  flexibility of 51, 64, 65
  X-ray diffraction of 102, 445–456, 457–468
time-resolved X-ray diffraction 391, 432
tropomyosin 64, 373, 395, 510, 569
troponin 64, 373, 395, 454, 457–470, 569, 609, 617
TV X-ray detector 393, 398, 411, 413, 414

visco-elasticity 172
viscosity
  effect on contraction 365–374
viscous force 600
V_{max} (maximum velocity of shortening)

effect of Ca on 360
in cardiac muscle 382–386
relation to Huxley model 357–358, 365
relation to V_0 348, 360–362
V_{max} (of actomyosin ATP-ase) 269, 273
V_0 (velocity of shortening at zero load)
see shortening
yielding
of cardiac muscle 242–247, 256–258
Young's modulus
of thin filaments 65, 67

RE
AP